OXFORD MATHEMATICAL

A. Belleni-Moranti: *Applied semigroups and evolution equations*
A.M. Arthurs: *Complementary variational principles* Second edition
M. Rosenblum and J. Rovnyak: *Hardy classes and operator theory*
J.W.P. Hirschfeld: *Finite projective spaces of three dimensions*
A. Pressley and G. Segal: *Loop groups*
D.E. Edmunds and W.D. Evans: *Spectral theory and differential operators*
Wang Jianhua: *The theory of games*
S. Omatu and J.H. Seinfeld: *Distributed parameter systems: theory and applications*
J. Hilgert, K.H. Hofmann, and J.D. Lawson: *Lie groups, convex cones, and semigroups*
S. Dineen: *The Schwarz lemma*
S.K. Donaldson and P.B. Kronheimer: *The geometry of four-manifolds*
D.W. Robinson: *Elliptic operators and Lie groups*
A.G. Werschulz: *The computational complexity of differential and integral equations*
L. Evens: *Cohomology of groups*
G. Effinger and D.R. Hayes: *Additive number theory of polynomials*
J.W.P. Hirschfeld and J.A. Thas: *General Galois geometries*
P.N. Hoffman and J.F. Humphreys: *Projective representations of the symmetric groups*
I. Györi and G. Ladas: *The oscillation theory of delay differential equations*
J. Heinonen, T. Kilpelainen, and O. Martio: *Non-linear potential theory*
B. Amberg, S. Franciosi, and F. de Giovanni: *Products of groups*
M.E. Gurtin: *Thermomechanics of evolving phase boundaries in the plane*
I. Ionescu and M. Sofonea: *Functional and numerical methods in viscoplasticity*
N. Woodhouse: *Geometric quantization* Second edition
U. Grenander: *General pattern theory*
J. Faraut and A. Koranyi: *Analysis on symmetric cones*
I.G. Macdonald: *Symmetric functions and Hall polynomials* Second edition
B.L.R. Shawyer and B.B. Watson: *Borel's methods of summability*
M. Holschneider: *Wavelets: an analysis tool*
Jacques Thévenaz: *G-algebras and modular representation theory*
Hans-Joachim Baues: *Homotopy type and homology*
P.D. D'Eath: *Black holes: gravitational interactions*
R. Lowen: *Approach spaces: the missing link in the topology–uniformity–metric triad*
Nguyen Dinh Cong: *Topological dynamics of random dynamical systems*
J.W.P. Hirschfeld: *Projective geometries over finite fields* Second edition
K. Matsuzaki and M. Taniguchi: *Hyperbolic manifolds and Kleinan groups*

David E. Evans and Yasuyuki Kawahigashi: *Quantum symmetries on operator algebras*
Norbert Klingen: *Arithmetical similarities: prime decomposition and finite group theory*
Isabelle Catto, Claude Le Bris, and Pierre-Louis Lions: *The mathematical theory of thermodynamic limits: Thomas–Fermi type models*
D. McDuff and D. Salamon: *Introduction to syplectic topology* Second edition
William M. Goldman: *Complex hyperbolic geometry*

Wavelets
An Analysis Tool

M. HOLSCHNEIDER
Centre National de la Recherche Scientifique (CNRS)
Centre de Physique Théorique
Marseille, France

CLARENDON PRESS · OXFORD

Oxford University Press, Great Clarendon Street, Oxford OX2 6DP
Oxford New York
Athens Auckland Bangkok Bogota Bombay Buenos Aires Calcutta Cape Town Chennai Dar es Salaam Delhi Florence Hong Kong Istanbul Karachi Kuala Lumpur Madrid Melbourne Mexico City Mumbai Nairobi Paris São Paulo Singapore Taipei Tokyo Toronto Warsaw
and associated companies in
Berlin Ibadan

Oxford is a registered trade mark of Oxford University Press

Published in the United States by
Oxford University Press Inc., New York

First published 1995
Reprinted 1995 (with corrections), 1998

A catalogue record for this book is available from the British Library

Library of Congress Cataloging in Publication Data

ISBN 0 19 853481 7 (Hbk.)
ISBN 0 19 850521 3 (Pbk.)

Printed in Great Britain on acid-free paper by
Bookcraft (Bath) Ltd, Midsomer Norton, Avon

PREFACE

This book is designed to be an easy-to-read, more or less self-contained introductory text about the theory of wavelets. It is intended for graduate students of mathematics and physics, as well as interested researchers from other fields, and engineers. Its main focus is the continuous wavelet transform and some of its applications. The intention is to fill a gap between the book by Ingrid Daubechies, whose focus is on orthonormal wavelet bases, and the more advanced books by Yves Meyer and Ronald Coifman. Hopefully this book will ease the access to such post-graduate literature. In this book only very elementary mathematics is needed. In addition, to ease the readability, I have opted to use sometimes a slightly non-rigorous but more intuitive language, for example in the discussion of the wavelet analysis of the Brownian motion.

In Chapter 1 we start by introducing the main object of interest which is the continuous wavelet transform. We shall discuss in great detail its various localization properties, in particular its mapping properties in Schwartz space and its dual (the space of tempered distributions), as well as in a Hilbert space context.

Chapter 2 is devoted to several sampling theorems linked to wavelet transforms. In particular we introduce the theory of frames. Several partial reconstruction formulas are discussed and the link to Calderón–Zygmund operators is made. In addition we present the continuous wavelet analysis over the circle.

Chapter 3 is entirely devoted to the construction of orthonormal wavelet bases and to multi-resolution analysis. We discuss this construction over the real numbers $\mathbb{R}$, the circle $\mathbb{T}$, the integers $\mathbb{Z}$, and the spaces $\mathbb{Z}/N\mathbb{Z}$. The last section contains a discussion of several algorithms to compute the continuous and discrete wavelet transforms efficiently.

Chapter 4 shows how wavelet analysis may be used to analyse the local self-similarity properties of fractals. We discuss how pointwise regularity may be read from the wavelet coefficients. As an example we apply this analysis to the Brownian motion, the non-differentiable function of Weierstrass and a function due to Riemann. We also give a short introduction to fractal dimensions and show how they can be measured in wavelet space.

Chapter 5 reformulates the main results of Chapters 1 and 3 using the language of group theory. As an application we discuss in the appendix the two dimensional wavelet transform in some detail. This is then used to solve the Radon inversion problem using a wavelet technique.

Chapter 6 gives a short discussion of the characterization of function spaces through their wavelet coefficients. It is clearly not complete but it should ease the access to the books of Triebel and Meyer discussing this subject in great detail.

The bibliography given at the end is clearly non-exhaustive and I refer to the 'wavelet literature survey' of Stefan Pittner, Josef Schneid, and Christoph W. Ueberhuber, that can be ordered by e-mail via pittner@uranus.tuwien.ac.at.

I would like to thank first of all Alex Grossmann, not only because among other things he introduced me to the interesting field of wavelet analysis but most of all because he created that kind of atmosphere which is a mixture between leisure and work, that made it a true pleasure to do science. My thanks (of another kind though) also go to Sabine, who, with her constant encouragement and warm moral support during the most critical periods of this work, has made this book possible.

Marseille M. H.
September 1994

CONTENTS

1
INTRODUCTION TO WAVELET ANALYSIS OVER $\mathbb{R}$

Wavelets comprise the family of translations and dilations of a single function called the *wavelet*. The name is due to the fact that every wavelet has to have at least some oscillations, and that it is localized. The wavelet transform is the set of scalar products of all dilated and translated wavelets with an arbitrary analysed function. This linear map is a partial isometry and its range is characterized by a reproducing kernel.

1 A short motivation

There are at least two approaches to wavelet analysis. The first is the interpretation of the wavelet transform as a time–frequency analysis tool. The second approach uses the wavelet analysis as a mathematical microscope. This second approach is closely linked to approximation theory.

1.1 Time–frequency analysis

Imagine yourself listening to a piano concert of Beethoven. Did you ever wonder how it is possible that somethings such as melodies, and hence music, exist? In more mathematical terms the miracle is the following: our ears receive a one-dimensional signal $p(t)$ over the real line $\mathbb{R}$ that describes how the local pressure varies with time t. However, this one-dimensional information is then somehow 'unfolded' into a two-dimensional time–frequency plane: the function over the real line $p(t)$ is mapped into a function over the time–frequency plane that tells us 'when' which 'frequency' occurs. Now, in the strict sense, this is a contradiction in itself. A pure frequency is given by the one-parameter family of complex exponentials $e_\omega = e^{it\omega}$. Consequently, it has no time point attached to it since it goes from $-\infty$ to $+\infty$. On the other hand a precise time point represented by the delta distribution $\delta(t)$ has all frequencies attached to it and hence no frequency can be associated with it. Therefore, neither $p(t)$ nor its Fourier transform $\hat{p}(\omega)$ give us sufficient information. The hearing process is thus based on some compromise between time localization and frequency localization. This is also true for wavelet transforms, as we will now see.

Recall that the Fourier transform of a function $s(t)$ over the real line is obtained by 'comparing' s with the one-parameter family of pure oscillations $e_\omega(t) = e^{i\omega t}$ by taking all possible scalar products, that is, for $s \in L^1(\mathbb{R})$ we have

$$s \mapsto Fs, \qquad (Fs)(\omega) = \langle e_\omega \mid s \rangle_{\mathbb{R}} = \int_{-\infty}^{+\infty} dt\, e^{-i\omega t}\, s(t),$$

where we have introduced the notation

$$\langle s \mid r \rangle_{\mathbb{R}} = \int_{-\infty}^{+\infty} dt\, \overline{s(t)}\, r(t),$$

whenever the integral converges absolutely. We usually write $\hat{s}$ for Fs. The Fourier transform of s is thus a function over the parameter ω, which may be interpreted as a frequency. We therefore call $\hat{s}(\omega)$ the *frequency content* or *frequency representation* of s, and s itself may be referred to as the *time representation.* It can be shown (e.g. Rudin 1992) that for $s \in L^1(\mathbb{R}) \cap L^2(\mathbb{R})$[1] the Fourier transform preserves the scalar product, and hence the energy, in the sense that

$$\langle s \mid w \rangle = \frac{1}{2\pi} \langle \hat{s} \mid \hat{w} \rangle, \qquad \int |s|^2 = \frac{1}{2\pi} \int |\hat{s}|^2. \tag{1.1.1}$$

Therefore, the Fourier transform may be extended to all of $L^2(\mathbb{R})$ by taking limits. More precisely for $s \in L^2(\mathbb{R})$ we may pick any sequence $s_n \in L^1(\mathbb{R}) \cap L^2(\mathbb{R})$ such that $\|s_n - s\|_2 \to 0$ as $n \to \infty$. Then there is a function in $L^2(\mathbb{R})$ not depending on the specific approximation sequence s_n, called $\hat{s} \in L^2(\mathbb{R})$, such that $\|\hat{s}_n - \hat{s}\|^2 \to 0$ as $n \to \infty$. This defines $\hat{s}$ almost everywhere. The *inverse Fourier transform* is given by the adjoint operator. For $r \in L^1(\mathbb{R}) \cap L^2(\mathbb{R})$ it reads

$$r \mapsto F^{-1}r, \quad (F^{-1}r)(t) = \frac{1}{2\pi} \int_{-\infty}^{+\infty} d\omega\, e^{i\omega}\, r(\omega), \quad FF^{-1} = F^{-1}F = \mathbb{1}.$$

It allows us to write s as a superposition of the elementary function e_ω if some regularity conditions on s are imposed, that is, for example, $s \in L^1(\mathbb{R})$ and $\hat{s} \in L^1(\mathbb{R})$ we have pointwise

$$s(t) = \frac{1}{2\pi} \int_{-\infty}^{+\infty} d\omega\, \hat{s}(\omega)\, e_\omega(t) = \frac{1}{2\pi} \int_{-\infty}^{+\infty} d\omega\, \hat{s}(\omega)\, e^{i\omega t}.$$

---

[1] As usual we denote by $L^p(\mathbb{R})$, $p > 0$ the space of functions for which

$$\|s\|_{L^p(\mathbb{R})} = \|s\|_p = \left\{ \int_{-\infty}^{+\infty} dt\, |s(t)|^p \right\}^{1/p} < \infty, \quad p < \infty.$$

For $p = \infty$ we set $\|s\|_{L^\infty(\mathbb{R})} = \|s\|_\infty = \text{ess sup}_{t \in \mathbb{R}} |s(t)|$, which means that $|s(t)| > \|s\|_{L^\infty(\mathbb{R})}$ only on a set of Lebesgue measure 0 and that for all $c < \|s\|_{L^\infty(\mathbb{R})}$ there is a set E of positive Lebesgue measure such that $|s(t)| \geq c$ for all $t \in E$.

To return to the problem of the hearing process the idea might be to replace the one-parameter family of pure oscillations e_ω by a two-parameter family of functions $g^{b,\omega}$, where b is a position parameter and ω corresponds to a frequency parameter. The transform from the function over the one-dimensional time–space $s(t)$ to the two-dimensional time–frequency space would be given by

$$\mathscr{T}s(b, \omega) = \langle g^{b,\omega} \mid s \rangle. \tag{1.1.2}$$

On way to obtain such a two-parameter family of time–frequency atoms would be to shift one basic function g with the help of translations in t and ω:

$$T_b\colon s(t) \mapsto s(t-b), \quad E_\omega\colon \hat{s}(\xi) \mapsto \hat{s}(\xi - \omega) \Leftrightarrow s(t) \mapsto e^{i\omega t}\, s(t)$$

upon setting

$$g^{b,\omega}(t) = E_\omega T_b g(t) = e^{ib\omega}\, g(t-b).$$

In this case the transform of (1.1.2) is called the windowed Fourier transform with respect to the window g:

$$\begin{aligned} s(t) \mapsto \mathscr{T}s(b,\omega) &= \int_{-\infty}^{+\infty} dt\, \bar{g}(t-b)\, e^{-i\omega t}\, s(t) \\ &= \frac{e^{-ib\omega}}{2\pi} \int_{-\infty}^{+\infty} d\xi\, e^{ib\xi}\, \bar{\hat{g}}(\xi - \omega) \hat{s}(\xi). \end{aligned}$$

Clearly, if g is localized around 0 with width Δ then the translated function $T_b g$ is localized around b with the same width. In the same sense if $\hat{g}$ is localized around 0, then the modulation $E_\omega g$ is localized around ω. Thus one cay say that the windowed Fourier transform is a time–frequency analysis where b is a time parameter and ω is a frequency parameter.

But there is a different way of changing the frequency localization of g and this brings us to wavelets. It is given by the dilation operator

$$D_a\colon g(t) \mapsto \frac{1}{a}\, g\left(\frac{t}{a}\right), \qquad \hat{s}(\omega) \mapsto \hat{s}(a\omega).$$

Indeed, suppose that the Fourier transform of g is localized around the central frequency $\omega_0 > 0$, say, with frequency width $\Delta\omega$. Then the dilated function $g_a = D_a g$ is localized around ω_0/a. At the same time the frequency width has changed from $\Delta\omega$ to $a^{-1}\Delta\omega$ such that $\Delta\omega/\omega$ has remained constant. See Figures 1.1 and 1.2. At the same time, because of the dilation, the time width Δt has changed to $a\Delta t$. However, the product $\Delta t \Delta\omega$ is constant (see Figure 1.3).

But if $\hat{g}$ is localized around $\omega_0 = 0$, the dilated function $\hat{g}(a\omega)$ is still localized around $\omega_0 = 0$ and the dilation operator does not change the position in frequency space but only the width of $\hat{g}$. Therefore, it is natural

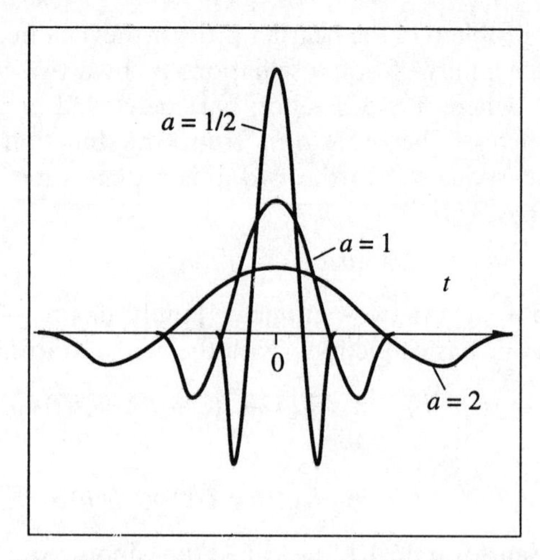

Fig. 1.1 The dilated wavelet $g_a = D_a g$ in the time representation.

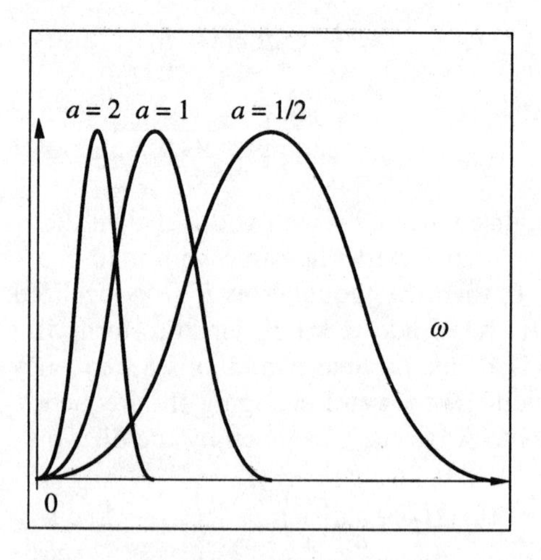

Fig. 1.2 The frequency localization of the dilated wavelet g_a. The translation will only change the phase of $\hat{g}$.

to require that g does not contain the 0 frequency or, amounts to the same, is of zero mean

$$\hat{g}(0) = 0 \Leftrightarrow \int_{-\infty}^{+\infty} dt\, g(t) = 0.$$

This, by the way, justifies the name 'wavelet' for such functions: a wavelet necessarily has some oscillations. Such a wavelet is never a mere bump.

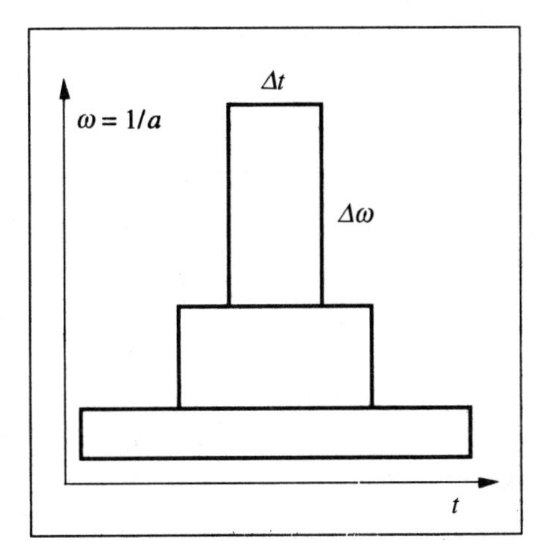

Fig. 1.3 The wavelet transform as time–frequency analysis. Every point (b, a) in the half-plane corresponds to a wavelet that is localized around b with time width $a\Delta t_0$ and around the central frequency ω_0/a with frequency width $\Delta\omega/\omega = \Delta\omega_0$. The area $\Delta\omega\Delta t$ is invariant.

The basic functions of wavelet analysis are thus obtained from a single function by dilation and translation:

$$g_{b,a} = T_b D_a g, \quad g_{b,a}(t) = \frac{1}{a} g\left(\frac{t-b}{a}\right).$$

Definition 1.1.1
A function $g \in L^1(\mathbb{R}) \cap L^\infty(\mathbb{R})$ with $\int g = 0$ is called a wavelet. The wavelet transform of a function $s \in L^p(\mathbb{R})$, $1 \leq p \leq \infty$, is defined as

$$\mathscr{W}_g s(b, a) = \langle g_{b,a} \mid s \rangle = \int_{-\infty}^{+\infty} dt \, \frac{1}{a} \bar{g}\left(\frac{t-b}{a}\right) s(t), \qquad b \in \mathbb{R}, a > 0.$$

We sometimes write $\mathscr{W}[g; s](b, a)$ for $\mathscr{W}_g s(b, a)$. The numbers $\mathscr{W}[g; s](b, a)$ will sometimes be called the *wavelet coefficients* of s with respect to the wavelet g.

The two-dimensional parameter space of wavelet analysis may be identified with the upper half-plane

$$\mathbb{H} = \{(b, a) : b \in \mathbb{R}, a > 0\}.$$

From what we have explained above, this may be interpreted as the time–frequency half-plane if we make the following identifications:

$$b \leftrightarrow \text{time},$$

$$\frac{\omega_0}{a} \leftrightarrow \text{frequency}.$$

The wavelet transform itself may be looked upon as a 'time–frequency analysis' with constant relative bandwidth $\Delta\omega/\omega = \text{const}$.

1.2 Wavelets and approximation theory

A second idea of wavelet analysis is to analyse arbitrary functions s over the real line $\mathbb{R}$ on different length scales. To fix the ideas suppose that $s \in L^p(\mathbb{R})$ for some $1 \leq p \leq \infty$. The first attempt to look at s at different length scales might be to look at smoothed versions of s. Therefore, pick a real-valued, non-negative $\phi \in C_c^\infty(\mathbb{R})$—the space of functions over the real line that are infinitely many times differentiable and that are compactly supported—and look at the family $\phi_a(t) = \phi(t/a)/a$, $a > 0$, of dilated versions of ϕ. For $\phi \neq 0$ we may scale the amplitude of ϕ in such a way that $\int \phi = 1$. Let us now look at the smoothed versions,

$$\sigma_a(b) = \phi_a * s(b) = \int_{-\infty}^{+\infty} dt \, \frac{1}{a} \phi\left(\frac{b-t}{a}\right) s(t),$$

of s. The convolution product of two functions is defined as usual as

$$s * r(t) = \int_{-\infty}^{+\infty} dt' \, s(t-t') r(t') = r * s(t).$$

Since the width of the support of ϕ_a is proportional to a, one might say that the smoothed version contains the details of s up to length scale a, where the scale is measured in units of the size of the support of $\phi = \phi_1$. The features of s that are on a smaller scale are smoothed out and are no longer visible in σ_a. As a becomes smaller, more and more details are added and eventually one reconstructs all of s. This heuristic arguments can be made precisely since we have the so-called 'approximation of the identity' lemma.

Lemma 1.2.1

The smoothed versions σ_a converge towards s in $L^p(\mathbb{R})$ provided that the amplitude of the smoothing bump ϕ is scaled such that $\int \phi = 1$. That is, we have, for $1 \leq p < \infty$,

$$\|\sigma_a - s\|_p \to 0 \quad \text{as } a \to 0. \tag{1.2.1}$$

If s is uniformly continuous then this holds for $p = \infty$, too.

A demonstration of this well-known fact can be found in Torchinsky (1980), for example.

The idea of wavelet analysis is to look at the details that are added if one goes from scale a to scale $a - da$ with $da > 0$ but infinitesimally small. Since σ_a contains all details of s *up to* scale a it follows that the difference $\sigma_{a-da} - \sigma_a$ consists of the details of s that are *at* length scale a. Going to the limit $da \to 0$ we are led to

$$\mathscr{W}(b, a) = -a\partial_a \sigma_a(b). \tag{1.2.2}$$

The factor a on the left is for convenience only. Now since ϕ is smooth and compactly supported we may exchange the derivation with the integration and obtain, thanks to the identity $-a\partial_a(\phi(t/a)/a) = (\phi(t/a) + (t/a)\phi'(t/a))/a$, that

$$\begin{aligned}\mathscr{W}(b, a) &= -a\partial_a \int_{-\infty}^{+\infty} dt\, \frac{1}{a}\phi\left(\frac{t-b}{a}\right)s(t)\\ &= \int_{-\infty}^{+\infty} dt\, \frac{1}{a} g\left(\frac{t-b}{a}\right)s(t)\\ &= g_a * s(b),\end{aligned}$$

with $g(t) = (t\partial_t + 1)\phi(t)$ and $g_a(t) = g(t/a)/a$. This expression looks similar to the one we used before in the definition of the approximations σ_a. However, one important thing has changed: ϕ, which was a bump—as expressed by $\int \phi = 1$—is replaced by g, which, rather, satisfies

$$\int_{-\infty}^{+\infty} dt\, g(t) = 0,$$

as can be seen by partial integration: $\int t\partial_t \phi = -\int \phi$. This also follows from the fact, verified by straightforward computation, that

$$\hat{g}(\omega) = \omega\partial_\omega \hat{\phi}(\omega).$$

Therefore, to obtain the details of s at length scale a we have to look at the convolution of s with the dilated version of a function g that is of 0-mean, that is, a wavelet. The convolutions $\mathscr{W}_a = g_a * s$ corresponding to the details of s at length scale a is the wavelet transform (Definition 1.1.3) of s with respect to the wavelet $\bar{g}(\cdot - t)$.

If we sum up all the details $\mathscr{W}_a = \mathscr{W}(\cdot, a)$ over all scales we might hope to recover s in the sense that

$$\int_\epsilon^\rho \frac{da}{a}\mathscr{W}_a = \int_\epsilon^\rho \frac{da}{a} g_a * s \to s \quad \text{in the limit } \epsilon \to 0, \rho \to \infty.$$

This can easily be seen if the wavelet g is derived from the smoothing bump g as given before in (1.2.2) and $s \in L^1(\mathbb{R}) \cap L^\infty(\mathbb{R})$ uniformly continuous.

Then, since $\mathcal{W}_a = -a\partial_a \sigma_a$, it follows that the above integral evaluates to $\sigma_\epsilon - \sigma_\rho$. The first term goes uniformly to s in the limit $\epsilon \to 0$ by the approximation of identity (1.2.1). The second term goes to 0 thanks to Hölder's inequality

$$|\langle s \mid r \rangle| \leq \|s\|_{L^p(\mathbb{R})} \|r\|_{L^q(\mathbb{R})}, \quad \frac{1}{p} + \frac{1}{q} = 1, \quad 1 \leq p, q \leq \infty.$$

Indeed, this implies that $\|\sigma_\rho\|_\infty \leq \|\phi_\rho\|_\infty \|s_1\|$, which goes to 0 as $\rho \to \infty$ since $\|\phi_\rho\|_\infty = \rho^{-1}\|\phi\|_\infty$.

Therefore, the wavelet transform allows us to unfold a function over the one-dimensional space $\mathbb{R}$ into a function over the two-dimensional half-plane $\mathbb{H}$ of positions and details (where is which detail generated?). It is this two-dimensional picture that has made a success of wavelet analysis.

Therefore, the parameter space $\mathbb{H}$ of the wavelet analysis may also be called the *position-scale* half-plane since if g is localized around 0 with width Δ then $g_{b,a}$ is localized around the position b with width $a\Delta$. The wavelet transform itself may now be interpreted as a mathematical microscope where we identify

$$b \leftrightarrow \text{position},$$
$$(a\Delta)^{-1} \leftrightarrow \text{enlargement},$$
$$g \leftrightarrow \text{optics}.$$

After this heuristic introduction we now want to study the properties of the wavelet transform in greater detail.

2 Some easy properties of the wavelet transform

We list some obvious properties of the wavelet transform for later reference.

Linearity The wavelet transform is a linear transformation—or linear operator—so that the superposition principle applies, that is, we have:

$$\mathcal{W}_g(s + r) = \mathcal{W}_g s + \mathcal{W}_g r, \qquad \mathcal{W}_g(\alpha s) = \alpha \mathcal{W}_g s,$$

for every function s and r and every complex number $\alpha \in \mathbb{C}$. With respect to the wavelet it is anti-linear:

$$\mathcal{W}[g + v; s] = \mathcal{W}[g; s] + \mathcal{W}[v; s], \qquad \mathcal{W}[\alpha g; s] = \bar{\alpha}\mathcal{W}[g; s].$$

Symmetry $g \leftrightarrow s$ If we exchange the roles of g and s we then have the following useful formula:

$$\mathcal{W}[g; s](b, a) = \frac{1}{a}\,\mathcal{W}[\bar{s}; \bar{g}]\left(-\frac{b}{a}, \frac{1}{a}\right). \tag{2.0.1}$$

Wavelet and parity For arbitrary functions the *parity* operator is defined by

$$s \mapsto Ps, \qquad (Ps)(t) = s(-t).$$

A function is called 'even' if $Ps = s$ and 'odd' if $Ps = -s$. The analogue for functions over the half-plane $\mathscr{T}$ is

$$\mathscr{T} \mapsto \mathscr{P}\mathscr{T}, \qquad (\mathscr{P}\mathscr{T})(b, a) = \mathscr{T}(-b, a).$$

By direct computation we can show that

$$\mathscr{W}[Pg; Ps] = \mathscr{P}\mathscr{W}[g; s], \qquad \mathscr{W}[Pg; s] = \mathscr{W}[g; Ps]. \tag{2.0.2}$$

Therefore, the wavelet transform of an even (odd) function with respect to an even (odd) wavelet is itself even, $\mathscr{P}\mathscr{W}_g s = \mathscr{W}_g s$. The wavelet transform of an even (odd) function with respect to an odd (even) wavelet is odd, $\mathscr{P}\mathscr{W}_g s = -\mathscr{W}_g s$.

3 Wavelet transform in Fourier space

We may write the action of dilation and translation in Fourier space as

$$\left.\begin{aligned} s(t) \mapsto s(t/a)/a &\Leftrightarrow \hat{s}(\omega) \mapsto \hat{s}(a\omega), \\ s(t) \mapsto s(t-b) &\Leftrightarrow \hat{s}(\omega) \mapsto \hat{s}(\omega)\, e^{-i\omega b}. \end{aligned}\right\} \tag{3.0.1}$$

Because of Parseval's equation for the Fourier transform of (1.1.1) we may rewrite the wavelet transform in frequency space. Since by (3.0.1) we have $\widehat{g_{b,a}}(\omega) = \hat{g}(a\omega)\, e^{-ib\omega}$ it follows that for $g, s \in L^2(\mathbb{R})$ we may compute the wavelet coefficients in Fourier space via

$$\mathscr{W}_g s(b, a) = \frac{1}{2\pi} \int_{-\infty}^{+\infty} d\omega\, \bar{\hat{g}}(a\omega)\, e^{ib\omega}\, \hat{s}(\omega). \tag{3.0.2}$$

The restriction $\mathscr{W}_g s(\cdot, a)$ of $\mathscr{W}_g s$ to one of the lines $a = c$ is called a *voice*. It is given by the convolution of s with the dilated analysing wavelet

$$\mathscr{W}_g s(\cdot, a) = \tilde{g}_a * s(\cdot), \text{ where } \tilde{g}_a(t) = \bar{g}(-t/a)/a.$$

Accordingly, since the convolution theorem states that for $r \in L^1(\mathbb{R})$, $s \in L^2(\mathbb{R})$, we have pointwise almost everywhere,

$$\widehat{r * s}(\omega) = \hat{r}(\omega)\hat{s}(\omega), \tag{3.0.3}$$

the Fourier transform of a voice reads

$$\widehat{\mathscr{W}_g s(\cdot, a)}(\omega) = \bar{\hat{g}}(a\omega)\hat{s}(\omega). \tag{3.0.4}$$

Because $a > 0$, the positive (negative) frequencies of g will only interact with the positive (negative) frequencies of s, that is, the wavelet transform does not mix the positive frequencies of the wavelet with the negative frequencies of the analysed function. It is therefore natural to treat the positive and negative parts separately and we have the following definition.

Definition 3.0.1

A function $s \in L^2(\mathbb{R})$ is called progressive iff[2] *its Fourier transform is supported by the positive frequencies only:*

$$\operatorname{supp} \hat{g} \subseteq \mathbb{R}_+.$$

It is called regressive if the time-reversed function $s(-t)$ is progressive, or, what amounts to the same iff its Fourier transform is supported by the negative frequencies only.

As usual we denote by $\operatorname{supp} s$ the support of a function s, that is, for continuous s it is the closure of set of points t where $s(t) \neq 0$. For more general s, $\operatorname{supp} s$ is the set of points where no open neighbourhood I exists on which $\langle \phi \mid s \rangle = 0$ for all $\phi \in C_c^\infty(I)$. As a complement of an open set the support is always closed.

The spaces of progressive and regressive functions in $L^2(\mathbb{R})$ are closed subspaces and we denote them by $H^2_+(\mathbb{R})$ and $H^2_-(\mathbb{R})$ respectively. The whole Hilbert space splits into an orthogonal sum:

$$L^2(\mathbb{R}) = H^2_+(\mathbb{R}) \oplus H^2_-(\mathbb{R}).$$

Since both spaces are closed it follows that the orthogonal projectors on the positive (negative) frequencies are continuous. In Fourier space they read

$$s \mapsto \Pi^+ s, \quad \hat{s}(\omega) \mapsto \Theta(\omega)\hat{s}(\omega); \qquad s \mapsto \Pi^- s, \quad \hat{s}(\omega) \mapsto \Theta(-\omega)\hat{s}(\omega),$$

where the Heaviside function Θ is defined as

$$\Theta(t) = \begin{cases} 0 & t < 0, \\ 1 & t \geq 0. \end{cases}$$

Because $\Pi^+ + \Pi^- = \mathbb{1}$ we may use these two projectors to split the analysing wavelet and the analysed function into a progressive and a regressive component. For the wavelet transform we obtain

$$\mathcal{W}[g; s] = \mathcal{W}[\Pi^+ g; \Pi^+ s] + \mathcal{W}[\Pi^- g; \Pi^- s].$$

Thus if g is a progressive wavelet, only the positive frequencies of the analysed function are 'seen' by the analysing wavelet we have

$$\mathcal{W}[g; s] = \mathcal{W}[g; \Pi^+ s] \quad \text{for } g \text{ progressive}.$$

[2] We write 'iff' instead of 'if and only if'.

From (3.0.4) it follows that, for progressive wavelets, each voice is a progressive function. In general, the analysis with a progressive wavelet means an *a priori* loss of information on s for a non-progressive $s \in L^2(\mathbb{R})$. However, if s is real valued, again no information is lost *a priori*. Indeed, a real-valued function is defined by its positive frequencies alone since it satisfies the Hermitian symmetry

$$\hat{s}(\omega) = \bar{\hat{s}}(-\omega)$$

and thus may be recovered from its progressive part $\Pi^+ s$:

$$s(t) = 2\Re \Pi^+ s(t) = \frac{1}{\pi} \Re \int_0^\infty d\omega \, \hat{s}(\omega) \, e^{i\omega t},$$

where $\Re z$ is the real part of $z \in \mathbb{C}$. Therefore, the wavelet transform of a real function s with respect to a real wavelet g may be expressed in terms of the associated progressive functions:

$$\mathscr{W}[g; s] = 2\Re \mathscr{W}[\Pi^+ g; \Pi^+ s] = 2\Re \mathscr{W}[g; \Pi^+ s] = 2\Re \mathscr{W}[\Pi^+ g; s]. \quad (3.0.5)$$

4 Co-variance of wavelet transforms

We have already encountered the dilation operator and the translation operator acting on functions over the real line via

$$T_b\colon s(t) \to s(t-b), \qquad D_a\colon s(t) \to s(t/a)/a.$$

Correspondingly, we have the dilation $\mathscr{D}_{a'}$, $a' > 0$, and translation operator $\mathscr{T}_{b'}$, $b' \in \Re$, acting on functions over the half-plane

$$\mathscr{R} \mapsto \mathscr{D}_{a'} \mathscr{R}, \qquad \mathscr{R}(b, a) \mapsto \frac{1}{a'} \mathscr{R}\left(\frac{b}{a'}, \frac{a}{a'}\right),$$

$$\mathscr{R} \mapsto \mathscr{T}_{b'} \mathscr{R}, \qquad \mathscr{R}(b, a) \mapsto \mathscr{R}(b - b', a).$$

The wavelet transform now satisfies the following co-variance property, which can be verified by direct computation:

$$\mathscr{W}[g; D_a s] = \mathscr{D}_a \mathscr{W}[g, s], \qquad \mathscr{W}[g; T_b s] = \mathscr{T}_b \mathscr{W}[g; s], \quad (4.0.1)$$

or, more explicitly,

$$\mathscr{W}[g; s(t - b')](b, a) = \mathscr{W}[g; s](b - b', a),$$

$$\mathscr{W}[g; s(t/a')](b, a) = \mathscr{W}[g; s](b/a', a/a').$$

This means that the wavelet transform of a dilated (translated) function is obtained by dilating (translating) the wavelet transform. The following co-variance holds with respect to the wavelet:

$$\mathcal{W}[g(t-b'); s](b, a) = \mathcal{W}[g; s](b + b'a, a),$$

$$\mathcal{W}[a'g(a't); s](b, a) = \mathcal{W}[g; s](b, a/a').$$

Actually, the dilation and the translation operators we have encountered are essentially defined via a dual action on the geometric support of the functions which is in this case either the real line or the half-plane. The dilation on the real line was obtained by re-scaling around the origin and shifting the real axis by an amount $1/a'$ and b' respectively, that is, we have $t \mapsto t/a'$ and $t \mapsto t - b'$. In the case of the half-plane $\mathbb{H}$ the analogous operations are

$$\text{re-scaling:} \qquad (b, a) \mapsto (b/a', a/a'), \qquad a' \in 0,$$

$$\text{shifting:} \qquad (b, a) \mapsto (b - b', a) \qquad b' \in \mathbb{R}.$$

In order to gain a little more geometric intuition for these transforms let us look at the *invariant* subsets of the half-plane, that is, those subsets of $\mathbb{H}$ that are mapped into themselves when an arbitrary re-scaling (shifting) is applied. Clearly, every straight line passing through the origin is mapped onto itself under all possible dilations and the same holds for any collection of such lines. Vice versa, every invariant subset must contain, together with the point (b, a), all its re-scaled points $\{(b/a', a/a'): a' > 0\}$. But this is the straight line passing through (b, a) and the origin $(0, 0)$. Therefore, the invariant sets for the re-scaling are all *cone-like* structures with top at the origin. A similar argument shows that the invariant subsets for the translations are the *strips* parallel to the real axis. See Figures 4.1 and 4.2 for illustrations.

Formally, every linear transform mapping functions over the real line to functions over the half-plane that satisfies the above co-variances is a formal wavelet transform. To see this we consider the delta function δ.[3] Recall that this generalized function acts on any continuous function s via the 'scalar-product' $\langle \delta \mid s \rangle = \delta(s) = s(0)$. Formally, every function can be written as a superposition of translated delta functions

$$s(t) = \int_{-\infty}^{+\infty} du\, \delta(t-u) s(u).$$

Therefore, thanks to the linearity of K and the translation co-variance, we have

$$Ks(b, a) = \int_{-\infty}^{+\infty} du\, (K\delta)(b-u, a) s(u).$$

[3] A. Grossmann, private communication.

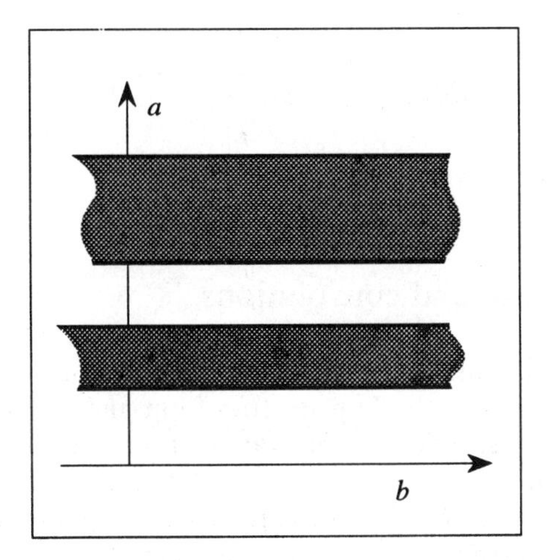

Fig. 4.1 An example of a translation invariant subset of the half-plane.

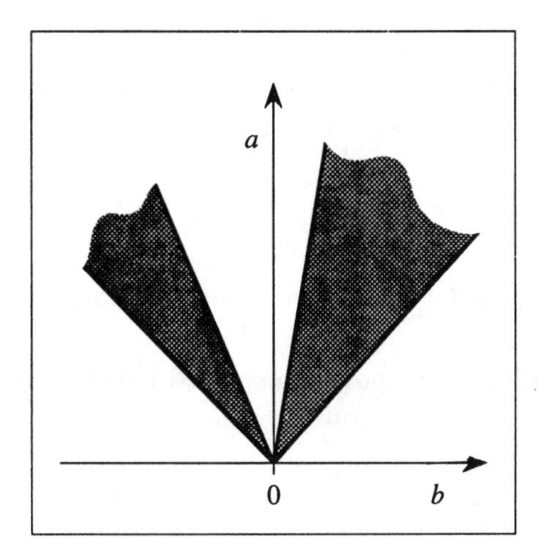

Fig. 4.2 A dilation invariant subset of the half-plane.

Now the delta function satisfies $D_a\delta = \delta$ in the sense that

$$\langle D_a\delta \mid s\rangle = \langle \delta \mid s(at)\rangle = s(0), \text{ which is again } \langle \delta \mid s\rangle.$$

By the assumed co-variance of K it follows that

$$(K\delta)(b, a) = (KD_a\delta)(b, a) = (K\delta)(b/a, 1)/a.$$

Therefore, if we set $\bar{g}(-b/a)/a$ to be the right-hand side of the last expression

we have

$$Ks(b, a) = \int_{-\infty}^{+\infty} du \frac{1}{a} \bar{g}\left(\frac{b-u}{a}\right) s(u),$$

which formally is a wavelet transform.

5 Voices, zooms, and convolutions

Before we pursue our analysis of the wavelet transform we now look at how this transform relates to the Laplace transform and the Mellin transform.

Recall that the restriction of the wavelet transform to one of the lines parallel to the real axis was called a *voice*, that is, the function $\mathscr{W}_g s(b, a)$ seen as a function of b for fixed a. In complete analogy we call a *zoom* the restriction of the wavelet transform to one of the lines passing through the point (0, 0) situated on the borderline of the half-plane, that is, the function $\lambda \mapsto \mathscr{W}_g(\lambda a, \lambda b)$ for some fixed $(b, a) \in \mathbb{H}$.

5.1 Laplace convolution

We recall the basic facts of Laplace convolution and Laplace transforms. The Laplace convolution is the convolution for functions over $\mathbb{R}_+$, that is, of causal functions. Let s, r be defined for $t \geq 0$, then

$$s * r(t) = r * s(t) = \int_0^t du\, s(t-u) r(u).$$

The Fourier transform may now be extended into the complex plane. It is now called the *Laplace* transform:

$$Ls(z) = L[s](z) = \int_0^\infty dt\, s(t)\, e^{-tz}. \tag{5.1.1}$$

Clearly, this is a Fourier transform with complex argument and on the imaginary it coincides with the Fourier transform

$$Fs(\omega) = Ls(i\omega).$$

The Laplace transform is well defined whenever s is locally integrable and if for some $\gamma \in \mathbb{R}$ we have, say,

$$\int_0^\infty dt\, |s(t)|\, e^{-\gamma t} < \infty.$$

In this case the Laplace transform is a holomorphic function in the open half-plane

$$\{z \in \mathbb{C} : \Re z > \gamma\}.$$

Indeed, for $\alpha > \gamma$ and $\Re z > \alpha$ we have $|s(t)\, e^{-tz}| \leq |s(t)|\, e^{-t\gamma}\, e^{-t(\alpha-\gamma)}$. Therefore, we may differentiate under the integral, thereby showing the analyticity. We will assume for the rest of the section that such a condition holds for all functions we consider. The inversion formula of the Laplace transform is given by

$$s(t) = \frac{1}{2i\pi} \int_{c-i\infty}^{c+i\infty} dz\, Ls(z)\, e^{tz},$$

where $c > \gamma$ such that the path of integration is completely inside the domain of analyticity of Ls. For the Laplace convolution the following convolution theorem holds:

$$v = s * r \Leftrightarrow Lv(z) = Ls(z) \cdot Lr(z),$$

where z should be in the intersection of the domains of analyticity of Ls and Lr. The translation and dilations behave under the Laplace transform as follows:

$$L[T_b r](z) = e^{bz}\, L[T_b r](z), \qquad L[D_a r](z) = L[r](az).$$

Let s be supported by $t \geq 0$ and let g be supported by $t \leq 0$. And assume that for some $\alpha > 0$ we have

$$\int_0^\infty dt\, |g(-t)|\, e^{\alpha t} < \infty \qquad \int_0^\infty dt\, |s(t)|\, e^{\alpha t} < \infty.$$

Then the wavelet transform of s with respect to g can be computed with the help of the Laplace transform:

$$L[\mathcal{W}_g s(\cdot, a)](z) = L[\tilde{g}](az) \cdot Ls(z),$$

$$\mathcal{W}_g s(b, a) = \frac{1}{2i\pi} \int_{-i\infty}^{+i\infty} dz\, L[\tilde{g}](az)\, e^{bz}\, Ls[z],$$

where $\tilde{g}(t) = \bar{g}(-t)$, as usual.

5.2 Scale convolution

For any two functions s and r defined for $t \geq 0$ we define their *scale convolution* or *logarithmic convolution* by

$$s *_l r(t) = \int_0^\infty \frac{da}{a}\, s\left(\frac{t}{a}\right) r(a) = r *_l s(t)$$

whenever this integral makes sense. The name 'logarithmic convolution' is justified by the fact that in the coordinate $u = \log t$ we have an ordinary convolution. The scale convolution is invariant under dilations

$$(D_a s) *_l r = s *_l (D_a r) = D_a(s *_l r).$$

Then the zooms can be seen as a scale convolution. Let $g_{b,-}$ and $g_{b,+}$ be the anti-causal ($t < 0$) and causal parts ($t > 0$) of $g(t - b)$ and let s_-, s_+ be the analogue split for s. By direct substitutions we can verify that

$$\mathscr{W}_g s(\lambda b, \lambda) = \bar{g}_{b,-} *_l s_-(\lambda) + \bar{g}_{b,+} *_l s_+(\lambda).$$

Therefore, the wavelet transform is a set of scale convolutions, indexed by a translation parameter. From the frequency integral of (3.0.2) it follows that for a progressive analysed function or a progressive analysing wavelet we have the following scale convolution over the frequency content of s:

$$\mathscr{W}_g s(\lambda b, \lambda) = \lambda^{-1} \bar{\hat{g}}_b *_l \hat{s}(1/\lambda),$$

where $\bar{\hat{g}}_b(\omega) = \bar{\hat{g}}(1/\omega)\, e^{ib/\omega}$.

5.3 Mellin transforms

The Mellin transform of a function s defined for $t \geq 0$ is defined as

$$Ms(z) = M[s](z) = \int_0^\infty \frac{dt}{t} t^z s(t).$$

It is well defined if

$$|s(t)| \leq c_1 t^\alpha \quad \text{for} \quad 0 < t \leq 1,$$

$$|s(t)| \leq c_2 t^{-\beta} \quad \text{for} \quad 1 < t < \infty,$$

for some $c_1, c_2 > 0$, and $\alpha < \beta$, in which case the Mellin transform is a holomorphic function in the strip $\{z \in \mathbb{C}: \alpha = \Re z < \beta\}$. It can be written as a Laplace transform by changing variables:

$$L[s(t)](z) = M[s(-\log t)](z), \qquad L[s(e^{-t})](z) = M[s(t)](z).$$

From this we deduce the inversion formula as

$$s(t) = \frac{1}{2i\pi} \int_{c-i\infty}^{c+i\infty} dz\, Ms(z) t^{-z},$$

where the path of integration should lie in the domain of analyticity of Ms, that is, $\alpha < c < \beta$ would work. For the same reason, the scale convolution is transformed by the Mellin transform into a pointwise multiplication

$$v = s *_l r \Leftrightarrow Mv(z) = Ms(z) \cdot Mr(z),$$

where z should be in the intersection of the respective domains of analyticity of Ms and Mr. The behaviour of the Mellin transform under various coordinate transforms in t space is given by

$$M[r(at)](z) = a^{-z}M[r(t)](z),$$

$$M[r(t^{\rho})](z) = \rho^{-1}M[r(t)](z/\rho),$$

$$M[t^{\lambda}r(t)](z) = M[r(t)](z+\lambda).$$

Therefore, we can write, for the Mellin transforms of the zooms of a wavelet transform of a causal function s with respect to an arbitrary wavelet g,

$$M[\mathcal{W}_g s(\lambda b, \lambda)](z) = -M[\bar{g}(\cdot - b)](1-z)\cdot M[s](z).$$

6 The basic functions: the wavelets

Let us return to the wavelet itself. In order to have good time resolution it is clear that wavelets should be localized in time. The time localization may be quantified by the following condition:

$$|g(t)| \leq c(1+|t|^2)^{-\nu/2},$$

with some $\nu > 1$ and some $c > 0$. The larger ν is, the better is the wavelet localized. In the same way we can quantify the localization in frequency space via

$$|\hat{g}(\omega)| \leq c(1+|\omega|^2)^{-\mu/2}, \qquad (6.0.1)$$

with some $\mu \geq 0$ and some $c > 0$. As an additional condition we have seen that at least the 0th moment of the wavelet should vanish. This can be generalized by requiring that some consecutive moments vanish, as expressed by

$$\int_{-\infty}^{+\infty} dt\, g(t)t^m = 0, \qquad m = 0, \ldots, n, \qquad (6.0.2)$$

for some integer $n \geq 0$. Here we have supposed that g is sufficiently localized for (6.0.2) to make sense (i.e. $(1+|t|)^m g \in L^1(\mathbb{R})$).

Condition (6.0.1) means, essentially, that some bounded derivatives of g exist.

Lemma 6.0.1

If $\hat{s} \in L^1(\mathbb{R})$ and $\omega^n \hat{s} \in L^1(\mathbb{R})$ then $s = \mathfrak{f}^{-1}\hat{s}$ is n times continuously differentiable, and the derivatives are given by

$$\frac{d^m}{dt^m} s(t) = \frac{1}{2\pi}\int_{-\infty}^{+\infty} d\omega\, (i\omega)^m \hat{s}(\omega)\, e^{i\omega t}, \quad m = 0, 1, \ldots, n. \qquad (6.0.3)$$

All derivatives are bounded and decay at infinity: $\lim_{|t|\to\infty} d^m s(t)/dt^m = 0$. On the other hand, suppose that $s \in L^1(\mathbb{R})$ is n-times differentiable with all *derivatives in* $L^1(\mathbb{R})$ *and suppose in addition that for* $0 \leq m \leq n-1$, *we have* $d^m s(t)/dt^m \to 0$ *as* $|t| \to 0$. *It then follows that* $\hat{s}$ *is localized in the sense that* $\omega^n \hat{s}(\omega) \to 0$ *in the limit* $|\omega| \to \infty$.

Note that we have to assume, in addition to $d^n s(t)/dt^n \in L^1(\mathbb{R})$, that these functions actually decay to 0 at infinity. This is a valid additional assumption since there are functions in $L^1(\mathbb{R})$ that do not decay at all, or that may even tend to infinity on a sequence of points $t_n \to \infty$.

As the result stated in the lemma is well known we will prove it here for the convenience of the reader. The proof is based on the well-known *Riemann–Lebesgue* lemma and the theorem of *dominated convergence* that we state without proof. A proof may be found in Torchinsky (1988).

Lemma 6.0.2 (*Riemann–Lebesgue*)
If $s \in L^1(\mathbb{R})$ *then* $\hat{s}$ *is a uniformly continuous function tending to* 0 *at infinity:*

$$\lim_{|\omega| \to \infty} s(\omega) = 0.$$

Recall that a function r is uniformly continuous if for all $\epsilon > 0$ there is a $\delta > 0$ such that the difference $|s(t) - s(u)| \leq \epsilon$ whenever $|t - u| \leq \delta$.

Theorem 6.0.3 (*Dominated convergence*)
Let $s_n \in L^1(\mathbb{R})$ *be a sequence of absolutely integrable functions that converges pointwise to a function* s:

$$\lim_{n \to \infty} s_n(t) = s(t),$$

for almost every $t \in \mathbb{R}$. *If in addition there is an integrable function* $r \in L^1(\mathbb{R})$ *that majorizes the family* s_n:

$$|s_n(t)| < r(t), \qquad \textit{for all } n,$$

then the limit function s *is integrable, too, and we may exchange the limit with the integration*

$$\lim_{n \to \infty} \int_{-\infty}^{+\infty} dt\, s_n(t) = \int_{-\infty}^{+\infty} dt \lim_{n \to \infty} s_n(t) = \int_{-\infty}^{+\infty} dt\, s(t).$$

Proof (of Lemma 6.0.1). We first show the first part of the lemma for $n = 1$. The general case will then follow by induction. From the Fourier inversion formula we have $s(t) = (1/2\pi) \int_{-\infty}^{+\infty} d\omega\, \hat{s}(\omega)\, e^{i\omega t}$. Consider now the finite difference for $u \neq 0$:

$$\frac{s(t+u) - s(t)}{u} = (2\pi)^{-1} \int_{-\infty}^{+\infty} d\omega\, \hat{s}(\omega)\, e^{i\omega t} \frac{e^{i\omega u} - 1}{u}. \qquad (6.0.4)$$

From the Taylor expansion of $e^{ix} = 1 + ix + O(x^2)$ we see that $(e^{ix} - 1)/x$ stays bounded as $x \to 0$. For $x \to \infty$ it decays anyway since it is majorized by $2/|x|$. We therefore have[4]

$$\left|\frac{e^{i\omega u} - 1}{u}\right| = |\omega|\left|\frac{e^{i\omega u} - 1}{|\omega| u}\right| \leq O(1)|\omega|,$$

and thus the integrand in (6.0.4) is bounded for all u by the absolutely integrable function $O(1)|\omega|\hat{s}(\omega)$. All this clearly holds for a subsequence $u_k \to 0$. In addition, the integrand converges pointwise as $u \to 0$ to the function $i\omega\hat{s}(\omega)\, e^{i\omega t}$. By the dominated convergence theorem, Theorem 6.0.3, we may exchange the limits and we obtain that the first part of (6.0.3) holds for $n = 1$. The continuity of ds/dt and its behaviour at infinity follows from Lemma 6.0.2.

Now suppose that $n > 1$. Then, by what we have shown, the derivative $ds(t)/dt$ of s satisfies the hypothesis of the theorem with $n - 1$ instead of n, and $i\omega\hat{s}(\omega)$ instead of $\hat{s}$. Continuing in this way we see that the lemma also holds for n.

We now want to show the second part of the lemma. For $n = 0$ there is nothing to prove since this is merely the Riemann–Lebesgue lemma again. Therefore, suppose $n \geq 1$. We may integrate by parts to obtain

$$\begin{aligned}\hat{s}(\omega) &= \int_{-\infty}^{+\infty} dt\, e^{-i\omega t}\, s(t) = -\frac{1}{i\omega}\int_{-\infty}^{+\infty} dt(\partial_t\, e^{-i\omega t})s(t) \\ &= -\frac{1}{i\omega} e^{-i\omega t}\, s(t)|_{-\infty}^{+\infty} + \frac{1}{i\omega}\int_{-\infty}^{+\infty} dt\, e^{-i\omega t}\, \partial_t s(t).\end{aligned}$$

This last expression equals $(-i\omega)^{-1}\int_{-\infty}^{+\infty} dt\, e^{-i\omega t}\, \partial_t s(t)$ since, by hypothesis, $s(t) \to 0$ as $|t| \to \infty$. We may continue in this way until we end up with

$$\hat{s}(\omega) = (i\omega)^{-n}\int_{-\infty}^{+\infty} dt\, e^{-i\omega t}\, \partial_t^n s(t).$$

On the right-hand side we recognize the Fourier transform of $\partial_t^n s(t)$, which is, by hypothesis, in $L^1(\mathbb{R})$. Therefore $\omega^n\hat{s}(\omega)$ is a function that goes to 0 at infinity by the Riemann–Lebesgue lemma. This shows that $\hat{s}(\omega) = o(\omega^{-n})$ as $|\omega| \to \infty$ and the proof is complete. □

[4] We use the notation

$$r(t) = O(s(t)), \qquad t \to t_0,$$

if for $|t - t_0|$ small enough $|r(t)/s(t)| < c$ with some constant $c < \infty$. In particular, $r(t) = O(1)$ means that $|r(t)| \leq c$ for t close enough to t_0. We write

$$r(t) = o(s(t)), \qquad t \to t_0,$$

if $\lim_{t \to t_0} r(t)/s(t) = 0$. In particular, $r(t) = o(1)$, $(t \to t_0)$ means that $r(t)$ tends to 0 as $t \to t_0$.

As follows directly from the definition of the Fourier transform for $s \in L^1(\mathbb{R})$ the conditions $\hat{s}(0) = 0$ and $\int s = 0$ are equivalent. The more general condition of (6.0.2) is equivalent to the fact that the wavelets do not contain the zero frequency, that is, their Fourier transform vanishes as ω tends to 0, as is shown by the following lemma.

Lemma 6.0.4
Condition (6.0.2) *is equivalent to*

$$\hat{g}(\omega) = o(\omega^n), \qquad (\omega \to 0),$$

whenever $g \in L^1(\mathbb{R})$ *and* $t^n g \in L^1(\mathbb{R})$.

Proof. From the previous lemma—upon exchanging the roles of $\hat{g}$ and g—we now see that $\hat{g}$ is n times continuously differentiable with derivatives $\hat{g}^{(m)}$, $m = 1, \ldots, n$, and that the derivatives are the Fourier transforms of $(it)^m g(t)$. Therefore, $i^m \int dt\, g(t) t^n = \hat{g}^{(n)}(0)$ and we can use Taylor's formula

$$\hat{g}(\omega) = \sum_{m=0}^{n} \hat{g}^{(m)}(0) \frac{\omega^m}{m!} + o(\omega^n), \qquad (\omega \to 0),$$

to conclude. □

According to the last condition a wavelet has some *oscillations*, which justifies the name 'wavelet'. A bump (e.g. $e^{-t^2/2}$) is not a wavelet. But the derivative of a bump is a wavelet.

We now consider some examples.

7 The real wavelets

These wavelets are, as the name implies, real-valued functions over the real line. As for any real-valued function the Fourier transform satisfies the *Hermitian* symmetry

$$\hat{g}(-\omega) = \bar{\hat{g}}(\omega).$$

Often, the real wavelets are either even, $g(-t) = g(t)$, or odd, $g(-t) = -g(t)$. In this case, the Fourier transform is real valued and even, or imaginary valued and odd respectively.

Example 7.0.1 (*A. Haar wavelet*)
Its explicit expression is given by (see Figure 7.1)

$$g(t) = \begin{cases} -1 & \text{for } 0 \leq t < 1/2 \\ +1 & \text{for } 1/2 \leq t < 1. \\ 0 & \text{otherwise} \end{cases}$$

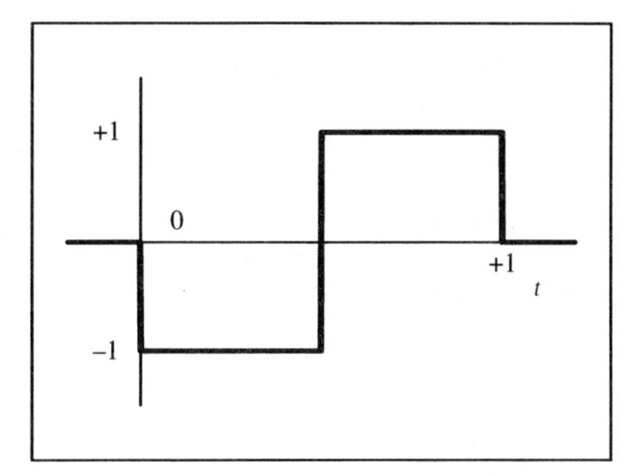

Fig. 7.1 Haar's wavelet in time representation.

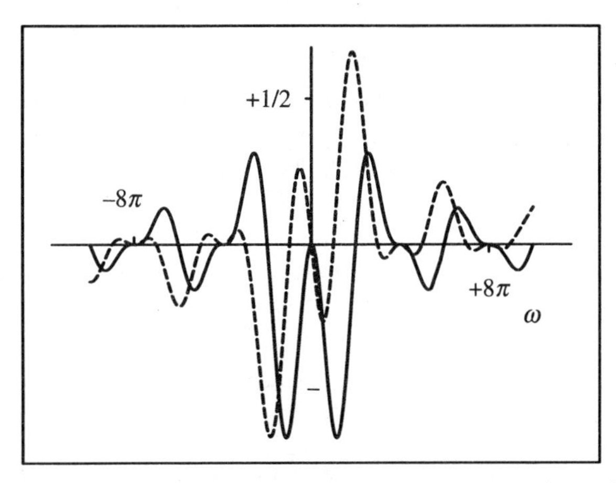

Fig. 7.2 Haar's wavelet in the frequency representation. Note the oscillations that produce a slow decrease of at most $1/\omega$. This is due to the low regularity of the time representation.

Obviously, this wavelet is extremely localized in time—it even has compact support—but it has only poor regularity—it is not even continuous. This is mirrored in a poor localization in frequency space (see Figure 7.2):

$$\hat{g}(\omega) = 2\,\frac{1 - \cos\omega}{\omega}\, e^{i(\omega + \pi)/2}.$$

This function was actually the first wavelet to be used in mathematics. It was introduced by Alfred Haar in 1906, where he also shows that the set $\{2^{j/2}g(2^j t - k)\}$ with $j, k \in \mathbb{Z}$, is an orthonormal basis of $L^2(\mathbb{R})$. The set of points $\{(2^j k, 2^j)\}$ forms a grid in the half-plane called the dyadic grid. It is enough to know the wavelet transform of $s \in L^2(\mathbb{R})$ with respect to g at all points of the dyadic grid to know it everywhere. This is an example of a sampling theorem in wavelet space that we will discuss in greater detail in later sections. We also will see in later sections that there is a whole family of more regular functions that have the same property, namely, that the sets of their dilates and translates with dilation and translation parameter in the dyadic grid form an orthonormal base of $L^2(\mathbb{R})$.

Example 7.0.2 (*The Poisson wavelets*)
Consider the function (see Figure 7.3)

$$g(t) = (t\partial_t + 1)P(t), \qquad P(t) = \frac{1}{\pi(1 + t^2)}. \tag{7.0.1}$$

In Fourier space this wavelet reads (see Figure 7.4)

$$\hat{g}(\omega) = |\omega|\, e^{-|\omega|}.$$

The analysis of a function s with respect to this wavelet is closely related to the boundary value problem of the Laplace operator. Recall that the Laplace operator acting on functions over the half-plane $\mathbb{H}$ reads

$$\Delta = \partial_b^2 + \partial_a^2.$$

A function $\mathcal{T} = \mathcal{T}(b, a)$ is called *harmonic* in the half-plane if it is two-times continuously differentiable at every point in the open half-plane and if it satisfies at

$$\Delta\mathcal{T}(b, a) = \partial_b^2\mathcal{T}(b, a) + \partial_a^2\mathcal{T}(b, a) = 0, \quad \text{at every point } (b, a) \in \mathbb{H}.$$

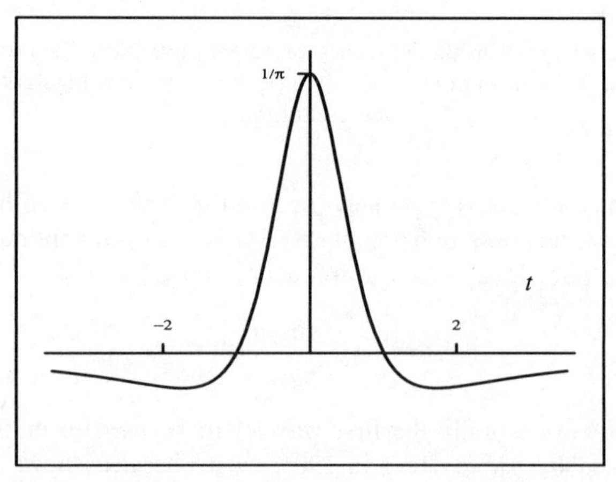

Fig. 7.3 The Poisson wavelet in the time representation.

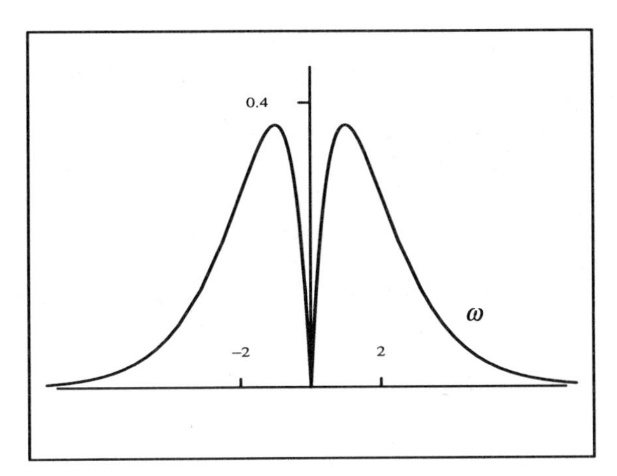

Fig. 7.4 The Poisson wavelet in the frequency representation. Note the singularity at $\omega = 0$.

Consider now the following boundary value problem: given a function s in $L^p(\mathbb{R})$ find a harmonic function $\mathscr{T}$ in the half-plane such that

$$\text{(i)} \int_{-\infty}^{+\infty} db\, |\mathscr{T}(b, a)|^p \leq c < \infty,$$

$$\text{(ii)} \ \mathscr{T}(\cdot, a) \to s(\cdot)\ (a \to 0) \quad \text{in } L^p(\mathbb{R}). \tag{7.0.2}$$

In general a second-order differential operators needs two boundary conditions (one for the value and a second one for its normal derivative). However, it turns out that the finiteness condition (i) ensures uniqueness and we have the following theorem due to Fatou. For a proof see Stein (1979).

Theorem 7.0.3

The initial value problem (7.0.2) *has exactly one solution for* $1 \leq p \leq \infty$. *It is given by*

$$\mathscr{T}(b, a) = P_a * s(b), \qquad P_a(t) = \frac{1}{a} P\left(\frac{t}{a}\right)$$

and P *as given by* (7.0.1). *In addition, we have for almost every* $b \in \mathbb{R}$ *that* $\mathscr{T}(b, a) \to s(b)$ *as* $a \to 0$.

Note that the following semigroup property holds:

$$P_\alpha * P_\beta = P_{\alpha+\beta}.$$

By direct computation one can verify that $\mathcal{T}$ is actually harmonic in the half-plane. The proof of the uniqueness, however, is a little more complicated. The kernel $P_a(b)$ is called the *Poisson* kernel and the function $\mathcal{T}$ in the theorem is called the *harmonic continuation* of s into the upper half-plane.

The wavelet transform with respect to the Poisson wavelet g is theorefore given by $\mathcal{W}_g s(b, a) = -\alpha \partial_a \mathcal{T}(b, a)$, where $\mathcal{T}$ is the harmonic continuation of s into the upper half-plane. This wavelet is actually a classic in harmonic analysis. Many theorems that we will prove for general wavelets have been proved before for the Poisson wavelet. For a general overview of the analysis of functions using this wavelet we refer to Stein (1979).

Example 7.0.4 (*D. Marr's wavelet*)

As we will now see, the speed of approach to equilibrium in a diffusion process is related to a wavelet transform. Let us adopt a physical language at this stage. Then $s(x)$ corresponds to a (one-dimensional) temperature profile. If no heat sources are present, this distribution will evolve according to the heat equation

$$(\partial_t - \partial_x^2)\mathcal{T}(x, t) = 0, \qquad t > 0,\ x \in \mathbb{R},$$

with the boundary condition $\mathcal{T}(x, 0) = s(x)$. This gives a family of smoothed functions $s_t = \mathcal{T}(\cdot, t)$ which will eventually tend to 0—the thermodynamic equilibrium can be seen in Figure 7.5. The speed of heat loss defines a function $\mathcal{R}$ over the half-plane:

$$\mathcal{R}(x, \sqrt{t}) = \partial_t \mathcal{T}(x, t).$$

This function has been used by D. Marr in the problem of edge detection

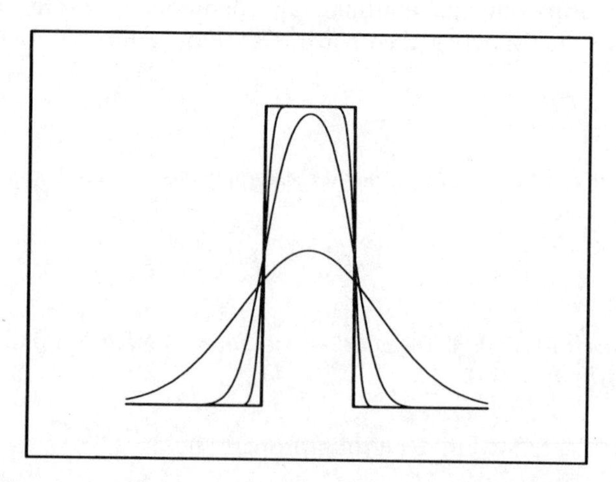

Fig. 7.5 The diffusion process starting from a characteristic function.

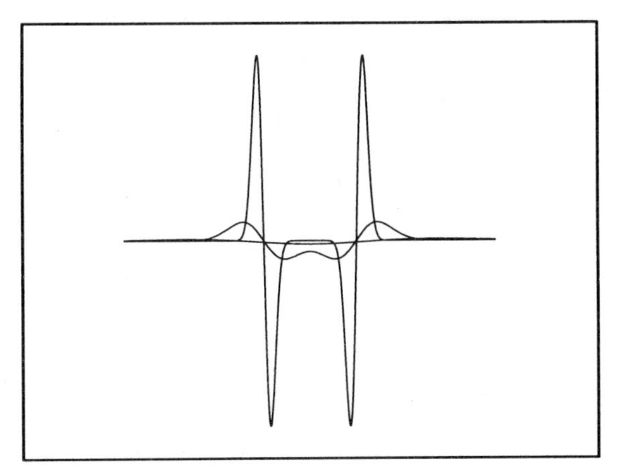

Fig. 7.6 The heat-loss rate. Note that the heat loss is most important at the sharp edges of the initial condition.

(Marr 1982). The idea is that the diffusion process will act strongest at the sharp corner of the function and hence the coefficients $\mathscr{R}(x, t)$ will be largest in the limit $t \to 0$ near the points x where $s(x)$ has a sharp transition. This is actually true, as can be seen in Figure 7.6. As is well known, the time evolution is given by convolution with the heat kernel:

$$\mathscr{T}(x, t) = K_t * s(x), \qquad K_t(x) = \frac{1}{\pi\sqrt{t}} e^{-x^2/t}.$$

Again a semigroup property holds:

$$K_t * K_u = K_{t+u}.$$

For the function $\mathscr{R}$ over the half-plane we therefore obtain (changing back to our standard b, a notation upon setting $b = t, a = \sqrt{t}$)

$$\mathscr{R}(b, a) = g_a * s(b), \qquad g(t) = \frac{1}{\pi}(t^2 - 1)\, e^{-t^2}, \; g_a(x) = g(t/a)/a,$$

which is again a wavelet transform. The first two moments of g vanish according to the decrease of $\hat{g}(\omega) = \omega^2 e^{-\omega^2}$ at $\omega \to 0$. In Figures 7.7 and 7.8 we illustrate this.

8 The progressive wavelets

In this section we will take a closer look at $H^2_+(\mathbb{R})$, the space of progressive functions. Recall that we defined this space as the closed subspace of $L^2(\mathbb{R})$

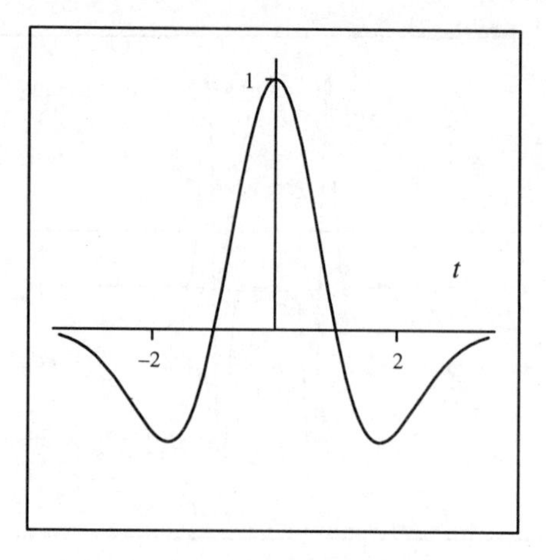

Fig. 7.7 D. Marr's wavelet in the time representation. Note that it looks like a smoothed second derivative.

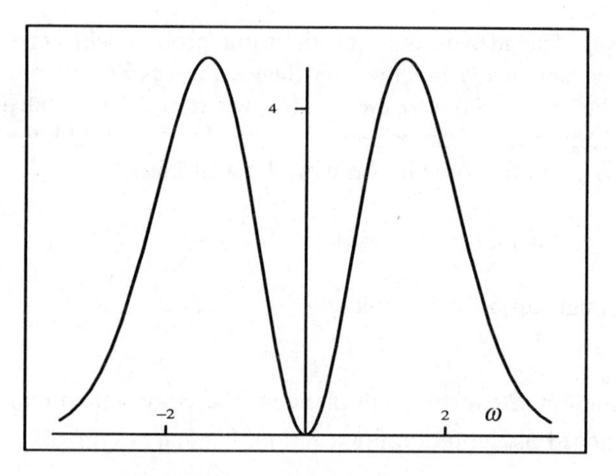

Fig. 7.8 Its Fourier space picture.

of functions having only positive frequencies. Accordingly, a progressive wavelet $g \in H^2_+(\mathbb{R})$ is a superposition of progressive exponentials

$$g(t) = \frac{1}{2\pi} \int_0^\infty d\omega \, \hat{g}(\omega) \, e^{i\omega t},$$

where the integral is to be taken in the mean square limit sense.

Since ω is never negative under the integral we may replace t by $z = t + ix$, where $x > 0$, and the integral is still convergent. It is even absolutely convergent since $|e^{i\omega z}| \leq e^{-\omega x}$. For the same reason we may derive with respect to z under the integral thereby showing that

$$g(z) = \frac{1}{2\pi} \int_0^\infty d\omega\, \hat{g}(\omega)\, e^{i\omega z}, \qquad z = t + ix,\ x > 0,$$

is an analytic function over the upper half-plane. In addition, since $|e^{-\omega x}| \leq 1$ under the integral the integrand is square integrable and it follows by Parseval's equation that we have

$$\int_{-\infty}^{+\infty} dt\, |g(t + ix)|^2 = \frac{1}{2\pi} \int_0^\infty d\omega\, |\hat{g}(\omega)|^2\, e^{-2\omega x} \leq c < \infty$$

for some constant not depending on $x > 0$. Note the similarity between this condition and the one we encountered in the case of harmonic functions over the half-plane. This is actually a complete characterization of $H^2_+(\mathbb{R})$ as is stated by Paley and Wiener's theorem (1934), to which we refer for a proof.

Theorem 8.0.1

Let s be a function in $H^2_+(\mathbb{R})$. Then there is a unique function $\mathcal{T}$, analytic in the upper half-plane such that

(i) $$\int_{-\infty}^{+\infty} dt\, |\mathcal{T}(t + ix)|^2 \leq c < \infty,$$

(ii) $$\lim_{x \to 0} \int_{-\infty}^{+\infty} dt\, |\mathcal{T}(t + ix) - s(t)|^2 = 0.$$

Vice versa, for any analytic function $\mathcal{T}$ over the upper half-plane satisfying (i) *there is a function $s \in H^2_+(\mathbb{R})$ such that s is the boundary value of $\mathcal{T}$ in the sense of* (ii).

Since $\int_0^\infty d\omega\, e^{-\omega x + i\omega t} = (x - it)^{-1}$ it follows that for $s \in H^2_+(\mathbb{R})$ the analytic continuation into the upper half-plane is given by the Cauchy kernel

$$s(t + ix) = C_x * s, \qquad C_x = x^{-1} C(\cdot/x), \qquad C(t) = \frac{1}{\pi(1 + it)}.$$

Again, a semigroup property holds for the Cauchy kernels

$$C_x * C_y = C_{x+y}.$$

The analogue holds for $H^2_-(\mathbb{R})$ where the upper half-plane has to be exchanged with the lower half-plane. A second possibility is to keep the upper half-plane but to exchange analyticity for anti-analyticity. Recall that this means that $\mathcal{T}(\bar{z})$ is analytic.

8.1 Progressive wavelets with real-valued frequency representation

A special set of progressive wavelets is the set of functions for which the Fourier transform is real valued. Here the real and the imaginary parts are given by

$$\mathfrak{R}g(t) = \frac{1}{4\pi}\int_{-\infty}^{+\infty} d\omega(\hat{g}(\omega) + \hat{g}(-\omega))\, e^{i\omega t} = \frac{1}{2\pi}\int_{0}^{\infty} d\omega\, \hat{g}(\omega)\cos(\omega t),$$

$$\mathfrak{I}g(t) = \frac{1}{4\pi}\int_{-\infty}^{+\infty} d\omega(\hat{g}(\omega) - \hat{g}(-\omega))\, e^{i\omega t} = \frac{1}{2\pi}\int_{0}^{\infty} d\omega\, \hat{g}(\omega)\sin(\omega t),$$

and thus the real part is an even function whereas the imaginary part is an odd function. They are related by a *Hilbert transform*

$$\mathfrak{R}g = -H\mathfrak{I}g, \qquad \mathfrak{I}g = H\mathfrak{R}g.$$

This transformation is essentially defined as the multiplication by $-i\, sign\, \omega = -i\omega/|\omega|$ in Fourier space:

$$H\colon \hat{s}(\omega) \mapsto -i\,\mathrm{sign}(\omega)\cdot\hat{s}(\omega). \tag{8.1.1}$$

Since in Fourier space H is multiplication by a bounded function it follows that H is continuous on $L^2(\mathbb{R})$. Now multiplication in Fourier space corresponds to convolution (3.0.3) and thus since

$$\int_{-\infty}^{+\infty} d\omega\, \mathrm{sign}(\omega)\, e^{-\epsilon|\omega| + it\omega} = \frac{2it}{t^2 + \epsilon^2},$$

we may use the approximations $e^{-\epsilon|\omega|}$ sign ω to sign ω to see that in the time representation the Hilbert transform reads

$$Hs(t) = \frac{1}{\pi}\lim_{\epsilon\to 0}\int_{-\infty}^{+\infty} du\, s(u)\frac{t-u}{(t-u)^2 + \epsilon^2},$$

where the convergence holds *a priori* only in $L^2(\mathbb{R})$ thanks to the dominated convergence. It actually holds pointwise almost everywhere (e.g. Stein 1979). Note that this is formally a convolution with $1/(\pi t)$.

From the representation of the Hilbert transform in Fourier space (8.1.1), we see that the orthogonal projectors $\Pi^+ : L^2(\mathbb{R}) \to H^2_+(\mathbb{R})$ and $\Pi^- : L^2(\mathbb{R}) \to H^2_-(\mathbb{R})$ on the progressive and regressive functions can be expressed as follows:

$$\Pi^+ = \tfrac{1}{2}(\mathbb{1} + iH), \qquad \Pi^- = \tfrac{1}{2}(\mathbb{1} - iH).$$

Example 8.1.1 (*The Cauchy wavelets*)

Actually, most explicit computations in this book makes use of them because of their nice algebraic and analytic properties. They are defined as (see

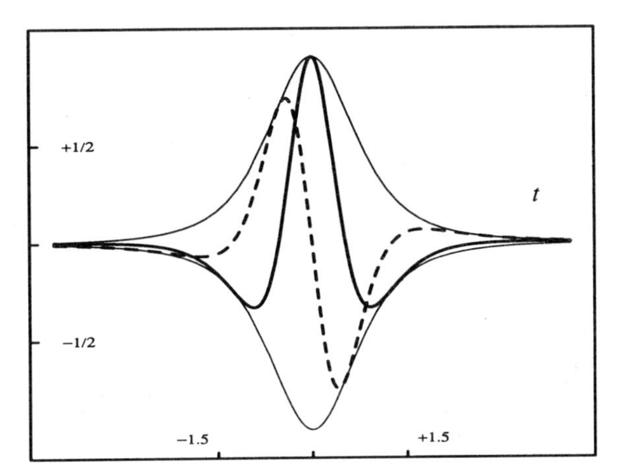

Fig. 8.1 The Cauchy wavelet for $\alpha = 2$ in time representation. The phase turns only finitely many times.

Figure 8.1)

$$g(t) = (2\pi)^{-1}\Gamma(\alpha + 1)(1 - it)^{-(1+\alpha)}, \qquad \alpha > 0,$$

where the gamma function is defined for $\Re z > 0$ by

$$\Gamma(z) = \int_0^\infty dt\, t^{z-1} e^{-t}.$$

These wavelets are of high regularity but have at most a polynomial decay at ∞. This is mirrored in the fast decrease of the Fourier coefficients and a lower regularity of the Fourier transform (at $\omega = 0$) (see Figure 8.2):

$$\hat{g}(\omega) = \begin{cases} \omega^\alpha e^{-\omega} & \text{for } \omega > 0, \\ 0 & \text{otherwise.} \end{cases}$$

Here we recognize that for $\alpha \in \mathbb{N}$ these wavelets are nothing but the αth derivative of the Cauchy kernel. Accordingly, we obtain the following interpretation of the wavelet transform of $s \in L^2(\mathbb{R})$ with respect to these wavelets: take the projection $\Pi^+ s$ of s into the space of progressive functions $H^2_+(\mathbb{R})$. This function can be extended to an analytic function over the upper half-plane. If we call this function $\mathscr{T}(z)$ we have ($\alpha \in \mathbb{N}$)

$$\mathscr{W}_g s(b, a) = a^\alpha \mathscr{T}^{(\alpha)}(b + i\alpha), \qquad \mathscr{T}^{(\alpha)}(z) = \partial_z^\alpha \mathscr{T}(z).$$

Thus the wavelet transform with respect to these wavelets is closely connected to the analysis of analytic functions over the half-plane. This observation is due to Paul (1985), who has used the wavelets in quantum mechanics.

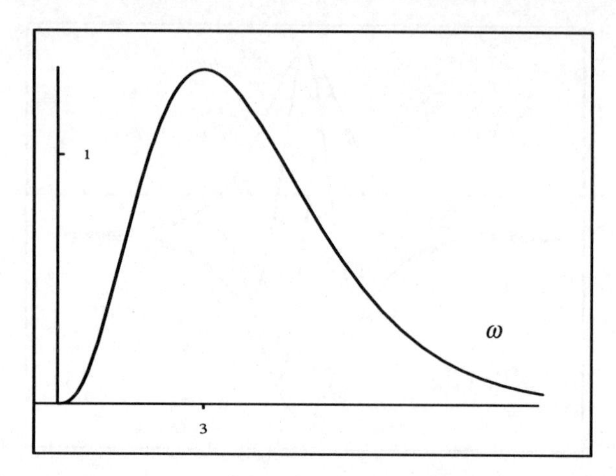

Fig. 8.2 The Cauchy wavelet for $\alpha = 2$ in frequency representation. Note the singular (=non-smooth) behaviour at $\omega = 0$.

These wavelets are complex valued and we may distinguish between the modulus $|g(t)|$ and the phase $\arg g(t)$. The phase is only defined up to an integer multiple of 2π. But on every open set on which $|g(t)| \neq 0$ it may be chosen to be a continuous function if g itself is continuous. At every point t where $g(t) = 0$ the phase is not defined. The phase speed is called the *instantaneous frequency*

$$\Omega(t) = \frac{d}{dt} \arg g(t).$$

This name is justified by the fact that the instantaneous frequency of the pure frequency $e^{i\omega t}$ is equal to ω.

By direct computation we can verify that for the Cauchy wavelets we have

$$|g(t)| = (1 + t^2)^{-(1+\alpha)/2}, \qquad \arg g(t) = (1 + \alpha) \arctan t.$$

Therefore, the phase turns $(1 + \alpha)/2$ times as t goes from $-\infty$ to $+\infty$.

Example 8.1.2 (*J. Morlet's* Gaussian 'wavelets')
These where first used in geophysical explorations (Grossman and Morlet 1985) and are at the origin of the development of wavelet analysis since this time. The wavelets are obtained by shifting a Gaussian function in Fourier space, or, what amounts to the same, by multiplying it by an exponential (see Figures 8.3 and 8.4):

$$g(t) = e^{i\omega_0 t} e^{-t^2/2}, \qquad \hat{g}(\omega) = \sqrt{2\pi}\, e^{-(\omega - \omega_0)^2/2}.$$

Strictly speaking, this function is not a progressive wavelet; it is not even a wavelet, because it is not of zero mean. However, the negative frequency components of g are small compared to the progressive component if

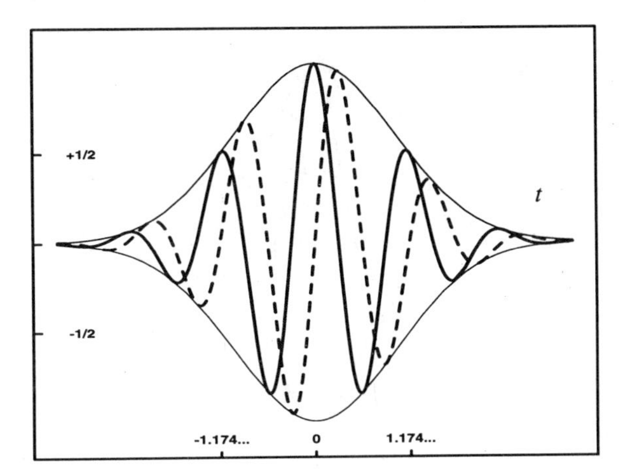

Fig. 8.3 J. Morlet's wavelet for $\omega_0 = 5.336\ldots$ in time representation. Note that the real part (solid line) and the imaginary part (dotted line) are oscillating around each other with constant phase speed. The modulus and negative modulus (thick solid line) are an envelope for these oscillations.

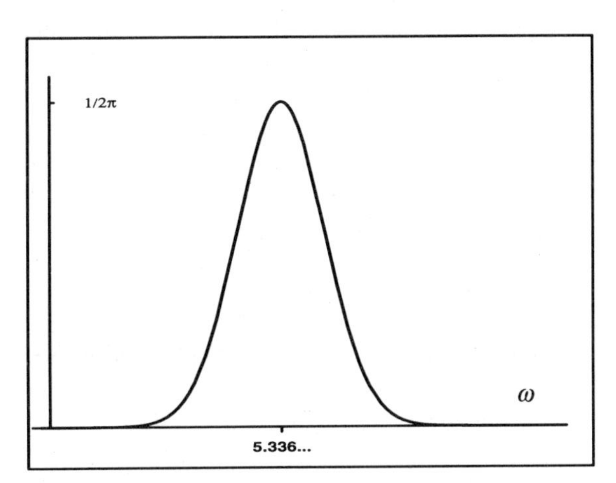

Fig. 8.4 The Fourier transform of J. Morlet's wavelet for $\omega_0 = 5.336\ldots$. Note that it does not vanish at the origin $\omega = 0$, but that it is numerically small.

$\omega_0 > 0$ is large enough. Therefore,

$$g(t) = e^{i\omega_0 t}\, e^{-t^2/2} + \eta(t),$$

where η is a small corrective term that makes g a progressive wavelet. The phase pulses—up to a correction—at a constant speed

$$\frac{d\Phi(t)}{dt} = \Omega(t) = \omega_0 + \rho(t).$$

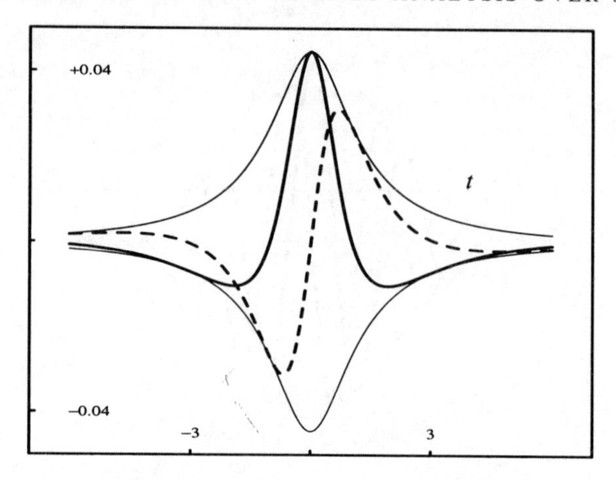

Fig. 8.5 The Bessel wavelet in the time representation.

In applications in signal processing it has been found that a particularly useful value for the internal frequency ω_0 is the one for which the real part touches the envelope $|g|$ at half its height. The smallest value is $5.336\ldots$.

Example 8.1.3 (*The Bessel wavelet*)

This is the first example of a family of wavelets that play a particular role in the analysis of distributions. The wavelets are very well localized in the time and frequency representations. The specific wavelet that we have in mind is given in Fourier space by (see Figure 8.6)

$$\hat{g}(\omega) = \begin{cases} e^{-(\omega + 1/\omega)} & \text{for } \omega > 0, \\ 0 & \text{for } \omega \leq 0. \end{cases}$$

In the time representation the wavelet is a *Bessel* function (see Figure 8.5):

$$g(t) = \frac{1}{\pi\sqrt{1 - it}} \mathrm{K}_1(2\sqrt{1 - it}).$$

This wavelet is very well localized in Fourier space, which is mirrored in the high regularity of g. The high regularity of the frequency representation implies high localization of g. Indeed, from the asymptotic behaviour of the Bessel function:

$$\mathrm{K}_1(z) = \sqrt{\frac{\pi}{2z}}\, e^{-z}\left(1 + O\left(\frac{1}{z}\right)\right), \qquad \Re z > 0,\ |z| \to \infty,$$

we obtain

$$g(t) = \frac{i}{2\sqrt{\pi}} t^{-1} e^{-(1+i)\sqrt{t/2}}(1 + O(t^{-1/2})), \qquad t \to +\infty.$$

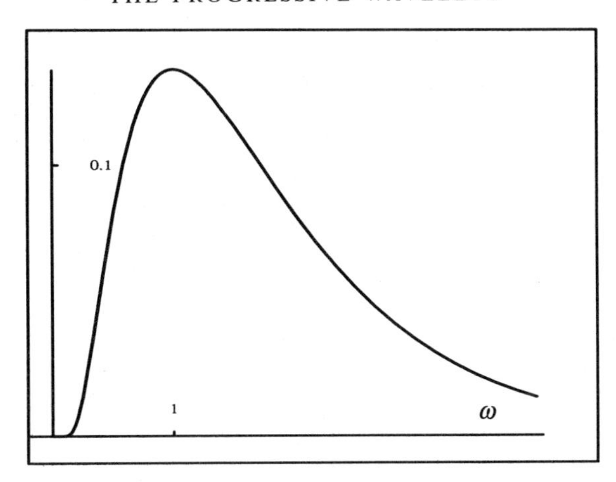

Fig. 8.6 The frequency representation of the Bessel wavelet. Note the fast decrease at $\omega = 0$.

The behaviour at $-\infty$ is obtained by complex conjugation. The phase turns an infinite number of times, with an instantaneous frequency that tends to 0 at large times

$$\frac{d\Phi(t)}{dt} \simeq \frac{1}{\sqrt{8t}}, \qquad |t| \to \infty .$$

Here we have introduced the following notation: we write $s(t) \simeq r(t)$, $t \to t_0$ if $\text{limit}_{t \to t_0} s(t)/r(t) = 1$, or, what amounts to the same, iff $s(t) = r(t)(1 + o(1))$, $t \to t_0$. All moments of g vanish because of the fast decrease of its Fourier transform at $\omega = 0$.

Note that a wavelet for which all moments exist and are 0 cannot be exponentially localized, since suppose that for some $\alpha > 0$ we had $|g(t)| \leq O(1)\, e^{-\alpha|t|}$. Then the Fourier transform of g would be analytic in a strip $\Im z \in (-\alpha, +\alpha)$ and thus if $\hat{g}$ is vanishing together with all its derivatives at $\omega = 0$ it follows, thanks to the analyticity, that $\hat{g} \equiv 0$ inside the whole strip of analyticity, and thus $g \equiv 0$, too. This in turn implies that a function of exponential decay is determined by all its moments.

8.2 Chirp wavelets

Now we discuss two progressive wavelets with non-real Fourier transform.

Example 8.2.1 (The linear-chirp wavelet)
Here we add a quadratic phase to the linearly growing phase of the previous

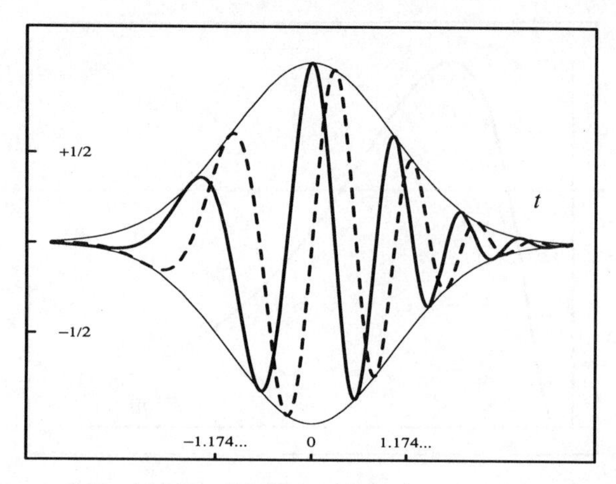

Fig. 8.7 The modulus (thick solid line), the real part (thin solid line), and the imaginary part (dotted line) of the linear chirp wavelet in the time representation. The instantaneous frequency increases visibly for large times.

example (see Figure 8.7):

$$g(t) = e^{i[\omega_0 + \beta t/2]t}\, e^{-t^2/2} + \eta(t).$$

The small correction η again ensures that g is a progressive wavelet. The modulus is again Gaussian. The instantaneous frequency is—up to a correction ρ due to η—linearly growing in time:

$$\frac{d\Phi(t)}{dt} = \Omega(t) = \omega_0 + \beta t + \rho(t).$$

In frequency space we have

$$\hat{g}(\omega) = \sqrt{\frac{2\pi}{1 - i\beta}}\, e^{-(\omega - \omega_0)^2/2(1 - i\beta)} + \hat{\eta}(\omega) \tag{8.2.1}$$

and thus the Fourier transform itself has a quadratic phase (see Figure 8.8). From the Fourier transform of g we see that the correction η is small if [5]

$$\omega_0 \gg \beta. \tag{8.2.2}$$

Therefore, the instantaneous frequency should only grow slowly compared to the central frequency ω_0.

[5] The sign '≫' means 'large compared to'.

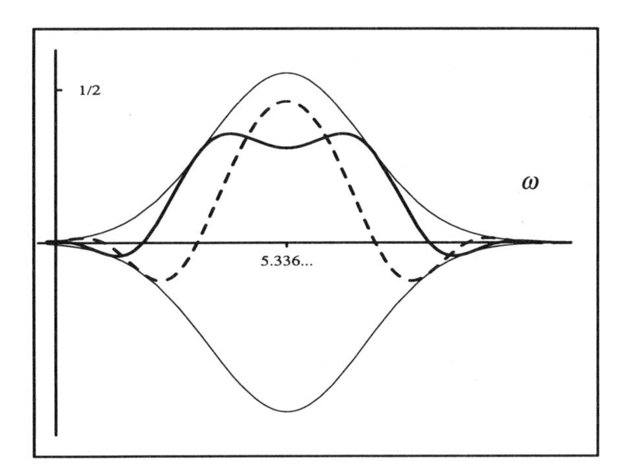

Fig. 8.8 The modulus (thick solid line), the real part (thin solid line), and the imaginary part (dotted line) of the linear chirp wavelet in the frequency representation. Note the zero frequency contribution, because (8.1.3) is not quite valid.

Example 8.2.2 (The asymptotic hyperbolic chirp wavelet)
Consider the Cauchy wavelet (example 8.1) where we take α complex valued, say $\alpha = \gamma + i\beta$ (see Figures 8.9 and 8.10):

$$g(t) = (2\pi)^{-1}\Gamma(\gamma + 1 + i\beta)(1 - it)^{-(1+\gamma+i\beta)}, \qquad \gamma > 0.$$

By direct computation we can verify that

$$M(t) = (2\pi)^{-1}|\Gamma(\gamma + i\beta)|\, e^{-\beta \arctan t}(1 + t^2)^{-(1+\gamma)/2},$$

$$\Phi(t) = \arg \Gamma(\gamma + 1 + i\beta) - \beta \log \sqrt{1 + t^2} + (1 + \gamma) \arctan t.$$

Note that for large $\beta > 0$ the positive times $t \gg 0$ are dumped down by a factor $e^{-\beta\pi/2}$, whereas the negative times are amplified by roughly $e^{\beta\pi/2}$. This explains the asymmetry in Figure 8.9. The phase turns infinitely many times. The instantaneous frequency asymptotically satisfies

$$\frac{d\Phi(t)}{dt} = \Omega(t) = -\frac{\beta t}{1 + t^2} + \frac{1 + \gamma}{1 + t^2} = -\frac{\beta}{t} + O\left(\frac{1}{t^2}\right), \qquad (|t| \to \infty),$$

which justifies the name 'asymptotic hyperbolic chirp'.

8.3 On the modulus of progressive functions

The Morlet wavelet was a progressive wavelet with a modulus that was essentially given by a Gaussian function. So the question is whether we can find a function in $H^2_+(\mathbb{R})$ whose modulus is exactly Gaussian. Unfortunately,

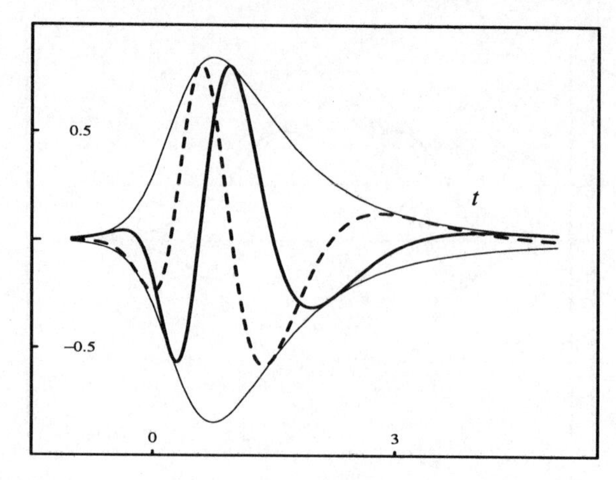

Fig. 8.9 The hyperbolic chirp wavelet in the time representation for $\alpha = 3 + 3i$.

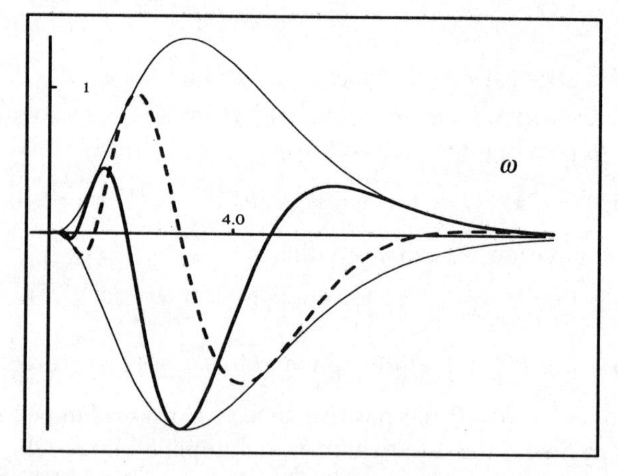

Fig. 8.10 The same wavelet in the frequency representation.

it turns out that this is not possible, as is shown by the following theorem due to Paley and Wiener (1934), to which we refer the reader for a proof.

Theorem 8.3.1
Let $s \in L^2(\mathbb{R})$ be a non-negative function $\neq 0$. Then there is a real-valued phase function $\phi(t)$ such that $r(t) = s(t)\, e^{i\phi(t)}$ satisfies $r \in H^2_+(\mathbb{R})$ if and only if

$$\int_{-\infty}^{+\infty} dt \, \frac{|\log s(t)|}{1+t^2} < \infty .$$

Therefore, no progressive function $s \in H^2_+(\mathbb{R})$ can have a modulus that decays monotonically as $t \to \infty$ as quickly as e^{-ct^2} for any choice of the constant $c > 0$.

Note that in any case the phase function is not unique. One possible realization is given by the following formal expression using the Hilbert transform:

$$\phi(t) = H \log s(t),$$

which states, as we have seen, that $\log s(t) + i\phi(t)$ is (formally) a progressive function. It thus has an analytic continuation into the upper half-plane. Its exponential $s(t)\, e^{i\phi(t)}$ is then again an analytic function over the half-plane, and it can be shown that it satisfies the condition of the Paley–Wiener theorem.

9 Some explicit analysed functions and easy examples

In this section we list some examples that reveal the main features of the wavelet transform. Although not all statements will be made mathematically precise, this section serves to enlarge our intuitions about wavelet transforms.

9.1 The wavelet transform of pure frequencies

These functions are related to the invariant strips in the half-plane. Consider the function $e_{\xi_0}(t) = e^{i\xi_0 t}$, which describes an oscillation with frequency ξ_0. The analysing wavelet should be absolutely integrable in order to have an everywhere well-defined wavelet transform.

Since e_{ξ_0} is a progressive function,[6] every voice of the wavelet transform will be a progressive function, too.

The pure oscillations are, up to a phase invariant under translations,

$$T_b\, e_{\xi_0} = e^{-i\xi_0 b}\, e_{\xi_0}.$$

Some of the main features of the wavelet transform of pure oscillations follow from this translation invariance and from the co-variance of the wavelet transform. Using the co-variance property of (4.0.1) we may write:

$$\mathcal{W}[g; e_{\xi_0}](b, a) = \mathcal{W}[g; e_{\xi_0}(t + b)](0, a).$$

From the invariance of e_{ξ_0} and the linearity of the wavelet transform it follows that

$$\mathcal{W}[g; e_{\xi_0}](b, a) = e^{-i\xi_0 \beta}\mathcal{W}[g; e_{\xi_0}](0, a).$$

---

[6] Strictly speaking, this holds only in the sense of distributions. For more details see Section 24.

Therefore every voice—the scale a is fixed to a constant—is a pure oscillation with the same frequency ξ_0 but with an amplitude and a phase that may vary from voice to voice. The whole transform is determined by the wavelet transform along a straight line passing through the border line of the half-plane, for example, $b = 0$.

Carrying out the integration

$$\mathscr{W}[g; e_{\xi_0}](0, a) = \int_{-\infty}^{+\infty} dt \frac{1}{a} \bar{g}\left(\frac{t}{a}\right) e^{i\xi_0 t} = \bar{\hat{g}}(a\xi_0)$$

we obtain

$$\mathscr{W}[g; e_{\xi_0}](b, a) = e^{i\xi_0 b} \bar{\hat{g}}(a\omega)$$

and thus the modulus and the phase of the wavelet transform read

$$M(b, a) = |\bar{\hat{g}}(a\xi_0)|, \qquad \Phi(b, a) = \arg \bar{\hat{g}}(a\xi_0) + \xi_0 b. \tag{9.1.1}$$

Therefore, the instantaneous frequency of every voice is imposed by the analysed pure oscillation[7]

$$\partial_b \Phi(b, a) = \xi_0.$$

The modulus of the wavelet transform is constant along every voice ($a =$ const.) and its behaviour along the line $b = 0$ is given by the modulus of the Fourier transform of the wavelet.

Suppose that $|\hat{g}|$ has its maximal value for $\omega = \omega_0$. Then the modulus of the wavelet transform takes its maximal value for the voice at scale

$$a = \omega_0/\xi_0. \tag{9.1.2}$$

This might indicate that ω_0/a is somehow related to a frequency. The localization around this central voice depends on the localization of the spectral envelope of g around ω_0. Therefore, the analysis of a pure oscillation with the help of the Haar wavelet of (7.0.1) wil give rise to a poor localization in the half-plane (see Figures 9.1 and 9.2), whereas, for example, the Bessel wavelet gives rise to a high localization around the scale (see Figures. 9.3 and 9.4).

Consider the lines of constant phase, that is, for example, the set of points in $\mathbb{H}$ where the phase Φ has a given value. If the phase of $\hat{g}$ is constant as, for example, if the wavelet has a real-valued Fourier transform, then these lines are the straight lines $b =$ const.

9.2 Chirp wavelets

If the Fourier transform of g itself has a phase that varies with ω, then the lines of constant phase are curved. As an example we consider the linear

[7] The phase itself is only defined modulo 2π, but if $\mathscr{W}_g s$ is differentiable, then the derivatives $\partial_b \Phi(b, a)$ and $\partial_a \Phi(b, a)$ are uniquely defined wherever $|M(b, a)| \neq 0$.

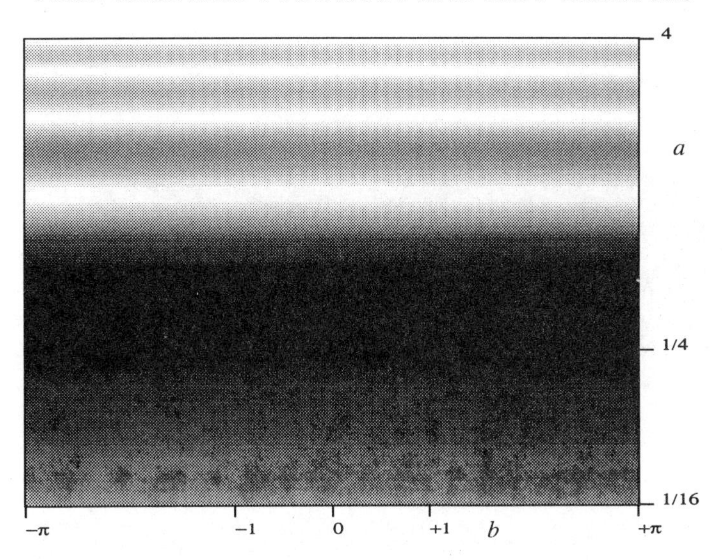

Fig. 9.1 The modulus of the wavelet transform of the pure oscillation with the help of Haar's wavelet. Black corresponds to high values whereas white are small values. Note that the maximum of the amplitude is at scale $a = 1/4$. However, there are oscillations for large scales due to the poor localization in frequency space of the Haar wavelet. Note that the scale-parameter axis is logarithmic.

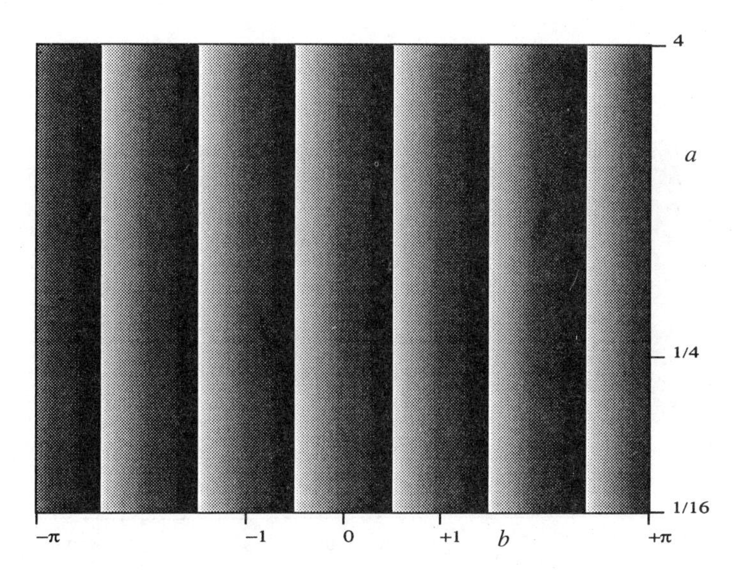

Fig. 9.2 The phase modulo 2π of the wavelet transform of a pure oscillation with the help of the Haar's wavelet. We use this convention that black is $+\pi$ and white corresponds to a phase of $-\pi$. Note that the phase turns at the speed of the analysed function.

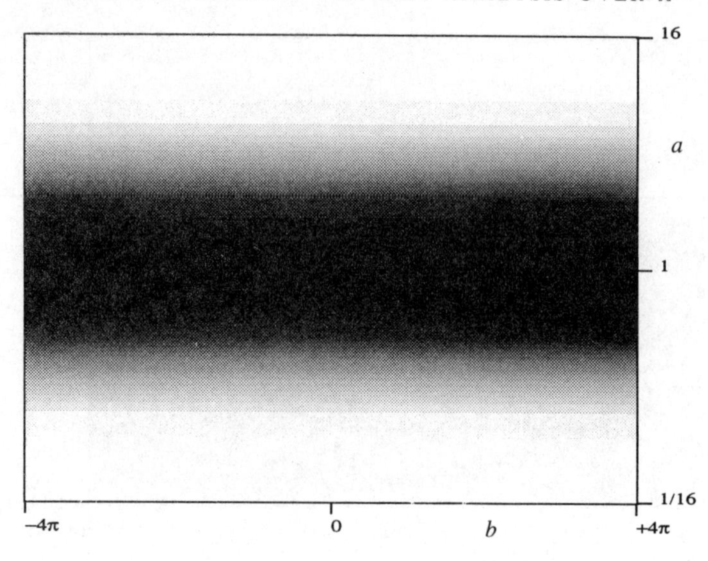

Fig. 9.3 The modulus of the wavelet transform of a pure oscillation with the help of the Bessel wavelet. Note that it is well localized around the central voice $a = \omega_0/\xi_0 = 1$.

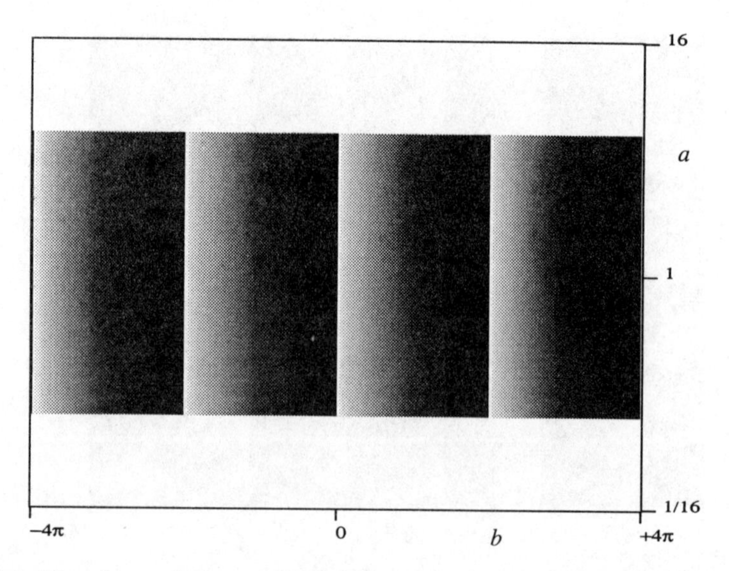

Fig. 9.4 The phase of the wavelet transform of a pure oscillation with the help of the Bessel wavelet. The phase pulses at a constant speed $\xi_0 = 1$, independent of the voice.

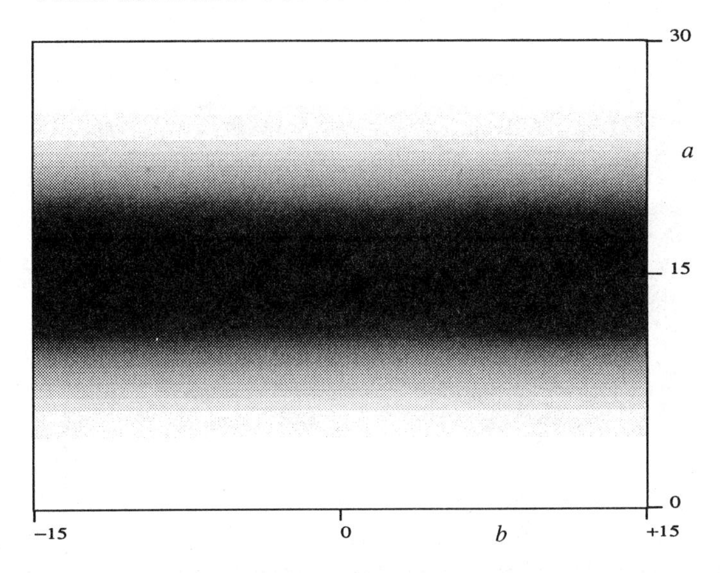

Fig. 9.5 The modulus of the wavelet analysis of pure frequency with the help of a linear chirp wavelet. Again the modulus is localized around the scale that corresponds to the analysed frequency.

chirp wavelet. From (8.2.1) and (9.1.1) we obtain that the lines of constant phase are given by

$$\frac{\beta\xi_0^2}{1+\beta^2}\left(a-\frac{\omega_0}{\xi_0}\right)^2 - 2\xi_0 b = \text{const.}$$

These lines are parabolas with the top at the scale $a = \omega_0\xi_0$. The curvature of the parabola at its top is proportional to ξ_0, as we have sketched in Figures 9.5 and 9.6. This result remains qualitatively true for every wavelet whose frequency representation has a quadratic phase.

9.3 The real oscillations

Consider

$$s(t) = \cos(\xi_0 t + \phi) = \tfrac{1}{2}(e^{-i\phi}\, e^{-i\xi_0 t} + e^{i\phi}\, e^{i\xi_0 t}).$$

The frequency content of s is localized at the frequencies $-\xi_0$ and ξ_0. The associated progressive function is the pure oscillation

$$\Pi^+ s(t) = e^{i\phi}\, e^{i\xi_0 t}/2.$$

For progressive wavelets the wavelet transform of s is given by the wavelet transform of the associated progressive function $\Pi^+ s$, which was treated in the previous example. A real-valued wavelet 'sees' instead the positive and

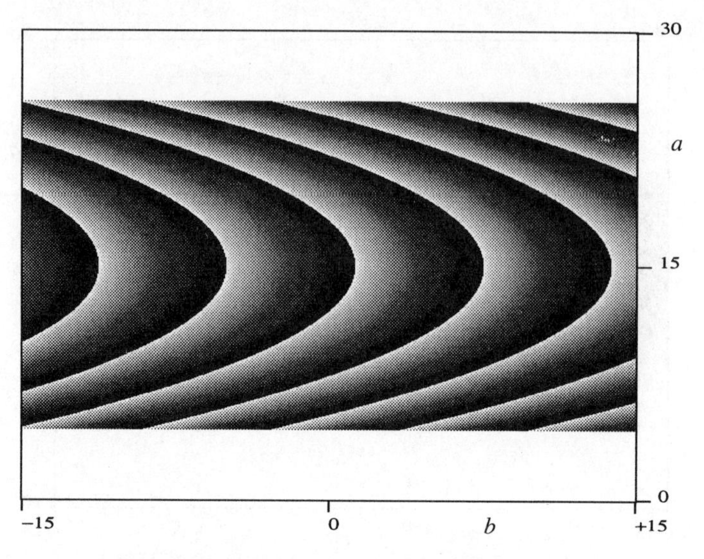

Fig. 9.6 The corresponding phase picture. Note that the curvature is most important at the scale corresponding to the analysed frequency. The phase speed for every voice is again given by the analysed frequency.

the negative frequency components. By (3.0.5) we obtain

$$\mathscr{W}[g; \cos(\xi_0 t + \phi)](b, a) = \Re(\hat{g}(a\xi_0)\, e^{i(b\xi_0 + \phi)}).$$

This implies that every voice is a real oscillation with frequency ξ_0 and a phase and an amplitude that varies with one voice to the other:

$$\mathscr{W}[g; \cos(\xi_0 t + \phi)](b, a) = A(a) \cos(\xi_0 b + \psi(a)),$$
$$A(a) = |\hat{g}(a\xi_0)|, \qquad \psi(a) = \arg \hat{g}(a\xi_0) + \phi.$$

See Figure 9.7 for a sketch.

9.4 The onsets

These functions are related to the dilation invariant regions in the half-plane. Consider a function that is 0 for negative times $t < 0$ and that suddenly starts more or less smoothly:

$$s(t) = |t|_+^\alpha = \Theta(t) t^\alpha, \qquad \alpha > -1, \tag{9.4.1}$$

where we have used the notation $|t|_\pm = (|t| \pm t)/2$. The bigger α is, the smoother is the transition at $t = 0$. In order to analyse these functions, the analysing wavelet g should be localized such that $t^\alpha g$ is still in $L^1(\mathbb{R})$. For instance,

$$|g(t)| \leq c(1 + |t|)^{-(1-\alpha+\epsilon)},$$

with some $\epsilon > 0$, will be enough.

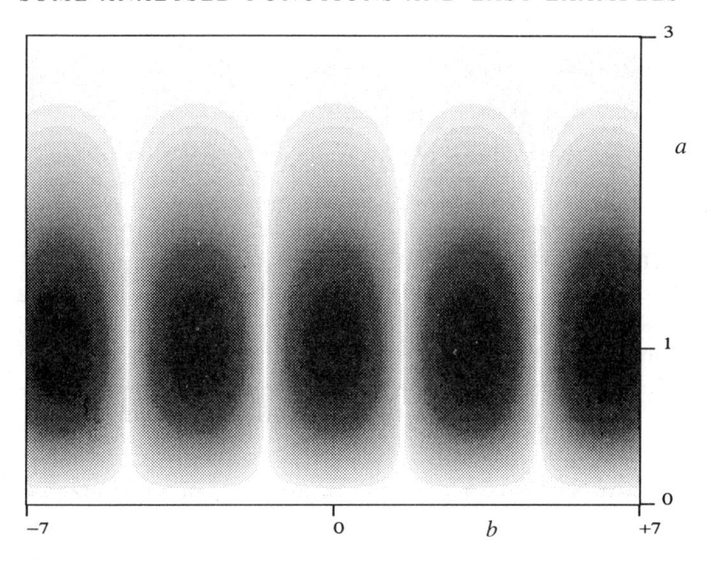

Fig. 9.7 The wavelet transform of a real oscillation with respect to D. Marr's real wavelet. The wavelet transform is real valued. It oscillates at the speed of the analysed function.

The onsets are—up to a scalar—dilation invariant, that is, they are homogeneous functions of degree α:

$$s(\lambda t) = \lambda^a s(t). \tag{9.4.2}$$

As for the pure oscillations, many of the general features of wavelet transforms of homogeneous functions follow from this invariance and the co-variance of the wavelet transform. From the homogeneity of s and (4.0.1) we have

$$\mathscr{W}[g; s](b, a) = \mathscr{W}[g; s(at)](b/a, 1) = a^\alpha \mathscr{W}[g; s](b/a, 1).$$

Therefore, the wavelet transform satisfies the same type of homogeneity as the analysed function ($\lambda > 0$):

$$\mathscr{W}[g; s](\lambda b, \lambda a) = \lambda^\alpha \mathscr{W}[g; s](b, a). \tag{9.4.3}$$

Because α is real valued, the phase of the wavelet remains, according to equation (9.4.3), unchanged along a line $b/a = b_c$ and thus every zoom has a constant phase. Therefore, the lines of constant phase converge towards the origin. The modulus of a zoom instead decreases with a power law that mirrors the type of singularity of the analysed function. This phenomenon has been observed for the first time by Grossmann (1984). In the context of signal analysis it has been used in Grossmann *et al.* (1987) and for the analysis of fractals in Holschneider (1988).

Carrying out the integrations we see that the modulus and the phase of the wavelet transform read in terms of the mellin transform of the shifted versions of g:

$$M(b, a) = a^{\alpha}|F(b/a)|, \qquad \Phi(b, a) = \arg F(b/a),$$

$$F(x) = M[g(\cdot - x)](\alpha + 1).$$

Consider as a first explicit example the wavelet transform of a delta function δ localized at the origin. It is not quite of the same kind as (9.4.1) but it satisfies the invariance equation of (9.4.2) with $\alpha = -1$. By direct computation we find

$$\mathcal{W}[g; \delta](b, a) = \frac{1}{a}\bar{g}\left(\frac{-b}{a}\right),$$

and thus the behaviour of a^{-1} of every zoom. In Figures 9.8 and 9.9 we have sketched the wavelet analysis of δ with the help of a Morlet wavelet.

To give an explicit example of a true onset we will use the Cauchy wavelet $g_\beta = (2\pi)^{-1}\Gamma(\beta + 1)(1 - it)^{-(1+\beta)}$ with $\beta > \alpha$. As we saw in Example 8.1.1, for this wavelet we may identify the half-plane with the complex upper half-plane by setting $z = b + ia$. Indeed, to give a different argument in order to see this, the complex conjugated, dilated and translated wavelet may be

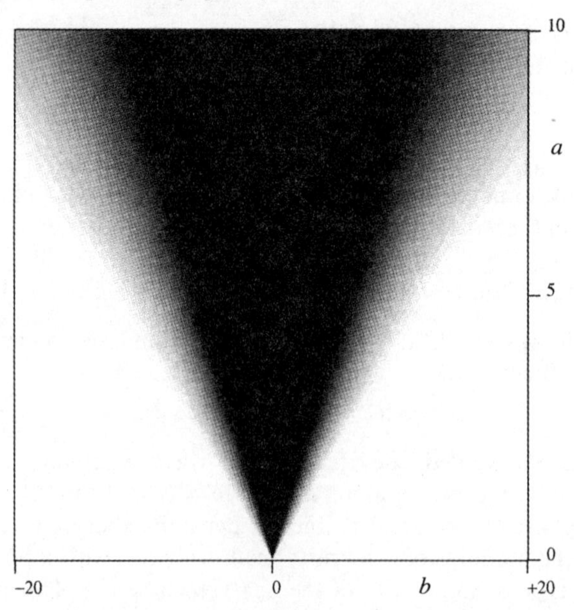

Fig. 9.8 The modulus of the wavelet transform of δ with respect to a Morlet wavelet. For representational reasons we re-scale each voice to take out the trivial factor $1/a$. Thus, $a|\mathcal{W}|$ is actually shown.

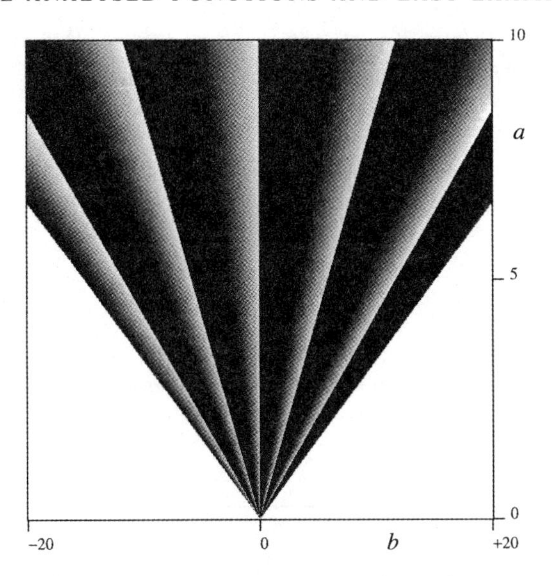

Fig. 9.9 The associated phase picture. Note that the lines of constant phase converge towards the point, where the singularity is located. We have used a cut-off: for small values of the modulus the phase is put to 0.

written as

$$\frac{1}{a} g_\beta\left(\frac{t-b}{a}\right) = (2\pi)^{-1} i^{1+\beta} \Gamma(1+\beta) a^\beta (b + ia - t)^{-1-\beta}$$

$$= (2\pi)^{-1} \Gamma(1+\beta) a^\beta \left(\frac{i}{z-t}\right)^{1+\beta}.$$

Therefore, the wavelet transform with respect to g_β is—up to the pre-factor a^β—an analytic function in $z = b + ia$.

In the case of the onsets we obtain by direct computation (compare Section 29)

$$\left.\begin{aligned} \mathscr{W}[g_\beta; |t|_+^\alpha](b, a) &= c a^\beta (-iz)^{\alpha-\beta}, \\ c = (2\pi)^{-1}\, e^{-i\pi(1+\alpha)/2} &\Gamma(1+\alpha)\Gamma(\beta-\alpha). \end{aligned}\right\} \tag{9.4.4}$$

In Chapter 4 we will use these kinds of wavelets to analyse the local scaling behaviour of fractal functions, that is, functions for which such kinds of singularity accumulate everywhere.

9.5 The wavelet analysis of a hyperbolic chirp

Consider the function

$$s(t) = (-it)^{i\alpha}.$$

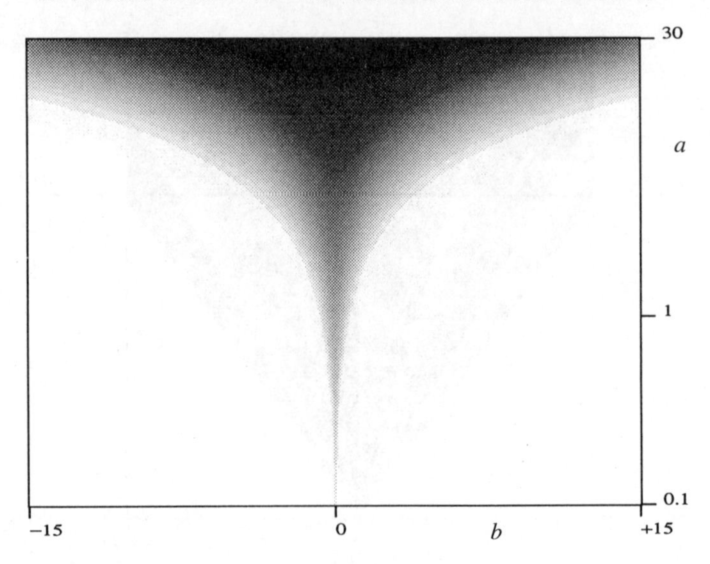

Fig. 9.10 The wavelet analysis of the onset $\sqrt{t}$ with the help of some Cauchy wavelet. Again, the re-scaled modulus: $\sqrt{a}|\mathscr{W}|$ is shown. Note that the a axis is logarithmic.

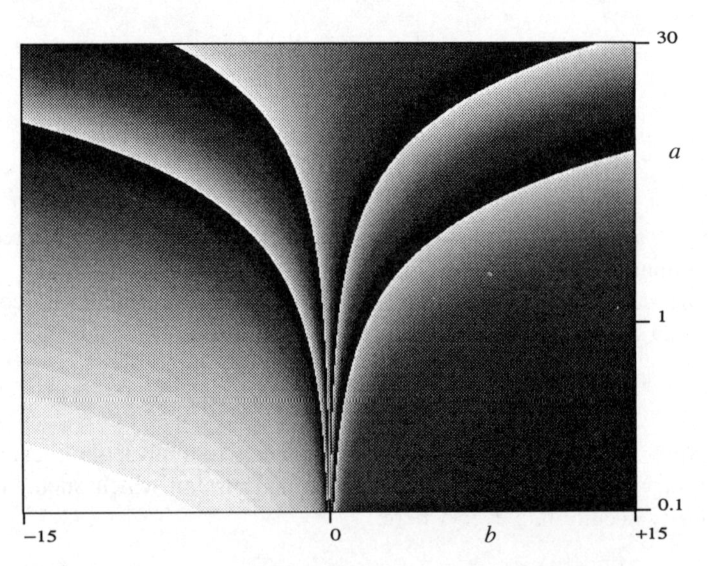

Fig. 9.11 Again, the lines of constant phase for each constant converge towards the point, where the singularity is located. The value of the phase of the zoom $b = 0$ is $-3\pi/4$ according to equation (9.4.4). The straight lines are curves due to the logarithmic scale in a.

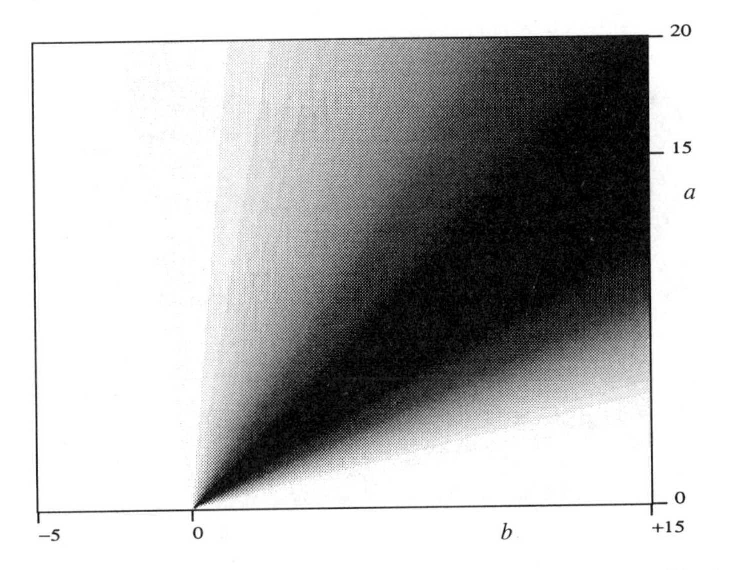

Fig. 9.12 The modulus of the wavelet analysis of a hyperbolic chirp with the help of a Cauchy wavelet. Note that the modulus is localized around the straight line $a = b$ that corresponds to the instantaneous frequency of the chirp.

This is a progressive function (at least in the sense of distributions). In addition, this function satisfies the discrete scaling invariance

$$s(\gamma t) = s(t), \qquad \gamma = e^{2\pi/\alpha},$$

which is mirrored in its wavelet transform via

$$\mathscr{W}_g s(\gamma b, \gamma a) = \mathscr{W}_g s(b, a).$$

The modulus of s is constant—with the exception of $t = 0$ where it jumps—and has a phase speed that decreases with α/t as $|t| \to \infty$:

$$M(t) = \tfrac{1}{2}\cosh\left(\frac{\alpha\pi}{2}\right) + \operatorname{sign}(t)\tfrac{1}{2}\sinh\left(\frac{\alpha\pi}{2}\right), \qquad \Omega(t) = \frac{\alpha}{t},$$

which is the reason for calling s a 'hyperbolic chirp'. The wavelet transform of the hyperbolic chirp with respect to the Cauchy wavelet g_μ is obtained using the analytic continuation in α of (9.4.4):

$$\mathscr{W}[g_\mu; (-it)^{i\alpha}](b, a) = ca^{\mu}(-iz)^{i\alpha-\mu},$$
$$c = (i\pi)^{-1}\sinh(\alpha\pi)\Gamma(1 + i\alpha)\Gamma(\mu - i\alpha).$$

Fig. 9.13 The phase picture of the wavelet transform of a hyperbolic chirp. Note that the lines of constant phase are logarithmic spirals turning around the point (0, 0).

For the modulus and the phase we obtain explicitly

$$M(b, a) = |c| a^{\mu} (b^2 + a^2)^{-\mu/2} \, e^{-\alpha \arctan a/b},$$

$$\Phi(b, a) = \arg c - \mu \arctan(a/b) + \alpha \log \sqrt{b^2 + a^2}.$$

The lines of constant phase are now logarithmic spirals turning around the point (0, 0), as can be seen in Figure 9.13. However, only that part of the spirals that lies in the upper half-plane is visible in the wavelet transform. These spirals are solutions of

$$\alpha \log(|z|) - \mu \arg z = c \text{ const} \Rightarrow |z| = c' \, e^{\arg z \mu/\alpha}.$$

Thus the density of spirals is determined by the quotient μ/α.

Let us now locate the points, where the modulus of a voice is maximal, that is, we look for the points in the half-plane where $\partial_b M(b, a) = 0$ and $\partial_b^2 M(b, a) < 0$. By direct computation we find that these points are located on the straight line

$$a = \frac{\mu}{\alpha} b. \tag{9.5.1}$$

The central frequency of the analysing wavelet g_μ is given by $\omega_0 = \mu$. If we identify, as in equation (9.1.2), the quantity μ/a with a frequency, we see from (9.5.1) that the modulus of the wavelet transform is located around the points in the upper half-plane, where the time corresponds to the position $b \leftrightarrow t$

and the inverse of the scale corresponds to the instantaneous frequency $\mu/a \leftrightarrow \Omega(t)$ of the analysed function. The phase speed of every voice on the line (9.5.1) corresponds to the instantaneous frequency of the analysed function if we again identify the position parameter b with the time t:

$$\partial_b \Phi(b, \mu b/\alpha) = \frac{\alpha}{b} = \Omega(b).$$

All this can be observed in Figures 9.12 and 9.13.

9.6 Interactions

In the next examples we want to combine the previous elementary examples to obtain more complex functions. Clearly, the superposition principle tells us that the wavelet transform of a superposition of functions is the superposition of the respective transforms. However, this does not apply to the modulus and the phase pictures because they are obtained by non-linear operations on the transform. Consider, for example, the superposition of two pure frequencies at frequencies ξ_0 and ξ_1. For the sake of simplicity we suppose that $\hat{g}$ is a real with a single maximum in ω_0 and monotonic to both sides. We then have

$$M(b, a) = |g(a\xi_0)\, e^{i\xi_0 b} + g(a\xi_1)\, e^{i\xi_1 b}|,$$

$$\Phi(b, a) = \arg\{g(a\xi_0)\, e^{i\xi_0 b} + g(a\xi_1)\, e^{i\xi_1 b}\}.$$

Thus for voices where $|g(a\xi_0)| \gg |g(a\xi_1)|$ the phase is determined by ξ_0, whereas in the inverse case the phase is essentially determined by ξ_1. Thus in the phase diagram we observe a transition from the phase speed ξ_0 to ξ_1. See Figures 9.15 and 9.17. For the modulus the same reasoning shows that it consists essentially of two strips if

$$|\hat{g}(\xi_0 \xi_1)| \ll 1, \quad \text{and} \quad |\hat{g}(\xi_1/\xi_0)| \ll 1.$$

Thus it depends on the frequency ratio[8] (not on the difference) whether or not two frequencies may well be separated through the wavelet transform. See Figures 9.14 and 9.16 for an illustration.

9.7 Two deltas

Now consider two delta functions located at t_0 and t_1. At large scale, the wavelet will not be able to distinguish both deltas, and modulus and phase are similar to the associated diagrams of one single delta located at the

---

[8] Note that music is also based on frequency ratios and not on frequency differences. An octave, for instance, is a ratio of 1:2.

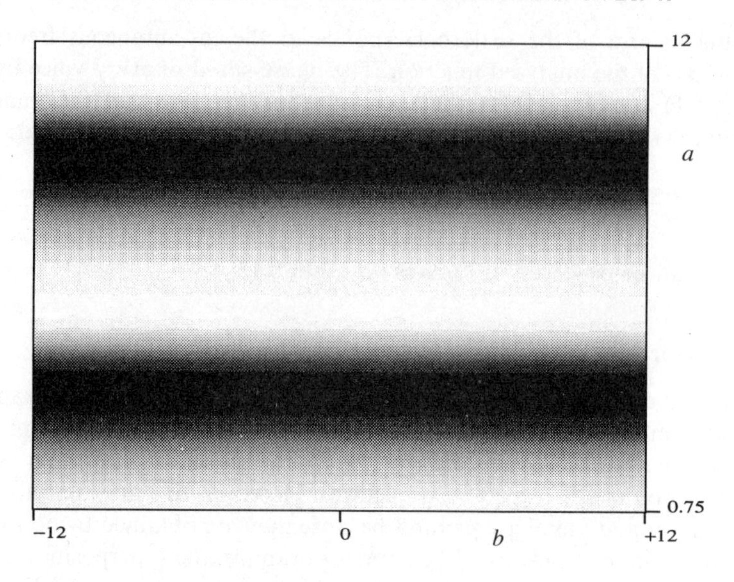

Fig. 9.14 The modulus of the wavelet analysis of the superposition of two pure frequencies with the help of the Morlet wavelet with internal frequency $\omega = 6$.

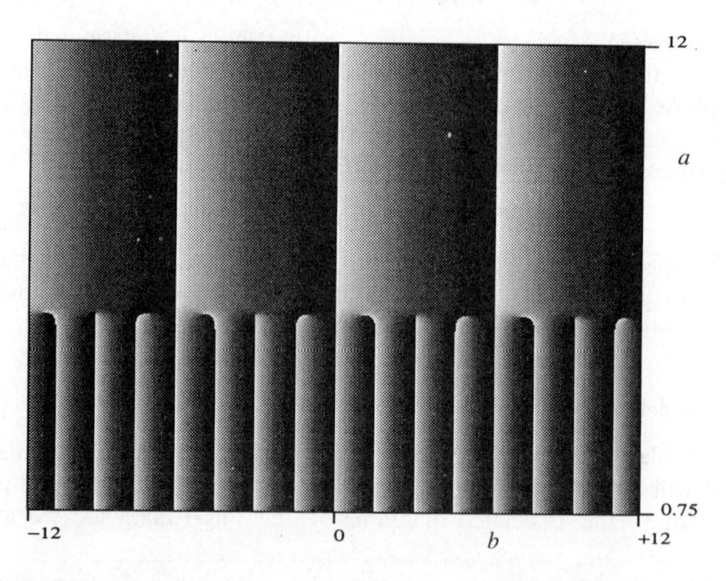

Fig. 9.15 The associated phase picture. Note the transition from one phase speed to the other through branching of the phase lines.

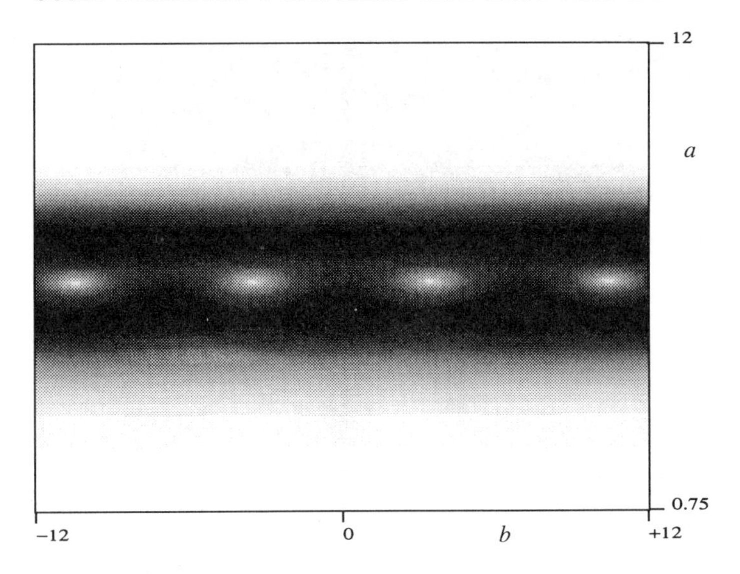

Fig. 9.16 The same as Figure 9.14 but the two frequencies have a frequency ratio closer to 1.

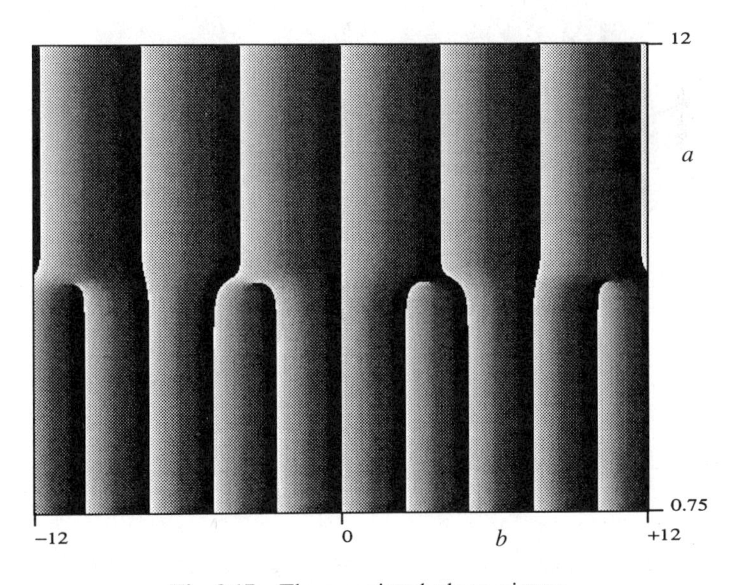

Fig. 9.17 The associated phase picture.

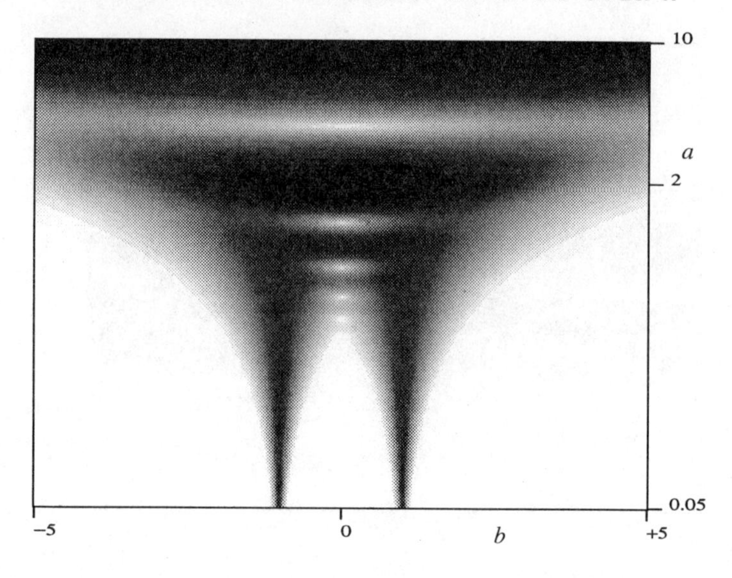

Fig. 9.18 Two delta functions at -1 and $+1$ analysed by some Morlet wavelet. The modulus diagram has been re-scaled again by the trivial factor a^{-1}.

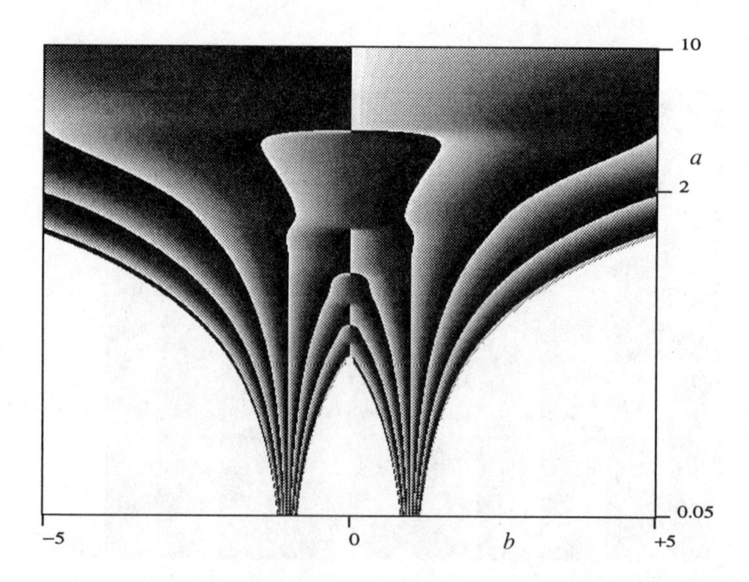

Fig. 9.19 The associated phase picture with cut-off. At small scale, again the phase lines eventually converge towards the respective locations of the singularities.

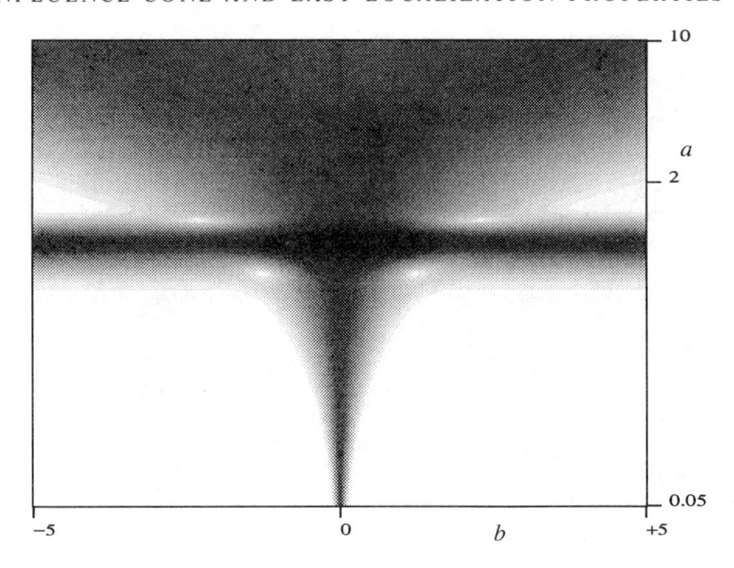

Fig. 9.20 Superposition of δ and pure frequency analysed by some Morlet wavelet. The modulus picture has been re-scaled again by the trivial factor a^{-1}.

midpoint. At small scale, however, the wavelet will be sufficiently localized to distinguish both functions. At an intermediate scale, a strong interaction can again be observed. (See Figures 9.18 and 9.19.)

9.8 Delta and pure frequency

To end this section consider a function that is a superposition of a delta function in time and a delta function in frequency. Again, an interaction zone can be observed, outside of which the features of the delta function and the pure frequency are qualitatively undisturbed. (See Figures 9.20 and 9.21.)

10 The influence cone and easy localization properties

Clearly, the wavelet transform of a bounded function $\|s\|_\infty < \infty$ with respect to an absolutely integrable wavelet, $\|g\|_1 < \infty$, is bounded by Hölder's inequality

$$|\mathcal{W}_g s(b, a)| \leq \|g\|_1 \, \|s\|_\infty . \tag{10.0.1}$$

If the function s is itself integrable, every voice is an integrable function by Young's inequality

$$\|\mathcal{W}_g s(b, a)\|_1 \leq \|g\|_1 \, \|s\|_1 . \tag{10.0.2}$$

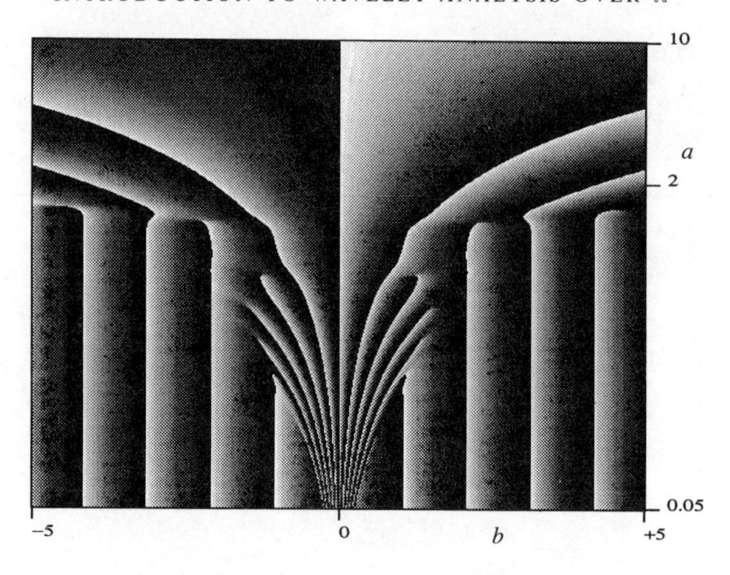

Fig. 9.21 The associated phase picture.

Recall that this inequality reads

$$\|s * r\|_w \leq \|s\|_p \|r\|_q$$

with

$$1 \leq p, q, w \leq \infty, \qquad 1 + \frac{1}{w} = \frac{1}{p} + \frac{1}{q}.$$

This shows that the localization of both—the wavelet and the analysed function—implies a localization of the wavelet transform. In the following we will clarify the dependence of the localization properties of the wavelet coefficients on the time localization of the analysed function and of the wavelet.

Consider the wavelet transform of the delta function localized at τ_0 with respect to the (continuous) wavelet g:

$$\mathscr{W}[g; \delta(t - \tau_0)](b, a) = \frac{1}{a}\bar{g}\left(\frac{\tau_0 - b}{a}\right).$$

Suppose that the wavelet is supported by the interval $[t_l, t_r]$. The dilated and translated versions $g_{b,a}$ of g are then supported by

$$[at_l + b; at_r + b].$$

The wavelet $g_{b,a} = a^{-1}g([\cdot - b]/a)$ interacts only with the delta if its position τ_0 is contained in this interval. The set of such points in the half-plane forms

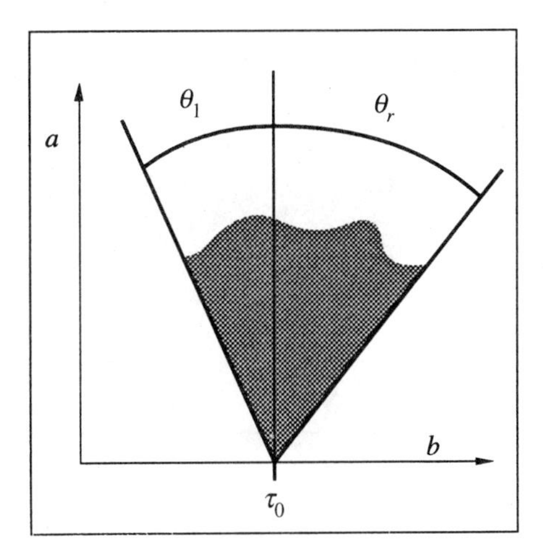

Fig. 10.1 The influence cone of compact supported wavelets. If the wavelet has an effective compact support this diagram rests qualitatively correct.

the so-called *influence cone* of the wavelet in τ_0:

$$I_g(\tau_0) = \{(b, a) \in \mathbb{H}: at_r \leq b - \tau_0 \leq at_l\}.$$

We note that the opening angle of the influence cone into the future direction $(b > \tau_0)$ is determined by the anti-causal part $(t \leq 0)$ of the wavelet $g(t)$:

$$\theta_r = \arctan(t_l),$$

whereas the opening angle to the past $(b < \tau_0)$ is determined by the causal part $(t \geq 0)$ of the wavelet

$$\theta_l = \arctan(t_r).$$

If a compactly supported function s with support $\operatorname{supp} s \subset [\tau_l, \tau_r]$ is analysed with the help of a compactly supported wavelet, every point in the support of s will give rise to its influence cone and thus the support of the wavelet transform will be contained in the superposition of all these influence cones. This is again a cone, but it is shifted towards the small scales such that its top is outside the half-plane in

$$a_0 = \frac{\tau_r - \tau_l}{t_l - t_r}, \qquad b_0 = p\tau_l + q\tau_r, \tag{10.0.3}$$

with $p = -(tr/(t_l - t_r))$ and $q = 1 - p$. Geometrically, the *a priori* support of the wavelet transform is the shadow of the support of the signal produced by a light source at (b_0, a_0) (see Figure 10.2).

The name 'influence cone' is justified by the fact that only inside the influence cone is the wavelet transform altered if we perturb the function s

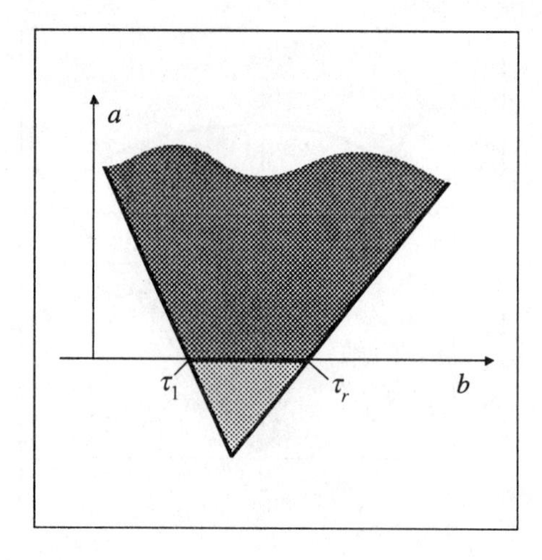

Fig. 10.2 The shifted influence cone associated with the interaction of compactly supported wavelets with compactly supported signals.

locally. More precisely, if two bounded functions s and r coincide, with the exception of a small neighbourhood of t_0,

$$s(t) = r(t), \qquad \text{for } |t - t_0| > \epsilon,$$

then the two wavelet transforms with respect to a compactly supported wavelet g coincide outside a neighbourhood of the influence cone of g in t_0:

$$\mathscr{W}_g s(b, a) = \mathscr{W}_g r(b, a)$$

whenever $(b - \epsilon, a) \notin I_g(t_0)$ and $(b + \epsilon, a) \notin I_g(t_0)$.

Indeed, the function $v = s - r$ is supported by $[t - \epsilon, t + \epsilon]$ and we must analyse the support of $\mathscr{W}_g v$. Now the set of points for which $(b - \epsilon, a) \notin I_g(t_0)$ or $(b + \epsilon, a) \notin I_g(t_0)$ is the shifted influence cone of (10.0.3) associated with the support of v.

As before, we say that the wavelet is effectively supported by the interval $[t_l, t_r]$ if the wavelet is so small outside this interval that its integral is small compared to the overall integral, or, what essentially amounts to the same, that we may decompose g into two parts

$$g = g_{\text{comp}} + g_{\text{small}},$$

with g_{comp} supported by $[t_l, t_r]$ and g_{small} with norm $\|g_{\text{small}}\|_1 \leq \epsilon \|g\|_1$. Suppose, further, that an analogue decomposition holds for s:

$$s = s_{\text{comp}} + s_{\text{small}},$$

where s_{comp} is supported by $[\tau_l, \tau_r]$, and $\|s_{\text{small}}\|_\infty \leq \|s\|_\infty$. Then the wavelet transform of s with respect to g splits into two parts

$$\mathscr{W}_g s = \mathscr{W}_{\text{comp}} + \mathscr{W}_{\text{small}},$$

where $\mathscr{W}_{\text{comp}} = \mathscr{W}_{g_{\text{comp}}} s_{\text{comp}}$ is supported by the shifted influence cone of (10.0.3), and $\mathscr{W}_{\text{small}}$ is small in the sense that

$$|\mathscr{W}_{\text{small}}(b, a)| \le (2\epsilon + \epsilon^2) \|g\|_1 \, \|s\|_\infty,$$

as follows by direct computation using the linearity in s and anti-linearity in g and the *a priori* estimate of (10.0.1).

11 Polynomial localization

We now wish to quantify the localization of the wavelet transform. Consider the function

$$\kappa_\alpha(t) = (1 + |t|^2)^{-\alpha/2}. \tag{11.0.1}$$

For $\alpha < 0$ this function is bounded from below and at infinity it grows as $|t|^{-\alpha}$. For $\alpha > 0$ this function decays with $|t|^{-\alpha}$ as $|t| \to \infty$. We say that a function s is *polynomially localized* if for some $\alpha > 0$ we have $\kappa_{-\alpha} s \in L^\infty(\mathbb{R})$ or, what amounts to the same, that s is majorized almost everywhere by κ_α:

$$|s(t)| \le c\kappa_\alpha(t) \quad \text{for almost all } t \in \mathbb{R}.$$

Therefore, the larger α is, the better is the localization of s. The set of polynomially localized functions forms an algebra, that is, if s and r are polynomially localized then the same is true for $s + r$ and sr. Specifically, if α and β are the respective localization exponents of s and r then the sum $s + r$ is polynomially localized with $\gamma = \min\{\alpha, \beta\}$ and the pointwise product sr is polynomially localized with $\alpha + \beta$. There is still another product for functions over $\mathbb{R}$ which is given by the convolution. Again, the set of polynomially localized functions is an algebra with respect to this multiplication, as is shown by the following lemma.

Lemma 11.0.1
Let κ_α be as in (11.0.1). *Then we have for $\alpha > 1$ that*

$$\kappa_\alpha * \kappa_\alpha(t) \le c\kappa_\alpha(t).$$

The constant c depends only on α.

The proof of this well-known result is postponed since we will prove a more general result soon. From the above result it follows that for polynomially localized g and s every voice of the wavelet transform is polynomially localized. More precisely, we have the following theorem.

Theorem 11.0.2

Let s and g be polynomially localized such that for some $\alpha > 1$ and some constant $c < \infty$ we have

$$|s(t)| \leq c\kappa_\alpha(t), \qquad |g(t)| \leq c\kappa_\alpha(t).$$

Then the wavelet transform of s with respect to g satisfies at

$$|\mathcal{W}_g s(b, a)| \leq c' \frac{1}{1+a} \kappa_\alpha\left(\frac{b}{1+a}\right),$$

with some $c' < \infty$ only depending on α.

Note that the lines $(b/(1 + a)) = \text{const}$ converge to the point $(-1, 0)$ outside the half-plane. Therefore, we re-discover the geometry of the shifted influence cone of (10.0.3).

The proof is again postponed to the next section, where we will give a more general result.

11.1 More precise results

It is sometimes useful to have sharp estimates of the localization exponents and to distinguish between the behaviour to the left and to the right. We therefore introduce the following functions:

$$\kappa_\alpha^+ = \begin{cases} \kappa_\alpha(t) & \text{for } t \geq 0, \\ 0 & \text{for } t < 0, \end{cases}$$

and $\kappa_\alpha^-(t) = \kappa_\alpha^+(-t)$. In addition, we write

$$\kappa_{\alpha,\beta}(t) = \kappa_\alpha^-(t) + \kappa_\beta^+(t).$$

This is a bounded function that behaves as $t^{-\alpha}$ to the right $(t \to +\infty)$ and as $|t|^{-\beta}$ to the left $(t \to -\infty)$.

Theorem 11.1.1

Suppose that α, β, α', $\beta' > 1$. We have,[9] *for $b \geq 0$,*

$$\mathcal{W}[\kappa_{\alpha,\beta}, \kappa_{\alpha',\beta'}](b, a) \sim \kappa_{\beta'}(a)\kappa_{\beta'}\left(\frac{b}{1+a}\right) + \frac{1}{a}\kappa_\alpha\left(\frac{1}{a}\right)\kappa_\alpha\left(\frac{b}{1+a}\right).$$

[9] We write $s(t) \sim r(t)$ for two real-valued functions iff there is a positive constant $c > 0$ such that $c^{-1} \leq s(t)/r(t) \leq c$ for all (considered) t.

For $b \leq 0$ we have

$$\mathscr{W}[\kappa_{\alpha,\beta}, \kappa_{\alpha',\beta'}](b, a) \sim \kappa_{\alpha'}(a)\kappa_{\alpha'}\left(\frac{b}{1+a}\right) + \frac{1}{a}\kappa_{\beta}\left(\frac{1}{a}\right)\kappa_{\beta}\left(\frac{b}{1+a}\right).$$

The elementary but tedious proof is given in the appendix.

Since $\kappa_\alpha + \kappa_\beta \sim \kappa_{\min\{\alpha,\beta\}}$ we obtain for the voice $a = 1$ the useful estimation

$$\kappa_{\alpha,\beta} * \kappa_{\alpha',\beta'} \sim \kappa_{\min\{\alpha,\alpha'\},\min\{\beta,\beta'\}} \tag{11.1.1}$$

and hence, in particular, Lemma 11.0.1.

Upon specifying $\alpha = \beta = \alpha' = \beta'$ we obtain Theorem 11.0.2.

In the general case, for every voice the decrease for $b \to +\infty\,(-\infty)$ is determined by the interaction of the anti-causal (causal) part of the wavelet and the causal (anti-causal) part of the analysed function. The less localized of both functions determines the localizaton of the voice. For fixed b we obtain at small scale ($|g| \leq O(\kappa_{\alpha,\beta})$, $|s| \leq O(\kappa_{\alpha',\beta'})$)

$$\limsup_{a \to 0} |\mathscr{W}_g s(b, a)| \leq c\kappa_{\alpha',\beta'}(b)$$

and therefore at small scales the effective support of the voice can essentially not be larger than the effective support of the analysed function. At large scale we have

$$\limsup_{a \to \infty} a|\mathscr{W}_g s(ba, a)| \leq c\kappa_{\beta,\alpha}(b),$$

and therefore at large scales the effective support of the voice is essentially given by the effective support of the dilated wavelet.

Therefore, the analysing wavelet should be more localized than the analysed function might be. Only, in this case, the localization of the wavelet transform reflects the localization of the analysed function. Note, however, that we have not shown so far that a certain localization of the wavelet transform implies a certain localization of the analysed function. And, as we shall see, this is not true in general.

12 The influence regions for pure frequencies

In this section we analyse the localization properties of the wavelet transform that are due to localization in frequency space.

A function s is called *band limited* if its Fourier transform is supported by a finite interval I. It is a *low-pass filter* if $0 \in I$. It is called *strip limited* or a *band-pass filter* if it is band limited and the frequency $\omega = 0$ is not contained in the support of $\hat{s}$. If s is strip limited, then there is a whole interval around the zero frequency where $\hat{s}$ vanishes.

Consider a progressive, strip-limited wavelet g. Its frequencies are supposed to be contained in the interval

$$\operatorname{supp} \hat{g} = [\omega_l; \omega_r],$$

with $0 < \omega_l < \omega_r < \infty$. The frequency content of the dilated and translated wavelet $g_{b,a}$ is then

$$\operatorname{supp} \widehat{g_{b,a}} = \left[\frac{\omega_l}{a}; \frac{\omega_r}{a}\right].$$

Consider the wavelet transform of a pure frequency e_{ξ_0}:

$$\mathcal{W}[g; e_{\xi_0}](b, a) = e^{i\xi_0 b}\bar{\hat{g}}(a\xi_0).$$

The wavelet is only interacting with the pure oscillation e_{ξ_0} if ξ_0 is contained in the support of its Fourier transform. The set of points (b, a) in the half-plane for which the wavelet is interacting with the pure frequency forms a strip parallel to the real axis:

$$I_g(\xi_0) = \left\{(b, a) \in \mathbb{H} : \frac{\omega_l}{\xi_0} \leq a \leq \frac{\omega_r}{\xi_0}\right\}.$$

This strip is called the *influence strip* of g at ω_0 (see Figure 12.1).

Consider now a strip-limited function s:

$$\operatorname{supp} \hat{s} \in [\xi_l, \xi_r], \qquad 0 < \xi_l < \xi_r < \infty.$$

Then superposing all influence strips of all frequencies contained in the analysed function we see that the wavelet transform is supported by the

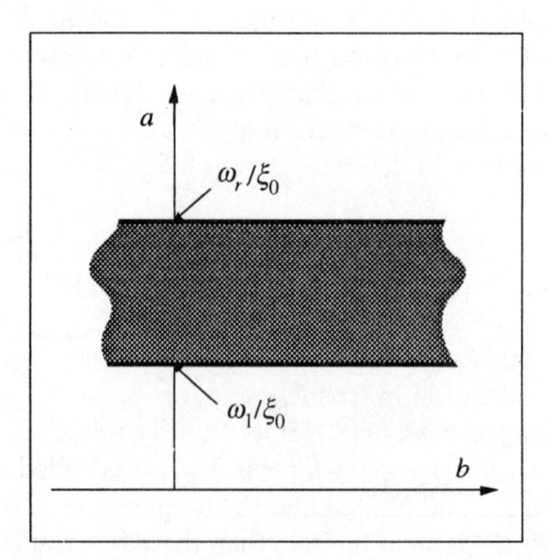

Fig. 12.1 The influence strip for strip-limited wavelets.

strip

$$\left\{(b, a) \in \mathbb{H} : \frac{\omega_l}{\xi_r} \leq a \leq \frac{\omega_r}{\xi_l}\right\}.$$

Here we note the important difference between the wavelet transform with respect to a wavelet (i.e. $\omega_l > 0$) and the one with respect to something bumplike (i.e. $\omega_l = 0$). Suppose that the analysed function is progressive and strip localized around a central frequency $\xi_0 > 0$. Whereas a wavelet reflects this localization by a wavelet transform that is supported by a strip around ξ_0/a, the bump will in general not allow us to distinguish between band-limited and strip-limited analysed function. For both types of analysed function, there is in general no small scale cut-off for the wavelet coefficients. The wavelet transform is, rather, supported by all scales smaller than the limit scale $a = \omega_r/\xi_l$.

In a general way the overall decay of the wavelet transform at small scale ($a \to 0$) is determined by the high frequency content of s and the behaviour of the wavelet around the zero frequency, that is, by the regularity of the analysed function and the number of vanishing moments of the wavelet. The decay of the wavelet transform for large scales is in turn determined by the high frequency behaviour of the wavelet and the zero frequency behaviour of the analysed function.[10] The following lemma quantifies these ideas precisely for the case of polynomial localization. We write

$$\phi_{\alpha,\beta}(\omega) = \frac{\omega^\alpha}{(1+\omega)^{\alpha+\beta}}.$$

For $\alpha, \beta > 0$ this is a function that tends to 0 like ω^α as $\omega \to 0$ and with $\omega^{-\beta}$ as $\omega \to \infty$.

We say that a progressive function s is *polynomial strip limited* (*band limited*) if its Fourier transform $\hat{s}$ is majorized by one of these functions $\hat{s} = O(\phi_{\alpha,\beta})$ with $\alpha, \beta > 0$.

Theorem 12.0.1

Let the wavelet g and the analysed function s be progressive. Suppose, further, that they are localized in frequency space such that for $\omega > 0$ we have

$$|\hat{g}(\omega)| \leq \phi_{\alpha,\beta}(\omega), \qquad |\hat{s}(\omega)| \leq \phi_{\alpha',\beta'}(\omega).$$

Then the wavelet transform is localized:

$$|\mathscr{W}_g s(b, a)| \leq c\phi_{\min\{\alpha, b'-1\}, \min\{\alpha'+1, \beta\}}(a),$$

[10] This duality is again an immediate consequence of equation (2.0.1) where the large scale behaviour of $\mathscr{W}_g s$—the wavelet transform of s with respect to g—is related to the small scale behaviour of $\mathscr{W}_s g$—the wavelet transform of g with respect to g.

with a constant c that does not depend on g or s. We suppose that $\alpha, \beta > 0$, $\alpha' > -1$, $\beta' > 1$, $\alpha \neq \beta' - 1$, $\alpha' \neq \beta - 1$.

As a special case note that for $\phi_\alpha = \phi_{\alpha,\alpha+1}$, $\alpha > 0$ we have

$$|\hat{s}(\omega)|, |\hat{g}(\omega)| \leq \phi_\alpha(\omega) \Rightarrow |\mathcal{W}_g s(b, a)| \leq O(1)\phi_{\alpha'}(a),$$

for all $\alpha' < \alpha$. Since

$$|\mathcal{W}_g s(b, a)| = \frac{1}{2\pi}\left|\int_0^\infty d\omega\, \bar{\hat{g}}(a\omega)\, e^{ib\omega}\hat{s}(\omega)\right| \leq \frac{1}{2\pi}\int_0^\infty d\omega\, |\hat{g}(a\omega)|\, |\hat{s}(\omega)|,$$

the theorem is an immediate consequence of the following fact for the scale convolutions:

$$s *_l r(t) = \int_0^\infty \frac{da}{a} s\left(\frac{t}{a}\right) r(a) = r *_l s(t),$$

Lemma 12.0.2
We have for $\alpha, \beta, \alpha', \beta' > 0$, $\alpha \neq \alpha'$, $\beta \neq \beta'$

$$\phi_{\alpha,\beta} *_l \phi_{\alpha',\beta'} \sim \phi_{\min\{\alpha,\alpha'\},\min\{\beta,\beta'\}}. \tag{12.0.1}$$

This means that the smallest of the absolute values of the localization exponents determines the localization of the scale convolution.

Proof. We introduce the inversion operator

$$s \mapsto Is, \qquad (Is)(t) = s\left(\frac{1}{t}\right).$$

The scale convolution satisfies the co-variance

$$Is *_l Ir = I(s *_l r), \tag{12.0.2}$$

as can be verified by direct computation.

We have to estimate

$$\phi_{\alpha,\beta} *_l \phi_{\alpha',\beta'}(t) = t^\alpha\left\{\int_0^1 + \int_1^\infty\right\}\frac{da}{a}\frac{a^{\alpha'-\alpha}}{(1 + t/a)^{\beta+\alpha}(1 + a)^{\beta'+\alpha'}} = X_1 + X_2.$$

Because $I\phi_{\alpha,\beta} = \phi_{\beta,\alpha}$ and because of (12.0.2) it is enough to show the lemma for $t \leq 1$, that is, we have to show that

$$X_1 + X_2 \sim t^{\min\{\alpha,\alpha'\}}, \qquad 0 \leq t \leq 1.$$

In X_1 we obtain $0 < a \leq 1$ and thus $1 + a \sim 1$. Changing $a \to t/a$ and calling it a again we obtain

$$X_1 \sim t^\alpha \int_0^1 \frac{da}{a}\frac{a^{\alpha'-\alpha}}{(1 + t/a)^{\beta+\alpha}} = t^{\alpha'}\int_0^{1/t}\frac{da}{a}\frac{a^{\alpha'+\beta}}{(1 + a)^{\beta+\alpha}}.$$

We distinguish two cases and obtain

$$X_1 \sim \begin{cases} t^{\alpha} & \alpha' > \alpha, \\ t^{\alpha'} & \alpha' < \alpha. \end{cases}$$

In X_2 we have $a > 1$ and thus $1 + a \sim a$ and $1 + t/a \sim 1$. Therefore,

$$X_2 \sim t^{\alpha} \int_1^{\infty} \frac{da}{a} a^{-\alpha-\beta'} \sim t^{\alpha},$$

since, by hypothesis on the exponents, the integral converges. □

13 The space of highly time–frequency localized functions

In this section we present the mathematical environment in which we will work in the following section. It is the space of highly localized and regular functions. We denote by $S_+(\mathbb{R})$ the set of progressive functions which are rapidly decreasing, and for which the Fourier transform is arbitrarily well polynomial strip localized, that is, the set of functions s with $\operatorname{supp} \hat{s} \subset \mathbb{R}_+$ for which for every localized exponent $\alpha > 0$ we have

$$\sup_{t \in \mathbb{R}} \kappa_\alpha^{-1}(t)|s(t)| + \sup_{\omega \geq 0} \phi_\alpha^{-1}(\omega)|\hat{s}(\omega)| < \infty,$$

where $\kappa_\alpha(t) = (1 + |t|^2)^{\alpha/2}$ and $\phi_\alpha(\omega) = \omega^\alpha(1 + \omega)^{-2\alpha-1}$. The left-hand side will be denoted by $\|s\|_{S_+(\mathbb{R});\alpha}$ or simply by $\|s\|_\alpha$ if the context is clear. It defines a seminorm since it satisfies at

$$\|\lambda s\|_{S_+(\mathbb{R}),\alpha} = \lambda \|s\|_{S_+(\mathbb{R}),\alpha},$$

$$\|s + r\|_{S_+(\mathbb{R}),\alpha} \leq \|s\|_{S_+(\mathbb{R}),\alpha} + \|r\|_{S_+(\mathbb{R}),\alpha}.$$

In addition, we have a hierarchy of seminorms, in the sense that $\|s\|_\alpha$ may always be compared to $\|s\|_\beta$:

$$\|s\|_\alpha \leq \|s\|_\beta \qquad \text{for all } \beta > \alpha.$$

The seminorms are actually norms since $\|s\|_\alpha = 0$ implies $s = 0$. A (coarsest) topology on $S_+(\mathbb{R})$ is given by imposing that all these norms be continuous. Accordingly, a base of neighbourhoods of the origin is given by

$$U_{\epsilon,\alpha} = \{s \in S_+(\mathbb{R}) \colon \|s\|_\alpha < \epsilon\}, \qquad \epsilon, \alpha > 0.$$

The $U_{\epsilon,\alpha}$ are convex and thus $S_+(\mathbb{R})$ is a locally convex space, that is, it has a base for its topology that consists of convex sets. Accordingly, a sequence of functions $s_n \in S_+(\mathbb{R})$ tends to 0 in (the topology of) $S_+(\mathbb{R})$ iff

$$\lim_{n \to \infty} \|s\|_{S_+(\mathbb{R});\alpha} \to 0 \qquad \text{for all } \alpha \geq 0.$$

Clearly, the continuous parameter α may be restricted to $\alpha \in \mathbb{N}$ without changing the topology and, accordingly, $S_+(\mathbb{R})$ has a countable base of neighbourhoods of 0. In addition, as we will show later, $S_+(\mathbb{R})$ is complete. To summarize, we have that the space $S_+(\mathbb{R})$ is a Fréchet space. Actually, we will see later that it is a closed subspace of the class of Schwartz $S(\mathbb{R})$. It consists of those functions in $S(\mathbb{R})$ that are progressive (see Section 19). The Bessel wavelet and all its dilates and translates are, for instance, in $S_+(\mathbb{R})$ and thus the set is not empty. It is again an algebra with respect to pointwise addition and convolution as product. Because of the high localization of s it follows that for any $n \in \mathbb{N}_0$ the nth derivative of $\hat{s}$ is continuous, and because for[11] $\omega \to 0^-$ we have $\partial_\omega^n \hat{s}(\omega) \to 0$, we have $\partial_\omega^n \hat{s}(0) = 0$. This shows, by Lemma 6.0.8, that for $s \in S_+(\mathbb{R})$ all moments vanish:

$$\int_{-\infty}^{+\infty} dt\, s(t)t^n = 0, \qquad n = 0, 1, 2, \ldots.$$

The image of $S_+(\mathbb{R})$ under the parity operator will be denoted by $S_-(\mathbb{R})$:

$$s \in S_-(\mathbb{R}) \Leftrightarrow s(-t) \in S_+(\mathbb{R}).$$

The direct sum of $S_+(\mathbb{R})$ and of $S_-(\mathbb{R})$ will be denoted by $S_0(\mathbb{R})$:

$$S_0(\mathbb{R}) = S_+(\mathbb{R}) \oplus S_-(\mathbb{R}).$$

In an analogous way, we wish to define high localization over the half-plane. The corresponding space of highly localized functions over the half-plane will be denoted by $S(\mathbb{H})$. It is the set of functions $\mathscr{T}$ over the half-plane for which, for every localization exponent $\alpha \geq 0$, we have

$$\sup_{(b,a) \in \mathbb{H}} (1 + a)\phi_\alpha^{-1}(a)\kappa_\alpha^{-1}\left(\frac{b}{1 + a}\right)|\mathscr{T}(b, a)| < \infty.$$

The left-hand side is again a norm and will be denoted by $\|\mathscr{T}\|_{S(\mathbb{H});\alpha}$ or simply $\|\mathscr{T}\|_\alpha$. Again, these norms are hierarchical in the sense that

$$\|\mathscr{T}\|_\alpha < \|\mathscr{T}\|_\beta \qquad \text{for } \beta > \alpha.$$

Again, the collection of norms induces a coarsest topology on $S(\mathbb{H})$ that makes them all continuous. This makes $S(\mathbb{H})$ a locally convex space. Since, again, we may restrict α to the countable set $\mathbb{N}_0$ the topology is generated by a countable family of seminorms. The space is obviously complete and hence $S(\mathbb{H})$ is a Fréchet space. A sequence $\mathscr{T}_n$ tends to 0 iff $\|\mathscr{T}_n\|_\alpha \to 0$ for all $\alpha > 0$.

---

[11] We write $t \to t_0^-$ for $t \to t_0$ and $t < t_0$, analogously for $t \to t_0^+$.

14 The inversion formula

The wavelet transform preserves all the information of the transformed function since there exists an (explicit) inversion formula which allows us to recover the function s from its wavelet transform. As we have seen, the wavelet transform is a set of convolutions. A single convolution has in general no inverse, or at least a stable[12] one, but the entirety of convolutions defining a wavelet transform allows stable inversion formulas.

Definition 14.0.1
A function h is called a reconstruction *wavelet for the analysing wavelet g if the two constants*

$$c_{g,h}^{+} = \int_0^{\infty} \frac{d\omega}{\omega} \bar{\hat{g}}(\omega)\hat{h}(\omega), \qquad c_{g,h}^{-} = \int_0^{\infty} \frac{d\omega}{\omega} \bar{\hat{g}}(-\omega)\hat{h}(-\omega), \quad (14.0.1)$$

are equal and finite:

$$c_{g,h}^{+} = c_{g,h}^{-}, \qquad 0 < |c_{g,h}^{+}| < \infty .$$

In the case of a progressive wavelet g we require h to satisfy only the positive frequency part of the previous definition:

$$c_{g,h} = \int_0^{\infty} \frac{d\omega}{\omega} \bar{\hat{g}}(\omega)\hat{h}(\omega), \qquad 0 < |c_{g,h}| < \infty .$$

Analogously for a regressive wavelet. If a wavelet is its own reconstruction wavelet we say that it is admissible. *In addition we suppose, unless otherwise stated, that all integrals are absolutely convergent.*

All admissible wavelets with $g \in L^1(\mathbb{R})$ have no zero frequency contribution or, what amounts to the same, they are of zero mean:

$$\hat{g}(0) = 0, \qquad \int_{-\infty}^{+\infty} dt\, g(t) = 0.$$

Indeed, otherwise the integral (14.0.1) could never converge since $\hat{g}$ is continuous. In order to keep the demonstration as simple as possible, we first show the inversion theorem for highly time frequency localized analysed functions. However, the formula holds for a much larger class of functions, as we will see later.

[12] By 'stable inversion' we mean that small perturbations of the transformed function will not alter the result of the inversion formula drastically.

Theorem 14.0.2

Let $h \in S_+(\mathbb{R})$ be a reconstruction wavelet for the wavelet $g \in S_+(\mathbb{R})$. Then we can recover a progressive function $s \in S_+(\mathbb{R})$ from its wavelet transform $\mathcal{W}_g s$ by means of the following double integral over the position-scale half-plane:

$$s(t) = \frac{1}{c_{g,h}} \int_0^\infty \frac{da}{a} \int_{-\infty}^{+\infty} db\, \mathcal{W}_g s(b, a) \frac{1}{a} h\left(\frac{t-b}{a}\right). \tag{14.0.2}$$

Clearly, the analogue theorem holds for $S_-(\mathbb{R})$ and $S_0(\mathbb{R})$. Thus the inverse transform allows us to write quite arbitrary functions over the real line as a superposition of elementary functions $h_{b,a}(t) = (1/a)h((t-b)/a)$. Each is weighted by the associated wavelet coefficient.

Proof. Calling the right-hand side of (14.0.2) $s^\#$ and using the Fourier representation (3.0.2) of the wavelet transform we may write

$$s^\#(t) = \int_0^\infty \frac{da}{a} \int_{-\infty}^\infty db\, \mathcal{W}_g s(b, a) \frac{1}{a} h\left(\frac{t-b}{a}\right)$$

$$= \frac{1}{2\pi} \int_0^{+\infty} \frac{da}{a} \int_{-\infty}^{+\infty} db \int_0^\infty d\omega\, \bar{\hat{g}}(a\omega)\, e^{ib\omega}\, \hat{s}(\omega) \frac{1}{a} h\left(\frac{t-b}{a}\right).$$

The double integral over b and ω converges absolutely by hypothesis on s, g, and h and we may exchange the integration. Carrying out, first, the integral over b we obtain the Fourier transform of $(1/a)h((t-b)/a)$, and thus

$$s^\#(t) = \frac{1}{2\pi} \int_0^\infty \frac{da}{a} \int_0^\infty d\omega\, \bar{\hat{g}}(a\omega) \hat{s}(\omega) \hat{h}(a\omega)\, e^{it\omega}.$$

The double integral over the scales and ω is absolutely convergent since all functions are arbitrarily well polynomial strip localized. Exchanging the integration again we obtain

$$s^\#(t) = \frac{1}{2\pi} \int_0^\infty d\omega\, e^{it\omega}\, \hat{s}(\omega) \int_0^\infty \frac{da}{a}\, \bar{\hat{g}}(a\omega) \hat{h}(a\omega).$$

The integral over the measures da/a is invariant under re-scalings $a \mapsto \omega a$. Morever exactly, we have for an arbitrary function r and $\omega \neq 0$

$$\int_0^\infty \frac{da}{a} r(\omega a) = \int_0^\infty \frac{da}{a} r(a \operatorname{sign} \omega)$$

whenever the integral converges absolutely. Therefore, by hypothesis on the reconstruction wavelet, the integral over the scales $\int_0^\infty (da/a)\bar{\hat{g}}(a\omega)\hat{h}(a\omega)$ is identically equal to $c_{g,h}$—with only the exception of $\omega = 0$—and thus we may write

$$s^{\#}(t) = c_{g,h} \frac{1}{2\pi} \int_0^{\infty} d\omega \, e^{it\omega} \hat{s}(\omega) = c_{g,h} s(t),$$

where we have used the inversion formula for the Fourier transform. □

14.1 Fourier transform in wavelet space

Because of the inversion formula we may express the Fourier transform—and at least formally any other linear transform—in terms of wavelet coefficients.

Theorem 14.1.1
Suppose that s and g are progressive, then

$$\hat{s}(\omega) = \frac{1}{c_{g,h}} \int_0^{\infty} \frac{da}{a} \int_{-\infty}^{+\infty} db \, \mathcal{W}_g s(b, a) \hat{h}(a\omega) \, e^{-ib\omega} \tag{14.1.1}$$

with a reconstruction wavelet h. We suppose in addition that g, h, and s are in $S_+(\mathbb{R})$.

Proof. First note that from (10.0.2) every voice is integrable. From the frequency representation of the voice (3.0.2) we see that the Fourier transform of a voice is given by

$$\int_{-\infty}^{+\infty} db \, e^{-i\omega b} \mathcal{W}_g s(b, a) = \bar{\hat{g}}(a\omega) \hat{s}(\omega).$$

Therefore, we have

$$\int_0^{\infty} \frac{da}{a} \int_{-\infty}^{+\infty} db \, \hat{h}(a\omega) \, e^{-ib\omega} \mathcal{W}_g s(b, a) = \hat{s}(\omega) \int_0^{\infty} \frac{da}{a} \bar{\hat{g}}(a\omega) h(a\omega) = c_{g,h} \hat{s}(\omega),$$

and therefore (14.1.1) holds. □

15 Reconstruction with singular reconstruction wavelets

Here we show some formal results without giving the proofs. These results will be justified in Chapter 2, Section 6. Throughout this section we suppose that s and g are progressive. The general case can be obtained as in (17.0.1). Suppose that for some $\beta \in \mathbb{R}$ we have

$$c_g = \int_0^{\infty} \frac{d\omega}{\omega} \bar{\hat{g}}(\omega) \, e^{-i\beta\omega}, \qquad 0 < |c_g| < \infty.$$

Then the translated delta function $\delta_\beta = \delta(t - \beta)$ is formally a reconstruction wavelet. Therefore, formally, we have

$$s(t) = c_g^{-1} \lim_{\epsilon \to 0, \rho \to \infty} \int_\epsilon^\rho \frac{da}{a} \mathscr{W}_g s(t - \beta a, a).$$

Thus the value of the analysed function in t is reconstructed by a simple summation over all scales along a straight line passing through the point $(t, 0)$ on the border line of the half-plane. For $\beta = 0$ we obtain Calderòn's reconstruction formula:

$$s(t) = c_g^{-1} \lim_{\epsilon \to 0, \rho \to \infty} \int_\epsilon^\rho \frac{da}{a} \mathscr{W}_g s(t, a), \qquad c_g = \int_0^\infty \frac{da}{a} \bar{\hat{g}}(a).$$

In this form the reconstruction formula is very intuitive. If we take a wavelet of the form as we have considered in the introduction as derived from a bump ϕ, then $\mathscr{W}_a s = -a\partial_a \phi_a * s$ has been interpreted as the details that are created at scale a. Now summing all these details up we recover s, as the above formula shows.

Suppose now that for some ξ we have

$$0 < |\hat{g}(\xi)| < \infty.$$

The the pure frequency e_ξ is formally a reconstruction wavelet for g. Carrying out the integration in Fourier space we obtain

$$\hat{s}(\omega) = \frac{1}{\bar{\hat{g}}(\xi)} \int_{-\infty}^{+\infty} db\, \mathscr{W}_g s\left(b, \frac{\xi}{\omega}\right) e^{-ib\omega}.$$

This shows again that the scale a is related to the inverse of a frequency.

16 The wavelet synthesis operator

It is natural to extend the inversion formula to arbitrary functions $\mathscr{T}$ over the half-plane, thus giving rise to a linear operator $\mathscr{M}_h$ mapping functions over the half-plane $\mathbb{H}$ to functions over the real line $\mathbb{R}$. We thus write $(h_{b,a} = T_b D_a h)$

$$\mathscr{T} \mapsto \mathscr{M}_h \mathscr{T}, \qquad (\mathscr{M}_h \mathscr{T})(t) = \int_0^\infty \frac{da}{a} \int_{-\infty}^{+\infty} db\, \mathscr{T}(b, a) h_{b,a}(t),$$

whenever the double integral is absolutely convergent. In particular, for $h \in L^1(\mathbb{R})$ it is well defined for all functions $\mathscr{T} \in S(\mathbb{H})$ because they are rapidly decaying as $|b|$ or $a + 1/a$ becomes large. The operator $\mathscr{M}_h$ is a wavelet synthesis since arbitrary functions are obtained by mere superposition of the wavelets $h_{b,a}$. Wavelet analysis with respect to an appropriate wavelet is one possible way of obtaining the expansion coefficients $\mathcal{T}(b, a)$. As we have seen these are by no means unique since many analysing wavelets will do the job.

We list some formal properties of the wavelet synthesis.

Linearity. The wavelet synthesis is a linear operator

$$\mathcal{M}_h(\mathcal{T} + \mathcal{R}) = \mathcal{M}_h\mathcal{T} + \mathcal{M}_h\mathcal{R}, \qquad \mathcal{M}_h(\alpha\mathcal{T}) = \alpha\mathcal{M}_h\mathcal{T}.$$

In contrast with the wavelet analysis it is also linear with respect to the synthesizing wavelet h:

$$\mathcal{M}_{h+g}\mathcal{T} = \mathcal{M}_h\mathcal{T} + \mathcal{M}_g\mathcal{T}, \qquad \mathcal{M}_{\alpha h}\mathcal{T} = \alpha\mathcal{M}_h\mathcal{T}$$

Wavelet and parity. Using the notation P for the parity operator acting on functions over $\mathbb{R}$, $s(t) \to s(-t)$ and $\mathcal{P}$ for functions over the half-plane, $\mathcal{T}(b, a) \to \mathcal{T}(-b, a)$, we have

$$\mathcal{M}[Ph; \mathcal{P}\mathcal{T}] = P\mathcal{M}[h; \mathcal{T}].$$

Fourier transform. By straightforward formal computation we can show that the Fourier transform of the synthesized function reads

$$\widehat{\mathcal{M}_h\mathcal{T}}(\omega) = \int_0^{\infty} \frac{da}{a} \int_{-\infty}^{+\infty} db\, \mathcal{T}(b, a)\hat{h}(a\omega)\, e^{-ib\omega}. \tag{16.0.1}$$

Co-variance. With respect to the synthesis coefficients the following co-variance holds:

$$\mathcal{M}[h; \mathcal{T}(b - b', a)](t) = \mathcal{M}[h; \mathcal{T}](t - b'),$$

$$\mathcal{M}[h; \mathcal{T}(\lambda b, \lambda a)](t) = \mathcal{M}[h; \mathcal{T}](\lambda t).$$

With respect to the synthesising wavelet nothing can be said *a priori* for general $\mathcal{T}$, only if $\mathcal{T}$ is the wavelet transform with respect to some g in which case the reconstruction is—up to some multiplicative constant—independent of h.

Convolutions and scale convolutions. The wavelet synthesis can be seen as a superposition of convolutions of every voice with the associated dilated filter $h_a = D_a h$:

$$\mathcal{M}_h\mathcal{T} = \int_0^{\infty} \frac{da}{a} \mathcal{T}(\cdot, a) * h_a. \tag{16.0.2}$$

If now the analysed function s, the wavelet $\tilde{g}$, and the reconstruction wavelet h are causal, then the wavelet transform $\mathcal{W}_g s$ is supported by the causal part of the half-plane, that is, every voice is causal. In this case the convolution in (16.0.2) is the Laplace convolution of Section 5.1.

The wavelet synthesis can also be seen as a superposition of scale convolutions. If we set $\mathscr{R}_b(a) = \mathscr{T}(ab, a)$ and $h_b(t) = h(t - b)$, we see that for $t > 0$ the wavelet synthesis reads

$$\mathscr{M}_h \mathscr{T}(t) = \int_{-\infty}^{+\infty} db\, h_b *_l \mathscr{R}_b(t).$$

The case where $t < 0$ follows from the behaviour of the wavelet synthesis under the parity operator.

17 Reconstruction without reconstruction wavelet

We have shown in Theorem 14.0.2 that on a suitable space of progressive functions and by symmetry for an analogue space of regressive functions we have, respectively,

$$\mathscr{M}_h \mathscr{W}_g = c^+_{g,h} \mathbb{1}, \qquad \mathscr{M}_h \mathscr{W}_g = c^-_{g,h} \mathbb{1}.$$

Upon separating the positive and negative frequencies we see that in the most general case—that is, if h is not necessarily a reconstruction wavelet for g in the sense that the constants $c^\pm_{g,h}$ are still finite but not equal—the inversion formula gives a mixture between the identity and the Hilbert transform:

$$\mathscr{M}_h \mathscr{W}_g = \tfrac{1}{2}(c^+_{g,h} + c^-_{g,h}) \mathbb{1} + \frac{i}{2}(c^+_{g,h} - c^-_{g,h}) H. \tag{17.0.1}$$

Consider the special case where one analyses a real-valued function s with the help of a progressive wavelet g. Upon reconstructing with either a real-valued or a progressive function h we obtain the associated progressive function

$$\mathscr{M}_h \mathscr{W}_g s = \tfrac{1}{2} c_{g,h}(s + iHs).$$

The real part of this reconstruction gives us back the analysed function, whereas the imaginary part is its Hilbert transform.

18 Localization properties of the wavelet synthesis

In the previous sections we have shown that localization in time and frequency space of the analysed function implies a certain localization of the wavelet coefficients. Now if the wavelet transform shows a certain localization over the half-plane, what can we say about the localization of the analysed function? A first answer is given by the following theorems, that discuss the relation between the localization of h, $\mathscr{T}$, and $s = \mathscr{M}_h \mathscr{T}$.

18.1 Frequency localization

Suppose, that the reconstruction wavelet is strip limited:

$$\operatorname{supp} \hat{h} \subseteq [\zeta_l; \zeta_r],$$

with $\zeta_l > 0$. The points in the half-plane that contribute to the value of $\hat{s}(\xi_0)$ lie in the influence strip of h at ξ_0. Therefore, the image of a function over the half-plane that is supported by a strip

$$\operatorname{supp} \mathscr{T} \subseteq \{(b, a) \in \mathbb{H}: a_{\min} \leq a \leq a_{\max}\}$$

is a strip-limited function

$$\operatorname{supp} \widehat{\mathscr{M}_h \mathscr{T}} \subseteq \left[\frac{\zeta_r}{a_{\min}}; \frac{\zeta_l}{a_{\max}}\right].$$

Suppose we analyse a strip-limited function s, $\operatorname{supp} \hat{s} \subset [\xi_l; \xi_r]$ with the help of a strip-limited wavelet g, supp, $\hat{g} \subset [\omega_l; \omega_r]$. The estimation given above for the support of $\hat{s}$ yields

$$\operatorname{supp} \hat{s} \subseteq \left[\omega_l \frac{\xi_l}{\zeta_r}; \omega_r \frac{\xi_r}{\zeta_l}\right].$$

Note that the support of $\hat{s}$ cannot be obtained by this estimation. We only obtain an interval that strictly contains the support of the analysed function.

The case of polynomial localization is treated by the following theorem. We again use the notations $\kappa_{\alpha,\beta}$ and $\phi_{\alpha,\beta}$ of Sections 11 and 12.

Theorem 18.1.1

Let $\mathscr{T}$ be a function over the half-plane that is localized as

$$\int_{-\infty}^{+\infty} db\, |\mathscr{T}(b, a)| \leq \phi_{\alpha,\beta}(a), \qquad \alpha, \beta > 0.$$

Suppose the synthesizing wavelet h is polynomially strip localized:

$$|\hat{h}(\omega)| \leq \phi_{\alpha',\beta'}, \qquad \alpha', \beta' > 0, \quad \alpha \neq \beta', \beta \neq \alpha'.$$

Then $\mathscr{M}_h \mathscr{T}$ is polynomially strip localized:

$$|\widehat{\mathscr{M}_h \mathscr{T}}(\omega)| \leq c\phi_{\min\{\beta,\alpha'\},\min\{\alpha,\beta'\}}(\omega).$$

Therefore, the lowest localization of h and $\mathscr{T}$ determines the localization of the synthesized function.

Proof. We can use formula (16.0.1) to estimate

$$|\widehat{\mathcal{M}_h \mathcal{T}}(\omega)| = \left| \int_0^\infty \frac{da}{a} \int_{-\infty}^{+\infty} db \, e^{-i\omega b} \mathcal{T}(b, a) \hat{h}(a\omega) \right|$$

$$\leq O(1) \int_0^\infty \frac{da}{a} \phi_{\alpha,\beta}(a) |\hat{h}(a\omega)|$$

$$= O(1) \int_0^\infty \frac{da}{a} \phi_{\beta,\alpha}(a) \phi_{\alpha',\beta'}\left(\frac{\omega}{a}\right).$$

The theorem follows from the estimation of (12.0.1). □

Note that we need a little more on the wavelet side than we obtained in Theorem (12.0.1). To prove the polynomial strip localization we needed in addition a little localization of the voices. For instance, a localization like

$$|\mathcal{W}_g s(b, a)| \leq \frac{1}{1+a} \kappa_\gamma \left(\frac{b}{1+a} \right) \phi_{\alpha,\beta}(a)$$

with some $\gamma > 1$ and $\alpha, \beta > 0$ will be enough to imply $|\hat{s}(\omega)| \leq c\phi_{\beta,\alpha}(\omega)$.

18.2 Time localization

Here more serious difficulties appear. Suppose the reconstruction wavelet h is compactly supported. Then the points in the upper half-plane that contribute to $\mathcal{M}_h \mathcal{T}(t)$ are situated in the influence cone $I_h(t)$ of h at t. On the other hand, suppose g and s to be compactly supported. The support of $\mathcal{W}_g s$ is contained in the shifted influence cone Γ. But for all t the influence cone $I_h(t)$ intersects Γ even if t is outside the support of s, and thus even for this highly localized situation no conclusion concerning the localization of the wavelet synthesis is possible.

Another way of saying essentially the same thing is the following. The points in the half-plane that are influenced by the behaviour of s on some interval I are in the shifted influence cone Γ. On the other side the points that are influenced only by this restriction of s to I are in some triangular region with base I, as shown in Figure 18.1.

An example may underline this difficulty. Consider the wavelet transform of the onset $|t|_+^\alpha$ with respect to the Cauchy wavelets with parameter β, as we did in (9.4.4). Every voice is polynomially localized with exponent $\beta - \alpha$ but the analysed function is not even bounded! The reason for this is, roughly speaking, that the large time behaviour is related to the local behaviour at $\omega = 0$, which is ignored by the wavelet transform since $\hat{g}$ is flat at $\omega = 0$.

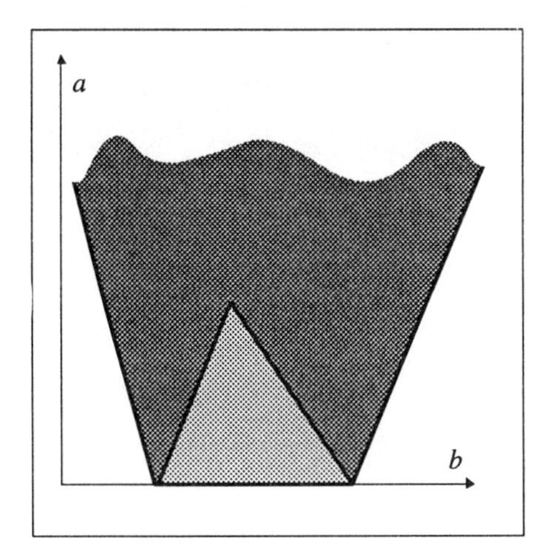

Fig. 18.1 The shifted influence cone (i.e. all points in the half-plane that may be influenced by the behaviour of the analysed function in some interval) and the exclusive triangular region (i.e. the points that are only influenced by the behaviour of the analysed function in some interval).

However, if the wavelet transform is at the same time polynomially strip localized we obtain some information about the localization of the analysed function.

Theorem 18.2.1
If $\mathscr{T}$ is localized as

$$|\mathscr{T}(b, a)| \leq \kappa_{\alpha,\beta}\left(\frac{b}{1+a}\right)\phi_{\gamma,\delta}(a), \qquad \alpha, \beta > 1,\ \gamma, \delta > 0,$$

and if h is polynomially localized, too:

$$|h(t)| \leq \kappa_{\alpha',\beta'}(t), \qquad \alpha', \beta' > 1,$$

then s is polynomially localized:

$$|s(t)| \leq c\kappa_{\min\{\alpha,\alpha',\delta\},\min\{\beta,\beta',\delta\}}(t).$$

Proof. We split the integral over the half-plane into two parts. We may do this because it is absolutely convergent:

$$\mathscr{M}_h\mathscr{T}(t) = \left\{\int_0^1 + \int_1^\infty\right\}\frac{da}{a}\mathscr{T}(\cdot, a) * h_a(t) = X_1 + X_2, \qquad h_a = h(\cdot/a)/a.$$

In the first term ($0 < a \leq 1$) we write

$$\mathscr{T}(b, a) \leq O(1) a^{\gamma} \kappa_{\alpha, \beta}(b).$$

Since $h_a * \kappa_{\alpha, \beta}$ can be formally identified with a wavelet transform of $\kappa_{\alpha, \beta}$ with respect to $\bar{h}(-t)$ we can use the estimation of (11.1.1) to write

$$|\mathscr{T}(\cdot, a) * h_a(t)| \leq O(1) a^{\gamma} (\kappa_{\alpha, \beta}(t) + a^{-1} \phi_\alpha(a) \kappa_{\alpha'}^{-}(t) \\ + a^{-1} \phi_\beta(a) \kappa_{\beta'}^{+}(t)) = X_3(t,a),$$

and therefore

$$X_1 \leq \int_0^1 \frac{da}{a} X_3(t, a) \leq O(1) \kappa_{\min\{\alpha, \alpha'\}, \min\{\beta, \beta'\}}(t).$$

In the second term ($a > 1$) we write

$$|\mathscr{T}(b, a)| \leq O(1) a^{-\delta} \kappa_{\alpha, \beta}\left(\frac{b}{a}\right).$$

We may now use the co-variance of the convolution under dilations ($D_a s * D_a r = D_a(s * r)$) and the estimation of (11.1.1) to write

$$|\mathscr{T}(\cdot, a) * h_a(t)| \leq O(1) a^{-\delta} \kappa_{\min\{\alpha, \alpha'\}, \min\{\beta, \beta'\}}\left(\frac{t}{a}\right).$$

Integrating over the scales we obtain

$$\begin{aligned} X_2 &\leq O(1) \int_1^\infty \frac{da}{a} a^{-\delta} \kappa_{\min\{\alpha, \alpha'\}, \min\{\beta, \beta'\}}\left(\frac{t}{a}\right) \\ &= O(1) |t|^{-\delta} \int_0^{|t|} \frac{da}{a} a^{\delta} \kappa_{\min\{\alpha, \alpha'\}, \min\{\beta, \beta'\}}(a \operatorname{sign} t) \\ &= O(1) \kappa_{\min\{\alpha, \alpha', \delta\}, \min\{\beta, \beta', \delta\}}(t). \end{aligned}$$

Both terms together prove the theorem. □

19 Wavelet analysis over $S_+(\mathbb{R})$

This section is intended to prove the following theorem.

Theorem 19.0.1

The spaces $S_+(\mathbb{R})$, and $S(\mathbb{H})$ are Fréchet spaces. The wavelet analysis and wavelet synthesis

$$\mathscr{W}: S_+(\mathbb{R}) \times S_+(\mathbb{R}) \to S(\mathbb{H}), \qquad (g, s) \mapsto \mathscr{W}_g s,$$
$$\mathscr{M}: S_+(\mathbb{R}) + S(\mathbb{H}) \to S_+(\mathbb{R}), \qquad (h, \mathscr{T}) \mapsto \mathscr{M}_h \mathscr{T},$$

are continuous maps.

Clearly, the same holds for $S_-(\mathbb{R})$ and $S_0(\mathbb{R})$, too. Recall that for V_1 and V_2 locally convex spaces the Cartesian product $V_1 \times V_2$ is again locally convex if it is equipped with the product topology, that is, the open sets are of the form $U_1 \times U_2$ with U_1 open in V_1 and U_2 open in V_2.

Recall, further, that a Fréchet space is a complete topological vector space with the topology generated by a countable family of semi-norms.

Proof. We have by definition $s \in S_+(\mathbb{R})$, $\mathscr{T} \in S(\mathbb{H})$,

$$\|s\|_\alpha = \sup_{t \in \mathbb{R}} \kappa_\alpha^{-1}(t)|s(t)| + \sup_{\omega > 0} \phi_\alpha^{-1}|\hat{s}(\omega)|,$$

$$\|\mathscr{T}\|_\alpha = \sup_{(b,a) \in \mathbb{H}} (1 + a)\phi_\alpha^{-1}(a)\kappa_\alpha^{-1}\left(\frac{b}{1 + a}\right)|\mathscr{T}(b, a)|,$$

where we have again used the weighting functions $\kappa_\alpha(t) = (1 + |t|^2)^{-\alpha/2}$ and $\phi_\alpha(a) = a^\alpha(1 + a)^{-2\alpha-1}$. The continuity of the respective transformations follows from the estimates,

$$\|\mathscr{W}_g s\|_\alpha \le c_\alpha \|g\|_{2\alpha}\|s\|_{2\alpha}, \qquad \|\mathscr{M}_h \mathscr{T}\|_\alpha \le c_\alpha \|h\|_\alpha \|\mathscr{T}\|_{\alpha+1},$$

that we are going to prove now.

The localization results of the previous section imply that we have

$$\sup_{t \in \mathbb{R}} \kappa_\alpha^{-1}(t)|\mathscr{M}_h \mathscr{T}(t)|$$
$$\le O(1) \sup_{t \in \mathbb{R}} \kappa_\alpha^{-1}(t)|h(t)| \sup_{(b,a) \in \mathbb{H}} (1 + a)\phi_\alpha^{-1}(a)\kappa_\alpha^{-1}\left(\frac{b}{1 + a}\right)|T(b, a)|$$

$$\sup_{\omega > 0} \phi_\alpha^{-1}(\omega)|\widehat{\mathscr{M}_h \mathscr{T}}(\omega)|$$
$$\le O(1) \sup_{\omega > 0} \phi_\alpha^{-1}(\omega)|h(\omega)| \sup_{(b,a) \in \mathbb{H}} (1 + a)\phi_{\alpha+1}^{-1}(a)\kappa_\alpha^{-1}\left(\frac{b}{1 + a}\right)|\mathscr{T}(b, a)|$$

and hence the second inequality holds.

To show the continuity of $\mathscr{W}$ we will use the theorems of Sections 11 and 12. Note that they can be rephrased as

$$\sup_{(b,a) \in \mathbb{H}} (1 + a)\kappa_\gamma^{-1}\left(\frac{b}{1 + a}\right)|\mathscr{W}_g s(b, a)| \le \left(\sup_{t \in \mathbb{R}} \kappa_\alpha^{-1}(t)|g(t)|\right)\left(\sup_{t \in \mathbb{R}} \kappa_\alpha^{-1}(t)|s(t)|\right),$$

$$\sup_{(b,a) \in \mathbb{H}} \phi_\alpha^{-1}(a)|\mathscr{W}_g s(b, a)| \le O(1)\left(\sup_{\omega > 0} \phi_\alpha^{-1}(\omega)|\hat{g}(\omega)|\right)\left(\sup_{\omega > 0} \phi_\alpha^{-1}(\omega)|\hat{s}(\omega)|\right).$$

We may now write

$$\left(\sup_{(b,a)\in\mathbb{H}} (1+a)\phi_\alpha^{-1}(a)\kappa_\alpha^{-1}\left(\frac{b}{1+a}\right)|\mathcal{W}_g s(b,a)|\right)^2$$
$$\leq \left(\sup_{(b,a)\in\mathbb{H}} (1+a)\kappa_\alpha^{-2}\left(\frac{b}{1+a}\right)|\mathcal{W}_g s(b,a)|\right)$$
$$\times \left(\sup_{(b,a)\in\mathbb{H}} (1+a)\phi_\alpha^{-2}(a)|\mathcal{W}_g s(b,a)|\right).$$

Now since $\kappa_\alpha^{-2}(t) = \kappa_{2\alpha}^{-1}(t)$ and $(1+a)\phi_\alpha^{-2}(a) \leq O(1)\phi_{2\alpha}^{-1}(a)$ the estimation follows.

We now show that all spaces are Fréchet spaces.

Clearly, $S(\mathbb{H})$ is a Fréchet space. Indeed, thanks to the hierarchical behaviour of the norms defined in Section 13 we may replace the continuous family of norms $\|\cdot\|_\alpha$, $\alpha > 0$, by the countable family of norms $\|\cdot\|_n$, $n \in \mathbb{N}$, without changing the topology. Now $S(\mathbb{H})$ is certainly complete since the localization passes to the limit without problems.

To show that $S_+(\mathbb{R})$ is a Fréchet space, too, let $s_n \in S_+(\mathbb{R})$ be a Cauchy sequence. Its wavelet transform $\mathscr{T}_n = \mathcal{W}_g s_n$ is a Cauchy sequence, too, since $\mathcal{W}_g$ is continuous. It thus has a limit, say $\mathscr{T}$. On the other hand, the inverse wavelet transform is continuous, too, and thus $s_n = \mathcal{M}_g\mathscr{T}_n \to \mathcal{M}_g\mathscr{T}$, where we have supposed, as we may, that g was admissible with $c_g = 1$. Now $\mathcal{M}_g\mathscr{T}$ is in $S_+(\mathbb{R})$ and we are done. □

19.1 Schwartz space

The problem is the following: the space $S_+(\mathbb{R})$ has been defined only by its localization in time and in frequency space. From the high localization of the Fourier transform it follows that the functions in $S_+(\mathbb{R})$ are infinitely many times differentiable. And vice versa, the Fourier transform is infinitely many times differentiable, too, thanks to the high localization in t. However, it is *a priori* not clear whether the derivatives $\partial^n s$ are localized or not. Here the wavelet transform helps as a powerful time frequency analysis tool that allows us to use localization properties of the time and the frequency representation jointly.

Lemma 19.1.1

The mappings $s \mapsto \partial^n s$, $n \in \mathbb{Z}$, and $s \mapsto t^m s$, $m \in \mathbb{N}$, are continuous from $S_+(\mathbb{R})$ onto itself.

Proof. Let g be the Bessel wavelet defined in Fourier space through $\hat{g}(\omega) = e^{-\omega - 1/\omega}$ for $\omega > 0$ and 0 otherwise. This wavelet and all its derivatives and primitives[13] are in $S_+(\mathbb{R})$. Thus the constants

$$c_n = \int_0^\infty \frac{d\omega}{\omega} (i\omega)^{-n} |\hat{g}(\omega)|^2$$

are finite and different from 0. This means that g is a reconstruction wavelet for the nth primitive $\partial^{-n} g$ of g. By partial integration we obtain

$$\mathscr{W}[(-\partial)^{-n} g; \partial^n s](b, a) = a^{-n} \mathscr{W}[g; s](b, a).$$

Therefore, the derivative is essentially multiplication with $1/a$ in wavelet space:

$$\partial^n s(t) = c_n^{-1} \mathscr{M}_g a^{-n} \mathscr{W}_g s.$$

Now multiplication by some power of a is a continuous operation in $S(\mathbb{H})$. And thus as a combination of continuous maps the ∂^n are continuous.

The second part of the proof makes use of the formula

$$\mathscr{W}[g; ts](b, a) = a\mathscr{W}[tg; s](b, a) + b\mathscr{W}[g; s](b, a),$$

which shows that the operator $s \mapsto ts$ corresponds essentially to multiplication by $b + a$ in wavelet space, which is again a continuous operation. We leave the details to the reader. □

Consider now the class of Schwartz, $S(\mathbb{R})$, which is the space of functions that decay together with all their derivatives faster than any polynomial. This space is a Fréchet space with an ascending chain of norms

$$\|s\|_{S(\mathbb{R}); n, m} = \|s\|_{n, m} = \sum_{n'=0}^{n} \sum_{m'=0}^{m} \sup_{t \in \mathbb{R}} |t^{n'} \partial_t^{m'} s(t)| < \infty.$$

Note that the multiplication by t and the derivation are continuous operations in $S(\mathbb{R})$.

As a consequence of Lemma 6.0.1 we have

Theorem 19.1.2

The Fourier transform is a continuous map from $S(\mathbb{R})$ onto $S(\mathbb{R})$.

The relation between $S_+(\mathbb{R})$ and $S(\mathbb{R})$ now becomes apparent.

Theorem 19.1.3

$S_+(\mathbb{R})$ ($S_-(\mathbb{R})$) is the closed subspace of $S(\mathbb{R})$ that consists of those functions in $S(\mathbb{R})$ whose Fourier transform is supported by the positive (negative) frequencies only.

---

[13] Clearly, the integration constant has to be choosen appropriately.

Then, clearly, $S_0(\mathbb{R})$ is the space of functions in $S(\mathbb{R})$ for which all moments vanish. It again is a closed subspace.

Proof. Let us denote by Σ the space of functions in $S(\mathbb{R})$ whose Fourier transforms are supported by the positive frequencies $\omega \geq 0$. Clearly, each function $s \in \Sigma$ is polynomially localized: $|s(t)| \leq c_\alpha(1 + t^2)^{-\alpha/2}$, for all $\alpha > 0$. Since $\hat{s}$ is again in $S(\mathbb{R})$ the Fourier transform is localized, too: $|\hat{s}(\omega)| \leq O(1)(1 + \omega^2)^{-\alpha/2}$ for all α. Now since it contains only positive frequencies, the Taylor development near 0 shows that $|\hat{s}(\omega)| \leq O(\omega^\alpha)$ as $\omega \to 0$ for all $\alpha > 0$. This shows that for $\omega > 0$ we have $|\hat{s}(\omega)| \leq c'_\alpha(\omega + 1/\omega)^{-\alpha}$ for all $\alpha > 0$ and hence $\Sigma \subset S_+(\mathbb{R})$. The constant c'_α can be estimated in terms of $\|\hat{s}\|_{S(\mathbb{R});0,m}$ and hence the embedding is continuous.

On the other hand, let $s \in S_+(\mathbb{R})$. As we have seen (Lemma 19.1.1), all its derivatives are also localized and hence $s \in S(\mathbb{R})$. In addition, the localization of $\partial^n s$ may be estimated in terms of $\|s\|_{S_+(\mathbb{R});\alpha}$ since, as we have seen, the derivative is a continuous operator in $S_+(\mathbb{R})$. This shows that the embedding is again continuous. □

19.2 The regularity of the image space

Until now we have only seen that the image of the wavelet transform is localized. In fact, it consists of smooth functions. Indeed, on the image of $\mathcal{W}_g$ the partial derivatives ∂_b and ∂_a are again essentially multiplication by a^{-1}:

$$\partial_b \mathcal{W}[g; s](b, a) = \mathcal{W}[g; \partial_t s](b, a) = -\frac{1}{a}\mathcal{W}[\partial_t g; s](b, a)$$

$$-a\partial_a \mathcal{W}[g; s](b, a) = \mathcal{W}[(t\partial_t + 1)g, s](b, a).$$

Note that this is the co-variance equation of the wavelet transform, but this time in infinitesimal language. Now, together with s, the functions ts and ∂s are in $S_+(\mathbb{R})$ and therefore $\partial_b^n(a\partial_a)^m \mathcal{W}_g s(b, a)$ is in $S(\mathbb{H})$. From now on let $S(\mathbb{H})$ be the space of functions for which

$$\|\mathcal{T}\|_{S(\mathbb{H});n,m,\alpha,\beta} = \sup_{(b,a)\in\mathbb{H}} |(a + 1/a)^\alpha(1 + b^2)^{\beta/2}\partial_b^n(a\partial_a)^m \mathcal{T}(b, a)| < \infty.$$

The left-hand side again defines a family of norms and the weakest topology in which they are continuous makes $S(\mathbb{R})$ a Fréchet space again. It is now evident that the wavelet transform is also continuous in this *a priori* finer topology. This will become apparent in the following section, where we show that the image of the wavelet transform may be characterized through a reproducing kernel (equation (20.0.1) below). From this it follows easily that localization in the image space implies localization of all partial derivatives.

20 The reproducing kernel

For a given wavelet, the wavelet transform $\mathscr{W}_g$ maps functions over the real line to functions over the half-plane. From this it is plausible that not every function over the half-plane is the wavelet transform with respect to g of some function over the real line. In this section we will characterize the image of wavelet transforms, that is, those functions over the half-plane that are the wavelet transforms of some function over the real line. It turns out that they have internal correlations as expressed by the presence of a reproducing kernel.

As we have seen, the wavelet transformation followed by an inverse wavelet transform with respect to a reconstruction wavelet gives back the analysed function. Let us consider this identity in wavelet space, that is, we consider the operator on functions over the half-plane

$$\Pi_{g,h} = \frac{1}{c_{g,h}} \mathscr{W}_g \mathscr{M}_h.$$

As we will see in the following theorem this operator characterizes the image of the wavelet transform.

Theorem 20.0.1

If $h \in S_+(\mathbb{R})$ is a reconstruction wavelet for $g \in S_+(\mathbb{R})$, then the operator $\Pi_{g,h}$ is a continuous projector onto the image of $\mathscr{W}_g$.

Proof. As composition of continuous maps it follows that $\Pi_{g,h}$ is a well-defined continuous map from $S(\mathbb{H})$ to $S(\mathbb{H})$. Clearly, for $\mathscr{T} \in S(\mathbb{H})$ we have that $\Pi_{g,h}\mathscr{T}$ is in the image of $\mathscr{W}_g$ since it is the wavelet transform of $c_{g,h}^{-1}\mathscr{M}_h\mathscr{T}$. On the other hand, suppose that $\mathscr{T}$ is in the image of $\mathscr{W}_g$. Then there is a function $s \in S_+(\mathbb{R})$ such that $\mathscr{T} = \mathscr{W}_g s$. We have $\Pi_{g,h}\mathscr{T} = c_{g,h}^{-1}\mathscr{W}_g(\mathscr{M}_h \mathscr{W}_g s) = \mathscr{W}_g s = \mathscr{T}$. This shows that

$$\Pi_{g,h}\Pi_{g,h} = \Pi_{g,h},$$

and thus $\Pi_{g,h}$ is a projector on the image of $\mathscr{W}_g$. □

If we write the projector explicitly we obtain, exchanging the integrations,

$$\begin{aligned}\Pi_{g,h}\mathscr{T}(b,a) &= c_{g,h}^{-1} \int_{-\infty}^{+\infty} dt\, \bar{g}_{b,a}(t) \int_0^{\infty} \frac{da'}{a'} \int_{-\infty}^{+\infty} db'\, \mathscr{T}(b',a') h_{b',a'}(t) \\ &= \int_0^{\infty} \frac{da'}{a'} \int_{-\infty}^{+\infty} db' \frac{1}{a'} P_{g,h}\left(\frac{b-b'}{a'}, \frac{a}{a'}\right) \mathscr{T}(b',a')\end{aligned}$$

with

$$P_{g,h}(b, a) = c_{g,h}^{-1}\mathcal{W}[g; h](b, a).$$

Therefore, we have the following corollary.

Corollary 20.0.2

The image of $S_+(\mathbb{R})$ under the wavelet transform $\mathcal{W}_g$ with $g \in S_+(\mathbb{R})$ is the closed subspace of functions in $\mathcal{T} \in S(\mathbb{H})$ that satisfy pointwise

$$\mathcal{T}(b, a) = \int_0^\infty \frac{da'}{a'} \int_{-\infty}^{+\infty} db' \frac{1}{a'} P_{g,h}\left(\frac{b-b'}{a'}, \frac{a}{a'}\right)\mathcal{T}(b', a'),$$

where h is a reconstruction wavelet for g.

Therefore, the function $P_{g,h}$ is called the reproducing kernel.

From the localization theorems it follows that for polynomially time–frequency localized wavelets the reproducing kernel is polynomially localized over the half-plane. Suppose now that $P_{g,h}$ is localized around the point (0, 1) in the upper half-plane. Then $(1/a')P_{g,h}[([b-b']/a'), (a/a')]$ seen as a function of (b', a') for fixed (b, a) is localized around (b, a). Therefore, the weighted mean with the reproducing kernel localized around the some point (b, a) gives back the value of $\mathcal{W}_g s$ at exactly this point. Therefore, the localization of the reproducing kernel indicates how well the wavelet transform separates the scales and the positions of the analysed function. Thus, for example, analysis with the help of the Haar wavelet does not separate the scales well (see Figure 20.1).

The reproduction kernel equation becomes even more suggestive if we introduce the convolution for functions over the half-plane. For any two functions in $\mathcal{T}, \mathcal{R} \in S(\mathbb{H})$, we define their convolution product via

$$\mathcal{T} * \mathcal{R}(b, a) = \int_0^\infty \frac{da'}{a'} \int_{-\infty}^{+\infty} db' \frac{1}{a'} \mathcal{T}\left(\frac{b-b'}{a'}, \frac{a}{a'}\right)\mathcal{R}(b', a').$$

Note that this convolution is not commutative any more but it is still associative. With this notation the reproducing kernel equation can be written as

$$\mathcal{T} \in \text{image } \mathcal{W}_g \Leftrightarrow \mathcal{T} = P_{g,h} * \mathcal{T}. \tag{20.0.1}$$

Thus, the image of the wavelet transform is stable under convolution over the half-plane with the reproducing kernel.

By direct computation we can verify that the following symmetry, which we state for later reference, holds

$$\frac{1}{a'} P_{g,h}\left(\frac{b-b'}{a'}, \frac{a}{a'}\right) = \frac{1}{a} \bar{P}_{h,g}\left(\frac{b'-b}{a}, \frac{a'}{a}\right). \tag{20.0.2}$$

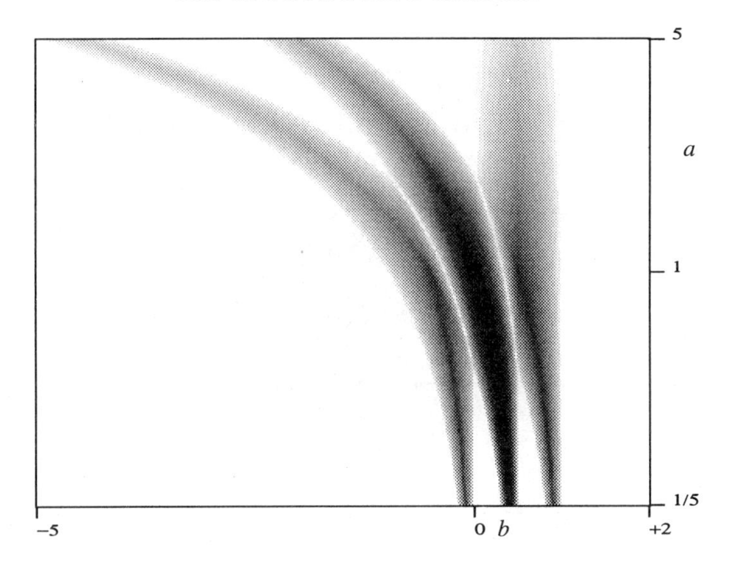

Fig. 20.1 The reproducing kernel of the Haar wavelet. Note the poor localization. Therefore, the analysis with this wavelet does not separate the scales well.

Example 20.0.3
Although these wavelets are not in $S_+(\mathbb{R})$ we consider them here as an illustration. The reproducing kernel of the Cauchy wavelets

$$g_\alpha(t) = \frac{\Gamma(1+\alpha)}{2\pi(1-it)^{1+\alpha}}$$

can easily be computed. In Fourier space we obtain ($z = b + ia$)

$$\begin{aligned}\mathscr{W}[g_\alpha; g_\beta](b, a) &= (2\pi)^{-1} a^\alpha \int_0^\infty d\omega\, \omega^{\alpha+\beta}\, e^{i(z+i)\omega} \\ &= (2\pi)^{-1}\Gamma(\alpha+\beta+1)a^\alpha(1-iz)^{-\alpha-\beta-1}.\end{aligned}$$

This function is polynomially localized around the point $z = i \leftrightarrow (0, 1)$:

The dilated and translated reproducing kernel can be written in the following concise form ($z = b + ia$, $\tau = b' + ia'$):

$$P[g_\alpha; g_\beta](z; \tau) = c\,\frac{(\Im z^\alpha)(\Im \tau^\beta)}{(z-\bar{\tau})^{\alpha+\beta+1}}, \qquad c = (2\pi)^{-1} i^{\alpha+\beta+1}.$$

20.1 The cross-kernel

Given the wavelet transform with respect to some wavelet g how can we obtain the wavelet transform with respect to h? In the case where g is

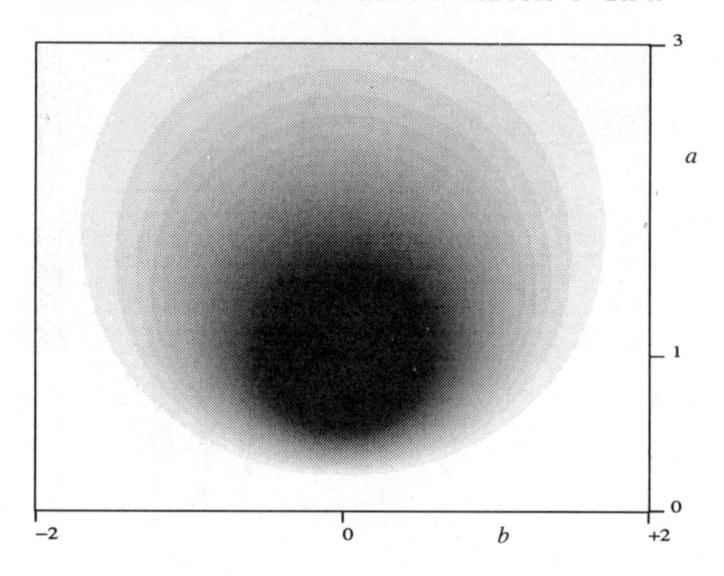

Fig. 20.2 The modulus of the reproducing kernel of the Cauchy wavelet for $\alpha = 6$. Note the localization around the point $(0, 1)$.

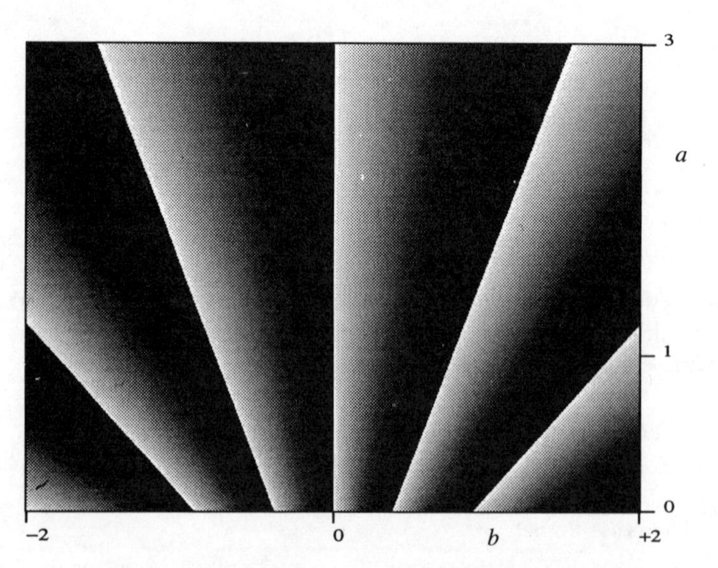

Fig. 20.3 The phase of the reproducing kernel. Note that the lines converge towards the point $(-1, 0)$ outside the half-plane.

admissible we might reconstruct with g and then analyse with h. Both operations together can again be written as a convolution over the half-plane

$$\mathscr{W}_h s = \Pi_{g \to h} * \mathscr{W}_g, \qquad \Pi_{g \to h} = \frac{1}{c_{g,g}} \mathscr{W}_h g. \tag{20.1.1}$$

This function is sometimes called the cross-kernel from g to h.

21 The wavelet transform of a white noise

The reproducing kernel can formally be interpreted as the minimal internal correlation of the wavelet coefficients. We only do formal computations. Consider a random function r over the real line. We note by $\mathbf{E}(r(t))$ the expectation of the random variable $r(t)$. The correlation of two time points, t, u, is the mean of the random function $\bar{s}(t)s(u)$:

$$\Phi(t, u) = \mathbf{E}(\bar{s}(u)s(t)).$$

A *white noise* s is a 'function' which is of zero mean, and for which different time points are completely un-correlated

$$\mathbf{E}(s(t)) = 0, \qquad \mathbf{E}(s(t)s(u)) = \delta(t-u).$$

Consider the wavelet transform of such a random function. It is itself a random function, but this time over the half-plane. The correlation between two points in the half-plane is given by

$$\Phi(b, a; b', a') = \mathbf{E}(\overline{\mathscr{W}_g s}(b', a')\mathscr{W}_g s(b, a)).$$

We exchange the integration formally with the mean over all realizations and obtain ($g_{b,a} = T_b D_a g$)

$$\begin{aligned}
\Phi(b, a; b', a') &= \mathbf{E}\left(\int_{-\infty}^{+\infty} dt\, g_{b,a}(t)s(t) \int_{-\infty}^{+\infty} du\, \bar{g}_{b',a'}(u)s(u)\right) \\
&= \int_{-\infty}^{+\infty} dt\, \bar{g}_{b,a}(t) \int_{-\infty}^{+\infty} du\, g_{b',a'}(u)\mathbf{E}(s(t)s(u)) \\
&= \int_{-\infty}^{+\infty} dt\, \bar{g}_{b,a}(t) \int_{-\infty}^{+\infty} du\, g_{b',a'}(u)\delta(t-u).
\end{aligned}$$

In the last equation we have used the correlation function of a white noise. Integrating over the delta function we see that the correlation function is

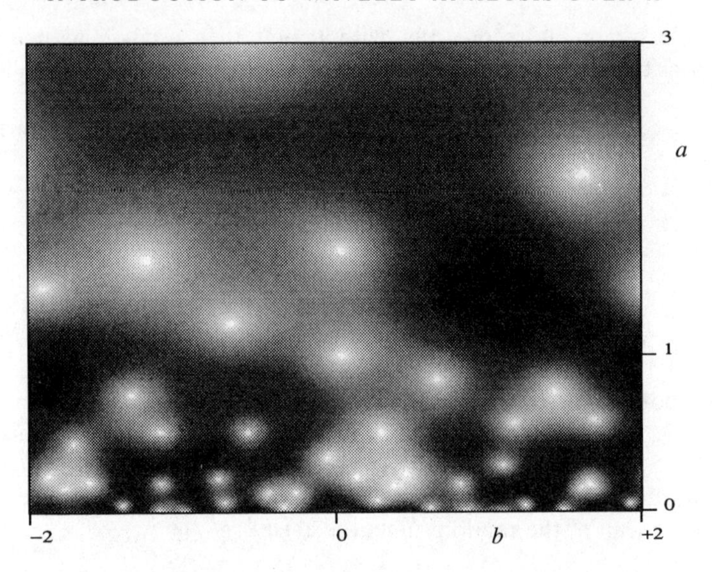

Fig. 21.1 The modulus of the wavelet transform of a white noise with respect to the Morlet wavelet. The regions of correlation change proportionally with the scale.

given by the reproducing kernel

$$\Phi(b, a; b', a') = c_g \frac{1}{a'} P_g\left(\frac{b - b'}{a'}, \frac{a}{a'}\right).$$

Therefore, even if the analysed function is completely un-correlated, the wavelet coefficients show a correlation because of the reproducing kernel property. The correlation depends on $(b - b')/a'$ and a/a'. Look at Figures 21.1 and 21.2, where the scale dependency of the correlations is visible.[14]

22 The wavelet transform in $L^2(\mathbb{R})$

In this and the following section we present the wavelet transform in a Hilbert space context. Recall that $H^2_+(\mathbb{R})$ is the closed subspace of $L^2(\mathbb{R})$ consisting of those functions whose Fourier transform is supported by the positive real axis $\omega \geq 0$. It will turn out that the wavelet analysis distributes the energy of the analysed function as measured by $\|s\|_2^2$ over the half-plane without

[14] Note that we did not quite explain this phenomenon. We only showed that, in the mean over many realizations, the correlation is given by the reproducing kernel. For one single realization of a white noise the argument is a little more tricky and beyond the scope of this section. In Chapter 4, Section 4 we will study the white noise in greater detail.

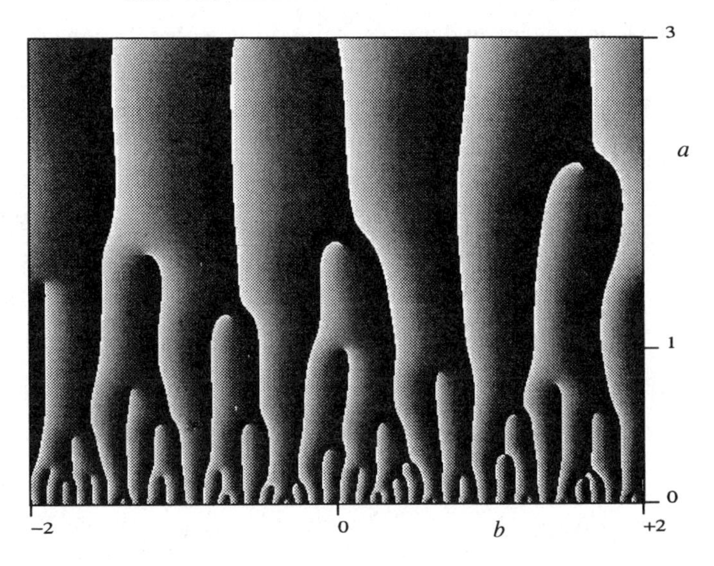

Fig. 21.2 The phase of the wavelet transform is a white noise. The phase turns on the average with a speed that is proportional to the length scale.

any loss. It therefore makes sense to speak of the distribution of the energy of s over the scales and positions. We now come to the details.

The scalar product for functions over the half-plane is defined by

$$\langle \mathcal{T} \mid \mathcal{R} \rangle_{\mathbb{H}} = \int_0^{\infty} \frac{da}{a} \int_{-\infty}^{+\infty} db \, \bar{\mathcal{T}}(b, a) \mathcal{R}(b, a)$$

whenever this expression make sense. Then $L^2(\mathbb{H})$ is the Hilbert space of functions for which $\|\mathcal{T}\|^2_{L^2(\mathbb{H})} = \|\mathcal{T}\|^2_2 = \langle \mathcal{T} \mid \mathcal{T} \rangle_{\mathbb{H}} < \infty$. We write $L^p(\mathbb{H})$ for the set of functions $\mathcal{T}$ over $\mathbb{H}$ for which

$$\|\mathcal{T}\| L^p(\mathbb{H}) = \|\mathcal{T}\|_p \leq \left\{ \int_0^{\infty} \frac{da}{a} \int_{-\infty}^{+\infty} db \, |\mathcal{T}(b, a)|^p \right\}^{1/p} < \infty .$$

In the case where $p = \infty$ we again set $\|\mathcal{T}\|_\infty = \text{ess sup}_{(b,a) \in \mathbb{H}} |\mathcal{T}(b, a)|$.

Then we have the following theorem.

Theorem 22.0.1

The wavelet synthesis $\mathcal{M}_g$ is the adjoint operator of $\mathcal{W}_g$, that is, we have

$$\langle \mathcal{M}_g \mathcal{T} \mid s \rangle_{\mathbb{R}} = \langle \mathcal{T} \mid \mathcal{W}_g s \rangle_{\mathbb{H}}. \tag{22.0.1}$$

We suppose that $s, g \in S_+(\mathbb{R})$, and $\mathcal{T} \in S(\mathbb{H})$.

Proof. The following integral is absolutely convergent and we may exchange the integrations:

$$\int_{-\infty}^{+\infty} dt \left\{ \int_0^\infty \frac{da}{a} \int_{-\infty}^{+\infty} db\, \mathcal{T}(b,a) \frac{1}{a} g\left(\frac{t-b}{a}\right) \right\} s(t)$$

$$= \int_0^\infty \frac{da}{a} \int_{-\infty}^{+\infty} db\, \mathcal{T}(b,a) \left\{ \int_{-\infty}^{+\infty} dt\, \frac{1}{a} \bar{g}\left(\frac{t-b}{a}\right) s(t) \right\}. \quad \square$$

The wavelet transform preserves the energy, as is shown by the following theorem.

Theorem 22.0.2

For $g, h \in S_+(\mathbb{R})$, and $s, r \in S_+(\mathbb{R})$ we have

$$\langle s \mid r \rangle_{\mathbb{R}} = c_{g,h}^{-1} \langle \mathcal{W}_h r \mid \mathcal{W}_g s \rangle_H. \tag{22.0.2}$$

In particular we have

$$\|s\|_{L^2(\mathbb{R})} = \frac{1}{c_g} \|\mathcal{W}_g s\|_{L^2(\mathbb{H})}.$$

Proof. We can use Theorems 14.0.2 and 22.0.1. From the inversion formula we have

$$\langle s \mid r \rangle_{\mathbb{R}} = c_{g,h}^{-1} \langle \mathcal{M}_h \mathcal{W}_g s \mid r \rangle_{\mathbb{R}}.$$

From the theorem on the adjoint operator we have

$$\langle s \mid r \rangle_{\mathbb{R}} = c_{g,h}^{-1} \langle \mathcal{W}_h s \mid \mathcal{W}_g r \rangle_{\mathbb{H}},$$

which proves the theorem. $\square$

In order to extend this result to all of $H^2_+(\mathbb{R})$ we need the following theorem.

Theorem 22.0.3

$S_+(\mathbb{R})$ is dense in $H^2_+(\mathbb{R})$.

Proof. The proof is completely standard but we will give it anyway. Indeed, let $\phi_\epsilon \in S_+(\mathbb{R})$, $0 < \epsilon < 1$, be such that its Fourier transform is identically equal to 1 on the interval $[\epsilon, 1/\epsilon]$ and that it goes monotonically to 0 such that the support of $\hat{\phi}_\epsilon$ is contained in $[0, 2/\epsilon]$. Further, let ψ be a $C^\infty(\mathbb{R})$ function that is equal to 1 on $[-1/\epsilon, +1/\epsilon]$ and that goes monotonically to 0 such that the support of ψ is contained in $[-2/\epsilon, +2/\epsilon]$. Let $s \in H^2_+(\mathbb{R})$ and consider

$$s_\epsilon = \phi_\epsilon * (\psi_\epsilon \cdot s).$$

For fixed $0 < \epsilon < 1$ we have $s_\epsilon \in S_+(\mathbb{R})$. Indeed, $\psi_\epsilon \cdot s$ is compactly supported; its Fourier transform is therefore a smooth function, which after multiplication by $\hat{\phi}_\epsilon$ yields a smooth function with support contained in $[\epsilon/2, 2\epsilon]$. In addition, we have

$$\lim_{\epsilon \to 0} \|s - s_\epsilon\|^2 = 0.$$

This can be seen by observing that $\psi_\epsilon \cdot s \to s$ in $L^2(\mathbb{R})$ since

$$\int_{-\infty}^{+\infty} dt\, |\psi_\epsilon(t) - 1|^2 \cdot |s(t)|^2 \le \int_{|t| \ge 1/\epsilon} dt\, |s(t)|^2 \to 0 \quad \text{as } \epsilon \to 0.$$

A similar argument in Fourier space allows us to conclude for the L^2-convergence of the convolution. In addition, the same argument shows that $\|\phi_\epsilon * r\|_2 \le c\|r\|_2$ with c independent on ϵ. Therefore, we have with $r_\epsilon = \psi_\epsilon \cdot s$

$$\|s - s_\epsilon\|_2 \le \|s - \phi_\epsilon * s\|_2 + \|\phi_\epsilon * (s - r_\epsilon)\|_2 \le \|s - \phi_\epsilon * s\|_2 + c\|s - r_\epsilon\|_2.$$

Both terms tend to 0, as we have seen. □

This last theorem shows that the wavelet transform with respect to a wavelet in $S_+(\mathbb{R})$ may be extended to an isometric linear operator from $H^2_+(\mathbb{R})$ to $L^2(\mathbb{H})$.

Theorem 22.0.4

There is a unique isometric operator $\mathscr{W}_g$ from $H^2_+(\mathbb{R})$ to $L^2(\mathbb{H})$ that coincides with $\mathscr{W}_g$ defined previously on the dense linear subset $S_+(\mathbb{R}) \subset H^2_+(\mathbb{R})$.

Proof. This is completely standard but we give the proof anyway. Indeed, let $s \in H^2_+(\mathbb{R})$. Then there is an approximating sequence $S_+(\mathbb{R}) \ni s_n \to s$ converging in $\mathbf{H}^2_+(\mathbb{R})$. By Lemma 22.0.2 it follows that $\mathscr{W}_g s_n$ is a Cauchy sequence in $L^2(\mathbb{H})$. Therefore, it converges in $L^2(\mathbb{H})$ to some function $\mathscr{T} \in L^2(\mathbb{H})$. We now define $\mathscr{W}_g s$ to be this function $\mathscr{T}$. We still have to check that this definition does not depend on the approximating sequence s_n. But this is easy since, let r_n be another approximating sequence, then $s_n - r_n$ converges to 0 in $H^2_+(\mathbb{R})$. It follows that

$$\|\mathscr{W}_g s_n - \mathscr{W}_g r_n\|_2^2 = \|\mathscr{W}_g(s_n - r_n)\|_2^2 \le c_g \|s_n - r_n\|_2^2 \to 0.$$ □

On the other hand we have defined the wavelet transform of $s \in H^2_+(\mathbb{R})$ with respect to $g \in H^2_+(\mathbb{R})$ pointwise via $\mathscr{W}_g s(b, a) = \langle g_{b,a} \mid s \rangle$. Both definitions of the wavelet transform coincide almost everywhere thanks to Fatou's lemma that we state here without proof. (A proof may be found in Torchinsky 1988.)

Theorem 22.0.5

Let w_n be a sequence of measurable functions on a measure space, then

$$\int d\mu(x) \liminf_{n \to \infty} w_n(x) \leq \liminf_{n \to \infty} \int d\mu(x) w_n(x),$$

where $d\mu$ is a positive Borell measure.

Clearly, both definitions of wavelet transforms agree (almost everywhere) for g and $s \in S_+(\mathbb{R})$. Now since $S_+(\mathbb{R})$ is dense in $H^2_+(\mathbb{R})$ it is enough to show that $s \mapsto Ks(b, a) = \langle g_{b,a} \mid s \rangle$ defines a bounded operator. Therefore, pick an approximating sequence $S_+(\mathbb{R}) \ni s_n \to s \in H^2_+(\mathbb{R})$ and consider $\mathscr{T}_n(b, a) = \langle g_{b,a} \mid s_n \rangle$. Upon replacing n by $n + N$ we may suppose that $\|s_n\|_2^2 \leq 2\|s\|_2^2$. By the continuity of the scalar product we have for each $(b, a) \in \mathbb{H}$ that $Ks(b, a) = \lim_{n \to \infty} \mathscr{T}_n(b, n)$. For each n we have $\|\mathscr{T}_n\|_2^2 = c_g \|s_n\|_2^2$ and, thanks to Fatou's lemma, we obtain for the limit $\|\lim_{n \to \infty} \mathscr{T}_n\|_2^2 \leq 2c_g\|s\|_2^2$, which shows that K is continuous and therefore $K = \mathscr{W}_g$.

We now want to give up the restriction that g is in $S_+(\mathbb{R})$. Suppose, therefore, that $s \in H^2_+(\mathbb{R})$ and that $g \in H^2_+(\mathbb{R})$ is admissible:

$$0 < c_g = \int_0^\infty \frac{d\omega}{\omega} |\hat{g}(\omega)|^2 < \infty.$$

Let us denote by A the set of admissible vectors in $H^2_+(\mathbb{R})$. It becomes a Hilbert space with respect to the scalar product and norm

$$\langle s \mid r \rangle_A = \int_0^\infty d\omega(1 + 1/\omega)\overline{\hat{s}(\omega)}\hat{r}(\omega),$$

$$\|g\|_A^2 = \int_0^\infty d\omega(1 + 1/\omega)|\hat{s}(\omega)|^2 = 2\pi\|s\|_2^2 + c_g^2.$$

By the same procedure of cutting and smoothing as in the proof of Theorem 22.0.3, we can find an approximating sequence $S_+(\mathbb{R}) \ni g_n \to g$ where the convergence holds in A, that is, we have simultaneously $g_n \to g$ in $H^2_+(\mathbb{R})$ and $c_{g_n - g_m} \to 0$ as $n, m \to \infty$. Therefore

$$\|\mathscr{W}_{g_m}s - \mathscr{W}_{g_n}s\|_2^2 = \|\mathscr{W}_{g_m - g_n}s\|_2^2 \leq c_{g_m - g_n}\|s\|_2^2,$$

which tends to 0. Thus this is a Cauchy sequence and there is a $\mathscr{T} \in L^2(\mathbb{H})$ with

$$\|\mathscr{W}_{g_n}s - \mathscr{T}\|_2 \to 0$$

which we define to be the wavelet transform of $s \in H^2_+(\mathbb{R})$ with respect to $g \in A$. This is well defined since, for any other sequence $h_n \to g$ in A, we have

$$\|\mathscr{W}_g s - \mathscr{W}_{h_n}s\|_2^2 = \|\mathscr{W}_{g - h_n}s\|_2^2 \leq c_{g - h_n}\|s\|_2^2 \to 0.$$

Again the same type of argument as before shows that this definition coincides almost everywhere with the previous one, namely, $\mathscr{W}_g s(b, a) = \langle g_{b,a} \mid s \rangle$.

Thus we have shown the following theorem, which is originally due to Grossmann and Morlet (1984).

Theorem 22.0.6
Let $g \in H^2_+(\mathbb{R})$ be admissible:

$$c_g = \int_0^\infty \frac{d\omega}{\omega} |\hat{g}(\omega)|^2, \qquad 0 < c_g < \infty.$$

Then the wavelet transform defined pointwise for $s \in H^2_+(\mathbb{R})$ as $\mathscr{W}_g s(b, a) = \langle g_{b,a} \mid s \rangle$ is in $L^2(\mathbb{H})$ and we have

$$\int_{-\infty}^{+\infty} db \int_0^\infty \frac{da}{a} |\mathscr{W}_g s(b, a)|^2 = c_g \int_{-\infty}^{+\infty} dt\, |s(t)|^2,$$

which means that $\mathscr{W}_g$ is a constant multiple of an isometry.

This shows in particular that the wavelet transform $\mathscr{W}: A \times H^2_+(\mathbb{R}) \to L^2(\mathbb{H})$, $(g, s) \mapsto \mathscr{W}_g s$ is a jointly continuous bi-linear mapping.

One may also extend equation (22.0.2) to all of $H^2_+(\mathbb{R})$.

Corollary 22.0.7
Let $g, h \in H^2_+(\mathbb{R})$ be admissible and let $s, r \in H^2_+(\mathbb{R})$. Then

$$\langle \mathscr{W}_h s \mid \mathscr{W}_g r \rangle_{\mathbb{H}} = c_{g,h} \langle s \mid r \rangle_{\mathbb{R}}, \qquad c_{g,h} = \int_0^\infty \frac{d\omega}{\omega} \hat{h}(\omega) \bar{\hat{g}}(\omega).$$

Note that $c_{g,h}$ is well defined thanks to the Schwarz inequality.

Proof. Again, an argument using approximating sequences will work. We leave the details to the reader. □

23 The inverse wavelet transform

The inverse wavelet transform with respect to an admissible wavelet can be extended to all functions in $\mathscr{T} \in L^2(\mathbb{H})$ by duality. Namely, we define $\mathscr{M}_g \mathscr{T} = s$ to be the unique element s in $L^2(\mathbb{R})$ such that

$$\langle s \mid r \rangle_{\mathbb{R}} = \langle \mathscr{T} \mid \mathscr{W}_g r \rangle_{\mathbb{H}}$$

for all functions $r \in L^2(\mathbb{R})$.

As an adjoint of a bounded map the wavelet synthesis itself is bounded.

Theorem 23.0.1
For $\mathcal{T} \in L^2(\mathbb{H})$ the energy of its wavelet synthesis with respect to an admissible wavelet $g \in H^2_+(\mathbb{R})$ is estimated by

$$\|\mathcal{M}_g \mathcal{T}\|_{H^2_+(\mathbb{R})} \leq c_g \|\mathcal{T}\|_{L^2(\mathbb{H})}.$$

This shows that the wavelet synthesis $\mathcal{M}: A \times L^2(\mathbb{H}) \to H^2_+(\mathbb{R}), (h, \mathcal{T}) \mapsto \mathcal{M}_h \mathcal{T}$ is a jointly continuous bi-linear mapping.

Proof. This is completely standard, but for the convenience of the reader we give the proof anyway. We have the following formula:

$$\|s\|_{L^2(\mathbb{R})} = \sup |\langle s \mid r \rangle|,$$

where the sup runs over all functions of unit norm $\|r\|_{L^2(\mathbb{R})} = 1$. For the inverse wavelet transform we therefore have

$$\begin{aligned} \|\mathcal{M}_g \mathcal{T}\|_{L^2(\mathbb{R})} &= \sup |\langle \mathcal{M}_g \mathcal{T} \mid r \rangle_{\mathbb{R}}| = \sup |\langle \mathcal{T} \mid \mathcal{W}_g r \rangle_{\mathbb{H}}| \\ &\leq \|\mathcal{T}\|_{L^2(\mathbb{H})} \|\mathcal{W}_g r\|_{L^2(\mathbb{H})} \leq c_g \|\mathcal{T}\|_{L^2(\mathbb{H})}. \end{aligned}$$ □

Again, the wavelet synthesis is the inverse of the wavelet analysis since it is the adjoint of an isometry. In the wavelet case this is just a rewriting of (22.0.7) and the definition of $\mathcal{M}_h$ as the adjoint of $\mathcal{W}_h$.

Theorem 23.0.2
We have for $s \in H^2_+(\mathbb{R})$

$$\mathcal{M}_h \mathcal{W}_g s = c_{g,h} s.$$

Again, we can also obtain this inversion by means of an integral over the half-plane.

Theorem 23.0.3
Let $h \in H^2_+(\mathbb{R})$ be admissible and let $\mathcal{T} \in L^2(\mathbb{H})$. Then let $I_j \subset \mathbb{H}$, $j \in \mathbb{R}$, be a non-decreasing family of closed, bounded sets of the open half-plane that converges to $\mathbb{H}$, that is, $I_j \subset I_{j'}$ for $j' \geq j$ and $\bigcup_j I_j = \mathbb{H}$. Then

$$\mathcal{M}_h \mathcal{T} = \lim_{j \to \infty} \int_{I_j} \frac{db\, da}{a} \mathcal{T}(b, a) h_{b,a}, \qquad h_{b,a}(t) = \frac{1}{a} h\left(\frac{t-b}{a}\right), \tag{23.0.1}$$

where the convergence holds in $L^2(\mathbb{R})$.

Proof. Let us call $\mathcal{M}_j$ the integral on the right-hand side of (23.0.1) for a given I_j. First observe that each $\mathcal{M}_j$ is a well-defined function in $H^2_+(\mathbb{R})$. It is obviously defined for almost every t. To see that it is of finite energy we pick $s \in L^2(\mathbb{R})$ arbitrary of unit length $\|s\| = 1$. Upon writing, using

Fubini's theorem,

$$\begin{aligned}\langle s \mid \mathcal{M}_j \rangle &= \int_{-\infty}^{+\infty} dt\, \bar{s}(t) \int_{I_j} \frac{db\, da}{a} \mathcal{T}(b, a) \frac{1}{a} h\left(\frac{t-b}{a}\right) \\ &= \int_{I_j} \frac{db\, da}{a} \mathcal{T}(b, a) \int_{-\infty}^{+\infty} dt \frac{1}{a} h\left(\frac{t-b}{a}\right) \bar{s}(t) \\ &= \int_{I_j} \frac{db\, da}{a} \mathcal{T}(b, a) \overline{\mathcal{W}_h s}(b, a),\end{aligned}$$

we see that this is bounded in absolute value by $c_h \|\mathcal{T}\|_{L^2(\mathcal{H})} \|s\|_{L^2(\mathcal{R})}$ since the wavelet transform with respect to h preserves the energy. Hence, by the Riesz representation theorem for finite j, $\mathcal{M}_j$ is a function in $L^2(\mathbb{R})$. It is obviously in $H^2_+(\mathbb{R})$ since h is.

To see that the $\mathcal{M}_j$ are a Cauchy net, we consider $\|\mathcal{M}_j - \mathcal{M}_{j'}\|_{L^2(\mathbb{R})}$ with $j \geq j'$. This equals

$$\sup |\langle s \mid \mathcal{M}_j - \mathcal{M}_{j'} \rangle|,$$

where the sup is over all $s \in L^2(\mathbb{R})$ with $\|s\| = 1$. As before, by Fubini's theorem the scalar product equals

$$\int_{I_j - I_{j'}} \frac{db\, da}{a} \mathcal{T}(b, a) \overline{\mathcal{W}_h s}(b, a),$$

which may be estimated by

$$\int_{I^c_{j'}} \frac{db\, da}{a} |\mathcal{T}(b, a)|\, |\mathcal{W}_h s(b, a)| \leq c_h \sqrt{\int_{I^c_{j'}} \frac{db\, da}{a} |\mathcal{T}(b, a)|^2},$$

where we have used Schwarz's inequality. As $j, j' \to \infty$, the integral tends to 0 since the integrand is in $L^1(\mathbb{H})$.

To show now that the limit is actually $\mathcal{M}_h \mathcal{T}$ is easy and is obtained by exchanging the integrals in $\langle s \mid \mathcal{M}_j \rangle$. We leave the details to the reader. □

The image of the wavelet transform may again be characterized by a reproducing kernel.

Theorem 23.0.4

The image of the wavelet transform with respect to $g \in H^2_2(\mathbb{R})$, admissible, is the closed subspace of functions $\mathcal{T} \in L^2(\mathbb{H})$ that satisfy pointwise at

$$\mathcal{T}(b, a) = \int_{-\infty}^{+\infty} db' \int_0^{\infty} \frac{da'}{a'} \frac{1}{a'} P_{g,h}\left(\frac{b-b'}{a'}, \frac{a}{a'}\right) \mathcal{T}(b', a'),$$

with $P_{g,h} = c_{g,h}^{-1} \mathcal{W}_g h$ *and* $h \in H^2_+(\mathbb{R})$ *any admissible reconstruction wavelet. For general* $\mathcal{T} \in L^2(\mathbb{H})$, *the right-hand side defines a projection onto the image of* $\mathcal{W}_g$. *For* $h = g$ *this projection is orthogonal.*

Proof. Because g and h are admissible, the operator $\Pi_{g,h} = \mathcal{W}_g \mathcal{M}_h$ is a well-defined bounded map from $L^2(\mathbb{H})$ to $L^2(\mathbb{H})$. Clearly, for $\mathcal{T} \in L^2(\mathbb{H})$ we have that $\Pi_{g,h}\mathcal{T}$ is in the image of $\mathcal{W}_g$ since it is the wavelet transform of $\mathcal{M}_h \mathcal{T}$. On the other hand, suppose that $\mathcal{T}$ is in the image of $\mathcal{W}_g$. Then there is a function s over the real line such that $\mathcal{T} = \mathcal{W}_g s$. Therefore, $\Pi_{g,h}\mathcal{T} = c_{g,h}^{-1} \mathcal{W}_g(\mathcal{M}_h \mathcal{W}_g s) = \mathcal{W}_g s = \mathcal{T}$. This shows that

$$\Pi_{g,h}\Pi_{g,h} = \Pi_{g,h},$$

and thus $\Pi_{g,h}$ is a projector on the image of $\mathcal{W}_g$. Suppose now that $h = g$. By (22.0.1) we see that $\Pi_{g,g}$ is self-adjoint.

As we know already, equation (22.0.2) holds in $S_+(\mathbb{R})$. Now we may pick approximating sequences g_n, h_n, and s_n in $S_+(\mathbb{R})$ converging in $H^2_+(\mathbb{R})$ to g, h, s, respectively, and in addition c_{g_n} and c_{g_n,h_n} converge to c_g and $c_{g,h}$. Now it follows that $\mathcal{T}_n = \mathcal{W}_{g_n} s_n$ converges in $L^2(\mathbb{H})$ and pointwise to $\mathcal{T} = \mathcal{W}_g s$, and $c_{g_n,h_n}^{-1} \mathcal{W}_{g_n} h_n$ converges to $P_{g,h}$ where the convergence holds again in $L^2(\mathbb{H})$ since the wavelet transform is jointly continuous. For fixed $(b, a) \in \mathbb{H}$ we may therefore go to the limit $n \to \infty$ in

$$\mathcal{T}_n(b, a) = \int_{-\infty}^{+\infty} db' \int_0^\infty \frac{da'}{a'} \frac{1}{a'} P_{g_n,h_n}\left(\frac{b - b'}{a}, \frac{a}{a'}\right) \mathcal{T}_n(b', a'),$$

and the theorem follows. □

24 The wavelet transform over $S'_+(\mathbb{R})$

We now want to discuss the wavelet transform of tempered distributions. These mathematical objects have typically very little regularity, as the example of δ suggests. In the wavelet transform we therefore expect to obtain a lot of energy at small scales. However, although δ is rough its wavelet transform $\mathcal{W}_g \delta(b, a) = \bar{g}(-b/a)/a$ is a smooth function in the interior of the half-plane. The roughness of δ only appears at the borderline where the coefficients explode. This is actually true in general and we will see that all tempered distributions correspond to smooth functions that are at most of polynomial growth in $|b|(a + 1/a)$. In addition, their action on a function in $S_+(\mathbb{R})$ can be written as an absolutely convergent integral over the half-plane.

We now come to the details. We recall the main properties of locally convex spaces and their duals. Essentially no proofs will be given and we refer the reader to Schaefer (1982).

Let X be a Fréchet space, that is, a complete locally convex space with the topology induced by some invariant metric. For every Fréchet space there is a countable family of seminorms $\|\cdot\|_\alpha$ on X that generate the topology of X. The space of linear continuous functionals,

$$\eta: X \to \mathbb{C}, \qquad \eta(\alpha s + \beta \omega) = \alpha\eta(s) + \beta\eta(\omega), \quad \alpha, \beta \in \mathbb{C},\ s, w \in X,$$

is called the dual space. Now continuity of η means that for some α we have

$$|\eta(s)| \le c\|s\|_\alpha, \qquad \text{for all } s \in X,$$

with some constant c that does not depend on s. The elements of X' separate points, that is, if for some $s, w \in X$ we have $\eta(s) = \eta(w)$ for all $\eta \in X'$ then it follows that $s = w$. Now every element s of X induces a linear functional on X' by setting

$$\rho_s: X' \to \mathbb{C}, \qquad \eta \to \eta(s).$$

The weak* topology on X' is defined to be the topology of pointwise convergence of these functionals, that is, a sequence $\eta_n \in X'$, tends to $\eta \in X'$ in the weak* topology iff for all $s \in X$ we have $\eta_n(s) \to \eta(s)$. A basis of neighbourhoods is then

$$U_\epsilon(s_1, \ldots, s_m) = \{\eta \in X': |\eta(x_p)| < \epsilon, p = 1, \ldots, m\},$$

where s_p are arbitrary points in X and $\epsilon > 0$. It can be shown that X' is again a locally convex space and its dual is X.

In the case where $X = S(\mathbb{R})$, or $S_\pm(\mathbb{R})$, or $S_0(\mathbb{R})$ we call the elements of X' tempered distributions.

Recall for later reference that the topology on $S_+(\mathbb{R})$ was generated by the seminorms

$$\|s\|_\alpha = \sup_{t\in\mathbb{R}}|s(t)|(1 + |t|^2)^{\alpha/2} + \sup_{\omega\ge 0}|\hat{s}(\omega)|\omega^{-\alpha}(1 + \omega)^{2\alpha+1}.$$

The corresponding locally convex space $S(\mathbb{H})$ over the half-plane was generated by the seminorms

$$\|\mathscr{T}\|_\alpha = \sum_{n\le\alpha}\ \sup_{(b,a)\in\mathbb{H}} |(1 + |b|^2)^{\alpha/2}(a + 1/a)^\alpha(a\partial_a)^n\partial_b^n\mathscr{T}(b, a)|.$$

The topology of the space $S(\mathbb{R})$ was generated by the seminorms

$$\|s\|_n = \sup_{t\in\mathbb{R}}|t^n\partial^n s|.$$

There is a natural continuous embedding of $S(\mathbb{R})$ into $S'(\mathbb{R})$ by setting

$$\iota: S(\mathbb{R}) \to S'(\mathbb{R}), \qquad s \mapsto \left(w \mapsto \int s\cdot w\right). \tag{24.0.1}$$

In Fourier space we have

$$\iota\colon \hat{s} \mapsto \left(\hat{w} \mapsto \int_{-\infty}^{+\infty} d\omega\, \hat{s}(-\omega)\hat{w}(\omega) \right).$$

Therefore, $s \in S_-(\mathbb{R})$ non-zero defines a linear non-zero linear functional on $S_+(\mathbb{R})$ under the embedding ι. This motivates the following definition.

Definition 24.0.1

The space of linear functionals on $S_-(\mathbb{R})$ will be denoted by $S'_+(\mathbb{R})$. The space of linear functionals on $S_+(\mathbb{R})$ will be denoted by $S'_-(\mathbb{R})$.

With this definition we have that

$$\iota\colon S_+(\mathbb{R}) \to S'_+(\mathbb{R}) \quad \text{and} \quad S_-(\mathbb{R}) \to S'_-(\mathbb{R}),$$

as it should be.

In the same way we may define a linear embedding of $S(\mathbb{H})$ into $S'(\mathbb{H})$ by setting

$$\iota\colon S(\mathbb{H}) \to S'(\mathbb{H}), \qquad \mathscr{T} \mapsto \left(\mathscr{R} \mapsto \int_{-\infty}^{+\infty} db \int_0^\infty \frac{da}{a} \mathscr{T}(b, a)\mathscr{R}(b, a) \right). \quad (24.0.2)$$

We now wish to discuss the relation between $S'_+(\mathbb{R})$ and $S'_-(\mathbb{R})$ and $S'(\mathbb{R})$. Clearly, any $\eta \in S'(\mathbb{R})$ defines a distribution in $S'_+(\mathbb{R})$ by restricting it to the closed subset $S_-(\mathbb{R})$. On the other hand, every distribution in $S'_+(\mathbb{R})$ can be extended to a distribution in $S'(\mathbb{R})$. This is an immediate consequence of the Hahn–Banach theorem that we state without proof. A demonstration may be found in Rudin (1991, page 58), for example.

Theorem 24.0.2 (Hahn–Banach)

Let V be a subspace of a locally convex space X and let η be a linear functional the satisfies, for some seminorm,

$$|\eta(s)| \le \|s\|, \qquad \text{for all } s \in V.$$

Then there is a functional $\chi \in X'$ that coincides with η on V and that satisfies

$$|\chi(s)| \le \|s\|, \qquad \text{for all } s \in X.$$

Note, however, that in view of this theorem the extension of $\eta \in S'_+(\mathbb{R})$ to $\chi \in S'(\mathbb{R})$ is in general not unique.

We now proceed by understanding the Fourier space picture of $S'_+(\mathbb{R})$. As we have seen before, the Fourier transform is a topological isomorphism of $S(\mathbb{R})$, that is, a continuous map with continuous inverse. The Fourier transform on $S'(\mathbb{R})$ is now defined by setting

$$\hat{\eta}(s) = \eta(\hat{s}), \qquad \eta \in S'(\mathbb{R}),\ s \in S(\mathbb{R}).$$

Again this is a topological isomorphism by duality. By exchanging the order of integration we can see that his is a true extension of the Fourier transform of $S(\mathbb{R})$ if we identify a function $s \in S(\mathbb{R})$ with a distribution in $S'(\mathbb{R})$ via $s(w) = \int sw$, as before. Since $S(\mathbb{R})$ is dense in $S'(\mathbb{R})$ this is the only possible continuous extension.

We introduce the notion of support for a distribution η in $S'(\mathbb{R})$. This is the complement of those points for which there is an open neighbourhood I such that for all $\phi \in C_c^\infty(I)$—the space of smooth functions with support in I—we have $\phi \cdot \eta = 0$. Equivalently, we may say that it is the complement of points that have a neighbourhood such that $\eta(\phi) = 0$ for all $\phi \in C_c^\infty(I)$. As the complement of an open set the support is a closed set. Again this notation coincides with the usual one in case s is a continuous function.

We still need the notion of quotient space of two topological vector spaces $V \subset X$. Consider the equivalence classes modulo V of an element $x \in X$, that is, the set of $y \in X$ such that $x - y \in V$. This set will be denoted by $x + V$. The set of equivalence classes modulo V will be denoted by X/V. It is again a vector space under addition of representatives and a topology is given as the finest topology that makes the natural map

$$\sigma: X \to X/V, \qquad x \mapsto x + V,$$

continuous; that is the open sets are of the form $U + V$ with U open in X.

As a last tool we need the so-called (absolute) polar of a subset $V \subset X$. It is defined as the set of vectors η in X' for which

$$|\eta(x)| \leq 1.$$

The polar is usually denoted by V°. In case of V being a subspace it follows that V° is a subspace of X'. It then consists of all vectors for which $\eta(x) = 0$ for all $x \in V$. It is therefore a closed subspace of X'. We now equip V with the topology inherited from X. By the Hahn–Bannach theorem every continuous functional $\eta: V \to \mathbb{C}$ can be lifted to X. It is defined up to some element of V°. Thus we can identify V' with X'/V°, at least algebraically. This identification also holds in the topological sense as is shown in the next theorem that we cite from Treves (1967, page 364).

Theorem 24.0.3

Let V be a subspace of a locally convex space X equipped with induced topology. Then

$$V' \simeq X'/V^\circ$$

iff V is closed in X. Here X' is equipped with weak topology.*

Applied to our situation this clearly means the following.

Theorem 24.0.4
We have

$$S'_+(\mathbb{R}) \simeq S'(\mathbb{R})/\{\eta \in S'(\mathbb{R}): \operatorname{supp} \hat{\eta} \subset \mathbb{R}_-\},$$
$$S'_-(\mathbb{R}) \simeq S'(\mathbb{R})/\{\eta \in S'(\mathbb{R}): \operatorname{supp} \hat{\eta} \subset \mathbb{R}_+\}.$$

The natural projections are given by

$$\sigma_+ : S'(\mathbb{R}) \to S'_+(\mathbb{R}), \qquad (\sigma_+\eta)(s) = \eta(s) \textit{ for all } s \in S_-(\mathbb{R}),$$
$$\sigma_- : S'(\mathbb{R}) \to S'_-(\mathbb{R}), \qquad (\sigma_-\eta)(s) = \eta(s) \textit{ for all } s \in S_+(\mathbb{R}).$$

24.1 Definition of the wavelet transform

For the sake of simplicity we will formulate everything for the analysing wavelet in $S_+(\mathbb{R})$ and distributions in $S'_+(\mathbb{R})$. All statements and defintions can easily be adapted to $g \in S_-(\mathbb{R})$ and $S_0(\mathbb{R})$.

For any two functions g, $s \in S_+(\mathbb{R})$ and $\mathscr{T} \in S(\mathbb{H})$ we have

$$\langle \mathscr{W}_g s \mid \mathscr{T} \rangle_{\mathbb{H}} = \langle s \mid \mathscr{M}_g \mathscr{T} \rangle_{\mathbb{R}}.$$

This, and identifications (24.0.1) and (24.0.2) motivates us to define the wavelet transform with respect to any wavelet $g \in S_+(\mathbb{R})$ of any distribution $\eta \in S'_+(\mathbb{R})$ as the distribution in $S'(\mathbb{H})$ that has the following action on any function $\mathscr{T} \in S(\mathbb{H})$:

$$\mathscr{W}_g \eta(\mathscr{T}) = \eta(\mathscr{M}_{\bar{g}} \mathscr{T}).$$

Since $g \in S_+(\mathbb{R})$ we have $\bar{g} \in S_-(\mathbb{R})$. Therefore, since $\mathscr{M}_{\bar{g}}$ is a continuous map from $S(\mathbb{H})$ to $S_-(\mathbb{R})$ and since $\eta \in S'_+(\mathbb{R})$ is continuous from $S_-(\mathbb{R})$ to $\mathbb{C}$ it follows that the whole is continuous and hence is a distribution in $S'(\mathbb{H})$.

This definition is a true extension of the previous definitions. To see this let us temporarily denote by $\mathscr{W}_g'$ the one we have just defined and by $\mathscr{W}_g$ the one we studied in the previous sections. We then have

$$\mathscr{W}_g' \iota_{S(\mathbb{R})} = \iota_{S(\mathbb{H})} \mathscr{W}_g,$$

as can be seen from

$$\begin{aligned}(\mathscr{W}_g' \iota_{S(\mathbb{R})} s)(\mathscr{T}) &= (\iota_{S(\mathbb{R})} s)(\mathscr{M}_{\bar{g}} \mathscr{T}) \\ &= \langle \bar{s} \mid \mathscr{M}_{\bar{g}} \mathscr{T} \rangle \\ &= \langle \mathscr{W}_{\bar{g}} \bar{s} \mid \mathscr{T} \rangle \\ &= \langle \overline{\mathscr{W}_g s} \mid \mathscr{T} \rangle \\ &= (\iota_{S(\mathbb{H})}(\mathscr{W}_g s))(\mathscr{T}).\end{aligned}$$

Since this is so, we will not distinguish any more between $\mathscr{W}_g$ and $\mathscr{W}_g'$.

We now define the inverse wavelet transform on $S'(\mathbb{H})$ with respect to $g \in S_+(\mathbb{R})$ via

$$(\mathcal{M}_g \Omega)(s) = \Omega(\mathcal{W}_{\bar{g}} s), \qquad \Omega \in S'(\mathbb{H}),\ s \in S_-(\mathbb{R}).$$

Again the synthesis $\mathcal{M}_g$ is a well-defined distribution in $S'_+(\mathbb{R})$ since as the composition of continuous map it goes continuously from $S_-(\mathbb{R})$ to $\mathbb{C}$. It is an extension of the previous definition since we have

$$\imath_{S(\mathbb{R})} \mathcal{M}_g = \mathcal{M}'_g \imath_{S(\mathbb{H})},$$

where we have used the analogue of the previous notations. Again we will not distinguish in the notation between the wavelet synthesis over $S(\mathbb{H})$ and that over $S'(\mathbb{H})$.

We are now ready for a series of theorems that describe the wavelet transform over $S'_+(\mathbb{R})$.

Theorem 24.1.1

The wavelet transform and the wavelet synthesis with respect to some fixed wavelets $g, h \in S_+(\mathbb{R})$:

$$\mathcal{W}_g : S'_+(\mathbb{R}) \to S'(\mathbb{H}), \qquad \eta \mapsto \mathcal{W}_g \eta,$$

$$\mathcal{M}_h : S'(\mathbb{H}) \to S'_+(\mathbb{R}), \qquad Y \mapsto \mathcal{M}_h Y,$$

are continuous in the respective weak topologies.*

Remark. Clearly, for a fixed analysed function the dependency on the wavelet is continuous, too. However, the bilinear map $(g, \eta) \mapsto \mathcal{W}_g \eta$ is not continuous but, rather, hypocontinuous (e.g. Treves 1967). However, if we limit ourselves to sequences (instead of generalized sequences) we still have that $g_n \to g$ and $\eta_n \to \eta$ implies $\mathcal{W}_{g_n} \eta_n \to \mathcal{W}_g \eta$, where the convergences take place in the respective topologies.

Proof. Fix $g \in S_+(\mathbb{R})$. We have to show that for any generalized sequence η_n, $n \in I$, from $\eta_n \to 0$ in $S'_+(\mathbb{R})$, it follows that $\mathcal{W}_g \eta_n \to 0$ in the topology of $S'(\mathbb{H})$. Now $\eta_n \to 0$ means that for all $s \in S_-(\mathbb{R})$ we have $\eta_n(s) \to 0$. But, therefore, for every $\mathcal{T} \in S(\mathbb{H})$ we have

$$\mathcal{W}_g \eta_n(\mathcal{T}) = \eta_n(\mathcal{M}_{\bar{g}} \mathcal{T}) \to 0,$$

which shows that $\mathcal{W}_g \eta_n \to 0$ in $S'(\mathbb{H})$.

The proof of the continuity of the inverse wavelet transform is completely analogous to the preceding proof, using this time the continuity of $\mathcal{W}_g$ on $S_\pm(\mathbb{R})$. □

Theorem 24.1.2

Suppose in addition that $h \in S_+(\mathbb{R})$ is a reconstruction wavelet for the analysing wavelet $g \in S_+(\mathbb{R})$. Then

$$c_{g,h}^{-1} \mathcal{M}_h \mathcal{W}_g = \mathbb{1}_{S'_+(\mathbb{R})}.$$

Proof. By direct application of the definitions of $\mathcal{W}_g$ and $\mathcal{M}_h$ we obtain, for every $\eta \in S_+(\mathbb{R})$ and every test function $s \in S_-(\mathbb{R})$,

$$(\mathcal{M}_h \mathcal{W}_g \eta)(s) = (\mathcal{W}_g \eta)(\mathcal{W}_{\bar{h}} s) = \eta(\mathcal{M}_{\bar{g}} \mathcal{W}_{\bar{h}} s).$$

Now $\bar{g}$ is also a reconstruction wavelet for $\bar{h}$ with the same constant $c_{g,h}$ and therefore the last expression equals $c_{g,h}\eta(s)$. □

Theorem 24.1.3

The image of $S'_+(\mathbb{R})$ under the wavelet transform with respect to $g \in S_+(\mathbb{R})$ is the closed subspace of $S'(\mathbb{H})$ of distributions Ξ that satisfy

$$\Xi = c_{g,h}^{-1} \mathcal{W}_g \mathcal{M}_h \Xi, \tag{24.1.1}$$

where $h \in S_+(\mathbb{R})$ is a reconstruction wavelet for g. For arbitrary $\Xi \in S'(\mathbb{H})$, the right-hand side defines a projection into the image of the wavelet transform.

Proof. The proof follows by duality and is left to the reader. □

We now want to show that the wavelet transform of a distribution may be identified with a smooth function over the half-plane of at most polynomial growth.

Theorem 24.1.4

Let $\eta \in S'_+(\mathbb{R})$, $g \in S_+(\mathbb{R})$. Then $\mathcal{W}_g \eta$ may be identified with

$$Y(b, a) = \eta(\bar{g}_{b,a}), \qquad g_{b,a}(t) = a^{-1} g(a^{-1}[t - b]),$$

in the sense that for all $\mathcal{T} \in S(\mathbb{H})$ we have

$$\mathcal{W}_g \eta(\mathcal{T}) = \int_0^\infty \frac{da}{a} \int_{-\infty}^{+\infty} db \; Y(b, a) \mathcal{T}(b, a). \tag{24.1.2}$$

The function Y is, together with all the derivatives of at most polynomial growth,

$$|\partial_b^n (a\partial_a)^m Y(b, a)| \leq O(1)(1 + |b|^2)^{\alpha/2}(a + 1/a)^\beta,$$

with some $\alpha, \beta \geq 0$ only depending on n, m. In addition, it satisfies the reproducing kernel equation

$$Y(b, a) = \int_0^\infty \frac{da'}{a'} \int_{-\infty}^{+\infty} db' \, \frac{1}{a'} P_{g,h}\left(\frac{b - b'}{a'}, \frac{a}{a'}\right) Y(b', a').$$

This equation holds pointwise. Vice versa, any function of at most polynomial growth satisfying this reproducing kernel equation is in the image of the wavelet transform where we again make the identification (24.1.2).

We therefore write $\mathcal{W}_g\eta$ for the function $Y(b,a) = \eta(\bar{g}_{b,a})$. The last statement shows that the action of $\eta \in S'_+(\mathbb{R})$ on any function $s \in S_-(\mathbb{R})$ can be written as an absolutely convergent scalar product over the half-plane

$$\eta(s) = \frac{1}{c_{g,h}} \int_0^\infty \frac{da}{a} \int_{-\infty}^{+\infty} db\, \mathcal{W}_g\eta(b,a)\mathcal{W}_{\bar{h}}s(b,a).$$

We can therefore say that somehow the wavelet transform de-singularized the distributions. The singular nature of the distribution can be found in the small scale ($a \to x$) and large ($|b| \to \infty$) position behaviour of its wavelet transform.

Proof. Let us consider the function over the half-plane that is pointwise defined via

$$Y(b,a) = \eta(\bar{g}_{b,a}), \qquad \bar{g}_{b,a}(\cdot) = \bar{g}([\cdot - b]/a)/a.$$

We want to show that this function is actually the wavelet transform of η in the sense of distributions. First let us look at the regularity of Y. We claim that Y is infinitely many times differentiable and that it is, together with all its derivatives, polynomially localized. The regularity follows from the next lemma.

Lemma 24.1.5

For dilations and translations the following derivation rules hold:

$$\partial_b(\eta(s(\cdot - b))) = \eta((\partial s)(\cdot - b)), \qquad a\partial_a(\eta(s(\cdot/a)/a)) = \eta((t\partial_t s)(\cdot/a)/a)$$

for every distribution $\eta \in S'(\mathbb{R})$ and every function $s \in S(\mathbb{R})$.

Clearly, the same holds for $\eta \in S'_+(\mathbb{R})$ and $s \in S_-(\mathbb{R})$, as can be seen by extending η to all of $S(\mathbb{R})$ by means of the Hahn–Banach theorem.

Proof. We only show the first equation, the second is analogue. We have

$$s(t - b + \epsilon) - s(t - b) = \epsilon s'(t - b) + \epsilon\rho_\epsilon(t),$$

with ρ_ϵ tending to 0 in $S(\mathbb{R})$ as $\epsilon \to 0$. Therefore, by linearity of η we have

$$\epsilon^{-1}(\eta(s(\cdot - b + \epsilon)) - \eta(s(\cdot - b))) = \eta(s'(\cdot - b)) + \eta(\rho_\epsilon).$$

Going to the limit $\epsilon \to 0$ we have obtained what we sought. □

The localization of $Y(b,a)$ and all its partial derivatives is a consequence of the next lemma. We use the notation $\kappa_\alpha(t) = (1 + t^2)^{\alpha/2}$ and $\phi_\alpha(a) = (1 + a)^{2\alpha+1}/a^\alpha$.

Lemma 24.1.6

For every distribution $\eta \in S'_+(\mathbb{R})$ and every function $s \in S_-(\mathbb{R})$ we have

$$|\eta(s(\cdot - b))| \leq c\kappa_\alpha(b), \qquad |\eta(s(\cdot/a)/a)| \leq c\phi_\alpha(a)$$

for some exponent $\alpha \geq 0$ and some constant $c > 0$ only depending on s and η.

Proof. It is elementary to see that the following estimations hold for $\alpha \geq 0$:

$$\left.\begin{aligned} \kappa_a(b + t) &\leq O(1)\kappa_\alpha(b)\kappa_\alpha(t), \\ \kappa_a(at) &\leq O(1)\kappa_\alpha(a)\kappa_\alpha(t), \\ \phi_a(at) &\leq O(1)\phi_\alpha(b)\phi_\alpha(t). \end{aligned}\right\} \tag{24.1.3}$$

Now we can write, since $|s(t)| \leq O(\kappa_{-\alpha}(t))$ and $|\hat{s}(\omega)| \leq O(\phi_{-\alpha}(\omega))$,

$$\begin{aligned} |\eta(s(\cdot - b))| \leq \|s(\cdot - b)\|_{S_+(\mathbb{R});\alpha} &\leq \sup_{t\in\mathbb{R}}|s(t - b)|\kappa_\alpha(t) + \sup_{\omega\geq 0}|\hat{s}(\omega)|\phi_\alpha(\omega) \\ &\leq O(1)\sup_{t\in\mathbb{R}}\kappa_{-\alpha}(t - b)\kappa_\alpha(t) + O(1) \\ &\leq O(1)\kappa_\alpha(b), \end{aligned}$$

for some $\alpha > 0$. Now for some $\alpha > 1$,

$$\begin{aligned} |\eta(s(\cdot/a)/a)| &\leq \|s(\cdot/a)/a\|_{S_+(\mathbb{R});\alpha} \\ &\leq O(1)\sup_{t\in\mathbb{R}}|\boldsymbol{\kappa}^\alpha(t)/s(t/a)/a + O(1)\sup_{\omega\geq 0}|\hat{s}(a\omega)|\phi_\alpha(\omega) \\ &\leq O(1)\sup_{t\in\mathbb{R}} a^{-1}\kappa_{-\alpha}(t/a)\kappa_\alpha(t) + O(1)\sup_{\omega\geq 0}\phi_{-\alpha}(a\omega)\phi_\alpha(\omega) \\ &\leq O(1)\phi_\alpha(a), \end{aligned}$$

and the lemma follows. □

Until now we have shown that $Y(b, a) = \eta(\bar{g}_{b,a})$ is a C^∞ function over the half-plane that is, together with all its derivatives, of at most polynomial growth in b and $(a + 1/a)$. We still have to show that $\mathscr{W}_g\eta$ corresponds to Y as claimed in the theorem. By definition we have

$$\mathscr{W}_g\eta(\mathscr{T}) = \eta(\mathscr{M}_{\bar{g}}\mathscr{T}), \qquad \mathscr{T} \in S(\mathbb{H}).$$

Now

$$\mathscr{M}_{\bar{g}}\mathscr{T} = \int_0^\infty \frac{da}{a}\int_{-\infty}^{+\infty} db\, \mathscr{T}(b, a)\bar{g}_{b,a}.$$

Upon replacing the integral by a truncated Riemannian sum we see that

$\mathscr{M}_{\bar{g}}\mathscr{T}$ may be approximated arbitrarily well by a finite sum ($\gamma = 1 + 1/N$):

$$\mathscr{M}_{\bar{g}}\mathscr{T} = \lim_{N\to\infty} s_N = \lim_{N\to\infty} \sum_{j=-N}^{N} \sum_{k=-N}^{N} \gamma^N \mathscr{T}(\gamma^j k, \gamma^j)\bar{g}_{\gamma^j k, \gamma^j}.$$

In addition, the convergence holds in $S_-(\mathbb{R})$ because of the strong localization of $\mathscr{T}$ and all its derivatives over the half-plane. Therefore, we may exchange the limits to write, with $Y(b, a) = \eta(\bar{g}_{b,a})$,

$$\mathscr{W}_g\eta(\mathscr{T}) = \lim_{N\to\infty} \eta(s_N) = \lim_{N\to\infty} \sum_{j=-N}^{N} \sum_{k=-N}^{N} \gamma^N Y(\gamma^j k, \gamma^j)\mathscr{T}(\gamma^j k, \gamma^j).$$

Now the right-hand side is a Riemannian sum that converges to

$$\int_0^\infty \frac{da}{a} \int_{-\infty}^{+\infty} db\, Y(b, a)\mathscr{T}(b, a),$$

and the first part of the theorem is proved.

We still have to prove the reproducing kernel equation. For all $\mathscr{T} \in S(\mathbb{H})$ we have just proved that

$$\langle \bar{Y} \mid \mathscr{T} \rangle_{\mathbb{H}} = \mathscr{W}_g\eta(\mathscr{T}) = c_{g,h}^{-1}(\mathscr{W}_g\mathscr{M}_h\mathscr{W}_g\eta)(\mathscr{T}).$$

By the definition of the wavelet transform and synthesis this equals

$$\mathscr{W}_g\eta(\mathscr{W}_{\bar{h}}\mathscr{M}_{\bar{g}}\mathscr{T}),$$

and finally by what we have proved, that is again a scalar product

$$c_{g,h}^{-1}\langle \bar{Y} \mid \mathscr{W}_{\bar{h}}\mathscr{M}_{\bar{g}}\mathscr{T} \rangle_{\mathbb{H}}.$$

This last expression reads explicitly

$$\int_0^\infty \frac{da}{a} \int_{-\infty}^{+\infty} db\, Y(b, a) \int_0^\infty \frac{da'}{a'} \int_{-\infty}^{+\infty} db' \frac{1}{a'} P_{\bar{h},\bar{g}}\left(\frac{b - b'}{a'}, \frac{a}{a'}\right)\mathscr{T}(b', a').$$

Since the polynomial growth of Y is compensated by the rapid decrease of the reproducing kernel, the integral is absolutely convergent and we may exchange the integrations and, using the symmetry of (20.0.2) we obtain

$$\begin{aligned}\langle \bar{Y} \mid \mathscr{T} \rangle_{\mathbb{H}} &= \int_0^\infty \frac{da'}{a'} \int_{-\infty}^{+\infty} db'\, \mathscr{T}(b', a') \int_0^\infty \frac{da}{a} \int_{-\infty}^{+\infty} db \frac{1}{a} \bar{P}_{g,h}\left(\frac{b' - b}{a}, \frac{a'}{a}\right)\bar{Y}(b, a)\\ &= \langle \bar{Y}' \mid \mathscr{T} \rangle_{\mathbb{H}},\end{aligned}$$

with

$$Y'(b', a') = \int_0^\infty \frac{da}{a} \int_{-\infty}^{+\infty} db \frac{1}{a} P_{g,h}\left(\frac{b' - b}{a}, \frac{a'}{a}\right)Y(b, a).$$

Since the holds for all $\mathscr{T}$ it follows that $Y' = Y$ almost everywhere. Since Y is continuous this holds everywhere. □

Corollary 24.1.7

The wavelet transform of $\eta \in S'_+(\mathbb{R})$ with respect to $g \in S_+(\mathbb{R})$ and that with respect to $h \in S_+(\mathbb{R})$ are linked via

$$\mathcal{W}_h\eta(b, a) = \int_0^\infty \frac{da'}{a'} \int_{-\infty}^{+\infty} db' \frac{1}{a'} \Pi_{g \to h}\left(\frac{b - b'}{a'}, \frac{a}{a'}\right) \mathcal{W}_g\eta(b', a'),$$

with $\Pi_{g \to h} = c_g^{-1} \mathcal{W}_h g$, and this equation holds pointwise.

Corollary 24.1.8

For any distribution $Y \in S'(\mathbb{H})$ the projection (24.1.1) *reads explicitly*

$$(\mathcal{W}_g \mathcal{M}_h Y)(b, a) = Y(\Pi_{b,a})$$

with

$$\Pi_{b,a}(\beta, \alpha) = \frac{1}{\alpha} \mathcal{W}_{\bar{h}} \bar{g}\left(\frac{\beta - b}{\alpha}, \frac{a}{\alpha}\right).$$

Again, the wavelet synthesis may be written as an integral over the half-plane.

Theorem 24.1.9

Let $Y \in S'(\mathbb{H})$ be a locally integrable function of at most polynomial growth. Then for $h \in S_+(\mathbb{R})$ we have

$$\int_\epsilon^\rho \int_{-\delta}^{+\delta} db\, Y(b, a) h_{b,a} \to \mathcal{M}_h Y$$

as $\epsilon \to 0$, $\rho, \delta \to \infty$, and the convergence takes place in the weak topology.*

Proof. The partial reconstruction over the region $K = \{\epsilon \le a \le \rho\} \times \{|b| \le \delta\}$ corresponds to multiplying Y by the characteristic function of K and to projecting it into the image of the wavelet transform with the help of a reproducing kernel P. It is easy to see that this family of functions is uniformly bounded by a polynomially bounded function over the half-plane. In addition, we have pointwise that $P * (K \cdot Y)(b, a) \to P * Y(b, a)$ as $K \to \mathbb{H}$. We may therefore use the dominated convergence theorem to conclude. □

Corollary 24.1.10

$S_+(\mathbb{R})$ is dense in $S'_+(\mathbb{R})$.

Proof. Take the inversion formula over a bounded region in the half-plane to obtain an approximation in $S_+(\mathbb{R})$. □

Remark. In Theorem 24.1.4 we have supposed that $h \in S_+(\mathbb{R})$. In this case any polynomial growth of Y is compensated by the reproducing kernel which is arbitrarily well polynomially localized. On the other hand, for a fixed Y we only need some localization. We may therefore choose h to be smooth but to have only a finite number of vanishing moments. In particular, we may choose h to be compactly supported. This technical remark will play a role in some demonstrations.

25 The wavelet transform on $S'(\mathbb{R})$

Recall the defintion of $S_0(\mathbb{R}) = S_+(\mathbb{R}) \oplus S_-(\mathbb{R})$. We define the wavelet transform of $\eta \in S_0'(\mathbb{R})$ with respect to $g \in S_0(\mathbb{R})$ and the inverse wavelet transform of $Y \in S'(\mathbb{H})$ with respect to $h \in S_0(\mathbb{R})$ as

$$\mathscr{W}_g \eta(\mathscr{T}) = \eta(\mathscr{M}_{\bar{g}} \mathscr{T}), \qquad \mathscr{M}_h Y(s) = Y(\mathscr{W}_{\bar{h}} s).$$

Obviously this is an extension of the previous definitions. The results of the previous section may be applied thanks to the following splitting. Let $g = g_- + g_+$ with $g_\pm \in S_\pm(\mathbb{R})$. Then

$$\mathscr{W}_g \eta = \mathscr{W}_{g_+} \sigma_+ \eta + \mathscr{W}_{g_-} \sigma_- \eta,$$

where the $\mathscr{W}_{g\pm}$ are the wavelet transforms on $S'_\pm(\mathbb{R})$ as defined in the previous section and the $\sigma_\pm$ are natural maps of $S'(\mathbb{R})$ onto $S'_\pm(\mathbb{R})$. The verification is straightforward and is left to the reader. The same holds trivially for the inverse wavelet transform

$$\mathscr{M}_h Y = \mathscr{M}_{h_+} Y + \mathscr{M}_{h_-} Y.$$

A last relation shows that the positive and negative frequency parts may be treated separately:

$$\mathscr{M}_{h_-} \mathscr{W}_{g_+} = \mathscr{W}_{g_+} \mathscr{M}_{h_-} = 0.$$

Since the sum $S_0(\mathbb{R}) = S_+(\mathbb{R}) \oplus S_-(\mathbb{R})$ is direct we obtain for the dual space

$$S_0'(\mathbb{R}) \simeq S'_+(\mathbb{R}) \oplus S'_-(\mathbb{R}).$$

Now the polar of $S_0(\mathbb{R})$ are those distributions in $S'(\mathbb{R})$ whose Fourier transform are supported by $\omega = 0$. This is precisely the space of polynomials $\mathscr{P}(\mathbb{R})$. We therefore have

$$S_0'(\mathbb{R}) \simeq S'(\mathbb{R})/\mathscr{P}(\mathbb{R}).$$

Consider now the space $S'(\mathbb{R})$ of all tempered distributions. We can still define the wavelet analysis by means of one of the following formulas ($g \in S_0(\mathbb{R})$, $\mathscr{T} \in S(\mathbb{H})$, $\eta \in S'$):

$$\mathscr{W}_g \eta(\mathscr{T}) = \eta(\mathscr{M}_{\bar{g}} \mathscr{T}), \qquad \mathscr{W}_g \eta(b, a) = \eta(\bar{g}_{b,a}).$$

It is a well-defined distribution in $S'(\mathbb{H})$ and again this is an extension of the previous definitions as a simple application of the Hahn–Banach theorem shows. However, the transformation is no longer invertible. This is because the wavelets do not see the polynomials. Therefore, a tempered distribution in $S'(\mathbb{R})$ is determined by its wavelet coefficients with respect to $g \in S_0$, admissible, up to some global polynomial. The wavelet transform factorizes out the polynomial part.

26 A class of operators

We now wish to introduce a class of operators that corresponds to manipulating the wavelet coefficients. Typically, if one is interested in one region of the position scale half-plane one would like to cut that region out by multiplying the wavelet coefficients by the characteristic function of that region. Now this kind of manipulation on the wavelet coefficients seems to be rather brutal since apparently it introduces singularities at the boundary of the region. In fact, it is not brutal at all since all operations in the half-plane are modulo the convolution with a reproducing kernel which in fact smoothes out the discontinuities.

To formalize this kind of manipulation let $\mathcal{T}$ be a locally integrable function over the half-plane that is of at most polynomial growth, that is, for some $\alpha > 0$ we have

$$|\mathcal{T}(b, a)| \leq O(1)(a + 1/a)^\alpha (1 + b^2)^{\alpha/2}$$

for some $\alpha \in \mathbb{R}$. Consider now the operator that acts as a multiplication operator in wavelet space:

$$K: s \mapsto \mathcal{M}_h(\mathcal{T} \cdot \mathcal{W}_g s).$$

From the reproducing kernel equation (20.0.1), it follows that K reads in wavelet space as ($\mathcal{R} = \mathcal{W}_g s$)

$$K: \mathcal{R} \mapsto P_{g,h} * (\mathcal{T} \cdot \mathcal{R}).$$

In this form the operator is known as the generalized Töplitz operator and $\mathcal{T}$ is called the symbol of the operator.

In some sense such an operator is approximately diagonal in the wavelet representation. Accordingly, since the polynomial localization over the half-plane of $\mathcal{T} \cdot \mathcal{R}$ may be estimated in terms of the localization of $\mathcal{T}$ it follows that we have the following theorem.

Theorem 26.0.1

The operator K maps $S_+(\mathbb{R})$ continuously into $S_+(\mathbb{R})$ and $S'_+(\mathbb{R})$ continuously into $S'_+(\mathbb{R})$.

26.1 The derivation operator and Riesz potentials

Consider now the Töplitz operator with symbol a^α, $\alpha \in \mathbb{R}$, that is,

$$K_\alpha : s \mapsto \mathcal{M}_h(a^\alpha \mathcal{W}_g s).$$

We want to show that this operator is essentially the αth derivation for $\alpha < 0$. For $\alpha > 0$ this is essentially the $|\alpha|$th primitive. Note that multiplication by a^{-1}, which corresponds to ∂, amplifies the wavelet coefficients that correspond to the small scale features of the analysed function. This is in line with the common experience that the derivative of a function is in general rougher than the function itself, that is, it has more small-scale features, whereas the primitive is smoother.

We now come to the details. We now work in a $S_0(\mathbb{R})$ context since this allows easier generalization to higher dimensions.

First observe the generalized Töplitz operators with symbol a^α are precisely those that are homogeneous under dilations.

Theorem 26.1.1

Let $B: S_0(\mathbb{R}) \to S_0'(\mathbb{R})$ be a continuous operator that satisfies the following co-variance properties:

$$T_b B = B T_b, \qquad D_a B = a^\alpha B D_a, \quad \alpha \in \mathbb{R}.$$

Then $B = K_\alpha$ for some wavelets g, h.

The exponent α will be called the degree of homogeneity of B.

Proof. Clearly, B is determined by its image of one single admissible function $g \in S_0(\mathbb{R})$ since by co-variance of the wavelet transform it is then known for all dilated and translated wavelets and hence for all functions in $S_0(\mathbb{R})$. More precisely, with $h = Bg$ by hypothesis in $S_0(\mathbb{R})$, we have

$$Bg_{b,a} = a^\alpha h_{b,a}$$

and hence by writing $s = c_g^{-1} \int_{-\infty}^{+\infty} db \int_0^\infty (da/a) \mathcal{W}_g s(b, a) g_{b,a}$ and passing the operator under the integral (which we may, thanks to the continuity of B and the convergence properties of the inversion formula),

$$B: s \mapsto \mathcal{M}_h a^\alpha \mathcal{W}_g s, \qquad h = c_g^{-1} Bg. \qquad \square$$

Obvious concrete examples of such homogeneous operators are the derivation operators ∂^n, which are of degree $-n$. Another example is the identity and the Hilbert transform H, which are both of degree 0. Still another example is given by the so-called Riesz potentials. They are defined via their action in Fourier space:

$$I_\alpha: S_0(\mathbb{R}) \to S_0(\mathbb{R}), \qquad \hat{s}(\omega) \mapsto |\omega|^{-\alpha}\hat{s}(\omega).$$

One easily verifies the following composition law:

$$I_\alpha I_\beta = I_{\alpha+\beta}.$$

In Fourier space we have, for $h, g, s \in S_0(\mathbb{R})$,

$$\widehat{K_\alpha s}(\omega) = \int_0^\infty \frac{da}{a} \int_{-\infty}^{+\infty} db\, a^\alpha \hat{h}(a\omega)\, e^{-ib\omega} \mathscr{W}_g s(b, a).$$

Now integrating over b we obtain the Fourier transform of a voice and thus

$$\widehat{K_\alpha s}(\omega) = \int_0^\infty \frac{da}{a} \int_{-\infty}^{+\infty} db\, a^\alpha \hat{h}(a\omega)\bar{\hat{g}}(a\omega)\hat{s}(\omega)$$

$$= |\omega|^\alpha \hat{s}(\omega) \int_0^\infty \frac{da}{a} a^\alpha \hat{h}(a \operatorname{sign} \omega)\bar{\hat{g}}(a \operatorname{sign} \omega).$$

Thus K_α can be written as the product

$$K_\alpha = (c_{1,\alpha}\mathbb{1} + c_{2,\alpha}H)I_\alpha,$$

with

$$c_{1,\alpha} + ic_{2,\alpha} = 2\int_0^\infty \frac{d\omega}{\omega} \omega^\alpha \hat{h}(\omega)\bar{\hat{g}}(\omega),$$

$$c_{1,\alpha} - ic_{2,\alpha} = 2\int_0^\infty \frac{d\omega}{\omega} \omega^\alpha \hat{h}(-\omega)\bar{\hat{g}}(-\omega).$$

As a particular case note that the derivative ∂^n may, up to some Hilbert transform, be written as a multiplication operator in wavelet space:

$$\partial^n = (c_1\mathbb{1} + c_2H)K_{-n}.$$

This shows that for $n \in \mathbb{Z}$ we may identify K_n—up to some combination of identity and Hilbert transform—with a multiple of the differentiation (for $n < 0$), or its inverse the integration (for $n > 0$). If we restrict ourselves to $S_+(\mathbb{R})$ or $S_-(\mathbb{R})$, the Hilbert transform is a constant multiple of the identity and the identification of a^{-n} with ∂^n holds is exactly. For $S_0(\mathbb{R})$ the wavelets g and h may again be chosen such that the identification holds without Hilbert transforms. Note that on $S_0(\mathbb{R})$ the inverse of ∂ is uniquely determined. In general, it is only defined up to a constant, but since there are no non-zero constants in $S_0(\mathbb{R})$ we have uniqueness.

26.2 Differentiation and integration over $S_0'(\mathbb{R})$

As we have said already, the Töplitz operator with polynomially bounded symbol is a weak* continuous operator since it transforms functions over the half-plane of at most polynomial growth into functions whose polynomial growth can be estimated in terms of the former ones.

In particular, the derivatives and primitives extend continuously to $S_0'(\mathbb{R})$. Note that this definition of derivation satisfies the following 'partial integration' rule, which can be taken as an equivalent definition:

$$\partial\eta(s) = -\eta(\partial s), \qquad \eta \in S_0'(\mathbb{R}), \quad s \in S_0(\mathbb{R}).$$

This definition of ∂ coincides with the usual one in the case where $\eta \in S_0'(\mathbb{R})$ is an absolutely continuous function. Recall that this means that $\sum |\eta(\beta_n) - \eta(\alpha_n)| \to 0$ whenever $\sum |\beta_n - \alpha_n| \to 0$. Then

$$\lim_{\epsilon \to 0} \epsilon^{-1}(\eta(t+\epsilon) - \eta(t))$$

exists almost everywhere and equals a function w that is locally in $L^1(\mathbb{R})$. Then the derivative of η in the sense of distributions coincides with w. Indeed, taken as a distribution the derivative of η is defined via $\partial\eta(s) = -\eta(\partial s)$ for all $s \in S_0(\mathbb{R})$. On the other hand, a standard partial integration which is valid for absolutely continuous functions show that $\int dt\, w(t)s(t) = -\int dt\, \eta(t)\, \partial s(t)$ for all $s \in S_0(\mathbb{R})$ and therefore both definitions coincide.

Suppose now that $s \in S_0'(\mathbb{R})$ is locally integrable. In this case the integration operator ∂^{-n} can be computed explicitly:

$$\partial^{-n}s(t) = \frac{1}{n!}\int_{t_0}^{t} du(t-u)^{n-1}s(u).$$

As a distribution in $S_0'(\mathbb{R})$ this definition does not depend on the origin of the integration t_0 since changing the origin will only add a global polynomial.

We give the following theorem as an application of these results.

Theorem 26.2.1

Every distribution in $S_0'(\mathbb{R})$ is the derivative (in the sense of distributions) of some function of at most polynomial growth. Or, vice versa, some primitive of $\eta \in S_0'(\mathbb{R})$ is a polynomially bounded function.

Clearly, the same holds for $S'(\mathbb{R})$ since we have only add some global polynomial.

Proof. Although this theorem is well known we will give a proof that uses the representation of distributions as smooth functions of at most polynomial growth over the half-plane. We first give a sufficient condition that a distribution over the half-plane corresponds via the wavelet synthesis to a polynomially bounded function over $\mathbb{R}$.

Lemma 26.2.2
Let $\mathcal{T} \in S'(\mathbb{H})$ be a locally integrable function such that

$$\rho(t) = \int_0^\infty \frac{da}{a} \int_{-\infty}^{+\infty} db\, |h_{b,a}(t)|\, |\mathcal{T}(b, a)|$$

is of at most weak polynomial growth: $(1 + |t|^2)^{-\alpha/2}\rho \in L^1(\mathbb{R})$ *for some* $\alpha > 0$. *Then* $\mathcal{M}_h\mathcal{T} \in S_0'(\mathbb{R})$ *is a function that coincides almost everywhere with*

$$w(t) = \int_{-\infty}^{+\infty} db \int_0^\infty \frac{da}{a} \mathcal{T}(b, a) \frac{1}{a} h\left(\frac{t-b}{a}\right).$$

Proof. Clearly, w is a locally integrable function of at most weak polynomial growth. It therefore defines a distribution in $S'(\mathbb{R})$. On the other hand, by definition we have $\mathcal{M}_h\mathcal{T}(s) = \mathcal{T}(\mathcal{W}_{\bar{h}} s)$ for all $s \in S_0(\mathbb{R})$. Explicitly, this reads

$$\mathcal{M}_h\mathcal{T}(s) = \int_{-\infty}^{+\infty} db \int_0^\infty \frac{da}{a} \mathcal{T}(b, a) \int_{-\infty}^{+\infty} dt \frac{1}{a} h\left(\frac{t-b}{a}\right) s(t).$$

By hypothesis the integral is absolutely convergent and we may exchange the order of integration. This shows that the action of $\mathcal{M}_h\mathcal{T}$ on all $s \in S_0(\mathbb{R})$ coincides with $\int ws$. This shows that $\mathcal{M}_h\mathcal{T} = w$. □

Now the proof is easy. Any distribution corresponds to a smooth function of at most polynomial growth over the half-plane. Call this function $\mathcal{T}$. We then split $\mathcal{T}$ into two parts. The first one, the small scales, coincides with $\mathcal{T}$ for $0 < a < 1$ and is 0 otherwise. The large scales $\mathcal{T}_2$ are such that $\mathcal{T}_1 + \mathcal{T}_2 = \mathcal{T}$. To see to what distributions these parts correspond we have to project them back into the image of the wavelet transform by convolution with a reproducing kernel P. It is an easy matter to verify that $P * \mathcal{T}_1$ has at most polynomial growth at small scales but decays rapidly at large scales. Therefore, some primitive—obtained by multiplying by a^m, m large enough—is a function that satisfies the conditions of the previous lemma. It therefore corresponds to a polynomially bounded function. On the other hand, the large scale part $P * \mathcal{T}_2$ decays rapidly at small scales and has at most polynomial growth at large scales. Therefore, some derivative—obtained through multiplication by a^{-m}, m large enough—is a function that again satisfies the conditions of the preceding lemma. It corresponds to a smooth function of at most polynomial growth. Together with this function all its primitives are smooth functions of at most polynomial growth. Both terms together give the desired function whose derivative is η. □

27 Singular support of distributions

It is not possible to characterize the support of a distribution in wavelet space by looking only at the small scale behaviour of the modulus of the wavelet coefficients if the analysing wavelet is in $S_0(\mathbb{R})$. The problem is that these wavelets cannot be compactly supported since otherwise their Fourier transform would be an entire function and could not vanish to all orders at $\omega = 0$. Therefore, the local decrease of the wavelet coefficients is the same whether locally the analysed function is 0 or merely a C^∞ function.

For a distribution we say that it is locally C^∞ if in some neighbourhood I we have, for all $\phi \in C_c^\infty(I)$, $\phi\eta \in C^\infty(I)$.

Definition 27.0.1
The complement of points that have a C^∞ regular neighbourhood is called the singular support of a distribution $\eta \in S'(\mathbb{R})$.

The singular support can actually be characterized in wavelet space, as the following theorem shows, which characterizes the complement of the singular support.

Theorem 27.0.2
Let $\eta \in S'(\mathbb{R})$. Then it is of C^∞ regularity in I, open (in the sense that $\phi\eta \in C^\infty(\mathbb{R})$ for all $\phi \in C^\infty(\mathbb{R})$ and compact support inside I), iff for some admissible wavelet $g \in S_0(\mathbb{R})$ we have

$$\mathcal{W}_g\eta(b, a) = O(a^m), \qquad a \to 0$$

for all $m > 0$, uniformly in b on all compact subsets K of I.

Proof. Suppose that η is as stated. Let $K \subset I$ be a compact set. Then we can find $\phi \in C^\infty(\mathbb{R})$ with support in I and containing K. In addition we may assume that $\phi = 1$ on some interval containing K. We now write $\eta = \phi\eta + (1 - \phi)\eta = \eta_1 + \eta_2$. By construction, $\eta_1 \in C^\infty(\mathbb{R})$. By Taylor's theorem, for every integral m large enough there is a polynomial P of degree $m - 1$ such that

$$\eta_1(b + t) = P(t) + \rho_b(t),$$

with $|\rho_b(t)| \leq O(t^m)$, uniformly in $t \in \mathbb{R}$ and $b \in K$. Now the wavelet in $S_0(\mathbb{R})$ has all moments vanishing and thus it does not see the polynomial part.

We therefore have uniformly in b:

$$|\mathscr{W}_g\eta_1(b, a)| \leq O(1) \int_{-\infty}^{+\infty} dt\, a^{-1}|g(t/a)|\, |t|^m$$

$$= O(a^m) \int_{-\infty}^{+\infty} dt\, |g(t)|\, |t|^m = O(a^m).$$

For the second term it is enough to observe that $\forall m$, $a^m\, \bar{g}_{b,a}(1-\phi)$ goes uniformly to 0 in $S(\mathbb{R})$ for $a \to 0$ and $b \in K$. Since $\mathscr{W}_g\eta_2(b, a) = \eta_2((1-\phi)\bar{g}_{b,a})$, the first part of the theorem follows.

To show the converse suppose that $\mathscr{T} = \mathscr{W}_g\eta$ satisfies the local decay as in the theorem. Let $K \subset I$ be a compact interval. Let $\alpha > 0$ be the distance of K from the border of I. The proof follows now splitting the function $\mathscr{T}$ into several parts. The first coincides with $\mathscr{T}$ on the small scales $0 < a < \alpha$ and it is 0 otherwise; the second is such that $\mathscr{T} = \mathscr{T}_1 + \mathscr{T}_2$. Now following the remark after Theorem 24.1.9 we may choose h to be compactly supported, say by $[-1, +1]$. The reconstruction of the large scales gives a function as smooth as h. Now $\mathscr{T}_1$ may be split again into $\mathscr{T}_3 + \mathscr{T}_4$ where $\mathscr{T}_3$ coincides with $\mathscr{T}_1$ in the union of the influence cones of h whose tops are in K. Clearly, upon exchanging the deviation with the integration we see that $\mathscr{M}_h\mathscr{T}_3$ is a function that is as smooth as h by the decay hypothesis on $\mathscr{T}$. On the other hand, $\mathscr{M}_h\mathscr{T}_4$ has support in the closure of the complement of K. The superposition of all terms shows that $\mathscr{M}_h\mathscr{T}$ is as smooth as h in K. Now h may be chosen arbitrarily smooth and the theorem follows since K was arbitrary. □

28 Bounded sets in $S_0(\mathbb{R})$ and $S_0'(\mathbb{R})$

For further reference in later sections we note that the bounded sets of $S_0(\mathbb{R})$ and $S_0'(\mathbb{R})$ may easily be characterized in wavelet space by some uniform localization. Recall that in a topological vector space Y a set X is bounded iff it is eventually absorbed by all neighbourhoods U of 0; in other words iff for all neighbourhoods $U \ni 0$ we have $X \subset tU$ for all large enough t. Equivalently (see e.g. Rudin 1991), a set X is bounded iff for all sequences $x_n \in X$ we have

$$\alpha_n x_n \to 0 \tag{28.0.1}$$

for all numerical sequences for which $\alpha_n \to 0$ as $n \to \infty$. For a Fréchet space generated by the seminorms p_n the bounded subsets $X \subset Y$ are precisely those for which each p_n is bounded on X.

Theorem 28.0.1

A set $X \subset S_0(\mathbb{R})$ is bounded iff there is some Π over the half-plane that decays faster than any polynomial in b and $(a + 1/a)$ such that

$$|\mathcal{W}_g x(b, a)| \leq \Pi(b, a)$$

for all $x \in X$. A set $X \subset S_0'(\mathbb{R})$ is bounded iff there is some function Π over the half-plane of at most polynomial growth in b and $(a + 1/a)$ such that

$$|\mathcal{W}_g x(b, a)| \leq \Pi(b, a),$$

for all $x \in X$.

Proof. A set $X \subset S_0(\mathbb{R})$ is bounded iff for all $\alpha > 0$

$$\sup_{x \in X} \|x\|_{S_0(\mathbb{R});\alpha} < \infty.$$

This is equivalent to the existence of constants c_α such that

$$|\mathcal{W}_g x(b, a)| \leq c_\alpha \kappa_\alpha^{-1}(b)\phi_\alpha^{-1}(a)$$

for all $x \in X$ with $\phi_\alpha(a) = (a + 1/a)^\alpha$ and $\kappa_\alpha(b) = (1 + |b|)^\alpha$. This proves the first part of the theorem.

The proof of the second half is by contradiction. Let $X \subset S_0'(\mathbb{R})$ be bounded and suppose that for all α and all c we can find some point (b, a) and some $x \in X$ such that $|\mathcal{W}_g x(b, a)| > c\kappa_\alpha(b)\phi_\alpha(a)$. In particular, we may choose $\alpha = c = n$, obtaining in that way a sequence of points (b_n, a_n) and distributions x_n. Consider now the sequence of translated and dilated wavelets $g_n = c_n g_{b_n, a_n}$ with $c_n = \kappa_n^{-1}(b_n)\phi_n^{-1}(a_n)$. We claim that this set is bounded in $S_0(\mathbb{R})$. To see this let $\Pi = |\mathcal{W}_g g|$. By the first part of the theorem we then have to show that the following expression is finite for all $m > 0$:

$$\sup_{n \in \mathbb{N}} \sup_{(b, a \in \mathbb{H})} \kappa_n^{-1}(b_n)\phi_n^{-1}(a_n)\kappa_m(b)\phi_m(a)\frac{1}{a_n}\Pi\left(\frac{b - b_n}{a_n}, \frac{a}{a_n}\right) < \infty.$$

Changing $a \to a_n a$ and $b \to a_n b + b_n$ and using the estimations

$$\kappa_\alpha(x + y) \leq \kappa_\alpha(x)\kappa_\alpha(y), \qquad \kappa_\alpha(|x|y) \leq \phi_\alpha(|x|)\kappa_\alpha(y),$$
$$\phi_\alpha(|x|\,|y|) \leq \phi_\alpha(|x|)\phi_\alpha(|y|),$$

we see that the above expression is majorized by

$$O(1) \sup_{n \in \mathbb{N}} \kappa_{m-n}(b_n)\phi_{m-n}(a_n)a_n^{-1}\phi_m(a_n) \sup_{(b, a) \in \mathbb{H}} \phi_m(a)\kappa_m(b)\Pi(b, a),$$

which is actually finite. But this is a contradiction. Indeed, suppose $Y \subset S_0(\mathbb{R})$ is a bounded set. Then $\{x(y): y \in Y, x \in X\}$ is bounded in $\mathbb{C}$. On the other hand, we have found a bounded set in $S_0(\mathbb{R})$ on which we have $|x_n(g_n)| = |\mathcal{W}_g x_n(b_n, a_n)| > n$, also a contradiction. The 'if' part follows from (28.0.1). Therefore the theorem is proved. □

29 Some explicit wavelet transforms of distributions

In this section we discuss some explicit distributions. We will mainly stick to homogeneous distributions of arbitrary order. All distributions are treated as distributions in $S_0'(\mathbb{R})$ or $S'_\pm(\mathbb{R})$.

29.1 The distributions $\ldots, \delta^{(-1)}, \delta, \delta', \delta'', \ldots, \delta^{(n)}$

For $n = 0, 1, \ldots$ we define the distribution $\delta^{(n)}$ as the evaluation of the nth derivative of s at the origin:

$$\delta^{(n)}(s) = (-\partial)^n s(0)$$

For $n = -1, -2, \ldots$ we use the integration operator I to write

$$\delta^{(n)}(s) = (-\partial)^n s(0) = \frac{(-1)^n}{n!} \int_{-\infty}^{0} dt\, t^{|n|} s(t) = \frac{1}{n!} \int_0^\infty dt\, t^{|n|} s(t).$$

The wavelet transform of these distributions now reads $n \in \mathbb{Z}$

$$\mathscr{W}[g; \delta^{(n)}](b, a) = a^{-n}(-1)^n(\partial^n \bar{g})(-b/a). \tag{29.1.1}$$

29.2 The distributions $|t|_-^\alpha, |t|_+^\alpha$

We now want to analyse the wavelet transform of the following functions:

$$|t|_+^\alpha = \begin{cases} t^\alpha & t > 0, \\ 0 & t \le 0, \end{cases} \qquad |t|_-^\alpha = \begin{cases} 0 & t \ge 0, \\ (-t)^\alpha & t < 0. \end{cases}$$

For $\Re\alpha > -1$ they are clearly distributions in $S_0'(\mathbb{R})$ if we define their action as $(s \in S(\mathbb{R}))$:

$$|t|_+^\alpha : s \mapsto \int_0^\infty dt\, t^\alpha s(t), \qquad |t|_-^\alpha : s \mapsto \int_{-\infty}^0 dt (-t)^\alpha s(t) = \int_0^\infty dt\, t^\alpha s(-t).$$

We will see that they are in $S_0'(\mathbb{R})$ for every $\alpha \in \mathbb{C}$ with the exception of $\alpha = -1, -2, \ldots$.

Recall that the *Mellin* transform essentially describes the action of $|t|_+^\alpha$ on a function s as a function of the exponent α:

$$Ms(\alpha) = M[s](\alpha) = \int_0^\infty dt\, t^{\alpha - 1} s(t).$$

By direct computation we obtain

$$\int_0^\infty dt \frac{1}{a} \bar{g}\left(\frac{t-b}{a}\right) t^\alpha = a^\alpha \int_0^\infty dt\, \bar{g}(t - b/a) t^\alpha$$

and

$$\int_{-\infty}^{0} dt \frac{1}{a} \bar{g}\left(\frac{t-b}{a}\right)(-t)^\alpha = a^\alpha \int_{0}^{\infty} dt\, \bar{g}(-t-b/a) t^\alpha$$

and thus the wavelet transform of $|t|_\pm^\alpha$ can be written as

$$\mathscr{W}[g; |t|_+^\alpha](b, a) = a^\alpha U_+(b/a), \qquad U_+(u) = M[\bar{g}(\cdot - u)](\alpha + 1),$$

$$\mathscr{W}[g; |t|_-^\alpha](b, a) = a^\alpha U_-(-b/a), \qquad U_-(u) = M[\bar{g}(u - \cdot)](\alpha + 1).$$

Actually, for a progressive or regressive wavelet, the two functions $U_\pm$ are not independent. For further use, in the following theorem we will prove a relation between U_+ and U_- for a large class of wavelets.

Lemma 29.2.1

If a progressive function r satisfies $r \in L^1(\mathbb{R}) \cap L^2(\mathbb{R})$, $t^{\alpha-1} r \in L^1(\mathbb{R})$ and $\hat{r} \in L^1(\mathbb{R}_+, d\omega/\omega^\alpha)$ then

$$M[r(-\cdot)](\alpha) = e^{-i\pi\alpha} M[r](\alpha)$$

and

$$M[r](\alpha) = \frac{1}{2\pi} \Gamma(\alpha)\, e^{i\pi\alpha/2} M[\hat{r}](1 - \alpha).$$

It follows at once that for regressive r with $r(-\cdot)$ satisfying the conditions of the theorem we have, instead,

$$M[r(-\cdot)](\alpha) = e^{i\pi\alpha} M[r](\alpha)$$

and

$$M[r](\alpha) = \frac{1}{2\pi} \Gamma(\alpha)\, e^{-i\pi\alpha/2} M[\hat{r}](1 - \alpha).$$

Proof. Because of the dominated convergence theorem we may write

$$M[r](\alpha) = \lim_{\lambda \to 0^+} \int_0^\infty dt\, t^{\alpha-1} e^{-\lambda t} r(t).$$

The Fourier transform of $|t|_+^{\alpha-1} e^{-\lambda t}$ is given by

$$\int_0^\infty dt\, t^{\alpha-1} e^{-\lambda t - i\omega t} = \frac{1}{(\lambda + i\omega)^\alpha} \int_0^\infty dt\, t^{\alpha-1} e^{-t}$$

$$= \frac{\Gamma(\alpha)}{(\lambda + i\omega)^\alpha}.$$

Writing the scalar product for every $\lambda > 0$ in frequency space we obtain

$$M[r](\alpha) = \lim_{\lambda \to 0^+} \frac{\Gamma(\alpha)}{2\pi} \int_0^\infty d\omega \frac{\hat{r}(\omega)}{(\lambda - i\omega)^\alpha}.$$

We may again use the theorem of dominated convergence to exchange the limit with the integral. Because $(\lambda - i\omega)^\alpha \to e^{-i\pi\alpha/2}\omega^{-\alpha}$ as $\lambda \to 0^+$ pointwise for every $\omega > 0$ we have

$$M[r](\alpha) = e^{i\pi\alpha/2}\frac{\Gamma(\alpha)}{2\pi}\int_0^\infty d\omega\, \omega^{-\alpha}\hat{r}(\omega).$$

This proves the second equation. On the other hand, for the same reasons as before we have

$$M[r(-\cdot)](\alpha) = \int_0^\infty dt\, t^{\alpha-1}s(t) = \int_{-\infty}^0 dt(-t)^{\alpha-1}r(-t)$$

$$= \lim_{\lambda\to 0^+}\int_{-\infty}^0 dt(-t)^{\alpha-1}\, e^{\lambda t}\, r(-t).$$

Passing in frequency space

$$|t|_-^{\alpha-1}\, e^{\lambda t} \to \Gamma(\alpha)(\lambda - i\omega)^{-\alpha}$$

and going to the limit $\lambda \to 0^+$ we find

$$M[Pr](\alpha) = \frac{\Gamma(\alpha)}{2\pi}\, e^{-i\alpha/2}\int_0^\infty d\omega\, \omega^{-\alpha}\hat{r}(\omega).$$

By comparing the two expressions we see that the lemma is proved. □

For a progressive wavelet g, the wavelet $\bar{g}$ is regressive and we obtain

$$U_+(t) = -e^{-i\pi\alpha/2}\frac{\Gamma(\alpha+1)}{2i\pi}\int_0^\infty d\omega\, \omega^{-\alpha}\, e^{i\omega t}\hat{\bar{g}}(\omega),$$

$$U_-(t) = -e^{i\alpha\pi}U_+(t).$$

Consider now the most general homogeneous distribution of degree α:

$$s_{\text{cusp}}(t) = c_-|t|_-^\alpha + c_+|t|_+^\alpha.$$

For the wavelet transform of s_{cusp} with respect to a progressive wavelet $g \in S_+(\mathbb{R})$ we obtain

$$\left.\begin{aligned} &\mathcal{W}[g; s_{\text{cusp}}](b, a) = a^\alpha U(b/a),\\ &U(u) = M[\bar{g}(\cdot - u)](\alpha+1)(c_+ - c_-\, e^{i\alpha\pi})\\ &\quad = \frac{\Gamma(\alpha+1)}{2i\pi}(c_+\, e^{-i\pi\alpha/2} - c_-\, e^{i\pi\alpha/2})\int_0^\infty d\omega\, \omega^{-\alpha-1}\hat{\bar{g}}(\omega)\, e^{i\omega u} \end{aligned}\right\}. \tag{29.2.1}$$

Note that for $\alpha = 0, 1, \ldots,$ and $c_+ = (-1)^\alpha c_-$ the wavelet transform of (29.2.1) vanishes, because then s_{cusp} is a polynomial.

Until now we have defined the distributions $|t|_\pm^\alpha$, for $\Re\alpha > -1$ only. The last expression ((29.2.1)) allows us to define the distributions $|t|_\pm^\alpha$ for all values of $\alpha \in \mathbb{C}$, $\Re\alpha < 0$, with the exception of $\alpha = -1, -2, \ldots$, where the gamma function Γ has simple poles, and the distributions $|t|_\pm^\alpha$ are no longer defined. For all other values of α, equation (29.2.1) gives a polynomial bounded function over the half-plane and hence defines a distribution in $S'_+(\mathbb{R})$. For a fixed test function $s \in S_-(\mathbb{R})$ the quantities $|t|_+^\alpha(s)$ and $|t|_-^\alpha(s)$ are holomorphic functions over the complex plane with simple poles at $\alpha = -1, -2, \ldots$.

To factor out the poles in $\alpha = -1, -2, \ldots$ we define, instead of $|t|_\pm$,

$$\chi_\pm^\alpha(t) = \frac{|t|_\pm^\alpha}{\Gamma(\alpha + 1)},$$

and the wavelet transform reads, for $g \in S_+(\mathbb{R})$,

$$\left.\begin{aligned} \mathscr{W}[g, \chi_\pm^\alpha](b, a) &= a^\alpha U_\pm(b/a), \\ U_\pm(t) &= \pm\frac{e^{\mp i\pi\alpha/2}}{2i\pi}\int_0^\infty d\omega\, \omega^{-\alpha}\, e^{i\omega t}\, \bar{\hat{g}}(\omega). \end{aligned}\right\} \tag{29.2.2}$$

These distributions are now defined for all $\alpha \in \mathbb{C}$ and $\chi_\pm^\alpha(s)$ is an entire function for $s \in S_-(\mathbb{R})$. Its action reads

$$\chi_\pm^\alpha(s) = c_g^{-1}\int_0^\infty \frac{da}{a}\int_{-\infty}^{+\infty} db\, a^\alpha U(b/a)\mathscr{W}_{\bar{g}}s(b, a).$$

For $n \in \mathbb{N}$ we obviously have, in the sense of $S'_+(\mathbb{R})$,

$$\chi_+^n = \delta^{(-n)}. \tag{29.2.3}$$

We claim that this is true for all $n \in \mathbb{Z}$. Indeed, from the Fourier space representation of the differentiation operator $\partial\colon \hat{s} \mapsto i\omega\hat{s}$ and from (29.2.2), it follows that

$$\mathscr{W}[g; \chi_+^{-n}](b, a) = a^{-n}(-1)^n\partial_t^{-n-1}\bar{g}(-b/a).$$

But this is the wavelet transform ((29.1.1)) of $\delta^{(-n)}$ and therefore (29.2.3) holds for all $n \in \mathbb{Z}$.

30 Extension to higher dimensions

Essentially, all the results that we have proved in the one-dimensional case carry over to higher dimensions. The proofs are all straightforward generalizations of the proofs given in the one-dimensional case and we limit ourselves to a very brief outline of the spaces involved. In Chapter 5, section 11 we will discuss the two-dimensional case in some detail.

The space $S(\mathbb{R}^n)$ is the space of smooth functions that decay at ∞, together with all their derivatives, faster than any polynomial. The closed subspace $S_0(\mathbb{R}^n) \subset S(\mathbb{R}^n)$ is the space of functions that have all moments vanishing, or, what amounts to the same,

$$|\hat{s}(k)| \leq O(|k|^n), \qquad |k| \to 0$$

for all $n \in \mathbb{N}$. The Fourier transform is defined as

$$\hat{s}(k) = \int d^n x \, e^{-i\langle k|x\rangle} \, s(x), \qquad s(x) = \frac{1}{(2\pi)^n} \int d^n k \, e^{i\langle k|x\rangle} \, \hat{s}(k).$$

There is no natural equivalent of $S_{\pm}(\mathbb{R})$ in higher dimensions. However, there is an equivalent of the Hilbert transform, as we shall see.

Now, first, the definition of the wavelet transform. Let $g \in S_0(\mathbb{R})$. The wavelet transform is now defined as

$$\mathscr{W}_g s(b, a) = \int d^n x \, \frac{1}{a^n} \, \bar{g}\left(\frac{x - b}{a}\right) s(x).$$

It is a function over the half-space $\mathbb{H}^n = \{(b, a) : b \in \mathbb{R}^n, a > 0\}$. It maps $S_0(\mathbb{R}^n)$ into the space of functions that are highly polynomially localized over the half-space $\mathbb{H}^n$. We call this space, together with its locally convex topology, $S(\mathbb{H}^n)$. The wavelet synthesis for $h \in S_0(\mathbb{R}^n)$ and $\mathscr{T} \in S(\mathbb{H}^n)$ reads

$$\mathscr{M}_h \mathscr{T}(x) = \int_0^\infty \frac{da}{a} \int d^n b \, \mathscr{T}(b, a) \frac{1}{a^n} \, h\left(\frac{x - b}{a}\right).$$

Again this is a continuous map. In the one-dimensional case the combination $\mathscr{M}_h \mathscr{W}_g$ was a combination of the identity and the Hilbert transform. In the higher-dimensional case we obtain the identity plus a continuous superposition of so-called Riesz transforms. Let $\xi \in \mathbb{R}^n$ be a unit vector $|\xi| = 1$, that is, a direction. Then the Riesz transform in the direction ξ is defined as

$$R_\xi : \hat{s}(k) \mapsto \hat{s}(k) \frac{\langle \xi \mid k \rangle}{|k|}$$

Along every one-dimensional ray passing through the origin this transforms acts like a Hilbert transform. However, its amplitude varies with the direction. The Riesz transform therefore is a kind of directed Hilbert transform.

Let us look at the wavelet transform in Fourier space:

$$\mathscr{W}_g s(b, a) = (2\pi)^{-n} \int d^n k \, \overline{\hat{g}(k)} \, e^{-i\langle k|x\rangle} \, s(k).$$

From this we see, as in the one-dimensional case, that

$$\widehat{\mathscr{M}_h \mathscr{W}_g s}(k) = f_{g,h}(k/|k|) \hat{s}(k),$$

where

$$f_{g,h}(\xi) = \int_0^\infty \frac{da}{a} \hat{h}(a\xi)\bar{\hat{g}}(a\xi), \qquad |\xi| = 1.$$

This Fourier space multiplier can be written as a superposition of Riesz transforms. Let $d^{n-1}\Omega$ denote the normalized measure, $\int d^{n-1}\Omega = 1$, on the unit sphere S^{n-1} in n dimensions; then we have

$$\mathcal{M}_h \mathcal{W}_g = c_{g,h}\mathbb{1} + \int_{S^{n-1}} d^{n-1}\Omega(\xi)(c_{g,h} - f_{g,h}(\xi))R_\xi,$$

$$c_{g,h} = \int_{S^{n-1}} d^{n-1}\Omega(\xi) f_{g,h}(\xi).$$

Therefore, a reconstruction wavelet for g should be such that $f_{g,h}(\xi)$ is independent of the direction. One way to ensure this is to suppose that all wavelets are radial functions, that is, they are functions of $|x|$ alone.

With this modification in mind all theorems that hold for $S_0(\mathbb{R})$, $S_0'(\mathbb{R})$, and $L^2(\mathbb{R})$ can be directly translated to the higher-dimensional case.

Appendix

31 Proof of Theorem 11.1.1

Proof. We need the following lemma.

Lemma 31.0.1
We have for $\beta > 1$, $\alpha > 1$,

$$I_{\alpha,\beta}(u) = \int_0^\infty dt(1+t)^{-\beta}(u+t)^{-\alpha} \sim \frac{u^{1-\alpha}}{1+u}$$

uniformly in $u > 0$. *And*

$$B_{\alpha,\beta}(u,v) = \int_0^1 dt(t+u)^{-\alpha}(v+1-t)^{-\beta}$$
$$\sim (1+u)^{-\alpha}(1+v)^{-\beta}[(1+u^{-1})^{\alpha-1} + (1+v^{-1})^{\beta-1}]$$

uniformly in $u, v > 0$.

Proof. Since

$$\int_0^\infty dt(1+t)^{-\beta}(u+t)^{-\alpha} = u^{1-\alpha-\beta}\int_0^\infty dt(u^{-1}+t)^{-\beta}(1+t)^{-\alpha},$$

we have $I_{\alpha,\beta}(u) = u^{1-\alpha-\beta} I_{\beta,\alpha}(1/u)$. We may therefore limit ourselves to $0 < u \le 1$, where we have to show that $I_{\alpha,\beta}(u) \sim u^{1-\alpha}$.

We split the integral into two parts:

$$\int_0^\infty dt(1+t)^{-\beta}(u+t)^{-\alpha} = \left\{\int_0^1 + \int_1^\infty\right\} dt(1+t)^{-\beta}(u+t)^{-\alpha}$$
$$= X_1 + X_2.$$

For X_1 we may write $(1+t)^{-\beta} \sim 1$ and thus

$$X_1 \sim \int_0^1 dt(u+t)^{-\alpha} = u^{1-\alpha}\int_0^{1/u} dt(1+t)^{-\alpha} \sim u^{1-\alpha},$$

since the last integral converges for $\alpha > 1$. In X_2 we may replace $(1+t)^{-\beta}$ by the equivalent expression $t^{-\beta}$ and we obtain

$$X_2 \sim \int_1^\infty dt\, t^{-\beta}(u+t)^{-\alpha}$$
$$= u^{1-\alpha-\beta}\int_{1/u}^\infty t^{-\beta}(1+t)^{-\alpha}.$$

Using $(1+t)^{-\alpha} \sim t^{-\alpha}$ for $t \geq 1$ once more we obtain that the last expression is equivalent to

$$\sim u^{1-\alpha-\beta}\int_{1/u}^\infty dt\, t^{-\alpha-\beta},$$

which is ~ 1 since $1-\alpha < 0$. The lemma now follows from $1 + u^{1-\alpha} \sim u^{1-\alpha}$ for $0 < u \leq 1$ and $\alpha > 1$.

To prove the second statement we write

$$B_{\alpha,\beta}(u, v) = \left\{\int_0^{1/2} + \int_{1/2}^1\right\} dt(t+u)^{-\alpha}(v+1-t)^{-\beta} = X_1 + X_2.$$

In X_1 we have $v + 1 + t \sim v + 1$ and thus

$$X_1 \sim (1+v)^{-\beta}\int_0^{1/2} dt(t+u)^{-\alpha} \sim (1+v)^{-\beta}\frac{u^{1-\alpha}}{(1+u)}.$$

In X_2 we have instead $t + u \sim 1 + u$ and thus

$$X_2 \sim (1+u)^{-\alpha}\int_{1/2}^1 dt(v+1-t)^{-\beta} \sim (1+u)^{-\alpha}\frac{v^{1-\alpha}}{1+v}$$

and the last estimation follows. □

Theorem 11.1.1 is an immediate consequence of the following lemma.

Lemma 31.0.2
The wavelet transform of κ_β^- with respect to κ_α^+ satisfies, for $\alpha, \beta > 1$,

$$\mathscr{W}[\kappa_\alpha^-; \kappa_\beta^+](b, a) \sim \left[\kappa_1\left(\frac{a}{b}\right) + \kappa_1\left(\frac{1}{b}\right)\right] \times \left[\kappa_\beta^+(a)\kappa_\beta^+\left(\frac{b}{1+a}\right) + \frac{1}{a}\kappa_\alpha^+\left(\frac{1}{a}\right)\kappa_\alpha^+\left(\frac{b}{1+a}\right)\right]. \qquad (31.0.1)$$

The wavelet transform of κ_β^+ with respect to κ_α^+ satisfies

$$\mathscr{W}[\kappa_\alpha^+; \kappa_\beta^+](b, a) \sim \frac{1}{1+a}\kappa_1\left(\frac{b}{1+a}\right)\left\{\kappa_{\alpha-1}^-\left(\frac{b}{a}\right) + \kappa_{\beta-1}^+(b)\right\}.$$

Proof. We now prove the first equation of the lemma. Suppose we had shown (31.0.1) for $a \leq 1$. From (2.0.1) we can write, for $a > 1$,

$$\mathscr{W}[\kappa_\alpha^-; \kappa_\beta^+](b, a) = \frac{1}{a}\mathscr{W}[\kappa_\beta^-; \kappa_\alpha^+]\left(\frac{b}{a}, \frac{1}{a}\right).$$

Using (31.0.1) supposed true for $a \leq 1$ this would prove the first part of the lemma.

For $a < 1$ we have to prove that

$$\mathscr{W}(b, a) \sim \begin{cases} 0 & \text{for } b \leq 0 \\ b/a & \text{for } 0 < b < a \\ 1 & \text{for } a \leq b \leq a + 1 \\ b^{-\beta} + a^{-1}(b/a)^{-\alpha} & \text{for } b > a + 1 \end{cases}$$

Clearly, the wavelet transform is supported by that part of the half-plane where $b \geq 0$. So the first equation is done.

We still need to estimate

$$\mathscr{W}(b, a) = \int_0^b dt\, \frac{1}{a}\kappa_\alpha^-\left(\frac{t-b}{a}\right)\kappa_\beta^+(t)$$

for $b > 0$. Upon using $\kappa_\beta^+(t) \sim (1+t)^{-\beta}$, $t \geq 0$, and $\kappa_\alpha^-(t) \sim (1-t)^{-\alpha}$, $t \leq 0$, we have, upon changing t to bt,

$$\mathscr{W}(b, a) \sim \int_0^b dt\, \frac{1}{a}(1 - [t-b]/a)^{-\alpha}(1+t)^{-\beta}$$
$$= b^{-\beta}\left(\frac{b}{a}\right)^{1-\alpha}\int_0^1 dt(1 + a/b - t)^{-\alpha}(b^{-1} + t)^{-\beta}.$$

We may now use the preceding lemma to conclude.

We now consider the second part. Again we may limit ourselves to $a \leq 1$ because of the symmetry of (2.0.1):

$$\mathscr{W}[\kappa_\alpha^+; \kappa_\beta^+](b, a) = \frac{1}{a} \mathscr{W}[\kappa_\beta^+; \kappa_\alpha^+]\left(\frac{b}{a}, \frac{1}{a}\right).$$

For $a \leq 1$ it remains to prove that

$$\mathscr{W}(b, a) \sim \begin{cases} b^{-\beta} & \text{for } b > 1 \\ 1 & \text{for } 0 \leq b \leq 1, \\ (1 - b)^{-1}(1 - b/a)^{1-\alpha} & \text{for } b < 0. \end{cases}$$

The case where $b \geq 0$ is obtained by

$$\begin{aligned} \mathscr{W} &\sim \int_b^\infty dt\, a^{-1}(1 + [t - b]/a)^{-\alpha}(1 + t)^{-\beta} \\ &= a^{-\beta} \int_0^\infty dt(1 + t)^{-\alpha}(t + [1 + b]/a)^{-\beta} \end{aligned}$$

where we may use (31.0.1) to write

$$\mathscr{W} \sim (1 + b)^{1-\beta}(1 + a + b)^{-1}$$

and we are done for $b \geq 0$.

Now consider $b < 0$. We may estimate

$$\begin{aligned} \mathscr{W}(b, a) &\sim \int_0^\infty dt\left(1 + \frac{t - b}{a}\right)^{-\alpha}(1 + t)^{-\beta} \\ &= a^{\alpha-1} \int_0^\infty dt(1 + t)^{-\beta}(a - b + t)^{-\alpha}. \end{aligned}$$

This integral is estimated by Lemma 31.0.1:

$$\mathscr{W}(b, a) \sim a^{\alpha-1} \frac{(a - b)^{1-\alpha}}{1 + a - b} = (1 - b)^{-1}\left[1 + \left(-\frac{b}{a}\right)\right]^{1-\alpha},$$

and the second part of the lemma is proved. □

2

DISCRETIZING AND PERIODIZING THE HALF-PLANE

Because of the reproducing kernel identity the wavelet coefficients have an internal correlation and are therefore not independent. So it is not surprising that knowledge of the wavelet transform on a subset of the half-plane is sufficient to reconstruct the analysed function. In the following sections we consider various partial reconstruction formulas associated with different subsets of the half-plane. In particular, we consider reconstruction over strips, over zooms and cones, and over various grids. In addition, we will construct a wavelet analysis over the circle. For the sake of simplicity we will formulate everywhere in one dimension. The reader should have no difficulty in generalizing the results to wavelet transforms of functions over $\mathbb{R}^n$.

1 Interpolation

Consider the following problem. Given N pairwise distinct points in the half-plane $(b_0, a_0), \ldots, (b_{N-1}, a_{N-1})$, and N complex numbers $\gamma_0, \ldots, \gamma_{N-1}$, can we find a function s over the real line, such that its wavelet transform with respect to a given wavelet g takes on the values γ_n at the points (b_n, a_n)

$$\mathscr{W}_g s(b_n, a_n) = \langle g_{b_n, a_n} \mid s \rangle = \gamma_n, \qquad n = 0, \ldots, N-1.$$

Equivalently, we may ask whether there is a function over the half-plane that satisfies at

(i) $\mathscr{T} \in \text{image } \mathscr{W}_g \Leftrightarrow \mathscr{T} = P_g * \mathscr{T}$,
(ii) $\mathscr{T}(b_n, a_n) = \gamma_n$.

If we can answer this question we know how to interpolate wavelet coefficients without leaving the image space of the wavelet transform. Clearly, we have to suppose that the functions g_{b_n, a_n} are linearly independent. This is a very weak assumption, however. Unfortunately, in this case there is now an infinity of ways to do this interpolation. Indeed, the space of square integrable functions over the half-plane line satisfying the reproducing kernel equation (i.e. that are in the image of the wavelet transform) is an infinite-dimensional, separable Hilbert space and the N linear equations are therefore satisfied on an infinite-dimensional affine subspace. We must therefore choose a particular solution. A canonical choice might be the function in the image space which has minimal energy as expressed by

(iii) for all $\mathscr{R} \in L^2(\mathbb{H})$ satisfying (i) and (ii) we have $\|\mathscr{R}\|_2 \geq \|\mathscr{T}\|_2$.

Conditions (i), (ii), and (iii) define unique by a function that can be constructed explicitly, as is shown by the following theorem, which, in the context of wavelet analysis, is due to Grossmann *et al.* (1985).

Theorem 1.0.1
Let $g \in H^2_+(\mathbb{R})$ be admissible and let $P_g = c_g^{-1}\mathscr{W}_g g$ be the associated reproducing kernel. Given N points $(b_n, a_n) \in \mathbb{H}$, $n = 0, \ldots, N-1$ let $\mathbf{M}_{n,m}$ be the $N \times N$ matrix defined by

$$\mathbf{M} = \mathbf{M}_{n,m} = \langle g_{b_n, a_n} \mid g_{b_m, a_m} \rangle = \frac{1}{a_m} P_g\left(\frac{b_n - b_m}{a_m}; \frac{a_n}{a_m}\right).$$

Suppose, further, that $\det \mathbf{M} \neq 0$. *Let $\gamma_0 \to \gamma_{N-1} \in \mathbb{C}$ be given. Then the interpolation problem* (i), (ii), *and* (iii) *has a unique solution $\mathscr{T}$. It is given by*

$$\mathscr{T}(b, a) = \sum_{n=0}^{N-1} \beta_n \frac{1}{a_n} P_g\left(\frac{b - b_n}{a_n}; \frac{a}{a_n}\right),$$

with $\beta = (\beta_0, \ldots, \beta_{N-1})$ given by

$$\beta = \mathbf{M}^{-1}\gamma \quad \Leftrightarrow \quad \gamma_n = \sum_{m=0}^{N-1} \mathbf{M}_{n,m}\beta_m.$$

This theorem is an immediate application of the following general lemma solving this optimization problem in a general vector space with scalar product.

Lemma 1.0.2
Let V be a complex pre-Hilbert space[1] *with scalar product $\langle \cdot \mid \cdot \rangle$ and norm $\|s\|^2 = \langle s \mid s \rangle$, and let $s_n \in V$, $n = 0, \ldots, N-1$ be linear independent vectors. Further, let N complex numbers γ_n, $n = 0, \ldots, N-1$ be given. Then the vector*

$$r = \sum_{n=0}^{N-1} \lambda_n s_n \tag{1.0.1}$$

with coefficients λ_n, which are the solution of

$$\sum_{n=0}^{N-1} \langle s_m \mid s_n \rangle \lambda_n = \gamma_m, \qquad m = 0, \ldots, N-1, \tag{1.0.2}$$

[1] We recall that a Hilbert space is a complete pre-Hilbert space.

is the unique vector in V that satisfies

(i) $\langle s_n \mid r \rangle = \gamma_n, \qquad n = 0, \ldots, N-1,$
(ii) *for all other vectors $v \neq r$ satisfying* (i) *we have* $\|v\| > \|r\|$.

Proof. First r is well defined since the vectors s_n are linearly independent and therefore the matrix $\mathrm{M}_{n,m} = \langle s_n \mid s_m \rangle$ is invertible. Then r satisfies (i) because from (1.0.1) and (1.0.2) it follows that

$$\langle s_n \mid r \rangle = \sum_{m=0}^{N-1} \langle s_n \mid s_m \rangle \lambda_m = \gamma_n.$$

Now let $v \neq r$ be another vector that satisfies (i). Then we claim that

$$\langle v - r \mid r \rangle = 0.$$

Indeed, using expansion (1.0.1) together with (1.0.2) for r, and using condition (i) for v, we obtain

$$\begin{aligned} \langle v - r \mid r \rangle &= \sum_{n=0}^{N-1} \langle v \mid s_n \rangle \lambda_n - \sum_{m=0}^{N-1} \sum_{n=0}^{N-1} \langle s_n \mid s_m \rangle \overline{\lambda_n} \lambda_m \\ &= \sum_{n=0}^{N-1} \overline{\gamma_n} \lambda_n - \sum_{m=0}^{N-1} \overline{\gamma_m} \lambda_m = 0. \end{aligned}$$

Therefore, we may write

$$\|v\|^2 = \|v - r + r\|^2 = \|v - r\|^2 + \|r\|^2 > \|r\|^2,$$

which proves the lemma. □

2 Reconstruction over voices

In chapter 1 we gave sufficient conditions under which the inversion formula

$$s = \frac{1}{c_{g,h}} \int_{\mathbb{H}} d\mu(b, a) \mathscr{W}_g s(b, a) h_{b,a}, \qquad d\mu(b, a) = db\, da/a,$$

$$h_{b,a}(t) = h([t - b]/a)/a$$

holds. We now want to know what kind of information we can obtain if we know the wavelet transform on some subset K of the half-plane. More explicitly, we consider for $K \subset \mathbb{H}$ the following partial reconstruction:

$$s^{\#} = \frac{1}{c_{g,h}} \int_K d\mu(b, a) \mathscr{W}_g s(b, a) h_{b,a}.$$

Equivalently, we may say that we want to analyse the Töpliz operators that we considered in Section 26 of the following form:

$$s^{\#} = c_{g,h}^{-1} \mathcal{M}_g(\chi_K \cdot \mathcal{W}_g s),$$

where χ_K is the characteristic function of K. In the limit case where K is of measure 0, χ_K will be taken to be a distribution with support in K. Note that in the case where K is some open set we have for the wavelet transform of the partial reconstruction

$$\mathcal{W}_g s^{\#}(b, a) = \int_K d\mu(b, a) \frac{1}{a'} P_{g,h}\left(\frac{b - b'}{a}, \frac{a}{a'}\right) \mathcal{W}_g s(b', a').$$

In the following sections we consider partial reconstruction over translation invariant regions—that is, strips in the half-plane—and dilation invariant regions—that is, cones.

2.1 One single voice

Since every strip in the half-plane is composed of lines parallel to the real axis we start with partial reconstruction over one single voice. Suppose we know the wavelet transform of $s \in H^2_+(\mathbb{R})$ with respect to an admissible wavelet g along a line parallel to the borderline of the half plane, that is, only one voice—say $f(b) = \mathcal{W}_g s(b, a)$—is known. We consider the partial reconstruction given by

$$s^{\#} = \int_{-\infty}^{+\infty} db\ \mathcal{W}_g s(b, a) h_{b,\alpha} = h_\alpha * \mathcal{W}_g s(\cdot, \alpha),$$

with $h_{b,\alpha} = h([\cdot - b]/a)/a$. In terms of the Töpliz operators this partial reconstruction may be written as

$$s^{\#} = \mathcal{M}_h(K \cdot \mathcal{W}_g s), \qquad K(b, a) = \delta(a/\alpha - 1).$$

The reason for choosing this approach rather than trying to generalize the interpolation of the previous section to the situation of infinitely many points is the following. Suppose that $|\hat{g}(\omega)| > 0$ for $\omega > 0$. Then s may 'in principle' be recovered from f since $\hat{s}(\omega) = \hat{f}(\omega)/\bar{\hat{g}}(\alpha\omega)$. Therefore, we cannot expect to obtain the analogue of the preceding interpolation results where an energy minimizing criteria could be used to select one solution among many possible ones since the unique solution is given by 'deconvolution'. However, this deconvolution may be very unstable whereas the partial reconstruction we now propose is stable, although perhaps not complete.

Since along a voice the wavelet transform is a convolution $\mathcal{W}_g s(\cdot, \alpha) = \tilde{g}_\alpha * s$, and since the inverse wavelet transform is a superposition of convolutions, it follows that

$$s^{\#} = r_\alpha * s \quad \text{with} \quad r_\alpha(t) = (h * \tilde{g})(t/\alpha)/\alpha, \qquad \tilde{g}(t) = \bar{g}(-t).$$

In Fourier space we have, accordingly,

$$\hat{r}_\alpha(\omega) = \hat{h}(\alpha\omega)\bar{\hat{g}}(\alpha\omega). \tag{2.1.1}$$

It might be interesting to note that partial reconstruction with the help over one voice is determined by the restriction of the reproducing kernel $P_{g,h} = c_{g,h}^{-1}\mathcal{W}_g h$ to the voice $a = 1$:

$$r_\alpha(t) = \frac{c_{g,h}}{\alpha} P_{g,h}\left(\frac{t}{\alpha}, 1\right).$$

In wavelet space partial reconstruction is nothing but a superposition of translated reproducing kernels

$$\mathcal{W}_g s^{\#}(b, a) = c_{g,h} \int_{-\infty}^{+\infty} db' \frac{1}{a'} P_{g,h}\left(\frac{b - b'}{a'}, \frac{a}{a'}\right) f(b'), \qquad f(b) = \mathcal{W}_g s(b, \alpha).$$

2.2 Infinitely many voices

With one single voice only a partial reconstruction is possible in general. Consider, therefore, the set of voices at scale $a = \alpha^n$, $n \in \mathbb{Z}$, with $\alpha > 1$. These scales are equidistant on a logarithmic scale $\log a$. And we now look at the partial reconstruction operator

$$s^{\#} = \mathcal{M}_h(K \cdot \mathcal{W}_g s), \qquad K(b, a) = \sum_{n=-\infty}^{+\infty} \delta(a/\alpha^n - 1).$$

More explicitly, the partial reconstruction along these voices reads

$$s^{\#} = \sum_{n=-\infty}^{+\infty} \int_{-\infty}^{+\infty} db\, \mathcal{W}_g s(b, \alpha^n) h_{b,\alpha^n}. \tag{2.2.1}$$

By mere superposition of the previous results we see that $s^{\#}$ is related to s by a 'Fourier multiplier'

$$\widehat{s^{\#}}(\omega) = \hat{r}(\omega)\hat{s}(\omega), \qquad \hat{r}(\omega) = \sum_{n=-\infty}^{+\infty} \hat{h}(\alpha^n\omega)\bar{\hat{g}}(\alpha^n\omega). \tag{2.2.2}$$

However, this filter r is not a smooth function since its Fourier transform is not decaying at infinity. As one can see in Figure 2.1 $\hat{r}$ is typically a bounded function that oscillates around some mean value. As we will see in Section 4, this kind of multiplier corresponds to a singular integral kernel known as a Calderòn–Zygmund integral operator.

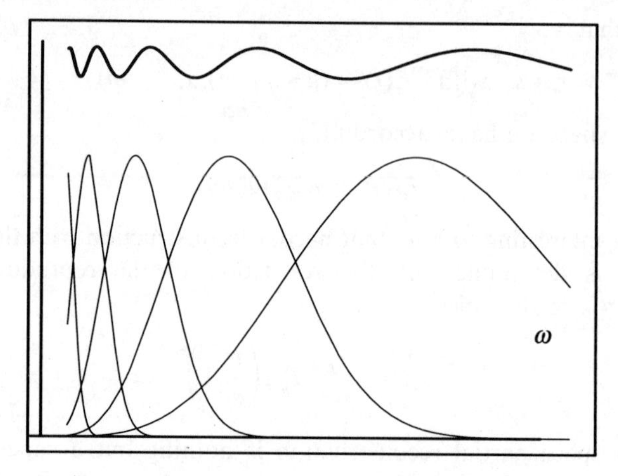

Fig. 2.1 The Fourier multiplier of the partial reconstruction over infinitely many voices.

If we define the operator $B: s \mapsto s^{\#}$, $\hat{s}(\omega) \mapsto \hat{r}(\omega)\hat{s}(\omega)$ with $h = g$, then it is immediately verified that B can only be bounded in $L^2(\mathbb{R})$ if g is admissible. Indeed, $\hat{r} \in L^{\infty}(\mathbb{R})$ implies that

$$\sum_{j \in \mathbb{Z}} |\hat{g}(\alpha^j \omega)|^2 \leq c < \infty .$$

Therefore, we can write, with some $c' > 0$,

$$c' \geq \int_1^{\alpha} d\omega \sum_{j \in \mathbb{Z}} |\hat{g}(\alpha^j \omega)|^2 = \sum_{j \in \mathbb{Z}} \int_{\alpha^j}^{\alpha^{j+1}} \frac{d\omega}{\alpha^j} |\hat{g}(\omega)|^2$$
$$\geq \alpha^{-1} \int_0^{\infty} \frac{d\omega}{\omega} |\hat{g}(\omega)|^2 ,$$

proving that g has to be admissible.

We now employ some elementary considerations in an $H^2_+(\mathbb{R})$ context. Note that the sum of (2.2.1) may be considered as a Riemann sum, namely, for ρ regular enough,

$$\int_0^{\infty} \frac{da}{a} \rho(a) = \lim_{\alpha \to 1} (\alpha - 1) \sum_{n=-\infty}^{+\infty} \rho(\alpha^n).$$

Therefore, if h is a reconstruction wavelet, then $(\alpha - 1)\hat{r}(\omega) \to c_{gh}$ as $\alpha \to 1$, and therefore $\widehat{(\alpha - 1)s^{\#}} \to c_{gh}\hat{s}$ and a complete reconstruction of s is obtained.

But even for α different from 1, $s^{\#}$ may be close to s provided the multiplier $\hat{r}$ is close to a constant. The following theorem gives an estimation of the quality of the approximation of the partial reconstruction.

Theorem 2.2.1

Let $g, h \in L^1(\mathbb{R}) \cap H^2_+(\mathbb{R})$ satisfy at

$$\sum_{j \in \mathbb{Z}} |\hat{g}(\alpha^j \omega)|^2 \leq \gamma < \infty, \qquad \sum_{j \in \mathbb{Z}} |\hat{h}(\alpha^j \omega)|^2 \leq \gamma < \infty,$$

with some $\gamma > 0$, and let $s \in H^2_+(\mathbb{R})$. Suppose that there is a constant ϵ, $0 \leq \epsilon < 1$, and another complex constant $c \in \mathbb{C}$ such that

$$\operatorname*{ess\,sup}_{\omega > 0} \left| c \sum_{n \in \mathbb{Z}} \hat{h}(\alpha^n \omega) \bar{\hat{g}}(\alpha^n \omega) - 1 \right| \leq \epsilon. \tag{2.2.3}$$

Then the following limit exists in $H^2_+(\mathbb{R})$:

$$s^{\#} = \lim_{N \to \infty} s_N = \lim_{N \to \infty} \sum_{n=-N}^{N} \int_{-\infty}^{+\infty} db\, \mathcal{W}_g s(b, \alpha^n) h_{b, \alpha^n}$$

and we have the following estimation:

$$\|s - cs^{\#}\|_2 \leq \epsilon \|s\|_2.$$

In the case where $h = g$ the hypothesis of the theorem is satisfied iff

$$0 < A = \operatorname*{ess\,inf}_{\omega > 0} \sum_{n \in \mathbb{Z}} |\hat{g}(\alpha^n \omega)|^2,$$

$$B = \operatorname*{ess\,sup}_{\omega > 0} \sum_{n \in \mathbb{Z}} |\hat{g}(\alpha^n \omega)|^2 < \infty,$$

in which case we can set

$$c = \frac{2}{B + A}, \qquad \epsilon = \frac{B - A}{B + A}. \tag{2.2.4}$$

Therefore, the smaller ϵ or the closer the constants A and B are to each other, the better is the reconstruction—at least in the square mean sense.

Proof. First observe that $s_N \in H^2_+(\mathbb{R})$ is well defined by Young's inequality. In Fourier space we have

$$\hat{s}_N(\omega) = \hat{r}_N(\omega) \hat{s}(\omega), \qquad \hat{r}_N(\omega) = \sum_{n=-N}^{N} \hat{h}(\alpha^n \omega) \bar{\hat{g}}(\alpha^n \omega).$$

By applying the Schwarz inequality to the sum we have

$$|\hat{r}_N(\omega)| \leq \sqrt{\sum_{n=-N}^{N} |\hat{h}(\omega)|^2} \sqrt{\sum_{n=-N}^{N} |\hat{g}(\omega)|^2} \leq \gamma.$$

Since, in addition, $\hat{r}_N(\omega) \to \hat{r}(\omega)$ pointwise with r given by (2.2.2), we may evoke the theorem of dominated convergence to conclude that s_N converges in the mean square sense to some $s^{\#}$. By hypothesis we have

$$\operatorname*{ess\,sup}_{\omega>0} |1 - c\hat{r}(\omega)| \leq \epsilon.$$

By Parseval's equation and Hölder's inequality we now have

$$\|s - c^{-1}s^{\#}\|_2^2 = \|\hat{s} - \widehat{cs^{\#}}\|_2^2 = \|(1 - c\hat{r})^2 \hat{s}^2\|_1 \leq \|1 - c\hat{r}\|_\infty^2 \|s\|_2^2$$
$$\leq \epsilon^2 \|s\|_2^2.$$

This proves the theorem. □

Again, the theorem may remain valid even for non-admissible wavelets provided the approximations s_N are well defined. For $h = \delta$, for example, we obtain the following approximation by summation over all voices:

$$s^{\#}(t) = \sum_{n \in \mathbb{Z}} \mathscr{W}(t, \alpha^n).$$

Condition 2.2.3 now has the following form:

$$\operatorname*{ess\,sup}_{\omega>0} \left| c \sum_{n \in \mathbb{Z}} \bar{\hat{g}}(\alpha^n \omega) - 1 \right| \leq \epsilon.$$

In the case of a wavelet with a real-valued Fourier transform we can again use (2.2.4) with the constants A and B given by

$$A = \operatorname*{ess\,inf}_{\omega>0} \sum_{n=-\infty}^{+\infty} \bar{\hat{g}}(\alpha^n \omega); \qquad B = \operatorname*{ess\,sup}_{\omega>0} \sum_{n=-\infty}^{+\infty} \bar{\hat{g}}(\alpha^n \omega).$$

3 An iteration procedure

Until now we have only obtained a partial reconstruction $s^{\#}$ that approximates s. As we will now see, this is only the first step in an approximation scheme that allows us to reconstruct s completely from knowledge of the voices at scale $a = \alpha^n$.

Because this procedure will be used in the sequel we state it in the following general form.

The set of all bounded operators for one Banach space B_1 to another Banach space B_2 is usually denoted by $\mathscr{B}(B_1, B_2)$. It is a Banach space in its own right with respect to the norm

$$\|K\|_{\mathscr{B}(B_1, B_2)} = \|K\| = \sup \frac{\|Ks\|_{B_2}}{\|s\|_{B_1}}$$

where the 'sup' runs over all vectors $s \in B_1$, where $s \neq 0$. In the case where $B_1 = B_2 = B$ we have in addition $\|K_1 K_2\| \leq \|K_1\| \, \|K_2\|$, which means that $\mathscr{B}(B) := \mathscr{B}(B, B)$ is a Banach algebra. After this preparation we are ready to state the following theorem.

Theorem 3.0.1
Let $K: B \to B$ be a bounded operator acting in a Banach space B that satisfies at

$$\|K - \mathbb{1}\|_{\mathscr{B}(B)} \leq \epsilon < 1.$$

Then K^{-1} exists and is in $\mathscr{B}(B)$. Let $s \in B$ be given. If we set

$$s_{n+1} = s_n + r_{n+1}, \qquad r_{n+1} = s_n - Ks_n,$$

with $s_0 = s$, then $s_n \to K^{-1}s$ as $n \to \infty$ and the difference may be estimated as

$$\|K^{-1}s - s_n\|_B \leq \frac{\epsilon^n}{1 - \epsilon} \|s\|_B.$$

With the help of theorem the complete reconstruction of s from its voices $\mathscr{W}_g s(\cdot, \alpha^j)$ can be obtained by applying it to the partial reconstruction operator

$$B: s \mapsto c^{-1} \sum_{n=-\infty}^{\infty} \int_{-\infty}^{+\infty} db \, \mathscr{W}_g s(b, \alpha^n) h_{b, \alpha^n},$$

with c as given by (2.2.3).

Proof. The proof is standard and will be reproduced here solely for the convenience of the reader.

Let $X_n: B \to B$ be defined by

$$X_n = \mathbb{1} + (\mathbb{1} - K) + (\mathbb{1} - K)^2 + \cdots (\mathbb{1} - K)^n$$

or, equivalently, $X_0 = \mathbb{1}$ and

$$X_{n+1} = \mathbb{1} + (\mathbb{1} - K)X_n = \mathbb{1} + X_n(\mathbb{1} - K). \tag{3.0.1}$$

Since $\|\mathbb{1} - K\| \leq \epsilon$ it follows that this series is a Cauchy sequence and hence converges in $\mathscr{B}(B)$ to some bounded operator X_∞. We have

$$\mathbb{1} - KX_n = \mathbb{1} - X_n K = X_{n+1} - X_n,$$

and therefore $KX_\infty = X_\infty K = \mathbb{1}$ and thus $K^{-1} = X_\infty$. From (3.0.1) we obtain the iteration procedure as stated in the theorem. Finally, from

$$\|K^{-1}s - s_n\|_B \leq \sum_{k=n+1}^{\infty} \|(\mathbb{1} - K)^k s\|_B \leq \frac{\epsilon^n}{1 - \epsilon} \|s\|_B,$$

we have the stated estimation. □

4 Calderón–Zygmund operators: a first contact

The partial reconstruction operators over the voices $a = \alpha^j$ that we have considered in the previous section belong to a class of operators that play a particular role in the theory of wavelet transforms: the Calderón–Zygmund singular integral operators. The prototype of these operators is the Hilbert transform

$$H: s \mapsto \lim_{\epsilon \to 0} \int_{-\infty}^{+\infty} du\, f_\epsilon(t-u)s(u), \qquad f_\epsilon(t) = \frac{t}{t^2+\epsilon^2}.$$

This transform is given as the limit of the family of operators K_ϵ acting as convolution operators with some filter f_ϵ. These filter functions f_ϵ satisfy at

$$\left.\begin{aligned} &\text{(i) } |\hat{f}_\epsilon(\omega)| \le c, \\ &\text{(ii) } |\partial f_\epsilon(t)| \le \frac{c}{|t|^2}, \end{aligned}\right\} \tag{4.0.1}$$

with some constant $c > 0$ that does not depend on ϵ.

Note that these operators behave well under dilation in the sense that $f_\epsilon(t/a)/a$ satisfies the same estimates as f_ϵ. It is for this reason that these operators are intimately connected to wavelet theory. The importance of these operators results from the fact that in spite of their bad localization—for example, the kernel of the Hilbert transform is only decaying as $1/|t|$—they preserve a lot of local regularity. For the moment we only cite a theorem whose proof can be found in Stein (1979, page 29).

Theorem 4.0.1
Let $f \in L^2(\mathbb{R})$ satisfy at

(i) $|\hat{f}(\omega)| \le c$,

(ii) *∂f continuous outside $\{0\}$ and* $|\partial f(t)| \le \dfrac{c}{|t|^2}$, $t \ne 0$.

Then for $s \in L^1(\mathbb{R}) \cap L^p(\mathbb{R})$ let us set

$$Ks = f * s.$$

Then for $1 < p < \infty$ there is constant c' that only depends on c and p such that

$$\|Ks\|_p \le c' \|s\|_p.$$

The relation between these kinds of operators and the partial reconstruction we have considered before is now made clear through the following theorem.

Theorem 4.0.2
Let $g, h \in S_+(\mathbb{R})$. Then for $\alpha > 1$ the partial reconstruction operator

$$s \mapsto Ks = \sum_{n=-\infty}^{\infty} \int_{-\infty}^{+\infty} db\, \mathcal{W}_g s(b, \alpha^n) h_{b,\alpha^n}$$

is a Calderón–Zygmund integral operator.

Proof. We will even show a little more. Consider partial reconstruction over a finite number of voices:

$$K_m: s \mapsto \sum_{n=-m}^{m} \int_{-\infty}^{+\infty} db\, \mathcal{W}_g s(b, \alpha^n) h_{b,\alpha^n}.$$

This can also be written as

$$K_m: s \mapsto f_m * s, \qquad f_m(t) = \sum_{n=-m}^{m} \alpha^{-n} \tilde{g} * h(t/\alpha^n), \qquad \tilde{g}(\cdot) = \bar{g}(-\cdot).$$

The first condition in (4.0.1) now follows directly from (2.2.2). Differentiating under the sum we obtain a stronger version of condition (ii), namely,

$$|\partial^n f_m(t)| \le \frac{c_n}{|t|^{1+n}}, \qquad n = 0, 1, \ldots,$$

with all constants c_n independent of m. We may therefore go to the limit $m \to \infty$ to conclude. □

In particular this shows that the partial reconstruction $s^\#$ may also be close to s with respect to the $L^p(\mathbb{R})$-norm if $1 < p < \infty$.

5 Reconstruction over strips

We will first give some formal results without checking questions of convergence. Suppose now that we know the wavelet transform over a strip $\epsilon < a < \rho$ in the scale-position half-plane. The application of the inversion formula over this region in the half-plane yields ($h_{b,a} = a^{-1}h([\cdot - b]/a)$, $h_a = h_{0,a}$)

$$s_{\epsilon,\rho} = \int_{\epsilon}^{\rho} \frac{da}{a} \int_{-\infty}^{+\infty} db\, \mathcal{W}_g s(b, a) h_{b,a} = \int_{\epsilon}^{\rho} \frac{da}{a} \mathcal{W}_g s(\cdot, a) * h_a,$$

and again the approximation $s_{\epsilon,\rho}$ is obtained through filtering, the filter being the superposition of the reconstruction filters over one voice (2.1.1):

$$s_{\epsilon,\rho} = r_{\epsilon,\rho} * s, \qquad r_{\epsilon,\rho}(t) = \int_\epsilon^\rho \frac{da}{a}\frac{1}{a} h * \tilde{g}\left(\frac{t}{a}\right), \tag{5.0.1}$$

with $\tilde{g}(\cdot) = \bar{g}(-\cdot)$. Actually, $r_{\epsilon,\rho}$ is the difference between two dilated filters,

$$r_{\epsilon,\rho} = r(\cdot/\epsilon)/\epsilon - r(\cdot/\rho)/\rho, \qquad r(t) = r_{1,\infty}(t) = \frac{1}{t}\int_0^t du\, h * \tilde{g}(u). \tag{5.0.2}$$

Indeed, supposing the integral in (5.0.1) is absolutely convergent, we may write

$$\begin{aligned} r_{\epsilon,\infty}(t) &= \int_\epsilon^\infty \frac{da}{a}\frac{1}{a} h * \tilde{g}\left(\frac{t}{a}\right) = \frac{1}{t}\int_0^{t/\epsilon} du\, h * \tilde{g}(u) \\ &= r_{1,\infty}(t/\epsilon)/\epsilon, \end{aligned} \tag{5.0.3}$$

and (5.0.2) follows from the obvious identity $r_{\epsilon,\rho} = r_{\epsilon,\infty} - r_{\rho,\infty}$. Therefore, the reconstruction filter $r_{\epsilon,\rho}$ is given in frequency space by

$$\widehat{r_{\epsilon,\rho}}(\omega) = \hat{r}(\epsilon\omega) - \hat{r}(\rho\omega).$$

Upon taking the Fourier transform of both sides of equation (5.0.3) we obtain the frequency representation of the filter r:

$$\hat{r}(\omega) = \int_1^\infty \frac{da}{a}\, \bar{\hat{g}}(a\omega)\hat{h}(a\omega) = \int_{|\omega|}^\infty \frac{da}{a}\, \bar{\hat{g}}(a \operatorname{sign} \omega)\hat{h}(a \operatorname{sign} \omega). \tag{5.0.4}$$

In particularly, it follows that $\int r = \hat{r}(0) = c_{g,h}$ if h is a reconstruction wavelet for g.

In general, r is essentially a low-pass filter. We may limit ourselves to the case of progressive functions and we therefore consider only positive frequencies. Suppose, therefore, that

$$\operatorname{supp} \hat{h}\bar{\hat{g}} \subseteq [\omega_l, \omega_r], \qquad \omega_l > 0.$$

Then from (5.0.4) it follows that r is band limited, and that

$$\operatorname{supp} r \subseteq [0, \omega_r].$$

However, r is not strip limited, and instead we have

$$\hat{r}(\omega) = c_{g,h}, \qquad 0 \le \omega \le \omega_l,$$

with the wavelet constant being given by Definition 14.0.1 of Chapter 1. However, the reconstruction filter $r_{\epsilon,\rho}$, as the difference of the two dilated low-pass filters $D_\epsilon r$ and $D_\rho r$, is strip limited (cf. Figure 5.1) and we have

$$\operatorname{supp} r_{\epsilon,\rho} \subseteq [\epsilon\omega_l, \rho\omega_r].$$

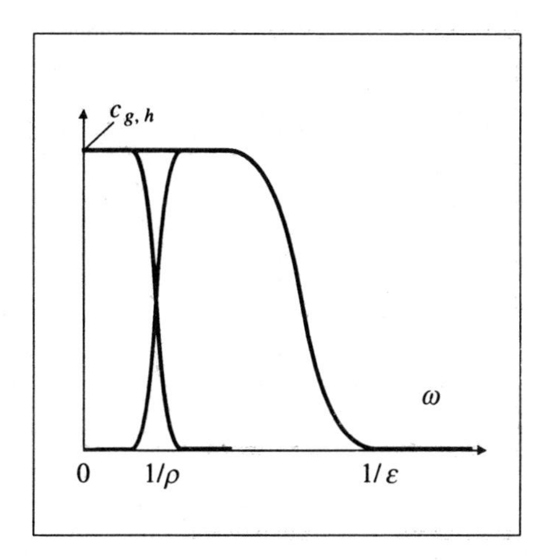

Fig. 5.1 The frequency content of the band-pass filter of the partial strip reconstruction as the difference between two dilated low pass filters.

As $\epsilon \to 0$ ($\rho \to \infty$) more and more high frequencies (low frequencies) of the analysed function are contained in the approximation $s_{\epsilon,\rho} = r_{\epsilon,\rho} * s$ and finally we get back all frequencies of s.

In the general case where the functions are neither progressive nor regressive the negative frequencies do not have in general, the same weight since, then,

$$r_{0,\infty}(\omega) = \begin{cases} c_{gh}^{+} & \omega > 0 \\ c_{gh}^{-} & \omega < 0 \end{cases} = \tfrac{1}{2}(c_{gh}^{+} + c_{gh}^{-}) + \tfrac{1}{2}(c_{gh}^{+} - c_{gh}^{-}) \operatorname{sign} \omega,$$

and we see once more that in general the reconstruction operator is a mixture between $\mathbb{1}$ and the Hilbert transform unless h is a reconstruction wavelet ($c_{gh}^{+} = c_{gh}^{-}$).

All this remains qualitatively true in the case where all functions are only polynomially strip localized, as is easiest seen by considering an example.

Example 5.0.1

Consider once more the family of Cauchy wavelets $g_\alpha = \Gamma(\alpha)/(1 - it)^{1+\alpha}$. The associated low-pass filter for an analysis with g_α and a partial reconstruction with g_β is given for $\omega > 0$ by

$$\hat{r}(\omega) = \int_{\omega}^{\infty} da\, a^{\alpha+\beta-1}\, e^{-2a}.$$

In the time representation we have

$$r(t) = \Re r(t) + i\Im r(t), \qquad \Im r = H\Re r,$$

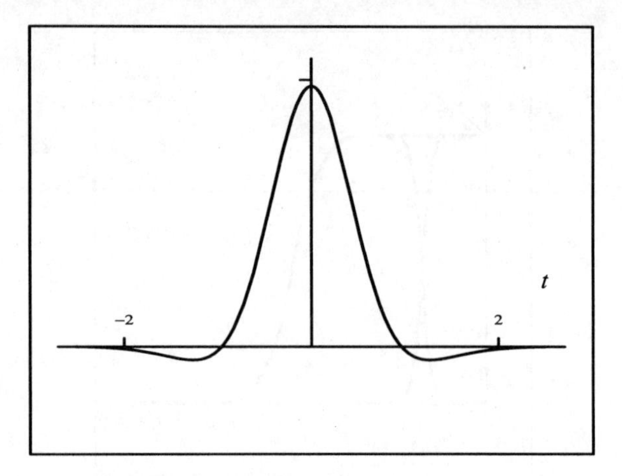

Fig. 5.2 The partial reconstruction filter for g_3 and g_5 in the time representation.

where H is the Hilbert transform. Now $\Re r$ can be computed as follows:

$$\begin{aligned}\Re r(t) &= \frac{1}{\pi}\int_0^\infty d\omega \cos(\omega t)\int_\omega^\infty \frac{da}{a}\, a^{\alpha+\beta}\, e^{-2a} \\ &= \frac{1}{\pi}\int_0^\infty d\omega \int_0^\infty \frac{da}{a}\, a^{\alpha+\beta}\, e^{-2a}\, \Theta(a-\omega)\cos(\omega t) \\ &= \frac{1}{\pi}\int_0^\infty da\, \frac{\sin(at)}{at}\, a^{\alpha+\beta}\, e^{-2a} \\ &= \frac{2\sin(\gamma \arctan(t/2))}{\gamma t(4+t^2)^{\gamma/2}},\end{aligned}$$

where Θ is the Heaviside function. In Figure 5.2 we have plotted the partial reconstruction filter. As we shall see, the oscillations of r are responsible for the so-called 'Gibbs' phenomenon.

As we have seen, the approximations $s_{\epsilon,\rho}$ tend towards s. In the following section we will have a closer look at the convergence in different function spaces.

6 The pointwise and uniform convergence of the inversion formula

This section may be skipped on first reading. We have already shown that the approximation

$$s_{\epsilon,\rho} = \int_\epsilon^\rho \frac{da}{a}\int_{-\infty}^{+\infty} db\, \mathscr{W}_g s(b,a) h_{b,a} = \int_\epsilon^\rho \frac{da}{a}\, h_a * \tilde{g}_a * s$$

converges in $L^2(\mathbb{R})$ towards the analysed function s in the limit where the strip $\epsilon < a < \rho$ becomes the entire half-plane. In this section we study the convergence in other functional settings. In particular, we give sufficient conditions for uniform convergence in $L^p(\mathbb{R})$ and for pointwise convergence.

Definition 6.0.1
A family of functions $r_\epsilon \in L^1(\mathbb{R}) \cap L^p(\mathbb{R})$, $1 \le p \le \infty$, $\epsilon > 0$, *is called an* $L^p(\mathbb{R})$ *approximation of the identity if*

(i) $\|r_\epsilon\|_{L^1(\mathbb{R})} < c < \infty$ *uniformly in* ϵ,

(ii) $\displaystyle\int_{-\infty}^{+\infty} dt\, r_\epsilon(t) = 1$ *for all* $\epsilon > 0$,

(iii) $\displaystyle\lim_{\epsilon \to 0} \int_{|t| \ge \eta} dt\, |r_\epsilon(t)|^p = 0$ *for all* $\eta > 0$.

Sometimes we replace condition (ii) *by the stronger one*

(ii′) $\displaystyle c_- = \int_{-\infty}^{0} dt\, r_\epsilon(t), \qquad c_+ = \int_{0}^{+\infty} dt\, r_\epsilon(t),$

for all $\epsilon > 0$, *where* $|c_-| + |c_+| < \infty$, $c_- + c_+ = 1$.

Example 6.0.2
For $r \in L^1(\mathbb{R})$ the family $r_\epsilon = r(\cdot/\epsilon)/\epsilon$ is an $L^1(\mathbb{R})$ approximation of the identity whenever $\int_{-\infty}^{+\infty} dt\, r(t) = 1$. Indeed, conditions (i) and (ii) are satisfied because the dilation does not change the overall integral. The last condition follows from

$$\int_{|t|>\eta} dt\, \frac{1}{\epsilon}\left| r\left(\frac{t}{\epsilon}\right)\right| = \int_{|t|>\eta/\epsilon} dt\, |r(t)|,$$

which tends to 0 with ϵ because $|r|$ is integrable. It even satisfies condition (ii′).

If, in addition, we have $s \in L^p(\mathbb{R})$ and

$$|r(t)| \le c|t|^{-1-\alpha}, \qquad \alpha > 0,$$

then $D_\epsilon r$ is an approximation of the identity in $L^p(\mathbb{R})$. Indeed,

$$\int_{\eta}^{\infty} dt\, \epsilon^{-p}\, |r(t/\epsilon)|^p \le c\epsilon^{1-p} \int_{\eta/\epsilon}^{\infty} dt\, t^{-p(1+\alpha)} = O(1)\epsilon^{\alpha p} \to 0$$

as $\epsilon \to 0$.

Because of the last two conditions r_ϵ becomes more and more concentrated around the origin as $\epsilon \to 0$ without changing the overall integral, and therefore we have formally $r_\epsilon \to \delta$. The lemmata below justify these considerations. We refer to Torchinsky (1988) for a proof.

Lemma 6.0.3
Let r_ϵ be an $L^1(\mathbb{R})$ approximation of the identity and let $s \in L^p(\mathbb{R})$ for some $1 \le p < \infty$. Then

$$\|r_\epsilon * s - s\|_{L^p(\mathbb{R})} \to 0$$

as $\epsilon \to 0$. If, in addition, s is uniformly continuous then the same estimate holds for $p = \infty$.

The following lemma shows that we even have pointwise convergence.

Lemma 6.0.4
Let r_ϵ be an $L^p(\mathbb{R})$ approximation of the identity $1 \le p \le \infty$, and let $s \in L^q(\mathbb{R})$ with $p^{-1} + q^{-1} = 1$. If for some $t \in \mathbb{R}$ the numbers $s(t+0)$ and $s(t-0)$ exist,[2] *and if they are finite, then*

$$\lim_{\epsilon \to 0} (r_\epsilon * s)(t) = c_+ s(t-0) + c_- s(t+0).$$

In particular, if s is continuous at t, then

$$\lim_{\epsilon \to 0} (r_\epsilon * s)(t) = s(t).$$

The constants $c_\pm$ are given by condition (ii′) *in Definition* 37.0.1.

6.1 Uniform convergence in $L^p(\mathbb{R})$, $1 < p < \infty$

We now wish to show that the approximation $s_{\epsilon,\rho}$ obtained from a partial reconstruction of s over a strip converge uniformly towards s. We give very weak sufficient conditions for the analysing wavelet g and the synthesizing wavelet h. It will turn out that neither of them needs to be regular. Only the joint function $\tilde{g} * h$ should satisfy certain properties. Therefore, the singular behaviour of either g or h may be compensated by the other. This has already been used in Section 15 where we used distributions as reconstruction wavelets. We will use it again in Chapter 5 where we invert the Radon transform with the help of a wavelet technique.

Suppose, now that $s \in L^p(\mathbb{R})$ with $1 \le p \le \infty$. In order to have a well-defined partial reconstruction $s_{\epsilon,\rho}$ in the sense that we may write

$$s_{\epsilon,\rho} = \int_\epsilon^\rho \frac{da}{a} h_a * (\tilde{g} * s) = \int_\epsilon^\rho \frac{da}{a} (h_a * \tilde{g}) * s = \left\{ \int_\epsilon^\rho \frac{da}{a} (h_a * \tilde{g}_a) \right\} * s,$$

[2] We write $s(t+0) = \lim_{u \to t, u > t} s(u)$, and $s(t-0) = \lim_{u \to t, u < t} s(u)$.

we assume that g and h are in one of the following two cases:

(i) $g, h \in L^1(\mathbb{R}) \cap L^2(\mathbb{R})$,
(ii) $g, h \in S'(\mathbb{R})$, one having compact support and $\tilde{g} * h \in L^1(\mathbb{R}) \cap C(\mathbb{R})$.

In both cases we have that $F = \tilde{g} * h$ is a continuous function, and the partial approximation reads

$$s_{\epsilon,\rho} = \int_\epsilon^\rho \frac{da}{a} F_a * s, \qquad F_a(\cdot) = F(\cdot/a)/a. \tag{6.1.1}$$

The following theorem then settles the question of uniform convergence in $L^p(\mathbb{R})$.

Theorem 6.1.1
Suppose that $F \in C(\mathbb{R})$ *satisfies at*

(i) $(2 + |\log|t||)\cdot F \in L^1(\mathbb{R})$,

(ii) $\int_0^\infty dt\, F(t) = \int_0^\infty dt\, F(-t) = 0$,

(iii) $c = \int_{-\infty}^{+\infty} dt \log(1/|t|)F(t) = \int_{-\infty}^{+\infty} \frac{dt}{|t|} \int_0^t du\, F(u \operatorname{sign} t) \neq 0$,

then for $s \in L^p(\mathbb{R})$, $1 < p < \infty$, *we have for* $s_{\epsilon,\rho}$ *given by* (6.1.1)

$$\lim_{\epsilon \to 0, \rho \to \infty} \|s - c^{-1} s_{\epsilon,\rho}\|_{L^p(\mathbb{R})} = 0.$$

In the case where F *and* s *are progressive, condition* (ii) *has to be replaced by*

(ii′) $\int_{-\infty}^{+\infty} dt\, F(t) = 0$.

Before we prove the theorem some comments are in order. The first condition on F is quite weak. Indeed, it is satisfied whenever we have a little localization, such as

$$|F(t)| \leq O(1)(1 + |t|^2)^{-\alpha/2}, \qquad \alpha > 1.$$

Conditions (i) and (ii) ensure that the partial reconstruction filter $r(t) = t^{-1} \int_0^t da\, F(a)$ is in $L^1(\mathbb{R})$, as we now show in a slightly more general way than is actually needed.

Lemma 6.1.2
If $v \in L^1(\mathbb{R}_+)$ *and* $\log t \cdot v \in L^1(\mathbb{R}_+)$, *and if* $\int_0^\infty dt\, v(t) = 0$, *then the function*

$$w(t) = \frac{1}{t} \int_0^t du\, v(u) = -\frac{1}{t} \int_t^\infty du\, v(u)$$

is in $L^1(\mathbb{R}_+)$ *and*

$$\int_0^\infty dt\, w(t) = \int_0^\infty dt \log(1/t) v(t).$$

Proof. We have to estimate

$$\int_0^\infty dt \left| \frac{1}{t} \int_0^t du\, v(u) \right| = \left\{ \int_0^1 + \int_1^\infty \right\} dt \left| \frac{1}{t} \int_0^t du\, v(u) \right| = X_1 + X_2.$$

It follows, exchanging the integrals, that

$$|X_1| \leq \int_0^1 dt \frac{1}{t} \int_0^t du\, |v(u)| = \int_0^1 du\, |v(u)| \int_0^1 dt \frac{1}{t}\, \Theta(t-u)$$
$$= -\int_0^1 du \log u |v(u)| < \infty,$$

showing that w is integrable at the origin. Using the fact that $\int_0^\infty dt\, v(t) = 0$ we may write, for $t > 0$,

$$w(t) = -\frac{1}{t} \int_t^\infty du\, v(u),$$

and thus, again exchanging the integrals,

$$|X_2| \leq \int_1^\infty dt\, |w(t)| \leq \int_1^\infty dt \frac{1}{t} \int_t^\infty du\, |v(u)|$$
$$= \int_1^\infty du\, |w(u)| \int_1^\infty dt \frac{1}{t}\, \Theta(u-t) = \int_1^\infty du \log u\, |v(u)| < \infty.$$

The value of the integral can be computed in the same way. □

The constant c as given in the theorem coincides with the previous definition of this constant, as is shown by the following two lemmas.

Lemma 6.1.3
For $v \in L^1(\mathbb{R})$ *set*

$$w(t) = \frac{1}{t} \int_0^t du\, v(u), \qquad t \in \mathbb{R}\backslash\{0\}.$$

If $w \in L^1(\mathbb{R})$, *too, then its Fourier transform is given by*

$$\hat{w}(\omega) = \int_{|\omega|}^\infty \frac{da}{a}\, \hat{v}(a \operatorname{sign} \omega)$$

if the integral is understood as $\lim_{\rho \to \infty} \int_{|\omega|}^{\rho}$.

In particular, for such v we have, upon setting $\omega = 0$,

$$\int_{-\infty}^{+\infty} dt \frac{1}{t} \int_0^t du\, v(u) = \lim_{\epsilon \to 0, \rho \to \infty} \int_\epsilon^\rho \frac{d\xi}{\xi} \hat{v}(\pm \xi).$$

Proof. By direct computation we can verify that ($\rho > 0$)

$$w(t) - \frac{1}{\rho} w\left(\frac{t}{\rho}\right) = \int_1^\rho \frac{da}{a} \frac{1}{a} v\left(\frac{t}{a}\right).$$

Therefore,

$$\hat{w}(\omega) - \hat{w}(\rho\omega) = \int_1^\rho \frac{da}{a} \hat{v}(a\omega) = \int_{|\omega|}^{\rho|\omega|} \frac{da}{a} \hat{v}(a \operatorname{sign} \omega).$$

For every $\omega \neq 0$ we have $\hat{w}(\rho\omega) \to 0$ as $\rho \to \infty$ by the Riemann–Lebesgue lemma, and on the right-hand side we have the stated limit. □

Proof (of Theorem 6.1.1). Upon, if necessary, replacing F by $F(t) + F(-t)$ in the progressive or regressive case we may assume that condition (ii) holds. As we saw in Section 35 we have

$$s_{\epsilon,\rho} = r_\epsilon * s - r_\rho * s, \qquad r_\alpha(\cdot) = r(\cdot/\alpha)/\alpha, \tag{6.1.2}$$

with

$$r(t) = \frac{1}{t} \int_0^t du\, F(u).$$

As we saw in Lemma 6.1.2 the approximation filter r is in $L^1(\mathbb{R})$, and its integral equals c.

The approximation of the identity lemma then shows that the first term in (6.1.2) tends to cs in $L^p(\mathbb{R})$.

For the second term we proceed as follows. Note that by the same hypothesis the approximation filter r is in $L^q(\mathbb{R})$ for all $1 \leq q \leq \infty$. Indeed, since $F \in L^1(\mathbb{R})$ it follows that at infinity we have

$$r(t) = \frac{1}{t} \int_0^t du\, F(u) \Rightarrow |r(t)| \leq \frac{\|F\|_1}{|t|}.$$

On the other hand we have that $|r(t)|$ stays bounded near 0 since F is continuous at 0. For every $\gamma > 0$ we can now split s into two terms, $s = v + w$, with $v \in L^1(\mathbb{R}) \cap L^p(\mathbb{R})$ and $\|w\|_p \leq \gamma$. By Young's inequality we obtain

$$\begin{aligned} \|r_\rho * s\|_p &\leq \|r_\rho\|_p \|v\|_1 + \|r_\rho\|_1 \|w\|_p \\ &\leq \rho^{1/p-1} \|r\|_p \|v\|_1 + \gamma \|r\|_1. \end{aligned}$$

And thus

$$\limsup_{\rho \to \infty} \|r_\rho * s\|_p \leq \gamma \|r\|_1,$$

which can be made arbitrarily small because γ was chosen arbitrarily. □

Remark. It is clear that the reconstruction cannot converge in $L^1(\mathbb{R})$ for all analysed functions $s \in L^1(\mathbb{R})$ since every partial reconstruction $s_{\epsilon,\rho}$ is of zero mean and the same must therefore hold for the limit.

6.2 Pointwise convergence in $L^p(\mathbb{R})$, $1 \leq p < \infty$

Here we have the following theorem.

Theorem 6.2.1
Let $s \in L^p(\mathbb{R})$ with $1 \leq p < \infty$ and let F be continuous and satisfy (ii) *respectively,* (ii′) *and* (iii) *of Theorem* 6.1.1. *Suppose that, instead of* (i) *we have*

(i′) $|F(t)| \leq O(1)|t|^{-\alpha-1}, \qquad \alpha > 0.$

Then for every $t \in \mathbb{R}$ where $s(t-0)$ and $s(t+0)$ exist and are finite we have

$$c^{-1}s_{\epsilon,\rho}(t) \to c_+ s(t-0) + c_- s(t+0), \qquad (\epsilon \to 0, \rho \to \infty),$$

with

$$c_+ = c^{-1}\int_0^\infty du \log(1/u)F(u), \qquad c_- = 1 - c_+$$

and c as given by Theorem 6.1.1. *In the case where all functions are progressive or regressive we have*

$$c_+ = c_- = \tfrac{1}{2}.$$

In particular, $s_{\epsilon,\rho}(t) \to s(t)$ at every point where s is continuous.

Proof. First note that $r(t) = (1/t)\int_0^t F \in L^q(\mathbb{R}) \cap L^1(\mathbb{R})$. Indeed, r is bounded near 0 since F is continuous and at infinity we have, by hypothesis,

$$|r(t)| \leq c|t|^{-\alpha-1}$$

since we have

$$|r(t)| \leq O(1)\frac{1}{|t|}\int_{|t|}^\infty du\, u^{-\alpha-1} = O(t^{-\alpha-1}) \qquad (|t| \to \infty).$$

By Example 6.0.2 the family $r_\epsilon = r(\cdot/\epsilon)/\epsilon$ is an $L^q(\mathbb{R})$ approximation of the identity. The first term in (6.1.2) tends to $s(t)$, as stated in the theorem because of the pointwise approximation of Lemma 6.0.4. For the second term we may use Hölder inequality to write

$$|r_\rho * s(t)| = |\langle \tilde{r}_\rho(\cdot - t) \mid s \rangle| \leq \|r_\rho\|_q \|s\|_p = \rho^{-1/p} \|r\|_q \|s\|_p,$$

which shows that $r_\rho * s(t) \to 0$ pointwise as $\rho \to \infty$.

The constants c_+ and c_- are computed with the help of Lemma 6.1.2 and the first part of the theorem is proved. In particular, in the progressive (regressive) case we may assume that F is even and thus $c_\pm = 1/2$. □

6.3 Pointwise convergence in $L^\infty(\mathbb{R})$

In the previous case we had to exclude the case where $s \in L^\infty(\mathbb{R})$. Actually, neither holds the convergence in the norm for the first term in equation (6.1.2) nor does the second term tend pointwise to 0 for $\rho \to \infty$. This will only be the case for a class of functions that are weakly oscillating around 0.

Definition 6.3.1
A locally integrable function s is weakly oscillating around 0 *iff*

$$\lim_{t \to \infty} \frac{1}{2t} \int_{y-t}^{y+t} du\, s(u) = 0$$

uniformly in y. The function s is weakly oscillating around c iff s − c is weakly oscillating around 0.

In particular, s is weakly oscillating around 0 iff

$$\sup_{t \in \mathbb{R}} |\chi_I * s(t)| = o(|I|), \qquad |I| \to \infty, \tag{6.3.1}$$

where $|I|$ denotes the length of the finite interval I, and χ_I denotes its characteristic function

$$\chi_I(t) = \begin{cases} 1 & \text{for } t \in I \\ 0 & \text{otherwise.} \end{cases}$$

Example 6.3.2
Every periodic function $s(t+1) = s(t)$ is weakly oscillating around $\int_0^1 dt\, s(t)$ if its restriction to one period is in $L^1([0,1])$.

For these functions we have the following lemma.

Lemma 6.3.3

If $s \in L^\infty(\mathbb{R})$ is weakly oscillating around c, then for $r \in L^1(\mathbb{R})$ with $\int_{-\infty}^{+\infty} dt\, r(t) = 1$ we have

$$\lim_{\rho \to \infty} |r_\rho * s(t) - c| = 0, \qquad r_\rho(\cdot) = r(\cdot/\rho)/\rho,$$

uniformly in $t \in \mathbb{R}$.

Proof. Because $\int_{-\infty}^{+\infty} dt\, r = 1$ we may suppose that $c = 0$. (If not, consider $s - c$ instead of s.) Given that $\epsilon > 0$ we find a finite number of disjoint intervals I_n and numbers c_n such that for $v = \sum c_n \chi_{I_n}$ we have

$$\|r - v\|_1 \leq \epsilon.$$

Therefore,

$$|r_\rho * s(t)| \leq |v_\rho * s(t)| + \epsilon \|s\|_\infty, \tag{6.3.2}$$

For every interval I_n we may use (6.3.1). The same is true for the finite superposition of such characteristic functions, v. Since ϵ was arbitrary, this shows that (6.3.2) can be made arbitrarily small uniformly in t, which proves the lemma. □

An application of this lemma to wavelet analysis yields the following theorem.

Theorem 6.3.4

Let F satisfy conditions (i), (ii) *respectively* (ii′) *and* (iii) *of Theorem* 6.1.1. *If s is weakly oscillating around d then*

$$c^{-1} s_{\epsilon,\rho}(t) \to c_+ s(t+0) + c_- s(t-0) - d$$

at every point where $s(t \pm 0)$ exist and are finite. The constants c and $c_\pm$ are again given as in Theorem 6.2.1.

Proof. By Lemma 6.1.2. the approximation filter (5.0.2) is in $L^1(\mathbb{R})$ and the theorem follows from Theorem 6.0.4 and Lemma 6.3.3. □

Corollary 6.3.5

If s is, in addition, uniformly continuous, we then have

$$\|s_{\epsilon,\rho} - s - d\|_\infty \to 0 \qquad (\epsilon \to 0, \rho \to \infty).$$

7 The 'Gibbs' phenomenon for $s_{\epsilon,\rho}$

In this section we show that at points where s has a finite discontinuity, the approximations

$$s_{\epsilon,\rho} = \int_{\epsilon}^{\rho} \frac{da}{a} h_a * \tilde{g}_a * s$$

may show oscillations with an amplitude that is essentially independent of ϵ and ρ. Therefore, although at all points outside the discontinuity we have pointwise convergence, the maximal difference between the approximation and the original function may remain unchanged. This phenomenon is well known in Fourier analysis where the truncated trigonometric sums show the same type of behaviour at any point of discontinuity of s. For the wavelet analysis it may happen that for a convenient choice of g and h this phenomenon disappears. For the sake of notational simplicity we assume that $c_{g,h} = 1$.

As we have seen, the approximation $s_{\epsilon,\rho}$ is given by the difference of two smoothing convolutions at 'scale' ϵ and ρ, namely,

$$s_{\epsilon,\rho} = r_\epsilon * s - r_\rho * s, \qquad r(t) = \frac{1}{t}\int_0^t \frac{d\tau}{\tau} h * \tilde{g}(\tau).$$

Throughout this section we suppose that $r \in L^1(\mathbb{R}) \cap L^\infty(\mathbb{R})$. Sufficient conditions on g and h have been given in the last section. If we now suppose in addition that $s \in L^1(\mathbb{R})$, in which case

$$r_\rho * s(t) \to 0 \qquad (\rho \to 0), \quad \text{uniformly for } t \in \mathbb{R}.$$

We may therefore limit ourselves to the analysis of the one-parameter family of approximations $s_{\epsilon,\infty} = r_\epsilon * s$ neglecting only some function $o(1)$ uniformly in $t \in \mathbb{R}$ and $\rho \to \infty$. We now specify s to be a function that represents the prototype of a function having some discontinuity at 0, namely,

$$s(t) = \chi_{[0,1]}(t) = \begin{cases} 1 & \text{for } 0 \le t \le 1, \\ 0 & \text{otherwise.} \end{cases}$$

Any function f having a discontinuity at some point t_0 can now be written as

$$f(t) = \alpha s(t - t_0) + \sigma(t), \qquad \text{with some } \alpha \in \mathbb{C},$$

where σ is now continuous in some neighbourhood of t_0. If $f \in L^1(\mathbb{R})$, then $\sigma \in L^1(\mathbb{R})$, too. By the lemma of the approximation of the identity the partial reconstruction of σ converges locally uniformly towards σ and we are left—by translation invariance—with the study of the approximations $s_{\epsilon,\infty} = r_\epsilon * s$ in a neighbourhood of 0. We may now write

$$s(t) = \tfrac{1}{2}(\text{sign}(t) + \text{sign}(1 - t)), \qquad \text{sign}(t) = \frac{t}{|t|},$$

and therefore with

$$2G(t) = r * \text{sign}(t) = 2\int_{-\infty}^{t} du\, r(u) - 1,$$

we have

$$s_{\epsilon,\infty}(t) = G\left(\frac{t}{\epsilon}\right) + G\left(\frac{1-t}{\epsilon}\right).$$

For $t \in [-1/2, +1/2]$ the second term tends uniformly towards 1/2 (recall that $c_{g,h} = \int r = 1$) and we are left with

$$s_{\epsilon,\infty}(t) = G(t/\epsilon) + 1/2 + o(1). \tag{7.0.1}$$

Coming back to the original function $s \in L^1(\mathbb{R})$ having an isolated discontinuity at t_0, we have proved the following theorem.

Theorem 7.0.1
Let $s \in L^1(\mathbb{R})$ be piecewise continuous to the left and to the right of t_0 with

$$s(t_0+0) - s(t_0-0) = 2\alpha \neq 0, \qquad s(t_0+0) + s(t_0-0) = 2\beta.$$

Then there is an open neighbourhood $I \ni t_0$ such that

$$\left.\begin{aligned} c_{g,h}^{-1} s_{\epsilon,\rho}(t) &= \beta + \alpha G([t-t_0]/\epsilon) + o(1), \\ (\epsilon \to 0,\ \rho &\to \infty)\ \textit{uniformly in}\ t \in I. \end{aligned}\right\} \tag{7.0.2}$$

Depending on the behaviour of G the approximations behave differently and we distinguish two cases.

7.1 Gibbs phenomenon

Suppose that there is at least one point $t > 0$ ($t < 0$) with

$$2G(t) > 1, \quad t > 0, \quad \text{or} \quad 2G(t) < -1, \quad t < 0,$$

respectively. Suppose r is continuous and has only a finite number of points where $r(t_n) = 0$. Then if t_1 and t_N are the smallest and the largest zero, respectively, of r the first of these conditions is equivalent to requiring that $r(t) < 0$ for $t > t_N$, respectively $t < t_1$. In this case the approximations $s_{\epsilon,\infty}$ show a behaviour known as the *Gibbs phenomenon*: (suppose $\alpha > 0$, $t_0 = 0$) although for every $t \neq 0$ the approximation $s_{\epsilon,\infty}(t)$ tends to $s(t)$, there is a constant $c > 0$ such that for every $\epsilon > 0$ there is at least one point $u > 0$ ($u < 0$) where $s_{\epsilon,\infty}(u) - s(u) > c$ ($<c$). To put it differently,

$$\lim_{\eta\to 0}\lim_{\epsilon\to 0}\sup_{0<t<\eta} s_{\epsilon,\infty}(t) > \lim_{\eta\to 0}\sup_{0<t<\eta}\lim_{\epsilon\to 0} s_{\epsilon,\infty}(t).$$

However, this 'over-shooting' is localized around the singularity, and from (7.0.2) it follows that the size of the domain, where the Gibbs phenomenon is important, is of the order of magnitude ϵ (see Figure 7.1).

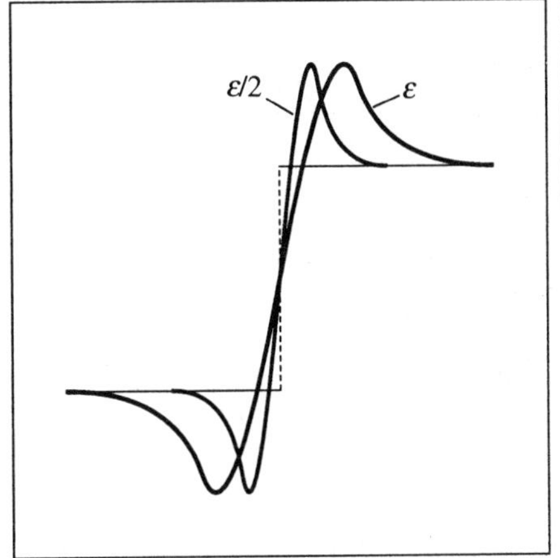

Fig. 7.1 The Gibbs phenomenon for the partial reconstruction $s_{\epsilon,\infty}$.

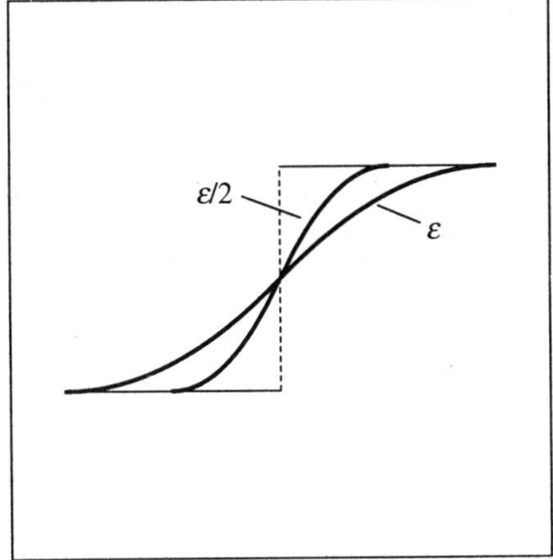

Fig. 7.2 The partial reconstruction $r_{\epsilon,\infty}$ without Gibbs phenomenon.

7.2 No Gibbs phenomenon

Suppose that

$$-1 \leq 2G(t) \leq 1.$$

In this case, as we see from (7.0.1), the principal term of the approximation $f_{\epsilon,\infty}$ is locally between $s(t_0 - 0)$ and $s(t_0 + 0)$. No Gibbs phenomenon as described previously can be observed. A sufficient but not necessary condition for this case is that $r(t) > 0$ (see Figure 7.2).

8 Reconstruction over cones

Much of what we have seen for the reconstruction over strips applies also to the situation where we reconstruct over a cone. Essentially, the Fourier transform has to be replaced by the Mellin transform. We will list only the main formulas and leave the details to the reader.

Throughout this section we suppose that the analysed function is causal, that is, $s(t) = 0$ for $t < 0$. The anti-causal case follows by symmetry.

Now suppose that the wavelet transform is only known in a cone-like region of the half-plane. These regions are dilation invariant. Therefore, it is no surprise that the partial reconstruction is related to the original function through a scale convolution which is dilation invariant. Recall that the scale convolution of two causal functions was defined as

$$s *_l r(t) = \int_0^\infty \frac{du}{u} s\left(\frac{t}{u}\right) r(u).$$

The wavelet synthesis may now be written as

$$\int_{-\infty}^{+\infty} d\beta \, h_\beta *_l \mathscr{R}_b(t), \qquad t > 0,$$

where $\mathscr{R}_\beta(\alpha) = \mathscr{W}_g s(\beta\alpha, \alpha)$. Suppose that only one zoom $\mathscr{R}_\beta(\cdot)$ is known. If we apply the inversion formula, the best we can do, we are led to consider the following function:

$$s^\#(t) = r_\beta *_l s, \qquad r(t) = \int_0^\infty da \, h(at - \beta)\bar{g}(a - \beta).$$

After a Mellin transform we therefore obtain

$$M[s^\#](z) = M[h(\cdot - \beta)](z) M[\bar{g}(\cdot - \beta)](1 - z) Ms(z).$$

Suppose now that we reconstruct over a cone-like region in the half-plane. That is, we have $\epsilon \leq \beta = b/a \leq \rho$. Again, we use the inversion formula over this cone region to obtain the following approximation:

$$s^\#(t) = \int_0^\infty \frac{da}{a} \int_{\epsilon a}^{\rho a} db \, \mathscr{W}_g s(b, a) h_{b,a}.$$

By mere superposition it follows that

$$s^\# = r_{\epsilon,\rho} *_l s, \quad \text{with}$$

$$M[r_{\epsilon,\rho}](z) = \int_\epsilon^\rho d\beta \, M[h(t - \beta)](z) M[\bar{g}(t - \beta)](1 - z).$$

9 The Poisson summation formula

In this section we present some material that will be used in the following sections. At the heart of all sampling theorems that we will use lies the Poisson summation formula, which states that sampling the time representation is related to periodization of the Fourier representation, and vice versa. All the results that we present here could be stated in general for Abelian groups. This is postponed to Chapter 5, Section 9. Here we prefer to state the theorems in a more explicit way.

9.1 Periodic functions

A function s is *periodic* with period 2π [3] iff

$$s(\omega + 2\pi) = s(\omega).$$

Example 9.1.1
The functions

$$e_n(\omega) = e^{i\omega n}, \qquad n = \cdots - 1, 0, +1 \ldots,$$

are periodic functions with period 2π.

To clarify the geometry we identify 2π-periodic functions over $\mathbb{R}$ with functions over the unit circle $\mathbb{T}$ with radius 1, and hence cirumference 2π. We use the notation $\mathbb{T}$ for the circle because the generalization of these results to more dimensions holds for the n-dimensional torus $\mathbb{T}^n$. This identification of functions is done in the obvious way, and we will not write the identification map explicitly. It should be clear from the context whether we are on the circle or on the real line. We identify integrals over the circle with integrals over one period:

$$\int_{\mathbb{T}} d\omega \equiv \int_0^{2\pi} d\omega.$$

The scalar product in the Hilbert space $L^2(\mathbb{T})$ of square summable functions over $\mathbb{T}$ of two functions r, s over the circle is given by

$$\langle r \mid s \rangle_{\mathbb{T}} = \int_{\mathbb{T}} d\omega\, \bar{r}(\omega) s(\omega) = \int_0^{2\pi} d\omega\, \bar{r}(\omega) s(\omega)$$

and the corresponding norm reads

$$\|s\|^2_{L^2(\mathbb{T})} = \langle s \mid s \rangle_{\mathbb{T}} = \int_0^{2\pi} d\omega\, |s(\omega)|^2,$$

where we have silently assumed that we identify functions that differ only on a set of measure 0. The analogous definition holds for the L^p spaces over $\mathbb{T}$.

Example 9.1.2
The scalar product of the 2π-periodic functions $e_n = e^{i\omega n}$, $n \in \mathbb{Z}$ is given by

$$\langle e_n \mid e_m \rangle_{\mathbb{T}} = 2\pi\, \delta_{n,m},$$

[3] For the sake of simplicity, all periodic functions over the real line that we consider in this section are 2π-periodic. All results can be extended trivially to arbitrary periods by using the appropriate commutation relations of the operators that we will introduce with the dilation operator D_a.

where $\delta_{n,m}$ is the *Kronecker symbol*

$$\delta_{n,m} = \begin{cases} 1 & \text{if } n = m, \\ 0 & \text{if } n \neq m. \end{cases}$$

9.2 The periodizing operator

There is a natural map of arbitrary—non-periodic—functions over the real line to functions over the circle, that is, to periodic functions over $\mathbb{R}$. This *periodization operator* is defined by

$$\Pi\colon L^1(\mathbb{R}) \to L^1(\mathbb{T}), \qquad (\Pi s)(\omega) = \sum_{n \in \mathbb{Z}} s(\omega + 2\pi n),$$

for almost all $\omega \in \mathbb{R}$.

The periodization operators formally allows us to write the scalar product over $\mathbb{R}$ of a function r with a periodic function s as a scalar product over the circle:

$$s(t + 2\pi) = s(t) \Rightarrow \langle \mathrm{r} \mid \mathrm{s} \rangle_{\mathbb{R}} = \langle \Pi r \mid s \rangle_{\mathbb{T}}. \tag{9.2.1}$$

Indeed, writing $\int_{-\infty}^{+\infty} = \sum_{n=-\infty}^{+\infty} \int_{2\pi n}^{2\pi(n+1)}$ and using the periodicity of s we obtain

$$\begin{aligned} \langle r \mid s \rangle_{\mathbb{R}} &= \int_{-\infty}^{+\infty} d\omega\, \bar{r}(\omega) s(\omega) \\ &= \sum_{n=-\infty}^{+\infty} \int_{2\pi n}^{2\pi(n+1)} d\omega\, \bar{r}(\omega) s(\omega) \\ &= \sum_{n=-\infty}^{+\infty} \int_0^{2\pi} d\omega\, \bar{r}(\omega + 2\pi n) s(\omega) \\ &= \int_0^{2\pi} d\omega \sum_{n=-\infty}^{+\infty} \bar{r}(\omega + 2\pi n) s(\omega) = \langle \Pi r \mid s \rangle_{\mathbb{T}}. \end{aligned}$$

Let us introduce the equally spaced delta-comb

$$Ш_\lambda(\omega) = \sum_{n \in \mathbb{Z}} \delta(\omega - n\lambda).$$

It is a sequence of delta functions situated at all points $\omega = n\lambda$. Then periodizing can (formally) be written as convolution with a delta-comb:

$$\Pi_{2\pi}\colon s \mapsto Ш_{2\pi} * s. \tag{9.2.2}$$

For the case where $s \in S(\mathbb{R})$ this holds in the sense of distributions.

9.3 Sequences and sampling

A function over the integral numbers $\mathbb{Z}$ is called a *sequence*. The numbers that make up a sequence l are denoted by $l(n)$, or by l_n, for $n \in \mathbb{Z}$. The scalar product of two sequences is given by

$$\langle l \mid v \rangle_{\mathbb{Z}} = \sum_{n \in \mathbb{Z}} \bar{l}(n) v(n).$$

It makes $L^2(\mathbb{Z})$, the space of square summable sequences with norm

$$\|l\|^2_{L^2(\mathbb{Z})} = \langle l \mid l \rangle_{\mathbb{Z}} = \sum_{n \in \mathbb{Z}} |l(n)|^2 < \infty,$$

a Hilbert space. The $L^p(\mathbb{Z})$ spaces are defined in an analogous way.

There is a natural map from continuous functions over the real line $\mathbb{R}$ to functions over the integral numbers $\mathbb{Z}$. This map is the perfect sampling operator with sampling period 1:

$$\Sigma\colon C(\mathbb{R}) \cap L^\infty(\mathbb{R}) \to L^\infty(\mathbb{Z}), \qquad s \mapsto \Sigma s, \quad (\Sigma s)(n) = s(n).$$

It maps a function over the real line to the sequence of the values of s at the points n, $n \in \mathbb{Z}$. This is well defined if s is continuous at the sample points.

Example 9.3.1

The sample sequence of the pure frequencies is

$$\Sigma e_\omega(n) = e^{i\omega n}.$$

Note, already, that we cannot distinguish e_ω from $e_{\omega + 2\pi}$ through its sampling sequence. This phenomenon is known as *aliasing* and is at the heart of the Shanon sampling theorem that follows later (Theorem 9.6.1).

Sampling a function s can also be seen as multiplication by the delta-comb $Ш_1$:

$$\Sigma\colon s \mapsto Ш_1\, s, \tag{9.3.1}$$

if we identify any sequence l with the modulated delta-comb

$$l \leftrightarrow \sum_{n \in \mathbb{Z}} l(n)\, \delta(t - n).$$

This is the reason for calling this operator the 'perfect' sampling operator. If we replace the delta distribution by some function $\varphi \in L^2(\mathbb{R})$ we obain the *imperfect sampling operators* with sampling function φ:

$$\Sigma_\varphi\colon L^2(\mathbb{R}) \to L^\infty(\mathbb{Z}), \qquad \Sigma_\varphi s(n) = \langle \varphi(\cdot - n) \mid s \rangle_{\mathbb{R}}.$$

That the imperfect sampling sequence is necessarily bounded follows from the Schwarz inequality. For later reference note that imperfect sampling can be written as smoothing with $\tilde{\phi} = \bar{\phi}(-\cdot)$ and subsequent perfect sampling:

$$\Sigma_\phi s = \Sigma(\tilde{\phi} * s).$$

9.4 The Fourier transform over the circle

The Fourier transform over the circle maps a function over the circle $\mathbb{T}$ to a function over $\mathbb{Z}$, that is, to a sequence. Let s be a function over the circle, that is, a periodic function with period 2π. Its Fourier transform is the sequence of Fourier coefficients with respect to the elementary functions e_n:

$$F^{\mathbb{T}}: L^2(\mathbb{T}) \to L^2(\mathbb{Z}), \qquad (F^{\mathbb{T}}s)(n) = \langle e_n \mid s \rangle_{\mathbb{T}} = \int_{\mathbb{T}} d\omega\, e^{-in\omega}\, s(\omega).$$

We usually write $\hat{s}$ instead of $F^{\mathbb{T}}s$ if no confusion is possible. The inverse Fourier transform maps a sequence l to a function over the circle:

$$F^{\mathbb{T}-1}: L^2(\mathbb{Z}) \to L^2(\mathbb{T}), \qquad F^{\mathbb{T}-1}l(\omega) = \frac{1}{2\pi} \sum_{n \in \mathbb{Z}} l(n)\, e^{in\omega}.$$

The inverse Fourier transform allows us to write s a a superposition of elementary functions e_n if some regularity conditions are imposed (e.g. $s \in L^1(\mathbb{T})$ and $F^{\mathbb{T}}s \in L^1(\mathbb{Z})$ will be sufficient that the following equation holds pointwise):

$$s = \frac{1}{2\pi} \sum_{n \in \mathbb{Z}} (F^{\mathbb{T}}s)(n)\, e_n, \qquad s(\omega) = \frac{1}{2\pi} \sum_{n \in \mathbb{Z}} \hat{s}(n)\, e^{in\omega}.$$

The scalar product of two functions can be computed in frequency space ($s, r \in L^2(\mathbb{T})$):

$$\langle s \mid r \rangle_{\mathbb{T}} = \int_{\mathbb{T}} d\omega\, \bar{s}(\omega) r(\omega) = \frac{1}{2\pi} \sum_{n \in \mathbb{Z}} \bar{\hat{s}}(n) \hat{r}(n) = \frac{1}{2\pi} \langle \hat{s} \mid \hat{r} \rangle_{\mathbb{Z}}. \qquad (9.4.1)$$

In particular, the Fourier transform over the circle preserves the energy:

$$\|s\|^2_{L^2(\mathbb{T})} = \frac{1}{2\pi} \|F^{\mathbb{T}}s\|^2_{L^2(\mathbb{Z})}.$$

For reasons of symmetry we introduce yet another operator, closely related to $F^{\mathbb{T}-1}$ by

$$F^{\mathbb{Z}}: L^2(\mathbb{Z}) \to L^2(\mathbb{T}), \qquad F^{\mathbb{Z}}l(\omega) = \sum_{n \in \mathbb{Z}} l(n)\, e^{-in\omega},$$

$$F^{\mathbb{Z}-1}: L^2(\mathbb{T}) \to L^2(\mathbb{Z}), \qquad s(n) = \frac{1}{2\pi} \int_0^{2\pi} d\omega\, s(\omega)\, e^{in\omega}.$$

Note that the sum that defines $F^{\mathbb{Z}}$ can be seen as a Riemannian sum that approximates the Fourier integral over $\mathbb{R}$ if we replace a continuous function by its sample values. It will therefore be referred to as the Fourier transform over $\mathbb{Z}$.

We end this subsection by recalling the well-known convolution theorems for periodic functions and for sequences. It states that the convolutions of a function s over $\mathbb{T}$ and a function over $\mathbb{Z}$,

$$s * r(\omega) = \int_{\mathbb{T}} d\omega' s(\omega - \omega') r(\omega'), \qquad l * v(n) = \sum_{k \in \mathbb{Z}} l(n-k) v(k),$$

reduce in Fourier space to pointwise multiplications,

$$F^{\mathbb{T}}(s * r)(n) = F^{\mathbb{T}} s(n) F^{\mathbb{T}} r(n), \qquad F^{\mathbb{Z}}(l * v)(\omega) = F^{\mathbb{Z}} l(\omega) F^{\mathbb{Z}} v(\omega).$$

9.5 The Poisson summation formula

The Poisson summation formula relates the Fourier transform of a function s over the real line to the sequence of Fourier coefficients of its periodization Πs. The relation is obtained through the sampling operator. More precisely, we have the following theorem.

Theorem 9.5.1
On $L^1(\mathbb{R})$ we have

$$\Sigma F = F^{\mathbb{T}} \Pi,$$

or, more explicitly, for $s \in L^1(\mathbb{R})$,

$$(F^{\mathbb{T}} \Pi s)(n) = \hat{s}(n),$$

and for $s \in L^1(\mathbb{R})$, $\hat{s} \in L^1(\mathbb{R})$,

$$\sum_{n \in \mathbb{Z}} s(\omega + 2\pi n) = \frac{1}{2\pi} \sum_{n \in \mathbb{Z}} \hat{s}(n)\, e^{in\omega}.$$

The last equation holds pointwise.

Formally, we have the following commutative diagrams which better illustrate the geometric relations between the function spaces involved. The second is trivially obtained from the first:

$$\begin{array}{ccc} (\mathbb{R}) & \xrightarrow{\Pi} & (\mathbb{T}) \\ \downarrow{\scriptstyle F} & & \downarrow{\scriptstyle F^{\mathbb{T}}} \\ (\mathbb{R}) & \xrightarrow{\Sigma} & (\mathbb{Z}) \end{array} \qquad \begin{array}{ccc} (\mathbb{R}) & \xrightarrow{\Sigma} & (\mathbb{Z}) \\ \downarrow{\scriptstyle F} & & \downarrow{\scriptstyle F^{\mathbb{Z}}} \\ (\mathbb{R}) & \xrightarrow{\Pi} & (\mathbb{T}) \end{array} \tag{9.5.1}$$

where $(\mathbb{R})$, $(\mathbb{T})$, $(\mathbb{Z})$ denote some suitable space of functions over $\mathbb{R}$, $\mathbb{T}$, $\mathbb{Z}$ respectively.

Proof. Since e_n is a periodic function with period 2π we may use Formula 9.2.1 to write

$$\hat{s}(n) = \langle e_n \mid s \rangle_{\mathbb{R}} = \langle e_n \mid \Pi s \rangle_{\mathbb{T}} = (F^{\mathbb{T}} \Pi s)(n),$$

which proves the theorem. □

Therefore, periodizing in the time representation (frequency representation) is equivalent to sampling in frequency space (time representation). Since periodizing a function amounts to convolving it with a delta-comb—equation (9.2.2)—and, since sampling a function is multiplication by a delta-comb—equation (9.3.1)—the Poisson summation formula may be restated as

$$\widehat{Ш_{2\pi}} = Ш_1,$$

where this equation holds in the sense of $S'(\mathbb{R})$.

9.6 Some sampling theorems

In general, neither the perfect nor the imperfect sampling operator has an inverse. However, there are spaces on which the sampling operator may be inverted. The best known is the space of band- or strip-limited functions for which the Shanon sampling theorem holds. We recall that a function is band or strip limited iff the support of its Fourier transform is contained in a finite interval, say $[\xi_l, \xi_r]$. Shanon's sampling theorem may now be stated as follows, which we prove as a first application of the Poisson summation formula.

Theorem 9.6.1

If $s \in L^2(\mathbb{R})$ is such that supp $\hat{s} \subset [-\pi, +\pi]$, *it follows that s is defined by its values at the points $t = n$, $n \in \mathbb{Z}$. An explicit reconstruction is given by*

$$s(t) = \sum_{n=-\infty}^{+\infty} s(n)\,\mathrm{sinc}(t-n), \qquad \mathrm{sinc}(t) = \frac{\sin(t)}{t}.$$

The convergence of the sum is in $L^2(\mathbb{R})$.

Before we come to the proof recall that the imperfect sampling operator with respect to ϕ reads

$$\Sigma_\phi s(n) = \langle \phi(\cdot - n) \mid s \rangle = \Sigma(\tilde{\phi} * s)(n).$$

Consequently, it reads in Fourier space

$$\Sigma_\phi \colon \hat{s} \mapsto \Pi(\bar{\hat{\phi}} \cdot \hat{s}) = \sum_{k \in \mathbb{Z}} \bar{\hat{\phi}}(\cdot + 2\pi k)\hat{s}(\cdot + 2\pi k).$$

The adjoint of the imperfect sampling operator is a reconstruction operator

$$\Sigma_\phi^*: l \mapsto \sum_{k \in \mathbb{Z}} l(k)\phi(\cdot - k).$$

In Fourier space it reads

$$\Sigma_\phi^*: F^{\mathbb{Z}} l(\omega) \to F^{\mathbb{Z}} l(\omega) \cdot F\phi(\omega).$$

Finally we obtain, for the combination of sampling with respect to some function $g \in L^2(\mathbb{R})$ and reconstructing with another function $h \in L^2(\mathbb{R})$,

$$\Sigma_h^* \Sigma_g: s \mapsto \sum_{k \in \mathbb{Z}} \langle g(\cdot - k) \mid s \rangle h(\cdot - k).$$

And in Fourier space we have

$$\Sigma_h^* \Sigma_g: \hat{s} \mapsto (\Pi(\bar{\hat{g}} \cdot \hat{s})) \cdot \hat{h}, \qquad \hat{s}(\omega) \mapsto \sum_{k \in \mathbb{Z}} \bar{\hat{g}}(\omega + 2\pi k)\hat{s}(\omega + 2\pi k)\hat{h}(\omega). \quad (9.6.1)$$

After this preparation the proof of Theorem 9.6.1 follows easily.

Proof. Consider the imperfect sampling operator with respect to sinc. We have to show that

$$\Sigma_{\text{sinc}}^* \Sigma s = s$$

if $s \in L^2(\mathbb{R})$ and $\operatorname{supp} \hat{s} \subset [-\pi, +\pi]$. Since $\widehat{\text{sinc}} = \chi_{[-\pi, +\pi]}$ this follows from

$$(\widehat{\Sigma_{\text{sinc}}^* \Sigma s})(\omega) = \chi_{[-\pi, +\pi]}(\omega) \sum_{k \in \mathbb{Z}} \hat{s}(\omega + 2\pi k) = \hat{s}(\omega).$$

The fact that the set of translates $\{\text{sinc}(\cdot - k)\}$ is an orthonormal family follows easily since its Fourier transform reads $\{\chi_{[-\pi, +\pi]}(\omega)\, e^{-ik\omega}\}$, $e_k = e^{ik\omega}$ is an orthonormal family on $[-\pi, +\pi]$, and $\Sigma s \in L^2(\mathbb{Z})$ since

$$\sum_n |s(n)|^2 = \frac{1}{2\pi} \sum_n \tilde{\hat{s}} * \hat{s}(2\pi n) = \frac{1}{2\pi} \tilde{\hat{s}} * \hat{s}(0) = \|s\|_2^2. \qquad \square$$

Applying this last theorem to a dilated and modulated version of s we obtain the following corollary.

Corollary 9.6.2
Suppose $s \in L^2(\mathbb{R})$ *and* $\operatorname{supp} \hat{s} \subset [\xi_l, \xi_r]$. *Then*

$$s(t) = \sum_{k \in \mathbb{Z}} s(\lambda k) r(t - \lambda k), \qquad r(t) = \lambda \operatorname{sinc}(\lambda t)\, e^{i\mu t},$$

where $\lambda = (\xi_r - \xi_l)/2$, $\mu = (\xi_r + \xi_l)/2$.

10 The continuous wavelet transform over $\mathbb{T}$

As a first application of the Poisson summation formula to wavelet analysis we will show how to construct a wavelet analysis over the one-dimensional torus that we identify with the circle $\mathbb{T}$. Here we follow essentially the presentation given in Holschneider (1990).

Clearly, there is no good dilation operator acting on functions over the circle and we must think of another way to construct a two-parameter family of functions over the circle. But here the periodization operator helps and we set, for any wavelet g over the real line,

$$g_a(t) = \Pi D_a g(t) = \frac{1}{a} \sum_{n \in \mathbb{Z}} g\left(\frac{t + 2\pi n}{a}\right)$$

and

$$g_{b,a}(t) = \Pi T_b D_a g(t) = g_a(t - b) = \frac{1}{a} \sum_{n \in \mathbb{Z}} g\left(\frac{t - b + 2\pi n}{a}\right) \qquad (10.0.1)$$

whenever this expression makes sense. An immediate application of the Poisson summation formula yields the following expression for the Fourier transform of $g_{b,a}$:

$$\hat{g}_{b,a}(n) = \int_{\mathbb{T}} dt\, g_{b,a}(t)\, e^{-int} = \hat{g}(an)\, e^{-ibn}.$$

Clearly, the Fourier transform on the left is the one over the circle, whereas the one on the right is over the real line.

The wavelet transform of a function over the circle $s \in L^p(\mathbb{T})$ with respect to the wavelet $g \in L^1$, $\int g = 0$, is then defined as

$$\mathcal{W}_g^{\mathbb{T}} s(b, a) = \int_{\mathbb{T}} dt\, \bar{g}_a(t - b) s(t) = \tilde{g}_a * s(b),$$

where $\tilde{g}(t) = \bar{g}(-t)$. The convolution is clearly a convolution over the circle. Since every voice is a convolution the wavelet transform is well defined thanks to Young's inequality. Note that the wavelet transform over $\mathbb{T}$, that we have just defined is nothing other than the usual wavelet transform over the real line of s if we identify the latter with a periodic function. Geometrically, the wavelet transform of a function over the circle $\mathbb{T}$ is a function over the half-cylinder $\mathbb{P} = \mathbb{R}^+ \times \mathbb{T}$. Functions over the half-cylinder may be identified in the obvious way with functions over the half-plane for which each voice is periodic with the same period.

We now come to the Fourier space picture. Suppose $g \in L^1(\mathbb{R})$ and $\int g = 0$. We then have $\hat{g}(0) = 0$. From the Poisson summation formula and the convolution theorem we obtain

$$\mathcal{W}_g^{\mathbb{T}} s(b, a) = \frac{1}{2\pi} \sum_{n \in \mathbb{Z}, n \neq 0} \bar{\hat{g}}(na)\, e^{inb}\, \hat{s}(n). \tag{10.0.2}$$

Clearly, the Fourier transform of g is the one over the real line, whereas $\hat{s}$ denotes the Fourier transform of s over the circle. Again, the positive and negative frequencies do not mix and we may split $L^2(\mathbb{T})$ according to

$$L^2(\mathbb{T}) = H^2_-(\mathbb{T}) \oplus \{1\} \oplus H^2_+(\mathbb{T}). \tag{10.0.3}$$

In parallel with the analysis of functions over the real line, we first analyse functions of very high regularity $S(\mathbb{T})$ that we identify with the 2π-periodic functions of C^∞ regularity. By duality we obtain the wavelet analysis of periodic distributions. Finally, we study the $L^2(\mathbb{T})$ theory. Since most proofs are along the same line as the theory developed in great detail in Chapter 1 we will be very brief in the proofs.

10.1 Wavelet analysis of $S(\mathbb{T})$ and $S'(\mathbb{T})$

A topology on $S(\mathbb{T})$ is given by the following norms that characterize the regularity of s:

$$\|s\|_{S(\mathbb{T});n} = \sup_{t \in \mathbb{T}} |\partial^n s(t)|. \tag{10.1.1}$$

We will denote by $S_0(\mathbb{T}) \subset S(\mathbb{T})$ the closed subspace of functions for which $\int_{\mathbb{T}} s = 0$. An equivalent topology to that of (10.1.1) is obtained in Fourier space through the decay of the Fourier coefficients: let $S(\mathbb{Z})$ be the Fréchet space of sequences that decay rapidly as expressed by the following norms:

$$\|l\|_{S(\mathbb{Z});m} = \sup_{n \in \mathbb{Z}} |n^m l(n)| < \infty.$$

Then the Fourier transform is a topological isomorphism between $S(\mathbb{T})$ and $S(\mathbb{Z})$.

We denote by $S'(\mathbb{T})$ the dual space of $S(\mathbb{T})$. The Fourier transform of $\eta \in S'(\mathbb{T})$ is defined as the sequence

$$\hat{\eta}(n) = \eta(e^{-int}).$$

By duality the Fourier transform is of at most polynomial growth, that is, we have for some $\alpha > 0$

$$|\hat{\eta}(n)| \leq O(1)(1 + |n|^2)^{\alpha/2}. \tag{10.1.2}$$

In analogy with the results of Section 24ff we define the wavelet transform of a distribution $\eta \in S'(\mathbb{T})$ through its action on the family of periodized wavelets

$$\mathcal{W}_g^{\mathbb{T}} \eta(b, a) = \eta(\bar{g}_{b,a}), \qquad g_{b,a} = \Pi T_b D_a g.$$

As in Chapter 1 it follows that the wavelet transform is a C^∞ function that is, together with all its derivatives, of at most polynomial growth in $a + 1/a$, uniformly in b. Clearly, the functions over the circle have a largest natural length scale, given by the length of the period. Therefore, the wavelet transform should not reveal any structure as a tends to ∞ regardless of what the analysed function is. This is actually the case, as is shown by the following theorem.

Theorem 10.1.1

Let $g \in S_0(\mathbb{R})$ and let $s \in S'(\mathbb{T})$. Then the wavelet transform decays, together with all its derivatives, at large scales as

$$|\partial_b^m (a\partial_a)^n \mathcal{W}_g^{\mathbb{T}} s(b, a)| \leq O(a^{-\alpha}), \qquad (a \to \infty),$$

for all $\alpha > 0$ uniformly in b.

The decay at small scale—if any—is again due to the regularity of the analysed function, that is, to its high-frequency behaviour.

Proof. We may use the Fourier space expansion (10.0.2) of the wavelet transform that is still valid for $g \in S_0(\mathbb{R})$ and $s \in S'(\mathbb{R})$. Therefore, passing to the absolute value and using (10.1.2) we obtain for all $\alpha > 0$ with some $\gamma \in \mathbb{R}$ in the limit $(a \to \infty)$

$$\begin{aligned} |a^\alpha \mathcal{W}_g^{\mathbb{T}} s(b, a)| &\leq O(1) \sum_{n \in \mathbb{Z} - \{0\}} a^\alpha |\hat{g}(na)| (1 + |n|^2)^{\gamma/2} \\ &\leq O(1) \sum_{n \in \mathbb{Z} - \{0\}} |n|^{\gamma + \alpha} \hat{g}(na) \leq O(1), \end{aligned}$$

thanks to the rapid decay of $\hat{g}$. The partial derivatives of $\mathcal{W}_g^{\mathbb{T}} s$ with respect to b and a are obtained as a superposition of wavelet transforms with respect to wavelets of the form $t^m \partial^n g$. These are again in $S(\mathbb{R})$ and the theorem is proved. □

Corollary 10.1.2

There are analysing wavelets such that $\mathcal{W}_g^{\mathbb{T}} s(b, a) = 0$ for all $s \in S'(\mathbb{T})$ and all $a \geq 1$.

Proof. Any $g \in S_0(\mathbb{R})$ with $\hat{g}$ supported by $[0, 1]$ will do. □

Therefore, the natural space of functions over the half-cylinder consists of functions of rapid decay at large scales. More precisely, let $S(\mathbb{P})$ be the space of functions over the half-cylinder for which the following seminorms are all finite:

$$\|\mathcal{T}\|_{S(\mathbb{P});n,l,m} = \sup_{(b,a)\in\mathbb{P}} |\partial_b^n(a\partial_a)^l a^m \mathcal{T}(b,a)| < \infty, \qquad n, l \in \mathbb{N}, m \in \mathbb{Z},$$

together with the locally convex topology that they induce.

We introduce the wavelet synthesis over the circle of a function $\mathcal{T} \in S(\mathbb{P})$ over the half-cylinder with respect to a family of periodized wavelets $h_{b,a} = \Pi T_b D_a h$ via

$$\mathcal{M}_h^{\mathbb{T}}\mathcal{T}(t) = \int_0^\infty \frac{da}{a} \int_{\mathbb{T}} db\, \mathcal{T}(b,a) h_{b,a}, \qquad h_{b,a} = \Pi T_b D_a h.$$

It can again be written as a superposition of convolutions of the voices:

$$\mathcal{M}_h^{\mathbb{T}}\mathcal{T} = \int_0^\infty \frac{da}{a} h_a * \mathcal{T}(\cdot, a).$$

If we identify functions over $\mathbb{P}$ with functions over $\mathbb{H}$ for which every voice is 2π-periodic we see that the $\mathcal{M}_h^{\mathbb{T}}$ may be written as $\mathcal{M}_h$. It is therefore clear that $\mathcal{M}_h$ is actually the inverse wavelet transform over the circle. But, as we have seen, we can only reconstruct functions modulo some polynomials. Now the only polynomials on the circle are the constant functions and thus the zero frequency is missing in the reconstruction. For the convenience of the reader we will now give a more explicit argument. Suppose for simplicity that g, h are functions in $S_0(\mathbb{R})$ and let h be a reconstruction wavelet (over $\mathbb{R}$) for g. Further, let $s \in C^\infty(\mathbb{T})$. Consider now the function $s^\#$ that we obtain when we analyse and reconstruct it through wavelet transforms:

$$s^\# = \int_0^\infty \frac{da}{a} h_a * \tilde{g}_a * s.$$

In Fourier space we obtain for $n \neq 0$, using the scaling invariance of the measure da/a as before,

$$\widehat{s^\#}(n) = \int_0^\infty \frac{da}{a} \bar{\hat{g}}(na)\hat{h}(na)\hat{s}(n) = \hat{s}(n) \int_0^\infty \frac{da}{a} \bar{\hat{g}}(a\,\text{sign}(n))\hat{h}(a\,\text{sign}(n)).$$

For $n = 0$, instead we obtain $\widehat{s^\#}(0) = 0$. Thus the term in $n = 0$ is missing and therefore,

$$c_{g,h}^{-1}\mathcal{M}_h^{\mathbb{T}}\mathcal{W}_g^{\mathbb{T}}s = s - \int_{\mathbb{T}} dt\, s(t), \qquad c_{g,h} = \int_0^\infty \frac{d\omega}{\omega} \bar{\hat{g}}(\omega)\hat{h}(\omega).$$

In the general case—if h is not necessarily a reconstruction wavelet for g—we again obtain a mixture between the identity and the Hilbert transform minus the mean value, where the Hilbert transform for functions on the circle is defined as

$$H\colon \hat{s}(n) \mapsto \begin{cases} -i\hat{s}(n) & \text{for } n > 0, \\ 0 & \text{for } n = 0, \\ i\hat{s}(n) & \text{for } n < 0. \end{cases}$$

In complete analogy with the theorems proved previously for the wavelet transform over the real line we have the following theorem.

Theorem 10.1.3

The wavelet transform and the wavelet synthesis on $S_0(\mathbb{T})$ are bi-continuous maps

$$\mathscr{W}^{\mathbb{T}}\colon S_0(\mathbb{R}) \times S_0(\mathbb{T}) \to S(\mathbb{P}), \qquad (g, s) \to \mathscr{W}_g^{\mathbb{T}} s,$$

$$\mathscr{M}^{\mathbb{T}}\colon S_0(\mathbb{R}) \times S(\mathbb{P}) \to S_0(\mathbb{T}), \qquad (h, \mathscr{T}) \to \mathscr{M}_h^{\mathbb{T}} \mathscr{T}.$$

For the wavelet transform of distributions with respect to a fixed $g \in S_0(\mathbb{R})$ we still have continuity[4]

$$\mathscr{W}_g^{\mathbb{T}}\colon S_0'(\mathbb{T}) \to S'(\mathbb{P}), \qquad s \mapsto \mathscr{W}_g^{\mathbb{T}} s,$$

$$\mathscr{M}_h^{\mathbb{T}}\colon S'(\mathbb{P}) \to S_0'(\mathbb{T}), \qquad \mathscr{T} \mapsto \mathscr{M}_h^{\mathbb{T}} \mathscr{T}.$$

In both cases the image of the wavelet transform is characterized by the reproducing kernel equation:

$$\mathscr{T} \in \text{image } \mathscr{W}_g \Leftrightarrow \mathscr{T}(b, a) = \int_0^\infty \frac{da'}{a'} \int_{\mathbb{T}} db' P_{g,h}(b, a; b', a') \mathscr{T}(b', a'),$$

with $h \in S_0(\mathbb{T})$ a reconstruction wavelet for g and

$$P(b, a; b', a') = \frac{1}{c_{g,h}} \langle g_{b,a} \mid h_{b',a'} \rangle. \tag{10.1.3}$$

The proof follows the same lines as the proof in Chapter 1 and is left to the reader.

From the frequency representation of the wavelet family we obtain the useful formula

$$P_{g,h}(b, a; b', a') = \sum_{n \neq 0} \bar{\hat{g}}(na) \hat{h}(na)\, e^{in(b - b')}.$$

Therefore, the reproducing kernel over the half-cylinder is the periodized reproducing kernel over the real line (denoted by $P'_{g,h}$):

$$P_{g,h}(b, a; b', a') = \sum_{k \in \mathbb{Z}} \frac{1}{a'} P'_{g,h}\left(\frac{b - b' + 2\pi k}{a'}, \frac{a}{a'}\right).$$

[4] As in the case of the wavelet transform over $\mathbb{R}$, the bi-linear mapping is hypocontinuous but not continuous.

10.2 The wavelet transform of $L^2(\mathbb{T})$

The wavelet transform over the circle conserves energy. More precisely, let us introduce the Hilbert space $L^2(\mathbb{P})$ of square integrable functions over the half-cylinder, with scalar product

$$\langle \mathcal{T} \mid \mathcal{R} \rangle_{\mathbb{P}} = \int_0^\infty \frac{da}{a} \int_{\mathbb{T}} db\, \bar{\mathcal{T}}(b, a) \mathcal{R}(b, a)$$

and norm

$$\|\mathcal{T}\|^2_{L^2(\mathbb{P})} = \langle \mathcal{T} \mid \mathcal{T} \rangle_{\mathbb{P}} = \int_0^\infty \frac{da}{a} \int_{\mathbb{T}} db\, |\mathcal{T}(b, a)|.$$

We denote by $H^2_+(\mathbb{T}) \subset L^2(\mathbb{T})$ the space of functions on $\mathbb{T}$ that contain only positive frequencies, that is, for which $\hat{s}(n) = 0$ for $n \leq 0$. Then $H^2_-(\mathbb{T})$ is the space of functions that contain only negative frequencies.

Theorem 10.2.1
The wavelet transform over $\mathbb{T}$ with respect to a progressive, admissible wavelet $g \in L^1(\mathbb{R}) \cap H^2_+(\mathbb{R})$, with $c_g = 1$,

$$\mathcal{W}_g^{\mathbb{T}} s\colon H^2_+(\mathbb{T}) \to L^2(\mathbb{P}), \qquad s \mapsto \mathcal{W}_g s,$$

is an isometry. The reproducing kernel equations (10.1.3) *is still valid pointwise. In the case where $g = h$ it is now an orthogonal projector on the image of the wavelet transform.*

Note that we have to suppose $g \in L^1(\mathbb{R})$ in order to have a well-defined periodized wavelet.

Proof. We limit ourselves to the dense subspaces $S_+(\mathbb{R})$ and $S_+(\mathbb{T})$ and leave the details of the limits to the reader. From (9.4.1) it follows, again using the scaling invariance of da/a, that

$$\|\mathcal{W}_g^{\mathbb{T}} s\|_{L^2(\mathbb{P})} = \frac{1}{2\pi} \sum_{n>0} \int_0^\infty \frac{da}{a} |\hat{g}(na\omega)|^2 \, |\hat{s}(n)|^2 = c_g \|s\|^2_{H^2_+(\mathbb{T})}. \qquad \square$$

In particular, the above theorem implies that we again have energy conservation

$$\|s\|^2_{H^2_+(\mathbb{T})} = \frac{1}{c_g} \|\mathcal{W}_g s\|^2_{L^2(\mathbb{P})}.$$

Clearly, the analogous theorem holds on $H^2_-(\mathbb{T})$, and the general case can be obtained from the splitting given by Formula (10.0.3).

11 Sampling of voices

In Section 33 we saw that it is enough to know the wavelet transform on a set of voices only in order to reconstruct the analysed function. In Theorem 9.6.1. and its corollary we saw that it is enough to know a strip-limited function on a discrete set of points to reconstruct it everywhere. We wish to use this idea to discretize the position-scale half-plane further.

To fix the ideas consider the wavelet g that is defined by

$$\hat{g}(\omega) = \chi_{[\pi, 2\pi]}(\omega) = \begin{cases} 1 & \text{for } \pi \leq \omega \leq 2\pi, \\ 0 & \text{otherwise.} \end{cases}$$

Consider the wavelet transform of any progressive function $s \in H^2_+(\mathbb{R})$. By Theorem 2.2.1 it is enough to know the set of voices $s_n = \mathcal{W}_g s(\cdot, 2^n)$ in order to reconstruct s completely. Every voice is a band-limited function supp $\hat{s}_n \subseteq [2^{-n}\pi, 2^{1-n}\pi]$, and therefore it may be sampled according to Shanon's sampling theorem, Theorem 9.6.1, without loss of information with a sampling frequency of 2^n. Therefore, it is enough to know the values of $\mathcal{W}_g s$ at a grid,

$$\Lambda = \{(n2^j, 2^j) \in \mathbb{H} : n, j \in \mathbb{Z}\},$$

in order to reconstruct the analysed function s. This specific subset of the half-plane is called the *dyadic* grid. See Figure 11.1.

In the following we will generalize this result in various ways to different wavelets and different grids. In particular, we will show that the image of the wavelet transform is in general a sampling space with respect to the

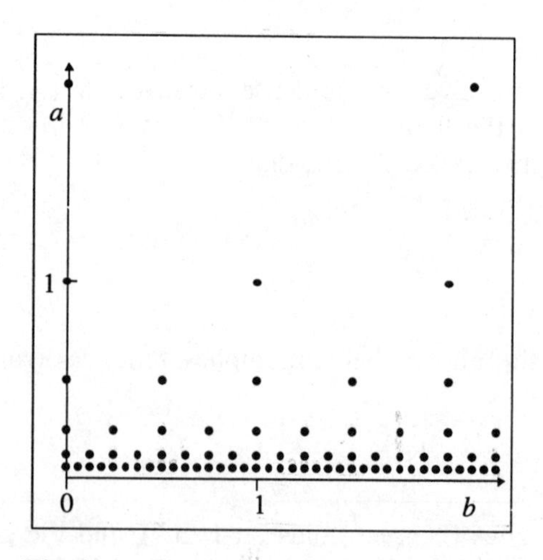

Fig. 11.1 The dyadic grid in the half-plane.

sampling operator associated with some grid. By this we understand the following: let $\Lambda \subset \mathbb{H}$ be a countable subset and let $\mu: \Lambda \to \mathbb{R}_+$ be a weight function. Then $L^2(\Lambda, \mu)$ is the Hilbert space of sequences whose scalar product and norm are given by

$$\langle l \mid v \rangle_\Lambda = \sum_{(\gamma \in \Lambda)} \mu(\gamma)\bar{l}(\gamma)v(\gamma), \qquad \|l\|_{L^2(\Lambda,\mu)} = \sum_{(\gamma \in \Lambda)} \mu(\gamma)|l(\gamma)|^2.$$

In this section we wish to give sufficient conditions on μ and Λ such that the sampling operator

$$\Sigma: \text{image } \mathscr{W}_g \to L^2(\Lambda, \mu), \qquad (\Sigma\mathscr{T})(b, a) = \mathscr{T}(b, a) \quad \text{for} \quad (b, a) \in \Lambda$$

may be inverted in a stable way, that is, it has a bounded left-inverse. In this case it is enough to know the wavelet transform on the grid Λ only in order to reconstruct it everywhere. However, it will turn out that, in general, the sampling operator is not surjective, that is, not every sequence over the grid in $L^2(\Lambda, \mu)$ can be obtained as a sample sequence of the wavelet transform of some function in $L^2(\mathbb{R})$. This is, loosely speaking, due to the internal correlations of the wavelets as expressed by the reproducing kernel. Only for special wavelets are the correlations between the wavelet coefficients at different points of a dyadic grid small enough to allow the sampling operator to be an isomorphism. This is the case of the so-called Riesz basis. In the case where the correlations disappear completely the associated wavelets are an orthonormal family. This is the main theme of the following chapter, where we show how to construct general orthonormal wavelet bases. In this chapter we are concerned with the weaker situation where the sampling operator is a continuous injective map.

The grids in the half-plane that we will use in the sequel are generalizations of the dyadic grid of the form

$$\Lambda = \Lambda_{\alpha,\gamma} = \{(n\alpha\gamma^j, \gamma^j) \in \mathbb{H}: n, j \in \mathbb{Z}\}' \qquad \gamma > 1, \alpha > 0. \tag{11.0.1}$$

Now sampling with the wavelet transform over the grid Λ means that we know $\mathscr{W}_g s(b, a)$ with $(b, a) \in \Lambda$ or, what amounts to the same, the set of scalar products

$$\{\langle g_{b,a} \mid s \rangle, (b, a) \in \Lambda\}.$$

This situation is common in the theory of frames and moments that we will present in the next section.

12 Frames and moments

Suppose we are given a countable family of vectors $\{g_n: n \in \mathbb{N}\}$ in a separable Hilbert space H. Suppose we know from a function $s \in H$ only the set of scalar products

$$\alpha(n) = \langle g_n \mid s \rangle, \qquad n \in \mathbb{N}.$$

The moments problem is to reconstruct s from the numbers $\alpha(n)$. Useful concepts when dealing with moments are the so-called frames and the closely related Bessel sequences.

Definition 12.0.1
A family $\{g_n\}$, $n \in \mathbb{N}$, of vectors in some separable Hilbert space is called a frame iff there are two constants A, B with $0 < A \leq B < \infty$ such that for every $r \in H$ we have

$$A\|r\|_H^2 \leq \sum_{n \in \mathbb{Z}} |\langle g_n \mid r \rangle|^2 \leq B\|r\|_H^2. \tag{12.0.1}$$

If $A = B$ the frame is called a tight frame.

The optimal constants A and B are called the *frame bounds*. Clearly, every orthonormal base of H is a tight frame, but the inverse is not necessarily true. Indeed, consider the one-dimensional vector space that is spanned by one single vector r. Then the set $\{r, r\}$ is a tight frame for this space, but it is clearly not an orthonormal base. However, a tight frame with $\|g_n\| = 1$ and frame bounds equal to 1 is an orthonormal base. (The proof is left to the reader.)

This situation is typical for frames. There are enough vectors to form a complete system, but in general there are too many to be independent.

Closely related to frames are the so-called Bessel sequences, which satisfy the upper frame bound.

Definition 12.0.2
A sequence $g_n \in H$ is called a Bessel sequence iff, for some finite $M > 0$,

$$\sum_{n \in \mathbb{N}} |\langle g_n \mid s \rangle|^2 \leq M \|s\|^2$$

for all $s \in H$.

Therefore, in the case of a Bessel sequence the imperfect sampling operator with respect to g_n,

$$K: H \to L^2(\mathbb{N}), \qquad s \mapsto \{\langle g_n \mid s \rangle\}, \quad n \in \mathbb{N},$$

is a bounded map from H into $L^2(\mathbb{N})$. If we want to stress the dependency on g_n we write $K_{\{g_n\}}$ or simply K_g. Its adjoint is given by the (*a-priori*) weakly convergent expression

$$K^*: \alpha \mapsto \sum_{k\in\mathbb{N}} \alpha(k) g_k.$$

In fact the sum converges strongly and unconditionally.[5]

Lemma 12.0.3

Let $\sum_{k\in\mathbb{N}} |\alpha(k)|^2 < \infty$. Then for any Bessel sequence g_n we have, for some $\eta \in H$,

$$\lim_{n\to\infty} \left\| \sum_{k\in I_n} \alpha(k) g_k - \eta \right\|_H = 0.$$

where $I_n \subset I_{n+1}$ is any growing family of finite subsets of $\mathbb{N}$ that tend to $\mathbb{N}$ as $n \to \infty$, that is, $\cup_{n\in\mathbb{N}} I_n = \mathbb{N}$.

Thanks to this theorem we may index the vectors in our frame by any countable index set and not only $\mathbb{N}$.

Proof. We have to estimate, with $m > n$,

$$\begin{aligned}
\left\| \sum_{k\in I_m} \alpha(k) g_k - \sum_{k\in I_n} \alpha(k) g_k \right\|^2 &= \left\| \sum_{k\in I_m\setminus I_n} \alpha(k) g_k \right\|^2 \\
&= \sup_{\|r\|=1} \left| \left\langle r \,\middle|\, \sum_{k\in I_m\setminus I_n} \alpha(k) g_k \right\rangle \right|^2 \\
&\le \sup_{\|r\|=1} \sum_{k\in I_m\setminus I_n} |\alpha(k)|^2 \sum_{k\in I_m\setminus I_n} |\langle r \mid g_k\rangle|^2 \\
&\le c \sum_{k\in I_m\setminus I_n} |\alpha(k)|^2.
\end{aligned}$$

This last expression tends to 0 as $n \to \infty$ and the sequence is a Cauchy sequence. □

We may thus consider the combination of sampling and reconstruction. In order to have a little more freedom we take a second Bessel sequence h_n for the reconstruction and we look at

$$B: H \to H, \qquad s \mapsto \sum_{k\in\mathbb{N}} \langle g_n \mid s\rangle h_n.$$

What we need now is that this operator be close to the identity. Therefore, we give the following definition.

[5] That is, the order of summation does not matter.

Definition 12.0.4

A Bessel family $h_n \in H$ is called a reconstructing family for the sampling family g_n—supposed to be a Bessel sequence again iff there are two constants $c \in \mathbb{C}$ and $0 \le \epsilon < 1$ such that

$$\left\| \sum_n c\langle g_n \mid r\rangle h_n - r \right\|_H^2 \le \epsilon \|r\|_H^2 \quad \text{for all } r \in H. \tag{12.0.2}$$

The pair of sequences g_n, h_n is called a sampling reconstruction pair.

Note that if h_n is a reconstructing family for g_n then g_n is also a reconstructing family for h_n with the same constants. This follows easily from the fact that an equivalent characterization of sampling reconstruction families is given by requiring that for all $s, r \in H$, we have

$$\left| c \sum_{k \in \mathbb{Z}} \langle g_n \mid r\rangle\langle s \mid h_n\rangle - \langle s \mid r\rangle \right| \le \sqrt{\epsilon}\|s\|_H \|r\|_H. \tag{12.0.3}$$

Indeed, suppose that (12.0.2) holds. Then we obtain (12.0.3) by the Cauchy–Schwarz inequality. On the other hand, since s and r are arbitrary we obtain (12.0.2) from (12.0.3).

An explicit scheme for reconstructing $s \in H$ from its samples $Ks \in L^2(\mathbb{N})$ is again obtained by iteration, as explained in Theorem 3.0.1. Explicitly, the iteration procedure reads for $v = v_0 = Ks$,

$$s_0 = cK_h^* v_0 = c \sum_{n \in \mathbb{N}} v(n) h_n, \qquad v_1(k) = K_g s_0(k) - v_0(k) = \langle g_k \mid s_0\rangle - v_0(k),$$

$$s_1 = cK_h^* v_1 = c \sum_{n \in \mathbb{N}} v_1(n) h_n, \qquad v_2(k) = K_g s_1(k) - v_1(k) = \langle g_k \mid s_1\rangle - v_1(k),$$

$\ldots,$

and $s_n \to s$ in the norm topology of H.

Theorem 12.0.5

The class of Bessel sequences having a reconstruction Bessel sequence coincides with the class of frames.

Proof. Let D be the left inverse of $K_h^* K_g$, which exists by Theorem 3.0.1. It follows by taking the adjoint that

$$\mathbb{1} = K_g^* K_h D^*.$$

In particular, K_g^* has a bounded right inverse $C = K_h D^*$. We may write

$$\|K_g s\| = \sup_{\|v\|_{L^2(\mathbb{N})} = 1} |\langle K_g^* v \mid s\rangle|.$$

Now putting $v = Cs/\|Cs\|$, the last expression is estimated from below by $\|C\|^{-1} > 0$. We have therefore shown that there is some constant $A > 0$ such that

$$\sum_{n \in \mathbb{N}} |\langle g_n \mid s \rangle|^2 \geq A \|s\|_H^2.$$

Therefore, a Bessel sequence that has a reconstructing Bessel sequence is a frame. And vice versa, as the next theorem shows a frame—which is clearly a Bessel sequence—has a Bessel sequence as the reconstruction sequence. Any reconstruction sequence is then a frame on its own, by symmetry.

Lemma 12.0.6

Suppose $\{g_n\}$ is a frame. Then there is a constant $0 \leq \epsilon < 1$ and another constant $c \in \mathbb{C}$ such that

$$\left\| \sum_n c \langle g_n \mid r \rangle g_n - r \right\|_H^2 \leq \epsilon \|r\|_H^2$$

for all $r \in H$. The relation between the frame bounds and a possible choice of constants is given by

$$c = \frac{2}{A + B} \quad \text{and} \quad \epsilon = \frac{B - A}{B + A}. \tag{12.0.4}$$

Proof. Suppose that the family g_n is a frame. The imperfect sampling–reconstruction operator $B = K^*K$ is a well-defined self-adjoint operator, and so is $cB - \mathbb{1}$. We obtain, for the norm of the operator, the following estimation by using (12.0.4) and (12.0.1):

$$\begin{aligned} \|cB - \mathbb{1}\|_H^2 &= \sup_{\|r\|_H = 1} \langle r \mid cBr - r \rangle \\ &= \sup_{\|r\|_H = 1} \left\{ c \sum_{n \in \mathbb{Z}} |\langle g_n \mid r \rangle|^2 - \|r\|_H^2 \right\} \leq \epsilon \|r\|_H^2. \end{aligned}$$

Therefore, the lemma holds with the constants as given in the lemma. □

This also proves Theorem 12.0.5. □

The sampling operator K with respect to a frame g_n can be approximately inverted by some reconstruction sequence. But even perfect reconstruction is possible. For this set

$$(K^*K)^{-1} g_n = g_n'.$$

This is again a frame since K^*K is bounded with a bounded inverse. We claim that

$$\sum_{n\in\mathbb{N}} \langle g_n \mid s\rangle g'_n = \sum_{n\in\mathbb{N}} \langle g'_n \mid s\rangle g_n = s. \tag{12.0.5}$$

We actually shall show a little more. Let K' be the sampling operator associated to g'_n. We then have

$$K' = K(K^*K)^{-1}.$$

Indeed,

$$K(K^*K)^{-1}s(k) = \langle g_k \mid (K^*K)^{-1}s\rangle = \langle (K^*K)^{-1}g_k \mid s\rangle = \langle g'_k \mid \cdot s\rangle.$$

Further, we have

$$(K')^*K = K^*K' = \mathbb{1},$$

which is, in concise form, equation (12.0.5). It follows easily from

$$\mathbb{1} = (K^*K)^{-1}K^*K = [K(K^*K)^{-1}]^*K = K'K$$

and

$$\mathbb{1} = K^*K(K^*K)^{-1} = K^*K'.$$

In general, the image of K is only a subspace of $L^2(\mathbb{N})$. A projector on its range is given by

$$K'K^* = K(K')^*. \tag{12.0.6}$$

Indeed, both sides are projectors since

$$K'K^*K'K^* = K'(K^*K')K^* = K'K^*,$$

$$K(K')^*K(K')^* = K[(K')^*K](K')^* = K(K')^*.$$

Since $K' = K(K^*K)^{-1}$ and since $(K^*K)^{-1}$ is surjective it follows that the range of K' and K coincide. We therefore have equality in (12.0.6). Since the operators are self-adjoint the projectors are orthogonal.

13 Some wavelet frames

Suppose, again, that we know the wavelet transform at the points of some grid $\Lambda \subset \mathbb{H}$ of type of (11.0.1), that is, only the set of numbers

$$\{\mathcal{W}_g s(b, a): (b, a) \in \Lambda\}, \qquad \Lambda = \{(n\alpha\gamma^j, \gamma^j): n, j \in \mathbb{Z}\}$$

is known. If we replace the inversion formula given by the double integral over the half-plane by a Riemannian sum we are led to consider, with

$\mu(b, a) = a$,

$$
\begin{aligned}
s^{\#} &= \sum_{(b,a)\in\Lambda} \mu(b, a)\mathscr{W}_g s(b, a) h_{b,a} \\
&= \sum_{j=-\infty}^{+\infty} \sum_{n=-\infty}^{+\infty} \gamma^j \mathscr{W}_g s(n\alpha\gamma^j, \gamma^j) h_{n\alpha\gamma^j, \alpha^j} \\
&= \sum_{j=-\infty}^{+\infty} \sum_{n=-\infty}^{+\infty} \gamma^j \langle g_{n\alpha\gamma^j, \alpha^j} \mid s \rangle h_{n\alpha\gamma^j, \alpha^j}.
\end{aligned} \tag{13.0.1}
$$

We now want to give some sufficient criteria that the collection $\{\sqrt{a} h_{b,a}: (b, a) \in \Lambda\}$ is a reconstruction sequence for $\{\sqrt{a} g_{b,a}: (b, a) \in \Lambda\}$. Since we find it more convenient to absorb the pre-factor $\sqrt{a}$ into some weight μ over Λ we say that $\{g_{b,a}, h_{b,a}\}$, $(b, a) \in \Lambda$ is a bi-frame over Λ with respect to the weight $\mu: \Lambda \to \mathbb{R}_+$.

We first need a sufficient condition that a family of dilated and translated functions is a Bessel sequence in order to give a well-defined meaning to the formula giving the tentative reconstruction $s^{\#}$ from the sampling values over some grid.

The condition given below is satisfied if the reproducing kernel of g has some localization over the half-plane. In order to simplify the notation we only show the theorem for the dyadic grid. The case of the more general grid follows trivially.

Theorem 13.0.1
Suppose that $g \in L^2(\mathbb{R})$ has a Fourier transform in $L^\infty(\mathbb{R})$ that satisfies, with some $\alpha > 0$ and $\beta > 1/2$, at

$$\hat{g}(\omega) = O(\omega^\alpha), \quad (\omega \to 0), \quad \hat{g}(\omega) = O(\omega^{-\beta}), \quad (|\omega| \to \infty).$$

Then $g_{j,k} = 2^{j/2} g(2^j \cdot - k)$ is a Bessel sequence.

Proof. Under the conditions on g we can find a $\epsilon > 0$ such that

$$\sum_{j\in\mathbb{Z}} |\hat{g}(2^j\omega)|^\epsilon \le c, \qquad \sum_{k\in\mathbb{Z}} |\hat{g}(\omega + 2\pi k)|^{2-\epsilon} \le c,$$

with some finite $c > 0$. From the Poisson summation formula and Parseval's equation we have, as usual,

$$
\begin{aligned}
\sum_{k\in\mathbb{Z}} |\langle g_{k,j} \mid s\rangle|^2 &= \int_0^{2\pi} d\omega\, 2^j |\widehat{\Sigma(\tilde{g} * D_{2^j} s)}(\omega)|^2 \\
&= \int_0^{2\pi} d\omega\, 2^j \left| \sum_{k\in\mathbb{Z}} \hat{s}(2^j[\omega + 2\pi k]) \bar{\hat{g}}(\omega + 2\pi k) \right|^2 \\
&= \int_0^{2\pi} d\omega\, 2^j \left| \sum_{k\in\mathbb{Z}} \hat{s}(2^j[\omega + 2\pi k]) \bar{\hat{g}}(\omega + 2\pi k)^{1-\epsilon/2} \bar{\hat{g}}(\omega + 2\pi k)^{\epsilon/2} \right|^2.
\end{aligned}
$$

The sum may be estimated using the Schwarz inequality and we obtain

$$\leq \int_0^{2\pi} d\omega\, 2^j \sum_{k\in\mathbb{Z}} |\hat{s}(2^j[\omega + 2\pi k])|^2\, |\hat{g}(\omega + 2\pi k)|^{\epsilon} \cdot \sum_{k\in\mathbb{Z}} |\hat{g}(\omega + 2\pi k)|^{2-\epsilon}.$$

The last sum is bounded by c and thus this is estimated by

$$c \int_0^{2\pi} d\omega\, 2^j \sum_{k\in\mathbb{Z}} |\hat{s}(2^j[\omega + 2\pi k])|^2\, |\hat{g}(\omega + 2\pi k)|^{\epsilon} = c \int_{-\infty}^{+\infty} d\omega\, |\hat{s}(\omega)|^2\, |\hat{g}(2^{-j}\omega)|^{\epsilon}.$$

Thus by summing over the scales we can use the estimate on g once more and we end up with

$$\sum_{j\in\mathbb{Z}} \sum_{k\in\mathbb{Z}} |\langle g_{j,k} \mid s\rangle|^2 \leq c \sum_{j\in\mathbb{Z}} \int_{-\infty}^{+\infty} d\omega\, |\hat{s}(\omega)|^2\, |\hat{g}(2^{-j}\omega)|^{\epsilon} \leq c^2 \|s\|^2,$$

and we are done. □

Consider now the 'sampling reconstruction' operator associated with the family of translates $g(\cdot - \alpha n)$ and $h(\cdot - \alpha n)$:

$$s \mapsto Bs = \sum_{n\in\mathbb{Z}} \langle g(\cdot - \alpha n) \mid s\rangle h(\cdot - \alpha n).$$

The reconstruction over the whole grid of (13.0.1) can now be written as

$$s^{\#} = \sum_{j\in\mathbb{Z}} D_{\gamma^j} B D_{\gamma^{-j}} s, \tag{13.0.2}$$

where D_a: $s(\cdot) \mapsto s(\cdot/a)/a$ is the dilation operator.

In Fourier space the sampling reconstruction operator B reads, as we have seen in (9.6.1),

$$B\colon \hat{s}(\omega) \mapsto \hat{h}(\omega) m(\omega), \qquad m(\omega) = \sum_{n=-\infty}^{+\infty} \bar{\hat{g}}(\omega + \lambda n) \hat{s}(\omega + \lambda n), \quad \lambda = 2\pi/\alpha.$$

Now suppose that g and h are strip limited with the support of their Fourier transforms contained in $[\omega_l, \omega_r]$ with $\omega_r - \omega_l \leq 2\pi/\alpha$. If we write

$$\widehat{Bs}(\omega) = \bar{\hat{g}}(\omega)\hat{h}(\omega)\hat{s}(\omega) + \hat{h}(\omega) \sum_{n\neq 0} \bar{\hat{g}}(\omega + 2\pi n/\alpha)\hat{s}(\omega + 2\pi n/\alpha)$$

we see that, by the support property of $\hat{g}$ and $\hat{h}$, the second term vanishes identically. Therefore, in this case we have

$$Bs = h * \tilde{g} * s.$$

Therefore, if there is a constant $0 \leq \epsilon < 1$ and another constant $c \in \mathbb{C}$ such

that

$$\operatorname*{ess\,sup}_{\omega>0} \left| c \sum_{n\in\mathbb{Z}} \hat{h}(\gamma^n\omega)\hat{g}(\gamma^n\omega) - 1 \right| \leq \epsilon,$$

then, in combination with Theorem 2.2.1 and decomposition (13.0.2), we have shown that $s^{\#}$ given by (13.0.1) is close to s, namely, $\|s - c^{-1}s^{\#}\|^2 \leq \epsilon\|s\|^2$. Therefore, the family of dilated and translated wavelets is a frame. Again, a perfect reconstruction can be obtained through iteration (cf. Theorem 3.0.1). But we may even have a perfect reconstruction in one step, as we will now show.

What is interesting is that for essentially any grid Λ of the form we have considered there are functions in S_0 whose translates and dilates with respect to Λ are a frame. It may even be a tight frame, as the next theorem shows.

Theorem 13.0.2
For every $\gamma > 1$ and for every $\alpha > 0$ there is a function $g \in S_0(\mathbb{R})$ such that the $\{\gamma^{j/2} g_{n\alpha^j,\gamma^j}\}$ are a tight frame.

In particular, this implies that

$$\sum_{j=-\infty}^{+\infty} \sum_{n=-\infty}^{+\infty} \gamma^j \langle g_{n\alpha^j,\gamma^j} \mid s\rangle g_{n\alpha^j,\gamma^j} = s$$

for all $s \in H^2_+(\mathbb{R})$.

Proof. Let β be some number that satisfies

$$\beta < \gamma\beta < \beta + \frac{2\pi}{\alpha}.$$

Let r be a function in the Schwartz class that satisfies at

(i) $\operatorname{supp} \hat{r} \subset [\beta, \beta + 2\pi/\alpha]$,

(ii) $\sum_{j\in\mathbb{Z}} |\hat{r}(\gamma^j\omega)|^2 > 0$.

By hypothesis on β such a function exists. Then set $g = h$ with

$$\hat{g}(\omega) = \hat{r}(\omega) \Big/ \sqrt{\sum_{j\in\mathbb{Z}} |\hat{r}(\gamma^j\omega)|^2}.$$

Clearly, $\hat{g} \in S_+(\mathbb{R})$ and the support of g is the same as that of r. By construction we have

$$\sum_{j\in\mathbb{Z}} |\hat{g}(\omega\gamma^j)|^2 \equiv 1,$$

and therefore by Theorem 2.2.1 the theorem is proved. □

We now come to a general estimation of the approximation due to Daubechies *et al.* (1986).

Theorem 13.0.3
Let $g, h \in L^1(\mathbb{R}) \cap H^2_+(\mathbb{R})$ be such that their translates and dilates with respect to $\Lambda = \{(n\alpha\gamma^j, \gamma^j): n, j \in \mathbb{Z}\}$ are Bessel sequences. Suppose, further, that there are constants $0 \leq \epsilon < 1$ and $c \in \mathbb{C}$ such that

$$\operatorname*{ess\,sup}_{\omega>0} \left| c \sum_{n\in\mathbb{Z}} \hat{h}(\gamma^n\omega)\bar{\hat{g}}(\gamma^n\omega) - 1 \right| \leq \epsilon.$$

Suppose, further, that

$$\rho = \sum_{k\neq 0} \delta(k) < \infty,$$

where

$$\delta(k) = \sup_{\omega\in\mathbb{R}} \sum_{j\in\mathbb{Z}} |\hat{g}(\gamma^j\omega + 2\pi k/\alpha)|\, |\hat{h}(\gamma^j\omega)|.$$

Then the following estimate holds for the reconstruction $s^\#$ given by 13.0.1, *namely,*

$$\|s - cs^\#\|_{L^2(\mathbb{R})} \leq \sqrt{\epsilon + \rho}\, \|s\|_{L^2(\mathbb{R})}.$$

Therefore, again for $\rho + \epsilon < 1$, a perfect reconstruction is possible through iteration, as shown in Theorem 3.0.1. In the special case where $g = h$ this theorem gives a sufficient condition that the family of translated and dilated wavelets $g_{b,a}$, where $(b, a) \in \Lambda$, is a frame with respect to the weight $\mu(b, a) = a$.

Proof. By an overall dilation we may assume that $\alpha = 1$ and that therefore the sampling points (b, a) in the half-plane are of the form $\{(k\gamma^j, \gamma^j)\}$, where $k, j \in \mathbb{Z}$. Let us denote by B the operator that maps s to its approximate reconstruction $s^\#$. The theorem is proved if we can find c and ϵ such that

$$|c\langle s \mid Br\rangle - \langle s \mid r\rangle| \leq \epsilon\langle s \mid r\rangle$$

for all $s, r \in H^2_+(\mathbb{R})$. Indeed, then the operator $cB - \mathbb{1}$ has an operator norm smaller than ϵ.

We are therefore led to consider

$$X_1 = \langle s \mid Br\rangle = \sum_{j\in\mathbb{Z}} \sum_{k\in\mathbb{Z}} \gamma^j \langle g_{k\gamma^j,\gamma^j} \mid r\rangle\langle s \mid h_{k\gamma^j,\gamma^j}\rangle.$$

For fixed j the sum over k in expression X_1 is nothing other than the scalar product of two sequences. These sequences can be obtained by applying the perfect sampling operator with sampling period γ^j to the filtered functions $\tilde{g}_{\gamma^j} * r$ and $h_{\gamma^j} * s$, where the single index again stands for the dilation $g_a = g(\cdot/a)a$. The scalar product of these two sequences can also be computed in Fourier space as an integral over one period, that is, over the interval $[0, 2\pi/\gamma^j)$. By the Poisson summation formula, the $2\pi/\gamma^j$-periodic Fourier transform of these perfect sample sequences is obtained by periodizing $\bar{\hat{g}}(\gamma^j\omega)\hat{r}(\omega)$ and $\hat{h}(\gamma^j\omega)\hat{s}(\omega)$. Altogether, we obtain

$$X_1 = \sum_{j\in\mathbb{Z}} \int_0^{2\pi/\gamma^j} d\omega \sum_{k\in\mathbb{Z}} \sum_{k'\in\mathbb{Z}} \bar{\hat{g}}(\gamma^j\omega + 2\pi k)\hat{r}(\omega + 2\pi k/\gamma^j)$$

$$\times\, \hat{h}(\gamma^j\omega + 2\pi k')\bar{\hat{s}}(\omega + 2\pi k'/\gamma^j).$$

By changing variables from ω to $\omega - 2\pi k'/\gamma^j$ and by replacing the sum over k by the sum over $k - k'$ we obtain, using $\sum_{k'\in\mathbb{Z}} \int_{k'2\pi/2^j}^{(k'+1)2\pi/2^j} d\omega = \int_{-\infty}^{+\infty} d\omega$,

$$X_1 = \sum_{j\in\mathbb{Z}} \int_{-\infty}^{+\infty} d\omega \sum_{k\in\mathbb{Z}} \bar{\hat{g}}(\gamma^j\omega + 2\pi k)\hat{r}(\omega + 2\pi k/\gamma^j)\hat{h}(\gamma^j\omega)\bar{\hat{s}}(\omega).$$

We now distinguish the term in $k = 0$ from the rest and we consider

$$X_2 = \int_{-\infty}^{+\infty} d\omega \left\{ \sum_{j\in\mathbb{Z}} \bar{\hat{g}}(\gamma^j\omega)\hat{h}(\gamma^j\omega) \right\} \hat{r}(\omega)\bar{\hat{s}}(\omega).$$

This is the main contribution to the reconstruction and may be treated as usual. By hypothesis on g and h we have

$$|cX_2 - \langle r \mid s\rangle| \le \epsilon|\langle r \mid s\rangle|.$$

We still have to estimate $X_3 = X_1 - X_2$. Here we have

$$X_3 = \sum_{k\neq 0} \int_{-\infty}^{+\infty} d\omega \sum_{j\in\mathbb{Z}} \bar{\hat{g}}(\gamma^j\omega + 2\pi k)\hat{r}(\omega + 2\pi k/\gamma^j)\hat{h}(\gamma^j\omega)\bar{\hat{s}}(\omega).$$

Passing the modulus and using the Cauchy–Schwarz inequality for the sum over j we obtain

$$|X_3| \le \sum_{k\neq 0} \int_{-\infty}^{+\infty} d\omega \left\{ \sum_{j\in\mathbb{Z}} |\bar{\hat{g}}(\gamma^j\omega + 2\pi k)|\, |\hat{h}(\gamma^j\omega)|\, |\hat{s}(\omega)|^2 \right\}^{1/2}$$

$$\times \left\{ \sum_{j\in\mathbb{Z}} |\bar{\hat{g}}(\gamma^j\omega + 2\pi k)|\, |\hat{h}(\gamma^j\omega)|\, |\hat{r}(\omega + 2\pi k/2^j)|^2 \right\}^{1/2}.$$

We now use the Cauchy–Schwarz inequality once more, but this time applied to the integral over ω to obtain

$$|X_3| \leq \sum_{k \neq 0} \left\{ \int_{-\infty}^{+\infty} d\omega \sum_{j \in \mathbb{Z}} |\bar{\hat{g}}(\gamma^j \omega + 2\pi k)| \, |\hat{h}(\gamma^j \omega)| \, |\hat{s}(\omega)|^2 \right\}^{1/2}$$

$$\times \left\{ \int_{-\infty}^{+\infty} d\omega \sum_{j \in \mathbb{Z}} |\bar{\hat{g}}(\gamma^j \omega + 2\pi k)| \, |\hat{h}(\gamma^j \omega)| \, |\hat{r}(\omega + 2\pi k/2^j)|^2 \right\}^{1/2}.$$

By changing the integration variable in the last parentheses from ω to $\omega - 2\pi/\gamma^j$ we have

$$|X_3| \leq \sum_{k \neq 0} \left\{ \int_{-\infty}^{+\infty} d\omega \sum_{j \in \mathbb{Z}} |\bar{\hat{g}}(\gamma^j \omega + 2\pi k)| \, |\hat{h}(\gamma^j \omega)| \, |\hat{s}(\omega)|^2 \right\}^{1/2}$$

$$\times \left\{ \int_{-\infty}^{+\infty} d\omega \sum_{j \in \mathbb{Z}} |\bar{\hat{g}}(\gamma^j \omega)| \, |\hat{h}(\gamma^j \omega - 2\pi k)| \, |\hat{r}(\omega)|^2 \right\}^{1/2}.$$

We now use the hypothesis on g and h to estimate

$$|X_3| \leq \sum_{k \neq 0} \delta(k)^{1/2} \|r\|_2 \, \delta(k)^{1/2} \, \|s\|_2 = \rho \|r\|_2 \, \|s\|_2.$$

This proves the theorem. □

13.1 Irregular sampling

We wish to end this section with a theorem that allows us to estimate the frame bounds—at least in principle—in the general case for irregular grids over the half-plane. The idea is to replace the reproducing kernel equation by an approximating Riemannian sum. This shows that the mean density of points where we have to sample without loss of information is related to the localization properties of the reproducing kernel $P_g = c_g^{-1} \mathcal{W}_g g$.

We first give a sufficient condition that $\{g_{b,a} : (b, a) \in \Lambda\}$ is a Bessel sequence with weight μ.

Theorem 13.1.1

Suppose there is a non-negative function $F \in L^1(\mathbb{H})$ such that

$$\sum_{(\beta,\alpha) \in \Lambda} \mu(\beta, \alpha)^2 \left| \frac{1}{a} P_g\left(\frac{b - \beta}{a}, \frac{\alpha}{a} \right) \frac{1}{\alpha} P_g\left(\frac{\beta - b'}{\alpha}, \frac{a'}{\alpha} \right) \right| \leq \frac{1}{a'} F\left(\frac{b - b'}{a}, \frac{a}{a'} \right). \quad (13.1.2)$$

Then

$$\sum_{(\beta,\alpha) \in \Lambda} \mu(\beta, \alpha)^2 |\mathcal{W}_g s(\beta, \alpha)|^2 \leq c_g \|F\|_{L^1(\mathbb{H})} \|s\|_{L^2(\mathbb{R})}^2.$$

Proof. For the sake of simplicity we write $d\nu(b, a)$ for $db\, da/a$ and $P(b, a; b', a')$ for the dilated and translated reproducing kernel. We have to estimate

$$\sum_{(\beta, \alpha) \in \Lambda} \mu(\beta, \alpha)^2 |\mathscr{W}_g s(\beta, \alpha)|^2,$$

which, by the reproducing kernel equation, equals

$$\sum_{(\beta, \alpha) \in \Lambda} \mu(\beta, \alpha)^2 \int d\nu(b, a) \int d\nu(b', a') \bar{P}(\beta, \alpha; b, a) \overline{\mathscr{W}_g s}(b, a) P(\beta, \alpha; b', a') \mathscr{W}_g s(b', a').$$

Passing to the modulus and exchanging the summation with the integration and replacing the sum over the grid by F, this is majorized by

$$\int d\nu(b, a) \int d\nu(b', a') |\mathscr{W}_g s(b, a)| \frac{1}{a'} F\left(\frac{b - b'}{a'}, \frac{a}{a'}\right) |\mathscr{W}_g s(b', a')|.$$

The desired estimated now follows from the Cauchy–Schwarz inequality, the isometric property of the wavelet transform, and the following lemma that we state as a result on its own. It is the analogue of Young's inequality for convolutions over the half-plane.

Lemma 13.1.2
Let $F \in L^1(\mathbb{H})$; then for $\mathscr{T} \in L^p(\mathbb{H})$ we have

$$\|F * \mathscr{T}\|_{L^p(\mathbb{H})} \leq \|F\|_{L^1(\mathbb{H})} \|\mathscr{T}\|_{L^p(\mathbb{H})}, \qquad 1 \leq p \leq \infty.$$

Proof. Let $\mathscr{R} \in L^q(\mathbb{H})$ with $p^{-1} + q^{-1} = 1$ be non-negative. We may also suppose that F and $\mathscr{T}$ are non-negative and we have to estimate

$$\int_{\mathbb{H}} d\nu(b, a) \int_{\mathbb{H}} d\nu(b', a') \mathscr{R}(b, a) \frac{1}{a'} \mathrm{F}\left(\frac{b - b'}{a'}, \frac{a}{a'}\right) \mathscr{T}(b', a')$$

$$= \int_{\mathbb{H}} d\nu(b, a) \int_{\mathbb{H}} d\nu(b', a') F(b, a) \mathscr{R}(a'b + b', a'a) \mathscr{T}(b', a').$$

The inner integral may be estimated, by Hölder's inequality, to be bounded by $\|\mathscr{T}\|_p \|\mathscr{R}\|_q$. The remaining integral is $\|F\|_1$ and the lemma follows by the Riesz representation theorem. □

The proof of Theorem 13.1.1 is finished. □

The condition can be used in practice if the reproducing kernel has a little localization and regularity so that we can replace the reproducing kernel equation applied to the kernel itself by a Riemannian sum. It can also be tested numerically.

The next theorem uses the same idea to give a sufficient condition for a grid together with a weight function to give rise to a bi-frame.

Theorem 13.1.3

Let $g, h \in L^1(\mathbb{R}) \cap H^2_+(\mathbb{R})$ be such that the reproducing kernel $P_{g,h} = c_{g,h}^{-1}\mathscr{W}_g h$ satisfies (13.1.1) with respect to some grid Λ and some weight function μ. Define

$$\mathscr{T}(b, a; b', a') = \sum_{(\beta,\alpha)\in\Lambda} \mu(\beta, \alpha)\frac{1}{\alpha} P_{g,h}\left(\frac{b-\beta}{\alpha}, \frac{a}{\alpha}\right)\frac{1}{a'} P_{g,h}\left(\frac{\beta - b'}{a'}, \frac{\alpha}{a'}\right).$$

Then $\{g_{b,a}, h_{b,a}\}$ is a bi-frame with respect to the weight μ if there is a constant $0 \leq \epsilon < 1$ and some $c \in \mathbb{C}$ such that we have

$$\left| c\mathscr{T}(b, a; b', a') - \frac{1}{a'} P_{g,h}\left(\frac{b-b'}{\alpha}, \frac{a}{a'}\right)\right| < \epsilon \frac{1}{a'} F\left(\frac{b-b}{a'}, \frac{a}{a'}\right), \quad (13.1.2)$$

with some function $F \in L^1(\mathbb{H})$ with $\|F\|_1 = 1$.

Proof. We have to show that

$$\left| c \sum_{(\beta,\alpha)\in\Lambda} \mu(\beta, \alpha)\langle g_{\beta,\alpha} \mid s\rangle\langle r \mid h_{\beta,\alpha}\rangle - \langle r \mid s\rangle\right| \leq \epsilon\|s\|_2\|r\|_2.$$

By the reproducing kernel equation for the wavelet coefficients we have

$$c \sum_{(\beta,\alpha)\in\Lambda} \mu(\beta, \alpha)\langle g_{\beta,\alpha} \mid s\rangle\langle r \mid h_{\beta,\alpha}\rangle$$

$$= c \sum_{(\beta,\alpha)\in\Lambda} \mu(\beta, \alpha) \int_0^\infty \frac{da}{a}\int_{-\infty}^{+\infty} db \frac{1}{a} P_{g,h}\left(\frac{\beta - b}{a}, \frac{\alpha}{a}\right)\mathscr{W}_g s(b, a)$$

$$\cdot \int_0^\infty \frac{da'}{a'}\int_{-\infty}^{+\infty} db' \frac{1}{a'} P_{g,h}\left(\frac{\beta - b'}{a'}, \frac{\alpha}{a}\right)\overline{\mathscr{W}_h r}(b', a').$$

Note that we have made use of the symmetry equation (20.0.2) in Chapter 1. Exchanging the summation with the integrals, which we may since everything converges absolutely, we obtain that the last expression equals

$$\int_0^\infty \frac{da}{a}\int_{-\infty}^{+\infty} db \int_0^\infty \frac{da'}{a'}\int_{-\infty}^{+\infty} db' c\mathscr{T}(b, a; b', a')\overline{\mathscr{W}}_g s(b, a)\overline{\mathscr{W}_h r}(b', a').$$

We may now replace $c\mathscr{T}$ by a reproducing kernel. The error is estimated by (13.1.2):

$$\epsilon \int_{\mathbb{H}} d\nu(b, a)|\mathscr{W}_g s(b, a)| \int_{\mathbb{H}} d\nu(b', a')a'^{-1}F([b - b']/a', a/a')|\mathscr{W}_g s(b', a')|.$$

To use the Schwartz inequality we have to estimate the $L^2(\mathbb{H})$ norm of the last integral. This can again be done with the help of Lemma 13.1.2. □

13.2 Calderón–Zygmund operators again

As we have already seen, partial reconstruction over infinitely many voices gives rise to a class of operators known as Calderòn–Zygmund operators. The point is that in terms of wavelet coefficients these operators do not alter the localization of the wavelet transforms. Loosely speaking, they map wavelets into possibly different wavelets. To give a more precise meaning to this we give the following definition.

Definition 13.2.1
Let $g_{b,a} = a^{-1}g([\cdot - b]/a)$ *with some admissible* $g \in S_0(\mathbb{R})$. *A linear operator* $\mathcal{T}$ *that satisfies at*

$$|\langle g_{b,a} \mid \mathcal{T} g_{b',a'}\rangle| \leq \frac{1}{a'}\Pi\left(\frac{b-b'}{a'}, \frac{a}{a'}\right)$$

with some function $\Pi \in S(\mathbb{H})$ *is called a highly regular Calderòn–Zygmund operator.*

Since, as we will see in Chapter 6, many function spaces are characterized by their localization properties in terms of wavelet coefficients it follows that the class of Calderòn–Zygmund operators is continuous in many different function spaces.

It is an easy matter to verify that the function g in the definition is unimportant in the sense that if the estimate holds for one admissible $g \in S_0(\mathbb{R})$ it holds for any others as well (with a possibly different Π). Therefore, the class of operators is well defined (see Chapter 6).

It is now no surprise that the partial reconstruction with respect to quite general grids is a Calderòn–Zygmund operator, provided the sum over the grid, together with the weight function, approximates the Riemann integral well enough. To be more precise, consider the operator

$$\mathcal{T}: s \mapsto \sum_{(\beta,\alpha)\in\Lambda} \mu(\beta,\alpha)\langle g_{\beta,\alpha} \mid s\rangle h_{\beta,\alpha}, \qquad g, h \in S_0(\mathbb{R}).$$

Let $P = |\mathcal{W}_g g|$ and $Q = |\mathcal{W}_g h|$. Then we have to estimate

$$\sum_{(\beta,\alpha)\in\Lambda} \mu(\beta,\alpha)\frac{1}{a}P\left(\frac{b-\beta}{a}, \frac{\alpha}{a}\right)\frac{1}{\alpha}Q\left(\frac{\beta-b'}{\alpha}, \frac{a'}{\alpha}\right).$$

Let us suppose that the grid, together with the weight function, is such that we may estimate the sum by the integral over the (β, α) half-plane. Since both P and Q are in $S(\mathbb{H})$, which is an algebra with respect to convolutions over the half-plane, it follows that in this case the sampling–reconstruction operator $\mathcal{T}$ is a highly localized Calderòn–Zygmund operator.

Appendix

14 A functional calculus

As we have seen, the wavelet transform has an inversion formula that converges very well in many function spaces. In particular, the analysing wavelet and the reconstruction wavelet may be different; one of them may even be singular provided the other is sufficiently regular. We will use this in Section 12 of Chapter 5 to invert the Radon transform. In this section we present an application where the analysing and the synthesizing wavelet differ in nature: the analysing wavelet is a scalar whereas the reconstruction wavelet is operator valued. This will be used to compute $s(A)$, where s if a function and A is a bounded operator acting in some Hilbert space H. We will follow Holschneider (1994b) and try to replace t by A in the inversion formula

$$s(t) = \frac{1}{c_{g,h}} \int_{-\infty}^{+\infty} db \int_0^{\infty} \frac{da}{a}\, \mathcal{W}_g s(b, a) h_{b,a}(t).$$

Thus we only have to know $h_{b,a}(A)$ for some suitable function h. But this kind of expression is given by the resolvent of A, $R_z(A) = (A - z)^{-1}$ since it can be written as $(z = b + ia)$

$$R_{\bar{z}}^n(A) = \frac{1}{(A - \bar{z})^n} = \frac{1}{a^{n+1}} \frac{1}{\left(\dfrac{A - b}{a} + i\right)^n} = a^{-n} h_{b,a}(A),$$

where $h(t) = [1/(t + i)^n]$. The function h is square integrable and its Fourier transform is supported by the positive frequencies only:

$$\hat{h}(\omega) = \begin{cases} -\dfrac{2\pi i}{(n-1)!}(-i\omega)^{n-1}\, e^{-\omega} & \text{for } \omega \geq 0, \\ 0 & \text{for } \omega < 0. \end{cases}$$

This is essentially the Cauchy wavelet. Suppose now that g is chosen in such a way that it is a reconstruction wavelet for h. Upon replacing $h_{b,a}(t)$ by $h_{b,a}(A) = R_{\bar{z}}(A)$ in the reconstruction formula we might hope to obtain a decent definition of $s(A)$. In order to avoid integration through the spectrum we have to suppose that it is contained in the closed upper half-plane. In order to obtain a norm-convergent integral we have to impose some regularity near the real axis. To be more precise, suppose that there is a natural number n such that in the lower half-plane $(\mathfrak{I}z < 0)$ we have

$$\|R_{b-i\alpha}^n(A)\| \leq O(1 + a^{-m}) \quad \text{for } a > 0 \tag{14.0.1}$$

uniformly in $b \in \mathbb{R}$. We say that A is of tempered type (in the lower half-plane).

Definition 14.0.1

Let $s, g \in S_+(\mathbb{R})$ and let g be such that

$$c_{g,n} = -\frac{2i\pi}{(n-1)!}\int_0^\infty \frac{d\omega}{\omega}(-i\omega)^{n-1}\bar{\hat{g}}(\omega)\, e^{-\omega}$$

is a non-zero finite number. Let A be of tempered type in the lower half-plane (14.0.1). *Then we define $s(A)$ to be*

$$s(A) = \frac{1}{c_{g,n}}\int_0^\infty da\, a^{n-2}\int_{-\infty}^{+\infty} db\, \mathcal{W}_g s(b,a) R^n_{b-ia}(A).$$

Clearly, the analogous definition is possible for functions in $S_-(\mathbb{R})$, in which case A should be of tempered type in the upper half-plane. The integral over the half-plane is well defined as norm-convergent integral since the norms of the resolvent are no stronger than the polynomial divergent and the wavelet transform $\mathcal{W}_g s$ compensates this well. Thus $s(A)$ is a bounded operator. We still have to check the dependency on g and n.

Theorem 14.0.2

The operator $s(A)$ does not depend on g or n.

Proof. It is enough to consider $n = 1$ since all other cases are obtained through mere partial integration, thanks to

$$\mathcal{W}_{\partial g} s(b,a) = -a\mathcal{W}_g \partial s(b,a) = a\,\partial_b \mathcal{W}_g s(b,a).$$

Consider the matrix element

$$\langle \psi \mid s(A)\phi\rangle = \frac{1}{c_{g,1}}\int_0^\infty \frac{da}{a}\int_{-\infty}^{+\infty} db\, \mathcal{W}_g s(b,a)\langle \psi \mid R_{b-ia}(A)\phi\rangle.$$

By hypothesis on the operator A the function $f(z) = \langle \psi \mid R_z(A)\phi\rangle$ is holomorphic in the lower half-plane ($\Im z < 0$), where it satisfies:

$$|f(z)| \le \frac{c}{|\Im z|^n}.$$

Therefore, the boundary value $\eta = \lim_{a\to 0} f(\cdot - ia)$ exists in the sense of distribution in $S'(\mathbb{R})$, and f may be recovered from its boundary by means of the Poisson kernel. With $\pi P(t) = 1/(t^2+1)$ we have (e.g. Beltrami and Wohlers, 1966).

$$f(b - ia) = P_a * \eta(b).$$

Now we may write

$$\begin{aligned} c_{g,1}\langle\psi \mid s(A)\phi\rangle &= \int_0^\infty \frac{da}{a}\int_{-\infty}^{+\infty} db\, \mathscr{W}_g s(b,a) f(b-ia) \\ &= \int_0^\infty \frac{da}{a}\int_{-\infty}^{+\infty} db(\tilde{g}_a * s(b))(P_a * \eta(b)) \\ &= \int_0^\infty \frac{da}{a}\tilde{g}_a * s * P_a * \eta\dagger(0); \qquad \eta\dagger(\cdot) = \eta(-\cdot) \\ &= s * \eta\dagger(0) \\ &= \eta(s). \end{aligned}$$

In the preceding equations we have used the fact that the inversion formula converges in $S(\mathbb{R})$ since $\tilde{g} * P \in S_+(\mathbb{R})$. □

We thus have, in addition, the following generalization of the spectral measures.

Corollary 14.0.3

Let A be of tempered type in the lower half-plane. Then for ever $\psi, \phi \in H$ there is a distribution $\eta_{\phi,\psi} \in S'_-(\mathbb{R})$ such that for every $s \in S_+(\mathbb{R})$ we have

$$\langle\psi \mid s(A)\phi\rangle = \eta_{\phi,\psi}(s).$$

Remark. For a given m in (14.0.1) the conditions on s may be relaxed considerably. All that is needed is essentially a local $C^{m+\epsilon}$ regularity since this implies a local small scale decrease of the wavelet coefficients of order $O(a^{m+\epsilon})$.

14.1 The case of self-adjoint operators

We now want to show that, at least for self-adjoint operators acting in some Hilbert space, the functional calculus we have defined so far coincides with the standard functional calculus (e.g. Reed and Simon (1972)). Let A be a self-adjoint operator with domain $D(A)$ acting on some Hilbert space. Since for self-adjoint A we have

$$\|R_{b\pm ia}(A)\| \le O(1/a),$$

the self-adjoint operators are all of real-tempered type. By the standard functional calculus it is easy to define $s(A)$ whenever $s \in L^\infty(\mathbb{R})$, by setting

$$s(A) = \int s(\lambda)\, dE_\lambda,$$

where E_λ is the spectral family of projection operators associated with A. In this sense the powers of the resolvent $R_z^n(A) = (A - z)^{-n}$ may be looked upon as a wavelet transform of the projection-valued measured E_λ, namely,

$$a^{n-1} R_{\bar{z}}^n(A) = \frac{1}{a} \int \frac{dE_\lambda}{\left(\dfrac{\lambda - b}{a} + i\right)^n}.$$

We now have the following theorem.

Theorem 14.1.1

Let $g \in S_+(\mathbb{R})$ be such that $c_{g,n}^+$ is non-zero. Let $s \in L^1(\mathbb{R}) \cap L^\infty(\mathbb{R})$ be continuous and contain only positive frequencies. Then for A a self-adjoint operator we have

$$s_{\epsilon,\rho}(A) = \frac{1}{c_{g,n}^\pm} \int_\epsilon^\rho \frac{da}{a} \int_{-\infty}^{+\infty} db\, \mathscr{W}_g s(b,a) R_{b-ia}^n(A) \to s(A),$$

as $\epsilon \to 0$, and $\rho \to \infty$. The convergence holds in the strong sense on $D(A)$, and $s(A)$ on the right-hand side is given by the standard functional calculus. If in addition s is uniformly continuous then the convergence holds in norm.

Proof. We may assume that $c_{g,n}^+ = 1$. By hypothesis, $\mathscr{W}_g s(\cdot, a) \in L^1(\mathbb{R})$, and thus the double integral is well defined for given ϵ and ρ, since it converges in the operator-norm. Now by exchanging the order of integration we have $s_{\epsilon,\rho}(A) = \int F_{\epsilon,\rho}(\lambda)\, dE_\lambda$ with

$$F_{\epsilon,\rho}(\lambda) = \int_\epsilon^\rho \frac{da}{a} \int_{-\infty}^{+\infty} db\, \mathscr{W}_g s(b,a) h_{b,a}(\lambda),$$

$$h_{b,a}(t) = h([t-b]/a)/a, \qquad h(t) = (t+i)^{-n}.$$

But since $s \in L^1(\mathbb{R})$ it is weakly oscillating and the theorem follows from the convergence behaviour of the inversion formula for wavelet transforms. □

Consider the family of functions $s_t(x) = e^{itx}$. Their wavelet transforms read $\mathscr{W}_g s_t(b,a) = \bar{\hat{g}}(at)\, e^{ibt}$. Therefore, we have formally[6]

$$e^{itA} = \frac{1}{c} \int_0^\infty da\, a^{n-2} \bar{\hat{g}}(at) \int_{-\infty}^{+\infty} db\, e^{itb}\, R_{b-ia}^n(A), \qquad (t > 0),$$

where $c = -2i\pi \int_0^\infty d(\omega/\omega)(-i\omega)^{n-1} \bar{\hat{g}}(\omega)\, e^{-\omega}$. The problem is that the b

[6] Note that for $t > 0$ only the lower half-plane for the resolvent has to be considered since the upper half-plane gives 0 anyway.

integration for unbounded A is not convergent. For bounded A, instead, it is convergent for $n \geq 2$ and the above formula is valid. In the following section we deal with the unbounded case. As we will see, a Cesaro type summation helps.

14.2 The function e^{itA}

In this section we will show how to apply the above formalism to the case of the function e^{itA}. Let A be a closed operator acting in some Hilbert space with dense domain $D(A)$. If there is a strongly continuous semigroup $F(t) \geq 0$, whose generator is iA, $dF(t)/dt = iAF(t)$, then a theorem of Hille and Yoshida shows that A is of tempered type in the lower half-plane. We write $F(t) = e^{itA}$. The resolvent can then be written in a strong sense as (see Hille and Philips (1957) for a proof)

$$\left.\begin{aligned} &R^n_{b-ia}(A)x = \frac{1}{i(n-1)!}\int_0^\infty d\tau\, e^{-i(b-ia)\tau}(-i\tau)^{n-1}F(\tau)x, \\ &a > 0, x \in D(A). \end{aligned}\right\} \tag{14.2.1}$$

Theorem 14.1.2

The semigroup generated by iA can be recovered with the help of the following Cesaro mean:

$$e^{itA}\,x = \lim_{\beta \to \infty} \int_0^\infty da\, a^{n-2} \int_{-\infty}^{+\infty} db\, \Phi(b/\beta)\bar{\hat{g}}(at)\, e^{ibt}\, R^n_{b-ia}(A)x, \qquad x \in D(A),$$

where Φ is any smooth function in C^∞ with compact support and $\Phi(0) = 1$ and $g \in S_+(\mathbb{R})$ normalized such that $c^+_{g,n} = 1$.

Proof. Inserting expression (14.2.1) in the above formula we have, after changing the order of integration for fixed β,

$$\int_0^\infty d\tau\, G_{\beta,t}(\tau)F(\tau)x,$$

where

$$G_{\beta,t}(\tau) = \frac{1}{i(n-1)!}\int_0^\infty \frac{da}{a}\int_{-\infty}^{+\infty} db\, \Phi(b/\beta)\bar{\hat{g}}(at)(-ia\tau)^{n-1}\, e^{-ar}\, e^{ib(t-\tau)}.$$

Carrying out the integration over b we obtain the Fourier transform of Φ and we have

$$G_{\beta,t}(\tau) = \frac{1}{i(n-1)!}\int_0^\infty \frac{da}{a}\,\bar{\hat{g}}(at)(-ia\tau)^{n-1}\, e^{-a\tau}\, \beta\hat{\Phi}(\beta[\tau - t]).$$

Since $\hat{\Phi}$ is in $S(\mathbb{R})$ and since $\int \hat{\Phi} = 2\pi\Phi(0) = 2\pi$ it follows easily that $G_{\beta,t}$ is an approximation of the identity, namely,

$$\text{(i) for all } \epsilon > 0 \lim_{\beta \to \infty} \int_0^{t-\epsilon} d\tau |G_{\beta,t}(\tau)| = \lim_{\beta \to \infty} \int_{t+\epsilon}^{\infty} d\tau |G_{\beta,t}(\tau)| = 0,$$

$$\text{(ii)} \lim_{\beta \to \infty} \int_{t-\epsilon}^{t+\epsilon} d\tau\, G_{\beta,t}(\tau) = 1,$$

$$\text{(iii) for all } \beta > 0 \int_0^{\infty} d\tau |G_{\beta,t}(\tau)| \leq c < \infty,$$

and the theorem follows by standard arguments which we recall for the convenience of the reader. By (ii) we may write

$$\begin{aligned} \lim_{\beta \to \infty} \int_0^{\infty} d\tau\, G_{\beta,t}(\tau) F(\tau) x &= F(t)x + \lim_{\beta \to \infty} \int_0^{\infty} d\tau\, G_{\beta,t}(\tau)(F(\tau)x - F(t)x) \\ &= F(t)x + \left\{ \int_0^{t-\epsilon} + \int_{t+\epsilon}^{\infty} + \int_{t-\epsilon}^{t+\epsilon} \right\} \\ &\qquad \times d\tau\, G_{\beta,t}(\tau)(F(\tau)x - F(t)x). \end{aligned}$$

In the first two integrals we use the fact that $\|F(\tau)\| \leq 1$ and (i) to show that these terms converge strongly to 0. In the third term we use (iii) and the continuity of F to see that its norm is $\leq n$, where η tends to 0 with ϵ. □

In particular, we may choose $\hat{g}$ to be supported by a small interval centred at $\omega = 1$, in which case the integral runs over a strip parallel to the real axis centred at a distance $1/t$. This shows explicitly that the behaviour of the resolvent at a distance $\Im z \sim 1/t$ from the real axis determines the behaviour of the unitary group generated by A for large times t.

3
MULTI-RESOLUTION ANALYSIS

In this chapter we present the general framework for the construction of an orthonormal basis of wavelets over the real line $\mathbb{R}$, the discretized real line $\mathbb{Z}$, the circle $\mathbb{T}$, and the discretized torus $\mathbb{Z}/N\mathbb{Z}$. As we saw in Section 13 of Chapter 2 there are wavelets g such that the wavelet transform

$$\mathscr{W}_g s(b, a) = \int_{-\infty}^{+\infty} dt \, \frac{1}{a} \bar{g}\left(\frac{t-b}{a}\right) s(t)$$

may be sampled over a dyadic grid, without loss of information. We obtained a special class of wavelets for which the reconstruction formula from the dyadic grid samples is particularly simple:

$$s = \sum_{j \in \mathbb{Z}} \sum_{n \in \mathbb{Z}} \mathscr{W}_g s(n2^j, 2^j) g(2^j \cdot - k) = \sum_{j \in \mathbb{Z}} \sum_{n \in \mathbb{Z}} \langle g_{j,n} \mid s \rangle g_{j,n},$$

where $g_{j,n}(t) = 2^{-j/2} g(2^{-j}t - n)$. However, this family of dilated and translated functions was not independent, in the sense that, in general, $\langle g_{j,n} \mid g_{j',n'} \rangle \neq \delta_{j,j'} \delta_{n,n'}$: the family of wavelets that we have found until now were a frame but not an orthonormal base. The aim of this section is to construct wavelets for which this family of dilated and translated wavelets is actually an orthonormal base, and hence the perfect sampling operator over the dyadic grid

$$\Sigma\colon \text{image } \mathscr{W}_g \to L^2(\mathbb{Z} \times \mathbb{Z}), \qquad T \mapsto 2^{j/2} T(2^j k, 2^j)$$

is an isometric isomorphism. The most appropriate tool for this construction is multi-resolution analysis. This is a sequence of spaces

$$V_j \subset V_{j+1},$$

each of which is just a re-scaled version of the former:

$$s \in V_j \Leftrightarrow s(2t) \in V_{j+1}.$$

In addition, V_0—and hence all V_j—has a translation invariant orthonormal basis. It turns out that the orthogonal complements $W_j = V_{j+1} \ominus V_j$ then also have a translation invariant orthonormal basis and the collection of all these basis functions is the desired wavelet basis.

1 Some sampling theorems

In this section we wish to study those spaces that have a translation invariant orthonormal basis. In the Shanon sampling theorem, Theorem 9.6.1, we have already encountered a typical example of such spaces. The space V consisting of those square integrable functions whose Fourier transform is supported by $[-\pi, +\pi]$ has an orthonormal basis that consists of the translates of one single function, namely, $\{\text{sinc}(\cdot - n)\}$. In addition, we have that

$$s \mapsto \langle \text{sinc}(\cdot - n) \mid s \rangle, \qquad s \in V, n \in \mathbb{Z}$$

defines a one-to-one correspondence between V and $L^2(\mathbb{Z})$. In order to give a formal definition let us introduce the shift operator for sequences:

$$v \mapsto Sv, \qquad Sv(n) = v(n-1).$$

Then we define the following.

Definition 1.0.1
A subspace V of $L^2(\mathbb{R})$ is called an equidistant sampling space if there is some function $g \in V$ such that the sampling operator

$$\Sigma_g \colon V \to L^2(\mathbb{Z}), \qquad s \mapsto \Sigma_g s(n) = \langle g(\cdot - n) \mid s \rangle$$

is an isomorphism[1] *that makes the following diagram commutative:*

$$\Sigma_g T_1 = S\Sigma_g \qquad \begin{array}{ccc} V & \xrightarrow{\Sigma_g} & L^2(\mathbb{Z}) \\ \downarrow{\scriptstyle T_1} & & \downarrow{\scriptstyle S} \\ V & \xrightarrow{\Sigma_g} & L^2(\mathbb{Z}) \end{array}$$

Note that sampling at $\mathbb{Z}$ may be easily replaced by sampling at $\alpha\mathbb{Z}$ by looking at $D_\alpha s$ instead of s.

1.1 Riesz bases

In order to understand the structure of sampling spaces we need the notion of a Riesz basis. For a general introduction to different concepts of basis we refer to Young (Young 1980). Recall that a Riesz basis $\{s_n\}$ of a (separable) Hilbert space $\mathscr{H}$ is a (Schauder) basis that is equivalent to an orthonormal basis $\{e_n\}$ in the sense that there is an isomorphism B such that $s_n = Be_n$ for all n. Since all infinite-dimensional separable Hilbert spaces are isomorphic we may choose $\{e_n\}$ in any infinite-dimensional separable Hilbert space, not necessarily in $\mathscr{H}$ itself. Therefore, $\{s_n\}$ is a Riesz basis for its closed linear

[1] An operator $K \colon X \to Y$ between two Banach spaces is an isomorphism iff K is bounded and bijective. From the open mapping theorem it then follows that K has a bounded inverse. Equivalently, K is an isomorphism iff it is surjective and if there are two finite, positive constants c_0 and c_1 such that

$$c_0 \|s\|_X \leq \|Ks\|_Y \leq c_1 \|s\|_X,$$

for all $s \in X$.

span $\mathscr{H}$ iff there are two constants $0 < c_1 \leq c_2 < \infty$ such that

$$c_1 \sum_{n \in \mathbb{Z}} |\alpha(n)|^2 \leq \left\| \sum_{n \in \mathbb{Z}} \alpha(n) s_n \right\|_{\mathscr{H}}^2 \leq c_2 \sum_{n \in \mathbb{Z}} |\alpha(n)|^2 \tag{1.1.1}$$

for all complex-valued sequences α, only a finite number of its entries being different from zero. Indeed, if we look at the image of the orthonormal base $\{\delta_n\}$ of $L^2(\mathbb{Z})$ under the operator

$$B: L^2(\mathbb{Z}) \to \mathscr{H}, \qquad \alpha \mapsto \sum_{n \in \mathbb{Z}} \alpha(n) s_n,$$

we see that (1.1.1) implies that B is an isomorphism and hence $s_n = B\delta_n$ form a Riesz basis for their closed linear span.

Vice versa, if B is an isomorphism then, by definition, (1.1.1) holds.

Equivalently, by looking at the adjoint of $K = B^*$ we see that $\{s_n\}$ is a Riesz basis iff the imperfect sampling operator

$$K: \mathscr{H} \to L^2(\mathbb{Z}), \qquad r \mapsto \langle s_n \mid r \rangle,$$

is an isomorphism between $\mathscr{H}$ and $L^2(\mathbb{Z})$.

With this remark we have the following theorem.

Theorem 1.1.1

A closed subspace $V \subset L^2(\mathbb{R})$ is a sampling space iff it has a translation invariant Riesz basis in the sense that there is some $g \in V$ such that the family of translates $\{g(\cdot - n): n \in \mathbb{Z}\}$ is a Riesz basis for V.

A last remark concerns the relation between frames and Riesz bases. Clearly, every Riesz basis is a frame, whereas the contrary obviously fails. However, it can be shown that if $\{s_n\}$ is a frame that is 'the least redundant possible', in the sense that whenever one single vector is removed the remaining set is no longer complete,[2] then the original frame is actually a Riesz basis. For a proof see Young (1980, page 188, Theorem 12).

Another way to see the difference is to look once more at the sampling operator $K: s \mapsto \langle s_n \mid s \rangle_{\mathscr{H}}$. If $\{s_n\}$ is a frame this operator is continuous and bounded from below—hence it is injective. In general, however, it is not surjective: not every sequence in $L^2(\mathbb{Z})$ can be obtained as a sampling sequence of some $s \in V$ due to the internal correlations of the s_n. In the case where $\{s_n\}$ is a Riesz basis the correlations are sufficiently low that K is surjective, and vice versa, K being bounded and bijective implies $\{s_n\}$ to be a Riesz basis.

[2] These frames are sometimes called 'exact frames'.

1.2 The Fourier space picture

We now wish to characterize the sampling spaces in Fourier space. We will need, besides the Fourier transform for functions over $\mathbb{R}$, that for sequences, for which we recall the definition:

$$F^{\mathbb{Z}}: L^2(\mathbb{Z}) \to L^2(\mathbb{T}), \qquad v \mapsto \sum_{k \in \mathbb{Z}} v(k)\, e^{-ik\omega}.$$

We will in general simply write $\hat{v}$ since from the context it is clear which Fourier transform is being used. Since translation of a function over $\mathbb{R}$ by n corresponds to multiplication by $e^{-in\omega}$ in Fourier space, it follows that the superposition of translates of one single $g \in L^2(\mathbb{R})$ with expansion coefficients v can be written

$$s = \sum_{n \in \mathbb{Z}} v(n) g(\cdot - n) \quad \Leftrightarrow \quad \hat{s}(\omega) = m(\omega)\hat{g}(\omega),$$

where $m(\omega)$ is the 2π-periodic function given by $m = \hat{v}$. This formula clearly holds if v has only a finite number of non-zero elements, or equivalently if m is a trigonometric polynomial. In this case we have

$$\|s\|_2^2 = 2\pi \int_{-\infty}^{+\infty} d\omega\, |m(\omega)|^2\, |\hat{g}(\omega)|^2 = 2\pi \int_0^{2\pi} d\omega\, |m(\omega)|^2 \sum_{n \in \mathbb{Z}} |\hat{g}(\omega + 2\pi n)|^2,$$

where we have used the periodicity of m. By taking limits we have proved the following theorem.

Theorem 1.2.1
Let $g \in L^2(\mathbb{R})$. Then the closed linear span consists of $\{g(\cdot - n)\}$ are exactly those functions that can be written as

$$\hat{s}(\omega) = m(\omega)\hat{g}(\omega),$$

with some 2π-periodic function in $L^2(\mathbb{T}, d\mu)$, where the measure μ is given by its $L^1(\mathbb{T})$-density

$$d\mu(\omega)/d\omega = \sum_{n \in \mathbb{Z}} |\hat{g}(\omega + 2\pi n)|^2.$$

Now, as the next theorem proves, $\{g(\cdot - n)\}$ is a Riesz basis iff this measure is equivalent to the Lebesgue measure.

Theorem 1.2.2
The family of translates $g(\cdot - n)$ is a Riesz basis for their linear span V iff there are two constants $0 < c_0 \leq c_1 < \infty$ such that

$$c_0 \leq \sum_{k \in \mathbb{Z}} |\hat{g}(\omega - 2\pi n)|^2 \leq c_1 \qquad (1.2.1)$$

for almost all $\omega \in \mathbb{R}$.

Proof. Let v be a finite sequence. Therefore $m(\omega) = \hat{v}(\omega)$ is a trigonometric polynomial. We then have, by periodicity of m and the unitarity of the Fourier transform,

$$\left\| \sum_{n\in\mathbb{Z}} v(n)g(\cdot - n) \right\|^2_{L^2(\mathbb{R})} = 2\pi \int_{-\infty}^{+\infty} d\omega\, |m(\omega)|^2\, |\hat{g}\omega|^2$$

$$= 2\pi \int_0^{2\pi} d\omega\, |m(\omega)|^2 \sum_{k\in\mathbb{Z}} |\hat{g}(\omega + 2\pi k)|^2. \quad (1.2.2)$$

Therefore, if (1.2.1) holds it follows that the mean $\sum_{k\in\mathbb{Z}} |\hat{g}(\omega + 2\pi k)|^2\, d\omega$ is equivalent to 1 and therefore, by the unitarity of the Fourier transform, $\|v\|^2_{L^2(\mathbb{Z})} = 2\pi\|m\|^2_{L^2(\mathbb{T})}$ is equivalent[3] to (1.2.2) and $\{g(\cdot - k)\}$ is a Riesz basis.

Vice versa, suppose now that $\{g(\cdot - n)\}$ is a Riesz basis for its linear span. By (1.1.1) there are two positive, finite constants such that

$$c_0\|v\|^2_{L^2(\mathbb{Z})} \le \left\| \sum_{n\in\mathbb{Z}} v(n)g(\cdot - n) \right\|^2_{L^2(\mathbb{R})} \le c_1\|v\|^2_{L^2(\mathbb{Z})}$$

for every finite sequence v. In Fourier space this reads ($\hat{v}(\omega) = F^{\mathbb{Z}}v(\omega) = m(\omega)$):

$$c_0 \int_0^{2\pi} d\omega\, |m(\omega)|^2 \le \int_0^{2\pi} d\omega\, |m(\omega)|^2 \sum_{n\in\mathbb{Z}} |\hat{g}(\omega + 2\pi n)|^2$$

$$\le c_1 \int_0^{2\pi} d\omega\, |m(\omega)|^2.$$

Since m is an arbitrary polynomial it may be chosen such that $|m(\omega)|^2$ is an approximation of the identity converging towards $\delta(\omega - \xi)$ with $\xi \in [-\pi, +\pi)$ arbitrary. In addition, since the sum is a function in $L^1(\mathbb{T})$ this approximation may be chosen[4] such that the convergence holds pointwise almost everywhere and (1.2.1) follows. □

Remark. The same argument shows that the existence of a finite c_1 in (1.2.1) is equivalent to the family of translates being a Bessel sequence.

---

[3] Two real valued functions or 'equivalent' if their quotient is bounded away from 0 and ∞.

[4] One way to obtain such an approximation is by periodizing an approximation over $\mathbb{R}$: let $\phi \in S(\mathbb{R})$ be such that its Fourier transform is compactly supported and that $\hat{\phi}(0) = 1$. Then

$$s_\epsilon(\omega) = \sum_{k\in\mathbb{Z}} \frac{1}{\epsilon}\phi\left(\frac{\omega + 2\pi k}{\epsilon}\right) \Leftrightarrow F^{\mathbb{T}}s_\epsilon(n) = \hat{\phi}(\epsilon n)$$

(by the Poisson summation formula) is an approximation of the identity, as can be seen if we identify functions over $\mathbb{T}$ with periodic functions over $\mathbb{R}$. In addition, each s_ϵ is a trongometric polynomial since its Fourier transform is a finite sequence.

1.3 Translation invariant orthonormal basis

Given that every sampling space has a translation invariant Riesz basis the question arises whether it even has a translation invariant orthonormal basis. The answer is affirmative, as is now proved.

Theorem 1.3.1
The family of translates $\{\phi(\cdot - n): n \in \mathbb{Z}\}$ *of one single function* $\phi \in L^2(\mathbb{R})$ *is an orthonormal set,* $\langle \phi(\cdot - n) \mid \phi(\cdot - m) \rangle = \delta_{n,m}$ *iff*

$$\sum_{k \in \mathbb{Z}} |\hat{\phi}(\omega + 2\pi k)|^2 = 1$$

for almost all $\omega \in \mathbb{R}$.

Proof. Clearly, $\phi(\cdot - n)$ is an orthonormal family iff

$$\langle \phi(\cdot - n) \mid \phi(\cdot - m) \rangle = (\tilde{\phi} * \phi)(m - n) = \delta_{n,m}.$$

Equivalently, we have, with the perfect sampling operator,

$$\Sigma(\tilde{\phi} * \phi) = \delta.$$

In Fourier space we may use the Poisson summation formula to obtain the equivalent expression

$$1 = \Pi(|\hat{\phi}|^2)(\omega) = \sum_{k \in \mathbb{Z}} |\hat{\phi}(\omega + 2\pi k)|^2. \qquad \square$$

Therefore, given a sampling space V with translation invariant Riesz basis generated by the translates of $g \in V$ we obtain an orthonormal base of V by setting

$$\hat{\phi}(\omega) = m(\omega)\hat{g}(\omega), \qquad |m(\omega)| = 1 \Big/ \sqrt{\sum_{k \in \mathbb{Z}} |\hat{g}(\omega + 2\pi k)|^2}.$$

Clearly, $m \in L^2(\mathbb{T})$ and therefore $\phi \in V$ and the span of their integer translates coincides with V. The orthonormality of the $\phi(\cdot - n)$ follows from the previous theorem. Note that ϕ is not unique since m is defined up to some phase function only. However, this is the only freedom we have in choosing ϕ.

This theorem is actually a special case of a general procedure to obtain an orthonormal basis from a Riesz basis (see Young 1980, page 48): let $\{g_n\}$ be a Riesz basis for $\mathscr{H}$. Then the sampling–reconstruction operator

$$B = \Sigma_g^* \Sigma_g : s \mapsto \sum_{k \in \mathbb{Z}} \langle g_n \mid s \rangle g_n$$

is strictly positive and bounded. It therefore has a unique, strictly positive square root $K = B^{1/2}$ which is again an isomorphism and we may consider $\phi_n = K^{-1} g_n$. As an isomorphic image of a Riesz basis it is again a Riesz basis and from

$$\langle \phi_n \mid \phi_m \rangle_{\mathscr{H}} = \langle B^{-1/2} g_n \mid B^{-1/2} g_m \rangle_{\mathscr{H}} = \langle g_n \mid B^{-1} g_m \rangle_{\mathscr{H}}$$
$$= \langle g_n \mid (\Sigma_g^* \Sigma_g)^{-1} g_m \rangle_{\mathscr{H}} = \langle \Sigma_g^{*-1} g_n \mid \Sigma_g^{*-1} g_m \rangle_{L^2(\mathbb{Z})}^2,$$

together with $\Sigma_g^* \delta_n = g_n$, it follows that this last expression equals $\delta_{n,m}$ and $\{\phi_n\}$ is an orthonormal basis.

1.4 Skew projections

Let V be a sampling space with translation invariant orthonormal base generated by ϕ. Consider the generalized sampling–reconstruction operator

$$\Sigma_h^* \Sigma_g : s \mapsto \sum_{k \in \mathbb{Z}} \langle g(\cdot - k) \mid s \rangle h(\cdot - k),$$

where we sample with the help of $g \in L^2(\mathbb{R})$ but reconstruct with $h \in V$.

We first check under which conditions this defines a projection operator. We will need a little regularity on g and h. Suppose that the $g(\cdot - k)$ are a Riesz basis for their linear span and suppose the same holds for h. Then, clearly, sampling with respect to g is bounded from $L^2(\mathbb{R})$ onto $L^2(\mathbb{Z})$. The reconstruction with respect to h is again bounded from $L^2(\mathbb{Z})$ to V, the linear span of the $h(\cdot - k)$. The combination $\Sigma_h^* \Sigma_g$ is therefore bounded from $L^2(\mathbb{R})$ onto V.

Let us examine the question when this is a projection operator onto V. Therefore, consider

$$s = \sum_{k \in \mathbb{Z}} \alpha(k) h(\cdot - k)$$

and let us look at $\Sigma_h^* \Sigma_g s$. We have

$$\Sigma_g s(k) = \sum_{k' \in \mathbb{Z}} M(k - k') \alpha(k'), \qquad M(k) = \langle g(\cdot - k) \mid h \rangle.$$

In the case of a projection the left-hand side must be equal to α for all $\alpha \in L^2(\mathbb{Z})$. This implies that $M = \delta$ and we have proved

Theorem 1.4.1

Suppose that $g, h \in L^2(\mathbb{R})$ are such that their translates are a Riesz basis for their linear span. Then the sampling–reconstruction operator is a projector onto the linear span of $\{h(\cdot - k)\}$ iff for almost all ω

$$\langle g(\cdot - k) \mid h \rangle = \delta(k) \Leftrightarrow \sum_{k \in \mathbb{Z}} \overline{\hat{g}(\omega + 2\pi k)} \hat{h}(\omega + 2\pi k) = 1. \qquad (1.4.1)$$

In particular if both g and h are in the same sampling space V then the above projector is an orthogonal projector—the space orthogonal to the span of integer translates of g, which is V, is mapped to 0—and thus the Riesz bases generated by the translates of g and h are a bi-orthogonal Riesz bases for V. Let ϕ generate a translation invariant orthonormal basis of V. Then suppose g and $h \in V$ generate a translation invariant Riesz basis of V. It follows that they are bi-orthogonal bases in V iff

$$\bar{\hat{g}}(\omega)\hat{h}(\omega) = |\hat{\phi}(\omega)|^2.$$

Indeed, the condition is sufficient because of Theorem 1.3.1. It is necessary because from $\hat{g} = m_g\hat{\phi}$ and $\hat{h} = m_h\hat{\phi}$ we obtain from 1.4.1 that $\bar{m}_g m_h = 1$.

1.5 Perfect sampling spaces

We now relax the condition $g, h \in L^2(\mathbb{R})$ and consider for later reference the extreme case where formally $g = \delta$ and the sampling operator with respect to g becomes the perfect sampling operator

$$\Sigma\colon s \mapsto \Sigma s(n) = s(n), \qquad n \in \mathbb{Z}.$$

Definition 1.5.1
A subspace V of $L^2(\mathbb{R})$ of continuous functions is called a perfect sampling space if the perfect sampling operator

$$\Sigma\colon V \to L^2(\mathbb{Z}), \qquad s \mapsto \Sigma s(n) = s(n)$$

is an isomorphism between V and $L^2(\mathbb{Z})$ that makes the following diagram commutative:

$$\Sigma T_1 = S\Sigma, \qquad \begin{array}{ccc} V & \xrightarrow{\Sigma} & L^2(\mathbb{Z}) \\ \downarrow{\scriptstyle T_1} & & \downarrow{\scriptstyle S} \\ V & \xrightarrow{\Sigma} & L^2(\mathbb{Z}) \end{array} \tag{1.5.1}$$

Example 1.5.2
Consider the space of continuous, piecewise, affine functions with knots at the integers. A function s in this space is defined if we know its value at the integers. The value at every intermediate point t with $n \le t < n + 1$ is then obtained by linear interpolation:

$$s(t) = (t - n)s(n) + (n + 1 - t)s(n + 1).$$

Consider the unique function h in this space that has $h(0) = 1$ and $h(n) = 0$ for $n \neq 0$. This is the triangular function

$$h(t) = \begin{cases} 1 - |t| & \text{for } |t| < 1, \\ 0 & \text{otherwise.} \end{cases}$$

A minute's reflection shows that

$$s(t) = \sum_{n \in \mathbb{Z}} s(n) h(t - n).$$

That the sampling operator $\Sigma: s \mapsto s(n)$ actually is an isomorphism is easy to see and is left to the reader (cf. Theorem 1.5.5 below).

This example actually reflects the main features of the perfect sampling space, as is shown by the following theorem.

Theorem 1.5.3

Every perfect sampling space V is generated by the translates $\{h(\cdot - n)\}$ of one single function h. This function is not unique. However, there is only one of these functions such that every function $s \in V$ can be developed according to

$$s(t) = \sum_{n \in \mathbb{Z}} s(n) h(t - n).$$

This function is uniquely determined by

$$h(n) = \delta(n) = \begin{cases} 1 & \text{for } n = 0, \\ 0 & \text{otherwise}. \end{cases} \tag{1.5.2}$$

for $n \in \mathbb{Z}$.

Note that we have, in complete formal analogy with the bi-orthogonal case,

$$\Sigma_h^* \Sigma = \mathbb{1}_V.$$

Proof. Clearly, $L^2(\mathbb{Z})$ is generated by the translates $\{S^n \delta : n \in \mathbb{Z}\}$. Since Σ is an isomorphism it follows that V is, according to the commutative diagram (1.5.1), also generated by the translates of $h = \Sigma^{-1}\delta$. But this is exactly the function h that is defined by (1.5.2). The translation invariance of V follows from that of $L^2(\mathbb{Z})$. □

The next lemma characterizes property (1.5.2) in Fourier space.

Lemma 1.5.4

Let h be a function with $\hat{h} \in L^1(\mathbb{R})$. Then h is continuous and $h(n) = \delta(n)$ iff

$$\sum_{n \in \mathbb{Z}} \hat{h}(\omega + 2\pi n) \equiv 1$$

for almost all ω.

Proof. This is an immediate application of the Poisson summation formula:

$$\Sigma h(n) = \delta(n) \Leftrightarrow \Pi \hat{h} = F^{\mathbb{Z}}\delta(n) \equiv 1.$$

This proves the lemma. □

Every perfect sampling space is a sampling space, as is shown by the following theorem.

Theorem 1.5.5

Every perfect sampling space V with generating function h has a translation invariant orthonormal base. A possible choice is given by

$$\phi(\omega) = m(\omega) \cdot \hat{h}(\omega), \qquad m(\omega) = 1 \Big/ \sqrt{\sum_{k \in \mathbb{Z}} |\hat{h}(\omega + 2\pi k)|^2},$$

where h is given by $\Sigma h = \delta$.

Note that the contrary is not true: in general on a sampling space the perfect sampling operator need not be an isomorphism, as we will soon see in examples.

Proof. From Theorem 1.5.3 it follows that every function in V can be written as a superposition of translates of h:

$$s = \sum_{k \in \mathbb{Z}} \alpha(k) h(\cdot - k),$$

and the $L^2(\mathbb{Z})$ norm of the expansion coefficient α is, by definition, equivalent to $\|s\|_{L^2(\mathbb{R})}$. This shows that $\{h(\cdot - k)\}$ is a Riesz basis and the theorem follows from Theorem 1.3.1 together with Theorem 1.2.2. □

The above theorem may be rephrased by saying that the projection of the translates of the δ distribution into V is a Riesz basis for V. More precisely, let ϕ be as in Theorem 1.5.5 Then ρ, the projection of δ into V reads in Fourier space:

$$\hat{\rho}(\omega) = \hat{\phi}(\omega) \sum_{k \in \mathbb{Z}} \bar{\hat{\phi}}(\omega + 2\pi k) = |m(\omega)|^2 \hat{h}(\omega) \sum_{k \in \mathbb{Z}} \bar{\hat{h}}(\omega + 2\pi k) = |m(\omega)|^2 \hat{h}(\omega).$$

Thus the perfect sampling operator acting on functions in V can also be replaced by the imperfect sampling operator with respect to $\rho \in V$:

$$\Sigma_\delta = \Sigma_\rho, \qquad \text{on } V.$$

This situation is similar to the situation we encountered in the continuous wavelet transform, where perfect sampling in the half-plane of a function in the image of the wavelet transform was the same as taking scalar products with a smooth function, namely, the reproducing kernel. In the case of a perfect sampling space we can reproduce in this way the value of s at least at all points $t \in \mathbb{Z}$.

1.6 Splines

We end this section by looking at the main example of a sampling space given by splines of arbitrary order. These spaces will be used in the sequel. Not all of them are perfect sampling spaces, as we will see, which underlines that the converse of Theorem 1.5.5 does not hold in general.

A spline of order n is a function that consists of pieces of polynomials on the intervals $[k, k+1)$ glued together such that the whole function is $n-1$-times continuously differentiable. The nth derivative may have finite jumps at the integral points $n \in \mathbb{Z}$. Let $\chi_{[0,1)}$ be the characteristic function of $[0, 1)$. Further, let

$$\beta = \beta_n = \overbrace{\chi_{[0,1)} * \chi_{[0,1)} * \cdots * \chi_{[(0,1)}}^{n+1\text{-times}}.$$

Then β is compactly supported, continuous for $n > 0$, and $n-1$-times continuously differentiable for $n > 1$. In addition, its restriction to each of the intervals $[m, m+1]$ coincides with a polynomial of degree n. So β is a spline of order n. We now fix the index n arbitrarily for the rest of this section so that we need not keep track of it in the notation. Let V be the space that is spanned by $\{\beta(\cdot - k)\}$, where $k \in \mathbb{Z}$. Since $\hat{\chi}_{[0,1)}(\omega) = (1 - e^{-i\omega})/i\omega$ it follows that

$$\hat{\beta}(\omega) = \left(\frac{1 - e^{-i\omega}}{i\omega}\right)^{n+1}.$$

Let us try to construct an orthonormal base for V. For this we are led to consider the periodization of $|\hat{\beta}|^2$

$$\begin{aligned} m(\omega) &= \sum_{k \in \mathbb{Z}} |\hat{\beta}(\omega + 2\pi k)|^2 \\ &= |1 - e^{-i\omega}|^{2n+2} \sum_{k \in \mathbb{Z}} \frac{1}{(\omega + 2\pi k)^{2n+2}} \\ &= \frac{2^n}{(2n-1)!} (1 - \cos(\omega))^{n+1} \partial_\omega^{2n} \frac{1}{\sin^2(\omega/2)}. \end{aligned}$$

As may be shown by induction we have $m \sim 1$ and therefore we may set

$$\varrho(\omega) = \frac{\hat{\beta}(\omega)}{\sqrt{m(\omega)}}$$

to obtain a translation invariant orthonormal base $\{\varrho(\cdot - k)\}$ for V.

1.7 Exponential localization

The orthonormal bases that we have constructed are all derived from some compactly supported function β. The orthogonalization procedure, however, destroys this sharp localization. However, as we will now show, the basis functions are still exponentially localized, since the expansion coefficients are exponentially decreasing.

Theorem 1.7.1
Let m be a 2π-periodic function that is analytic in some strip $|\Im z| \leq c$. Then m is the Fourier transform of a sequence that satisfies at

$$\|e^{c|n|}s\|_{\infty} < \infty.$$

Proof. Because of the periodicity and the analyticity of m we shift the path of integration into the complex plane

$$\int_0^{2\pi} m(\omega)\, e^{in\omega} = \int_0^{2\pi} m(\omega \pm ic)\, e^{in(\omega \pm ic)}.$$

Going to the modulus we obtain the desired estimation. □

1.8 Perfect sampling spaces of spline functions

Now that we have shown that the spline functions are naturally associated with some sampling space the question is whether they are also perfect sampling spaces. Equivalently, can we also find an elementary generating function $h \in V$ such that $h(k) = \delta(k)$?

Suppose that V were a perfect sampling space. Then $h \in V$ must satisfy at

(i) $\hat{h}(\omega) = \kappa(\omega)\hat{\beta}(\omega)$, with some $\kappa \in L^2(\mathbb{T})$,
(ii) $\sum_{k \in \mathbb{Z}} h(\omega + 2\pi k) \equiv 1$.

Therefore, we are led to consider

$$\eta(\omega) = \sum_{k \in \mathbb{Z}} \hat{\beta}(\omega + 2\pi k).$$

If there is such an h in V we would have

$$\kappa(\omega) = \eta(\omega)^{-1}.$$

We now distinguish two cases.

n odd

Suppose, first, that n is odd. We then obtain by induction (we leave the details to the reader)

$$\eta(\omega) = e^{-i(n+1)\omega/2} \frac{2^{n+1}}{(n-1)!} \sin^{n+1}(\omega/2) \partial_\omega^{n-1} \frac{1}{\sin^2(\omega/2)}.$$

Again $\eta \sim 1$ and therefore h is obtained as

$$\hat{h}(\omega) = \frac{\hat{\beta}(\omega)}{\eta(\omega)}.$$

n even

Now consider the case where n is even. In this case we are led to

$$\eta(\omega) = i^{n+1}(1 - e^{-i\omega})^{n+1} \frac{1}{n!} \partial_\omega^n \frac{1}{\tan(\omega/2)}.$$

Therefore, $\eta(0) = 0$, as can be shown by induction. This implies that $1/\eta$ cannot be in $L^2(\mathbb{T})$ and thus there is no elementary generating function for V if n is even.

2 Sampling spaces over $\mathbb{Z}$, $\mathbb{T}$, and $\mathbb{Z}/N\mathbb{Z}$

In this section we list the analogue of the previous results for sampling spaces over $\mathbb{Z}$, $\mathbb{T}$, and $\mathbb{Z}/N\mathbb{Z}$. Nothing mathematically new happens and essentially no proofs will be given since they follow the same lines as the previous proofs. This section merely serves as a reference for later sections.

2.1 Sampling spaces over $\mathbb{Z}$

Consider an equidistant sampling space over $\mathbb{R}$ with sampling frequency $\nu = 1$. If, instead, we sample this space with a higher frequency, say $\nu = 2$, so that we take as data the evaluation at all points $\mathbb{Z}/2$, we obtain obtain a subspace of $L^2(\mathbb{Z})$ because the samples are not independent. Indeed, we may drop every second element of the sampling sequence without loss of information: we could still reconstruct the unsampled function we started from and then oversample it again. We now want to characterize those subspaces of $L^2(\mathbb{Z})$ where undersampling (or decimation) is possible.

Let us introduce the subsampling operator, or decimation operator,

$$\Sigma_N^{\mathbb{Z}}\colon L^2(\mathbb{Z}) \to L^2(\mathbb{Z}), \qquad \Sigma_N^{\mathbb{Z}} v(m) = v(Nm).$$

It drops all but every Nth element of the sequence v.

In order to state the Poisson summation formula on $\mathbb{Z}$ we still need a periodization operator for functions over the torus, that is, for periodic functions with period 2π. This is achieved by

$$\Pi_N^{\mathbb{T}}: L^1(\mathbb{T}) \to L^1(\mathbb{T}), \qquad s \mapsto \frac{1}{N} \sum_{k=0}^{N-1} s(\omega/N + 2\pi k/N).$$

By straightforward computation, which we leave to the reader, we obtain the following theorem.

Theorem 2.1.1

On $L^1(\mathbb{Z})$ we have

$$F^{\mathbb{Z}} \Sigma_N^{\mathbb{Z}} = \Pi_N^{\mathbb{T}} F^{\mathbb{Z}}$$

or, more explicitly, for $v \in L^1(\mathbb{Z})$ we have

$$\Sigma_N^{\mathbb{Z}}: \hat{v} \mapsto \frac{1}{N} \sum_{k=0}^{N-1} \hat{v}(\omega/N + 2\pi k/N).$$

Again, the relations become clearer by using a diagram

$$\begin{array}{ccc} L^1(\mathbb{Z}) & \xrightarrow{\Sigma_N^{\mathbb{Z}}} & L^1(\mathbb{Z}) \\ \downarrow F^{\mathbb{Z}} & & \downarrow F^{\mathbb{Z}} \\ A(\mathbb{T}) & \xrightarrow{\Pi_N^{\mathbb{T}}} & A(\mathbb{T}) \end{array}$$

where we have written $A(\mathbb{T})$ for the Wiener algebra over $\mathbb{T}$ given by the image of $L^1(\mathbb{Z})$ under the Fourier transform. Therefore, sampling again corresponds via a Fourier transform to periodizing.

Closely related to the perfect sampling operator, we define the imperfect sampling operator with respect to $g \in L^2(\mathbb{Z})$ as

$$\Sigma_{g,N}^{\mathbb{Z}}: v \mapsto \langle g(\cdot - k) \mid v \rangle_{\mathbb{Z}}.$$

It can again be written as perfect sampling of a smoothed version of v:

$$\Sigma_{g,N}^{\mathbb{Z}} v = \Sigma_N^{\mathbb{Z}}(\tilde{g} * v), \qquad \tilde{g}(\cdot) = \bar{g}(-\cdot).$$

Accordingly, we have in Fourier space

$$\Sigma_{g,N}^{\mathbb{Z}}: \hat{v}(\omega) \mapsto \frac{1}{N} \sum_{k=0}^{N-1} \bar{\hat{g}}(\omega/N + 2\pi k/N) \hat{v}(\omega/N + 2\pi k/N).$$

Definition 2.1.2

A subspace V of $L^2(\mathbb{Z})$ is called an equidistant sampling space with decimation factor N and smoothing filter $g \in L^2(\mathbb{Z})$ iff

$$\Sigma_{g,N}^{\mathbb{Z}}: V \to L^2(\mathbb{Z}), \qquad v \mapsto \langle g(\cdot - k) \mid v \rangle_{\mathbb{Z}}$$

is an isomorphism that makes the following diagram commutative:

$$\Sigma_{g,N}^{\mathbb{Z}} S^N = S\Sigma_{g,N}^{\mathbb{Z}}, \qquad \begin{array}{ccc} V & \xrightarrow{\Sigma_{g,N}^{\mathbb{Z}}} & L^2(\mathbb{Z}) \\ \downarrow S^N & & \downarrow S \\ V & \xrightarrow{\Sigma_{g,N}^{\mathbb{T}}} & L^2(\mathbb{Z}) \end{array}$$

In the case where $g = \delta$ we speak of a perfect sampling space.

Note that we did not assume $g \in V$. If, actually, $g \in V$ then $\{g(\cdot - kN): k \in \mathbb{Z}\}$ is a Riesz basis of V, as before. In every case we have that $\{h(\cdot - kN: k \in \mathbb{Z})\}$ is a Riesz basis for V where $h = (\Sigma_{g,N}^{\mathbb{Z}})^{-1}\delta$. For a perfect sampling space h is the unique sequence in V that satisfies at

$$h(0) = 1, \qquad h(mN) = 0, \quad \text{for } m \neq 0,$$

and we have the following explicit reconstruction of an arbitrary sequence $v \in V$ from its decimated sequence:

$$v(m) = \sum_{k \in \mathbb{Z}} v(kN) h(m - kN).$$

The following theorems are in complete analogy with the case of sampling spaces in $L^2(\mathbb{R})$ and no proofs will be given.

Theorem 2.1.3

Let $N \in \mathbb{N}$. The closed linear span of $\{g(\cdot - kN): k \in \mathbb{Z}\}$, where $g \in L^2(\mathbb{Z})$, are those sequences $s \in L^2(\mathbb{Z})$ that can be written as

$$\hat{s}(\omega) = m(N\omega)\hat{g}(\omega),$$

with some $m \in L^2(\mathbb{T}, d\mu)$, where the positive measure μ reads

$$d\mu(\omega) = \sum_{k=0}^{N-1} |\hat{g}(\omega/N + 2\pi k/N)|^2 \, d\omega.$$

Theorem 2.1.4

Let $N \in \mathbb{N}$. A family of sequences $\{g(\cdot - kN): k \in \mathbb{Z}\}$, where $g \in L^2(\mathbb{Z})$ is a Riesz basis for their linear span iff there are two positive, finite constants such that

$$0 < c_0 \leq \sum_{n=0}^{N-1} |\hat{g}(\omega - 2\pi k/N)|^2 \leq c_2 < \infty$$

for almost all $\omega \in \mathbb{T}$.

By looking at $\Sigma_N^{\mathbb{Z}}(\tilde{\phi} * \phi)$ we see that the following theorem holds.

Theorem 2.1.5

A family of sequences $\{\phi(\cdot - kN): k \in \mathbb{Z}\}$ with $\phi \in L^2(\mathbb{Z})$ is an orthonormal basis for their linear span iff

$$\sum_{n=0}^{N-1} |\hat{\phi}(\omega - 2\pi k/N)|^2 = N$$

for almost all $\omega \in \mathbb{T}$.

Accordingly, a translation invariant orthonormal basis $\{\phi(\cdot - kN)\}$ can be constructed from a Riesz basis by setting

$$\hat{\phi}(\omega) = m(\omega)\hat{g}(\omega), \qquad |m(\omega)| = 1 \Big/ \sqrt{\frac{1}{N}\sum_{n=0}^{N-1} |\hat{g}(\omega - 2\pi k/N)|^2}.$$

2.2 Skew projections

The adjoint of the sampling operator reads

$$(\Sigma_{g,N}^{\mathbb{Z}})^*: v \mapsto \sum_{k\in\mathbb{Z}} v(k)g(\cdot - Nk)$$

and we consider the combination

$$(\Sigma_{h,N}^{\mathbb{Z}})^* \Sigma_{g,N}^{\mathbb{Z}}: v \mapsto \sum_{k\in\mathbb{Z}} \langle g(\cdot - Nk) \mid v\rangle_{\mathbb{Z}} h(\cdot - Nk).$$

Theorem 2.2.1

Let V be a sampling space over $\mathbb{Z}$ with a translation invariant Riesz base generated by h. Suppose in addition that $\{g(\cdot - Nk): k \in \mathbb{Z}\}$ is a Riesz basis for its linear span. Then

$$(\Sigma_{h,N}^{\mathbb{Z}})^* \Sigma_{g,N}^{\mathbb{Z}}$$

is a projection onto V iff for almost all ω

$$\langle g(\cdot - kN) \mid h\rangle = \delta(k) \Leftrightarrow \sum_{k=0}^{N-1} \bar{\hat{g}}(\omega + 2\pi k/N)\hat{h}(\omega + 2\pi k/N) = N.$$

In the case where both h and g lie in the same sampling space V the above projector is orthogonal (cf. Section 1.4). In particular, this means that $h(\cdot - Nk)$ and $g(\cdot - Nk)$ are a bi-orthogonal system of Riesz bases for V:

$$\langle g(\cdot - Nk) \mid h\rangle = \delta(k).$$

2.3 Oversampling of sampling spaces

The main example of sampling spaces over $\mathbb{Z}$ is given by oversampling some sampling space over $\mathbb{R}$. To explain what we mean let X be a sampling space over $\mathbb{R}$ with sampling rate 1 and orthonormal base generated by g. Further, let Y be a sampling space on $\mathbb{R}$ with sampling period N, that is, Y has an orthonormal basis of the form $\{h(\cdot - Nk)\}$. Let us suppose that

$$Y \subset X.$$

Then we have the following lemma.

Lemma 2.3.1

Let $V = \Sigma_g Y$. It follows that V is a sampling space in $L^2(\mathbb{Z})$ with orthonormal basis $\{\phi(\cdot - kN)\}$ generated by

$$\phi(k) = \Sigma_g h(k) = \langle g(\cdot - k) \mid h \rangle_{\mathbb{R}}.$$

Proof. The sampling operator Σ_g restricted to $Y \subset X$ is an isometry. It follows that the image of $\{h(\cdot - kN)\}$ is an orthonormal basis for $V = \Sigma_g Y$. From the commutation relation of the sampling operator with the translations along $\mathbb{Z}$ it follows that this set is just given by the translates of $\phi = \Sigma_g h$. □

2.4 Sampling spaces over $\mathbb{T}$

We now look at the image of a sampling space over $\mathbb{R}$ under the periodization operator. It turns out that these spaces are in general sampling spaces over $\mathbb{T}$ in the sense that it is enough to know $s(2\pi k/N)$, where $k = 0, 1, \ldots, N-1$, to reconstruct s everywhere.

Consider the perfect sampling operator $\Sigma_N^{\mathbb{T}}$ that sends (continuous) functions over $\mathbb{T}$ into a sequence over the discrete circle $\mathbb{Z}/N\mathbb{Z}$ by

$$\Sigma_N^{\mathbb{T}}: s(\omega) \mapsto \{s(2\pi k/N)\}, \qquad k \in \mathbb{Z}.$$

Functions over the discrete torus $\mathbb{Z}/N\mathbb{Z}$ may be identified with periodic sequences of period N. By abuse of language we will denote these sequence spaces by $L^2(\mathbb{Z}/N\mathbb{Z})$ or simply by $(\mathbb{Z}/N\mathbb{Z})$.

We can make $L^2(\mathbb{Z}/N\mathbb{Z})$ a Hilbert space by setting

$$\langle v \mid r \rangle_{\mathbb{Z}/N\mathbb{Z}} = \sum_{k=0}^{N-1} \bar{v}(k) r(k), \qquad \|v\|^2_{L^2(\mathbb{Z}/N\mathbb{Z})} = \sum_{k=0}^{N-1} |v(k)|^2.$$

The discrete Fourier transform is then a map from the space of functions over $\mathbb{Z}/N\mathbb{Z}$ into itself defined by

$$F^{\mathbb{Z}/N\mathbb{Z}}: L^2(\mathbb{Z}/N\mathbb{Z}) \mapsto L^2(\mathbb{Z}/N\mathbb{Z}), \qquad F^{\mathbb{Z}/N\mathbb{Z}} v(n) = \sum_{m=0}^{N-1} v(m)\, e^{-2i\pi nm/N}.$$

We write $\hat{v}$ instead of $F^{\mathbb{Z}/N\mathbb{Z}} v$ if no confusion is possible. Again, it preserves the scalar product and hence the norm:

$$\langle v \mid s \rangle_{\mathbb{Z}/N\mathbb{Z}} = \frac{1}{N} \langle \hat{v} \mid \hat{s} \rangle_{\mathbb{Z}/N\mathbb{Z}}, \qquad \|v\|^2_{L^2(\mathbb{Z}/N\mathbb{Z})} = \frac{1}{N} \|\hat{v}\|^2_{L^2(\mathbb{Z}/N\mathbb{Z})}.$$

The inverse discrete Fourier transform is, up to a scalar multiple, given by the adjoint

$$(F^{\mathbb{Z}/N\mathbb{Z}})^{-1}: L^2(\mathbb{Z}/N\mathbb{Z}) \to L^2(\mathbb{Z}/N\mathbb{Z}), \qquad v(n) \mapsto \frac{1}{N} \sum_{m=0}^{N-1} v(m)\, e^{2i\pi nm/N}.$$

The (cyclic) convolution

$$s * r(n) = \sum_{k=0}^{N-1} s(n-k) r(k),$$

reduces in Fourier space to a pointwise multiplication as usual:

$$v = s * r \Rightarrow \hat{v}(k) = \hat{s}(k) \cdot \hat{r}(k).$$

In order to state the Poisson summation formula we need a periodization operator for sequences mapping sequences over $\mathbb{Z}$ into sequences over $\mathbb{Z}/N\mathbb{Z}$. A natural candidate is given by

$$\Pi_N^{\mathbb{Z}}: v(n) \mapsto \sum_{m \in \mathbb{Z}} v(n + mN).$$

It is well defined whenever $v \in L^1(\mathbb{Z})$.

As before, sampling over the torus is related to periodizing the sequence of Fourier coefficients:

$$F^{\mathbb{T}} s(n) = \int_{\mathbb{T}} d\omega\, s(\omega)\, e^{-i\omega n}.$$

Recall that the $A(\mathbb{T})$ are those functions whose Fourier transform is absolutely summable.

Theorem 2.4.1

On $A(\mathbb{T})$ and $L^1(\mathbb{Z})$, respectively, we have

$$\Pi_N^{\mathbb{Z}} F^{\mathbb{T}} = \frac{2\pi}{N} F^{\mathbb{Z}/N\mathbb{Z}} \Sigma_N^{\mathbb{T}}, \qquad F^{\mathbb{Z}/N\mathbb{Z}} \Pi_N^{\mathbb{Z}} = \Sigma_N^{\mathbb{T}} F^{\mathbb{Z}},$$

or, more explicitly, we have for $s \in A(\mathbb{T})$

$$\frac{2\pi}{N} \sum_{m=0}^{N-1} s\left(\frac{2\pi m}{N}\right) e^{-2i\pi nm/N} = \sum_{m \in \mathbb{Z}} \hat{s}(n + mN).$$

Again, the geometry becomes clearer if we write the following commutative diagrams, the second being an easy consequence of the first:

$$\begin{array}{ccc} A(\mathbb{T}) & \xrightarrow{\Sigma_N^{\mathbb{T}}} & (\mathbb{Z}/N\mathbb{Z}) \\ \big\downarrow F^{\mathbb{T}} & & \big\downarrow \frac{2\pi}{N} F^{\mathbb{Z}/N\mathbb{Z}} \\ L^1(\mathbb{Z}) & \xrightarrow{\Pi_N^{\mathbb{Z}}} & (\mathbb{Z}/N\mathbb{Z}) \end{array} \qquad \begin{array}{ccc} L^1(\mathbb{Z}) & \xrightarrow{\Pi_N^{\mathbb{Z}}} & (\mathbb{Z}/N\mathbb{Z}) \\ \big\downarrow F^{\mathbb{Z}} & & \big\downarrow F^{\mathbb{Z}/N\mathbb{Z}} \\ A(\mathbb{T}) & \xrightarrow{\Sigma_N^{\mathbb{T}}} & (\mathbb{Z}/N\mathbb{Z}) \end{array}$$

In analogy with sampling over the real line we introduce the imperfect sampling operator associated with some $\phi \in L^2(\mathbb{T})$:

$$\Sigma_{\phi,N}^{\mathbb{T}}: L^2(\mathbb{T}) \to L^2(\mathbb{Z}/N\mathbb{Z}), \qquad s(\omega) \mapsto \langle \phi(\cdot - 2\pi k/N) \mid s \rangle_{\mathbb{T}}.$$

It can be written as smoothing followed by perfect sampling:

$$\Sigma_{\phi,N}^{\mathbb{T}} s = \Sigma_N^{\mathbb{T}}(\tilde{\phi} * s), \qquad \tilde{\phi}(\cdot) = \bar{\phi}(-\cdot).$$

In the case where $\phi = \delta$ we agree that $\Sigma_{\delta,N}^{\mathbb{T}} = \Sigma_N^{\mathbb{T}}$ is the perfect sampling operator. We then define the following.

Definition 2.4.2

We call $V \subset L^2(\mathbb{T})$ a sampling space over $\mathbb{T}$ if there is some $\phi \in V$ such that the sampling operator $\Sigma_{\phi,N}^{\mathbb{T}}: V \mapsto L^2(\mathbb{Z}/N\mathbb{Z})$ is an isomorphism that is compatible with the translations:

$$\Sigma_N^{\mathbb{T}} T_{2\pi/N} = S\Sigma_{\phi,N}^{\mathbb{T}}, \qquad \begin{array}{ccc} V & \xrightarrow{\Sigma_{\phi,N}^{\mathbb{T}}} & L^2(\mathbb{Z}/N\mathbb{Z}) \\ \big\downarrow T_{2\pi/N} & & \big\downarrow S \\ V & \xrightarrow{\Sigma_{\phi,N}^{\mathbb{T}}} & L^2(\mathbb{Z}/N\mathbb{Z}) \end{array}$$

In the case where $\phi = \delta$ we speak of a perfect sampling space.

Equivalently, we may say that V has a translation invariant basis.[5]

In Fourier space the linear spans of $\{g(\cdot - 2\pi k/N)\}$ are trivially those functions that can be written as

$$\hat{s}(n) = m(n)\hat{g}(n), \qquad m(n) = m(n+N). \tag{2.4.1}$$

The collection $\{g(\cdot - 2\pi k/N)\}$ is a basis for their span iff

$$\sum_{k \in \mathbb{Z}} |\hat{g}(n + kN)|^2 \neq 0$$

for all $n \in \mathbb{Z}/N\mathbb{Z}$. Indeed, this is equivalent to the fact that the dimension of the space of functions of the form of (2.4.1) is N. The collection $\{g(\cdot - 2\pi k/N)\}$ is an orthonormal family iff

$$\sum_{k \in \mathbb{Z}} |\hat{g}(n + kN)|^2 = \frac{2\pi}{N},$$

as can be seen by looking at $\Sigma_N^{\mathbb{T}}(\tilde{g} * g) = \delta$ in Fourier space.

[5] Since V is finite dimensional every basis is a Riesz basis.

2.5 Periodizing a sampling space over $\mathbb{R}$

As we have said already, sampling spaces over $\mathbb{T}$ can be obtained by periodizing a sampling space over $\mathbb{R}$. Let X be a sampling space over $\mathbb{R}$ with translation invariant orthonormal base $\{\phi_k = (\cdot - 2\pi k/N)\}$. Let V be the set of functions $\sum_{k \in \mathbb{Z}} s(\omega + 2\pi k)$ with $s \in X$, where s can be written as a finite superposition of the ϕ_k.

Theorem 2.5.1
If $\phi \in L^1(\mathbb{R}) \cap L^2(\mathbb{R})$ then V is an N-dimensional sampling space over $\mathbb{T}$. The translates of

$$\rho(\omega) = \Pi\phi(\omega) = \sum_{k \in \mathbb{Z}} \phi(\omega + 2\pi k)$$

generate an orthonormal basis for V.

Proof. In Fourier space we have, thanks to the Poisson summation formula,

$$\hat{\rho}(n) = F^{\mathbb{T}}\rho(n) = F\phi(n), \qquad n \in \mathbb{Z}.$$

Since the $2\pi\mathbb{Z}/N$-translates of ϕ are an orthonormal set we have $\sum_{k \in \mathbb{Z}} |\hat{\phi}(\omega + kN)|^2 = 2\pi/N$. Thus we obtain

$$\sum_{k \in \mathbb{Z}} |\hat{\rho}(n + kN)|^2 = \sum_{k \in \mathbb{Z}} |\hat{\phi}(n + kN)|^2 = 2\pi/N$$

and thus the translates of ρ are an orthonormal family. It is clearly complete since the translates of ϕ span X. □

The analogue holds if V is a perfect sampling space.

2.6 Periodizing a sampling space over $\mathbb{T}$

It is no surprise that the periodization of a sampling space over $\mathbb{T}$ again yields a sampling space with lower dimension, however. More precisely, let X be an N-dimensional sampling space and let $N = kN'$ with all numbers in $\mathbb{N}$. Then $\Pi_k^{\mathbb{T}} X$ is an N'-dimensional sampling space. To see this it is enough to look in Fourier space. Let ϕ be the generator of a translation invariant orthonormal base of X. Then $\rho = \sqrt{k\Pi_k^{\mathbb{T}}}\,\phi$ generates a translation invariant orthonormal base for V as follows from

$$\sum_{m \in \mathbb{Z}} |\hat{\rho}(n + mN')|^2 = \sum_{m \in \mathbb{Z}} |\hat{\phi}(nk + mN)|^2 = \frac{2\pi}{N}.$$

2.7 Sampling spaces over $\mathbb{Z}/N\mathbb{Z}$

We end this section with the most discrete space. Consider now the sampling operator over $\mathbb{Z}/N\mathbb{Z}$ for periodic sequences

$$\Sigma_K^{\mathbb{Z}/N\mathbb{Z}}: (\mathbb{Z}/N\mathbb{Z}) \to (\mathbb{Z}/N'\mathbb{Z}), \qquad \Sigma_K^{\mathbb{Z}/N\mathbb{Z}} v(n) = v(Kn).$$

Now K, N, and N' cannot be independent. Clearly, they must be compatible with the various periodicities, that is,

$$N = KN'.$$

In Fourier space sampling is again periodizing. We introduce the periodizing operator

$$\Pi_K^{\mathbb{Z}/N\mathbb{Z}}: (\mathbb{Z}/N\mathbb{Z}) \mapsto (\mathbb{Z}/N'\mathbb{Z}), \qquad \Pi_K^{\mathbb{Z}/N\mathbb{Z}} v(n) = \frac{1}{K} \sum_{p=0}^{K-1} v(n + pN'), \quad N = KN'.$$

By direct computation we verify the Poisson summation formula

$$F^{\mathbb{Z}/N'\mathbb{Z}} \Sigma_K^{\mathbb{Z}/N\mathbb{Z}} = \Pi_K^{\mathbb{Z}/N\mathbb{Z}} F^{\mathbb{Z}/N\mathbb{Z}}, \qquad \begin{array}{ccc} (\mathbb{Z}/N\mathbb{Z}) & \xrightarrow{\Sigma_K^{\mathbb{Z}/N\mathbb{Z}}} & (\mathbb{Z}/N'\mathbb{Z}) \\ \downarrow F^{\mathbb{Z}/N\mathbb{Z}} & & \downarrow F^{\mathbb{Z}/N'\mathbb{Z}} \\ (\mathbb{Z}/N\mathbb{Z}) & \xrightarrow{\Pi_K^{\mathbb{Z}/N\mathbb{Z}}} & (\mathbb{Z}/N'\mathbb{Z}) \end{array}$$

Or, more explicitly, we have

$$\Sigma_K^{\mathbb{Z}/N\mathbb{Z}}: \hat{v}(n) \mapsto \sum_{m=0}^{K-1} \hat{v}(n + mN').$$

The imperfect sampling operator is defined as

$$\Sigma_{\phi,K}^{\mathbb{Z}/N\mathbb{Z}}: (\mathbb{Z}/N\mathbb{Z}) \to (\mathbb{Z}/N'\mathbb{Z}), \qquad v(n) \mapsto \langle \phi(\cdot - Kn) \mid v \rangle_{\mathbb{Z}/N\mathbb{Z}},$$

which can also be written as

$$\Sigma_{\phi,K}^{\mathbb{Z}/N\mathbb{Z}} v = \Sigma_K^{\mathbb{Z}/N\mathbb{Z}} (\tilde{\phi} * v).$$

In complete analogy with the previous cases we give the following definition.

Definition 2.7.1

A subspace $V \subset L^2(\mathbb{Z}/N\mathbb{Z})$ is called an equidistant sampling space with decimation factor K if the sampling operator

$$\Sigma_{\phi,K}^{\mathbb{Z}/N\mathbb{Z}}: V \mapsto L^2(\mathbb{Z}/N'\mathbb{Z}), \qquad N = KN'$$

is an isomorphism that is compatible with the shift

$$S^1 \Sigma_{\phi,K}^{\mathbb{Z}/N\mathbb{Z}} = \Sigma_{\phi,K}^{\mathbb{Z}/N\mathbb{Z}} S^K.$$

All sampling spaces can be described in this situation. Again, a sampling space has a translation invariant basis $\{h(\cdot - mK): m \in \mathbb{Z}/N'\mathbb{Z}\}$. Consider the linear span X of the translates $\{g(\cdot - mK)\}$ of some sequence g. In Fourier space it consists of those sequences for which

$$\hat{s}(n) = m(n)\hat{g}(n), \qquad m(n + N') = m(n).$$

Let

$$\mu(n) = \sum_{m=0}^{K-1} |g(n + mN')|^2.$$

The dimensions of X are equal to the number of non-zero elements of μ:

$$\dim X = \#\{m: \mu(m) > 0, 0 \le m < N'\}.$$

Clearly, a sampling space with decimation factor $K = N/N'$ has dimension N' and $\mu(m) > 0$ for all m in this case.

A set of translated sequences $\{v(\cdot - mN)\}$ is an orthonormal set iff

$$\sum_{m=0}^{K-1} |\hat{v}(n + mN')|^2 \equiv K.$$

Therefore, a translation invariant orthonormal basis is obtained from a translation invariant basis by means of

$$\hat{v}(n) = m(n)\hat{g}(n), \qquad |m(n)| = \sqrt{K}\left\{\sum_{p=0}^{K-1} |\hat{g}(n + pN')|^2\right\}^{-1/2}.$$

3 Quadrature mirror filters in $L^2(\mathbb{Z})$

In this section we want to show that, roughly speaking, orthogonal complements of sampling spaces are again sampling spaces. To be more precise, we give the following theorem.

Theorem 3.0.1

Let $V \subset L^2(\mathbb{Z})$ be a sampling space with decimation factor 2. It follows that W defined by

$$L^2(\mathbb{Z}) = V \oplus W$$

is again a sampling space with decimation factor 2. If $\{\alpha(\cdot - 2k)\}$ is an orthonormal basis for V then $\{\beta(\cdot - 2k)\}$ is one for W where

$$\beta(k) = (-1)^k \bar{\alpha}(1 - k).$$

In particular, we have that

$$\{S^{2k}\alpha, S^{2k}\beta: k \in \mathbb{Z}\} \quad \text{is an orthonormal basis of } L^2(\mathbb{Z}). \tag{3.0.1}$$

This gives rise to the following definition.

Definition 3.0.2

A pair of sequences $\alpha, \beta \in L^2(\mathbb{Z})$ satisfying (3.0.1) is called a QMF-system (Quadrature Mirror Filter system).

The theorem is therefore proved if we can show that for every $\alpha \in L^2(\mathbb{Z})$ whose $2\mathbb{Z}$-translates are an orthonormal family there is a sequence $\beta \in L^2(\mathbb{Z})$ such that (α, β) is a QMF-system. We now see how this is possible.

Quadrature mirror filters were introduced in Estaband and Galand (1977) as a signal-processing technique that allows down-sampling and perfect reconstruction of sampling sequences of sound signals. This technique is now commonly known as 'sub-band coding'.

By the very definition of QMF-systems we have the following theorem.

Theorem 3.0.3

A pair of sequences $\alpha, \beta \in L^2(\mathbb{Z})$ is a QMF-system iff

$$U: L^2(\mathbb{Z}) \to L^2(\mathbb{Z}) \oplus L^2(\mathbb{Z}), \qquad v \mapsto [v_0, v_1] = [\Sigma_2^{\mathbb{Z}}(\tilde{\alpha} * v), \Sigma_2^{\mathbb{Z}}(\tilde{\beta} * v)] \quad (3.0.2)$$

is an isometry.

The reason for the name 'quadrature mirror filter' becomes clear when we rewrite the QMF condition in Fourier space. Recall the definition of the Fourier transform of a sequence:

$$F^{\mathbb{Z}}: L^2(\mathbb{Z}) \to L^2(\mathbb{T}), \qquad v(n) \mapsto \sum_{k \in \mathbb{Z}} v(n)\, e^{-in\omega}.$$

The orthonormality of a family $\{\alpha(\cdot - 2k)\}$, respectively, $\{\beta(\cdot - 2k)\}$, can be rephrased as (Theorem 2.1.5)

$$|\hat{\alpha}(\omega)|^2 + |\hat{\alpha}(\omega + \pi)|^2 = 2, \qquad |\hat{\beta}(\omega)|^2 + |\hat{\beta}(\omega + \pi)|^2 = 2.$$

And looking at $\Sigma_2^{\mathbb{Z}}(\tilde{\alpha} * \beta) = 0$ in Fourier space we see that

$$\langle \alpha(\cdot - 2k) \mid \beta(\cdot - 2k') \rangle = 0 \Leftrightarrow \bar{\hat{\alpha}}(\omega)\hat{\beta}(\omega) + \bar{\hat{\alpha}}(\omega + \pi)\hat{\beta}(\omega + \pi) \equiv 0.$$

We have thus shown the 'only if' part of the following theorem.

Theorem 3.0.4

Two sequences $\alpha, \beta \in L^2(\mathbb{Z})$ are a QMF-system iff the following matrix

$$\frac{1}{\sqrt{2}} \begin{bmatrix} \hat{\alpha}(\omega) & \hat{\alpha}(\omega + \pi) \\ \hat{\beta}(\omega) & \hat{\beta}(\omega + \pi) \end{bmatrix} \qquad (3.0.3)$$

is unitary for almost all $\omega \in \mathbb{T}$.

Proof. Only the sufficiency of condition (3.0.3) remains to be proved. Suppose now that $\alpha, \beta \in L^2(\mathbb{Z})$ are such that the above matrix is unitary. Then by the Poisson summation formula $\{\alpha(\cdot - 2k), \beta(\cdot - 2k)\}$ is an orthonormal set. To see that it is complete observe that it is enough to show that the sampling–splitting operator U as given in (3.0.2) is injective. This operator reads in Fourier space

$$U: \hat{v}(\omega) \mapsto \begin{bmatrix} \hat{v}_0(\omega) \\ \hat{v}_1(\omega) \end{bmatrix} = \begin{bmatrix} \bar{\hat{\alpha}}(\omega) & \bar{\hat{\alpha}}(\omega + \pi) \\ \bar{\hat{\beta}}(\omega) & \bar{\hat{\beta}}(\omega + \pi) \end{bmatrix} \begin{bmatrix} \hat{v}(\omega) \\ \hat{v}(\omega + \pi) \end{bmatrix}.$$

It is therefore injective since the matrix is unitary. □

3.1 Completing a QMF-system

Proof (of Theorem 3.0.1). We now come back to our original problem of showing that the orthogonal complement of a sampling space with decimation factor 2 is again a sampling space. In view of the preceding theorem we are left with the problem of filling a unitary 2×2 matrix of which the first row is given. This is easily done by means of

$$\hat{\beta}(\omega) = e^{-i\omega}\, \bar{\hat{\alpha}}(\omega + \pi).$$

In sequence space we have explicitly

$$\beta(k) = (-1)^k \bar{\alpha}(1 - k).$$

Therefore, the complement of the sampling space V in $L^2(\mathbb{Z})$ is a sampling space on its own and Theorem 3.0.1 is proved. □

3.2 Complements over $\mathbb{R}$

Suppose now that we are given two sampling spaces X and Y over the real line $\mathbb{R}$. Suppose that X may be sampled at all integral positions $k \in \mathbb{Z}$, whereas Y may even be sampled at a lower rate, at $k \in 2\mathbb{Z}$, that is, X has a translation invariant orthonormal base $\{x(\cdot - k)\}$ and Y has one of the form $\{y(\cdot - 2k)\}$. Intuitively, X is on a smaller scale than Y. It is therefore natural to consider the situation where $Y \subset X$. In this case we can look at the orthogonal complement W of Y in X defined through

$$X = Y \oplus W.$$

We then have the following theorem as before.

Theorem 3.2.1

The space W is a sampling space, too, with sampling rate 2. An explicit *orthonormal base of W is given by* $\{w(\cdot - 2k)\}$ *with*

$$w = \sum_{k \in \mathbb{Z}} \beta(k) x(\cdot - k), \qquad \beta(k) = (-1)^k \overline{\Sigma_x y(1-k)}.$$

Proof. By hypothesis the sampling operator Σ_x is an isometric isomorphism from Y into $L^2(\mathbb{Z})$. Let $\alpha = \Sigma_x y$. Then, consequently, $\{\alpha(\cdot - 2k)\}$ is an orthonormal set. It can be complemented by a sequence β such that (α, β) is a QMF-system. Set $w = (\Sigma_x)^{-1}\beta$. It follows that $\{y(\cdot - 2k), w(\cdot - 2k)\}$ is an orthonormal basis of X. Since $\{y(\cdot - 2k)\}$ is an orthonormal basis of Y it follows that $W = X \ominus Y$ is a sampling space with $\{w(\cdot - 2k)\}$ as the orthonormal basis. □

3.3 QMF over $\mathbb{Z}/N\mathbb{Z}$ and complements over $\mathbb{T}$

In the same way we define QMF over $\mathbb{Z}/N\mathbb{Z}$ to be any pair of sequences (α, β) such that

$$\{S^{2k}\alpha, S^{2k}\beta \colon k = 0, 1, \ldots, N/2 - 1\}$$

is an orthonormal basis of $L^2(\mathbb{Z}/N\mathbb{Z})$. It can be shown as before that this is equivalent to

$$\begin{bmatrix} \hat{\alpha}(n) & \hat{\alpha}(n + N/2) \\ \hat{\beta}(n) & \hat{\beta}(n + N/2) \end{bmatrix}$$

being unitary for all n. Therefore, as before, the complement of a sampling space $V \subset L^2(\mathbb{Z}/N\mathbb{Z})$ with decimation factor 2 is again a sampling space. If $\{\alpha(\cdot - 2k)\}$ is an orthonormal basis for V then $\{\beta(\cdot - 2k)\}$ is an orthonormal basis for $W = L^2(\mathbb{Z}/N\mathbb{Z}) \ominus V$, with

$$\beta(n) = (-1)^n \bar{\alpha}(1 - n).$$

Again, by using the imperfect sampling operator one can show as before that the complement of two sampling spaces over $\mathbb{T}$, one having half the dimension of the other, is again a sampling space. More precisely, let X be a sampling space over $\mathbb{T}$ with orthonormal basis $\{x(\cdot - 2\pi k/N)\}$, and let $V \subset X$ have a basis $\{y(\cdot - 4\pi k/N)\}$. It follows that $W = X \ominus V$ is a sampling space, too, with orthonormal base $\{w(\cdot - 4\pi k/N)\}$, where

$$w = \sum_{k=0}^{N-1} \beta(k) x(\cdot - 2\pi k/N), \qquad \beta(k) = (-1)^k \bar{\alpha}(1-k), \qquad \alpha = \Sigma_{x,N}^{\mathbb{T}} y.$$

4 Multi-resolution analysis over $\mathbb{R}$

After all this preparation we are ready for the main objective of this chapter. We start with a definition that will be explained by examples.

Definition 4.0.1

A multi-resolution analysis of $L^2(\mathbb{R})$ is an ascending chain of closed subspaces $V_j \subset L^2(\mathbb{R})$, $j \in \mathbb{Z}$:

$$\{0\} \subset \cdots \subset V_j \subset V_{j+1} \subset \cdots \subset L^2(\mathbb{R})$$

such that

(i) $\bigcap_{j \in \mathbb{Z}} V_j = \{0\}, \qquad \bigcup_{j \in \mathbb{Z}} V_j$ *is dense in* $L^2(\mathbb{R})$,

(ii) $s \in V_j \Rightarrow s(t/2) \in V_{j+1}$,

(iii) V_0 *is a sampling space with sampling frequency* 1.

This definition is due to Meyer and Mallat (1990). Let P_j be the orthogonal projector of V_j. Then the first condition states that the successive projections $s_j = P_j s$ of $s \in L^2(\mathbb{R})$ tend to s in the square mean as $j \to \infty$.

By conditions (ii) the space V_{j+1} is just a scaled version of V_j and therefore functions in V_{j+1} have in general more small-scale features than do functions in V_j, as examples will show. Thus every function s may be approximated by a sequence of functions that contains more and more small scale features.

The last condition essentially says that the functions in V_0 are at a scale not smaller than 1. More precisely, V_0 has a translation invariant orthonormal base generated by some $\phi \in V_0$, as we have shown in Section 1. Every function of V_0 can be written as

$$s = \sum_{n \in \mathbb{Z}} v(n)\phi(\cdot - n)$$

and the sequence $v \in L^2(\mathbb{Z})$ is uniquely determined by the sampling operator $v = \Sigma_\phi s$. The multi-resolution analysis is therefore completely determined by this single function ϕ, which in turn is determined by the multi-resolution analysis up to a Fourier multiplier of modulus 1.

By condition (ii) all other approximation spaces V_j are sampling spaces, too, but with a sampling frequency of 2^{-j}. This again underlines the idea that the functions in V_j may have details up to scale 2^{-j}.

A multi-resolution analysis is therefore the discrete analogue of the partial reconstruction

$$s_{\epsilon,\infty} = \int_{\varepsilon}^{\infty} \frac{da}{a} \int_{-\infty}^{+\infty} db\, \mathscr{W}_g s(b,a) g_{b,a},$$

in the sense that the projection of $P_j s$ of s into the approximation space corresponds to the partial reconstruction $s_{2^{-j},\infty}$ containing all details of s up to scale 2^{-j}.

In the continuous case we have seen that the details that are added if we go from the approximation s_ϵ to $s_{\epsilon - d\epsilon}$ give rise to the continuous wavelet transform. In the same spirit we now look at the details that are added if we go from one approximation $P_j s$ to the next finer one $P_{j+1}s$. For this purpose we look at the orthogonal complements W_j defined through

$$V_{j+1} = V_j \oplus W_j.$$

Then we have the following orthonormal decomposition of $L^2(\mathbb{R})$:

$$L^2(\mathbb{R}) = \overline{\bigoplus_{j \in \mathbb{Z}} W_j},$$

thanks to condition (i).

Since each V_j is a sampling space with sampling distance 2^{-j} it follows from Theorem 3.2.1 that W_j is a sampling space, too, with the same sampling frequency. In particular, W_0 has an orthonormal base $\{\psi(\cdot - k)\}$. Since by construction the W_j are obtained by dilating W_0 we have proved the following theorem due to St. Mallat and Y. Meyer (Mallat 1989; Meyer 1990).

Theorem 4.0.2

For every multi-resolution analysis $\{V_j\}$, there is a (up to a 2π-periodic Fourier phase) unique function ψ—called the wavelet—such that the family

$$\{\psi_{n,j} = 2^{j/2}\psi(2^j t - n) \colon j, n \in \mathbb{Z}\}$$

is an orthonormal basis of $L^2(\mathbb{R})$. Let $\{\phi(\cdot - k)\}$ be an orthonormal base for V_0. The wavelet ψ may be constructed explicitly from the generating function ϕ using the formula

$$\psi(t/2)/2 = \sum_{k \in \mathbb{Z}} \beta(k)\phi(t-k), \quad \textit{with} \quad \beta(k) = (-1)^k \bar{\alpha}(1-k),$$

$$\alpha(k) = \Sigma_\phi D_2 \phi(k) = \frac{1}{2}\int_{-\infty}^{+\infty} dt\, \bar{\phi}(t-k)\phi(t/2).$$

In view of Section 3 we call (α, β) the associated QMF-system of the multi-resolution analysis. In Section 8 we consider the inverse problem, namely, the question under which conditions a QMF-system generates a multi-resolution analysis.

4.1 Localization and regularity of ψ

By the explicit construction of ψ out of ϕ we have the following general principle on how the regularity and the localization of ψ and ϕ are connected.

Theorem 4.1.1

Suppose that the generating function ϕ is compactly supported (exponentially localized, polynomially localized), that is,

$$|\phi(t)| \le O(1)\, e^{-c|t|}, \quad \textit{or} \quad |\phi(t)| \le O(1)(1 + |t|^2)^{-n/2}.$$

Then the same type of localization holds for ψ if the Fourier phase is chosen as in the previous theorem. If ϕ is n-times continuously differentiable and if all its derivatives have the same localization (compact support, exponentially decaying, polynomially decaying), then ψ has the same regularity and localization.

Proof. The compact support of ϕ implies that the sampling sequence $\alpha = \sqrt{2}\Sigma_\phi D_2\phi$ has compact support, too. The filter β that makes (α, β) a QMF-pair may be chosen in such a way that it is also of finite length. Therefore the reconstruction $\Sigma_\phi^*\beta$ is again of compact support. All other cases of localization may be treated in the same way. The analogue holds for the regularity. □

5 Examples of multi-resolution analysis and wavelets

It is now time to consider some explicit constructions of wavelets from multi-resolution analysis.

5.1 The Haar system

Let $V_0 \subset L^2(\mathbb{R})$ be the space of piecewise constant functions generated by the translates of $\phi = \chi_{[0,1)}$. And let V_j be the scale of spaces obtained by dilating V_0 by powers of 2. It is immediate that $\{V_j\}$ is a multi-resolution analysis. The associated QMF reads

$$\alpha(0) = 1, \quad \alpha(1) = 1, \quad \alpha(k) = 0 \text{ otherwise},$$

and, accordingly, the complementary filter β reads

$$\beta(0) = -1, \quad \beta(1) = +1, \quad \beta(k) = 0 \text{ otherwise}.$$

Going back to functions over $\mathbb{R}$ we have

$$\psi(t) = \begin{cases} -1 & \text{for } 0 \leq t < 1/2 \\ +1 & \text{for } 1/2 < t \leq 1 \\ 0 & \text{otherwise,} \end{cases}$$

which clearly is A. Haar's wavelet.

5.2 Spline wavelets

The following construction is due to P. G. Lemarie (Lemarie 1988). Let $n \in \mathbb{N}$. As we saw in Section 1.6 the space V_0 of square integrable spline functions of order n with nodes at $\mathbb{Z}$ is a sampling space with sampling period 1. Let V_j be the dilated versions of V_0. It is now an easy matter to verify that $\{V_j\}$ is a multi-resolution analysis. Indeed, the inclusion property (ii) follows trivially and the completeness property (i) follows from standard arguments about approximating functions in $L^2(\mathbb{R})$ by splines (e.g. Chui 1992). For the uniqueness property we reason as follows. On V_0 we have

$$\|s\|_\infty \leq c\|s\|_2.$$

Indeed, V_0 has a Riesz basis of the form $\rho(\cdot - k)$, where ρ is continuous and compactly supported. The right-hand side of the above equation is equivalent to the $L^2(\mathbb{Z})$ norm of the sequence of expansion coefficients α in

$$s = \sum_{k \in \mathbb{Z}} \alpha(k)\rho(\cdot - k).$$

Because ρ is compactly supported and continuous we have $\|s\|_{L^\infty(\mathbb{R})} \leq O(1)\|\alpha\|_{L^\infty(\mathbb{Z})}$ and hence the above estimation. By simple dilation it follows that

$$\|s\|_\infty \leq c2^j\|s\|_2.$$

Therefore, if s is in all V_j it is 0.

The wavelet can now be computed explicitly from the generating function ϕ. Since ϕ is of exponential localization it follows that the wavelets are exponentially localized, too. For an explicit expression we refer, for example, to Chui (1992).

5.3 Band-limited functions

Let $V_j \subset L^2(\mathbb{R})$, the space of functions s whose Fourier transform is supported by $[-\pi 2^j, +\pi 2^j]$. Again, $V_j \subset V_{j+1}$ and all other properties of the multi-resolution approximation are easily verified. Accordingly, the wavelet is given in Fourier space by the characteristic function of $[-\pi, -\pi/2] \cup [+\pi/2, +\pi]$.

5.4 Littlewood–Paley analysis

This is a smoothed version of the former construction due to Y. Meyer (1990): Consider a function $\phi \in S(\mathbb{R})$ that satisfies

(i) $\operatorname{supp} \hat{\phi} \subset [-4\pi/3, +4\pi/3]$,
(ii) $\hat{\phi}(\omega) \equiv 1, \ \omega \in [-2\pi/3, +2\pi/3]$,
(iii) $\hat{\phi}(\omega) = \hat{\phi}(-\omega)$,
(iv) $0 \leq \hat{\phi}(\omega) \leq 1$,
(v) $\hat{\phi}^2(\omega) + \hat{\phi}^2(2\pi - \omega) \equiv 1, \ \omega \in [-2\pi, +2\pi]$.

It follows that ϕ generates a multi-resolution analysis. The associated wavelet now satisfies

$$\hat{\psi}(\omega) = e^{-i\omega}\sqrt{\phi^2(\omega/2) - \phi^2(\omega)}$$

and therefore $\psi \in S(\mathbb{R})$, too. In addition we have

$$\text{supp}\,\hat{\psi} \subset [-8\pi/3, -2\pi/3] \cup [+2\pi/3, +8\pi/3]$$

and therefore ψ is strip limited. In particular, all moments of ψ vanish. Thus ψ is rapidly decaying but it cannot be exponentially localized since it is in $S_0(\mathbb{R})$.

6 The partial reconstruction operator

Given a sequence of sampling spaces V_j that satisfy the inclusion property $V_j \subset V_{j+1}$, one still has to check the completeness and uniqueness condition (i) of Definition 4.0.1. Let ϕ be the generator of a translation invariant orthonormal basis of V_0. Then uniqueness and completeness can be stated in terms of the projection operators on V_j:

$$P_j = D_{2^j}^{-1} P_0 D_{2^j}, \qquad P_0 s = \sum_{k \in \mathbb{Z}} \langle \phi(\cdot - k) \mid s \rangle \phi(\cdot - k),$$

as

$$\lim_{j \to \infty} P_j = \mathbb{1}, \qquad \lim_{j \to -\infty} P_j = 0,$$

where the limits hold strongly. The following theorem gives sufficient conditions to cover all practical cases. For later reference we will prove the theorem in a far more general setting, including the case of skew projections. Therefore, let ϕ and ρ be generators of some translations invariant Riesz basis for their respective linear spans and consider the operators

$$P_j = D_{2^j}^{-1} P_0 D_{2^j}, \qquad P_0 s = \sum_{k \in \mathbb{Z}} \langle \rho(\cdot - k) \mid s \rangle \phi(\cdot - k).$$

Note that they need no longer be projectors.

Theorem 6.0.1
Under the above conditions we always have

$$\lim_{j \to -\infty} P_j s = 0, \qquad s \in L^2(\mathbb{R}).$$

Remark. The proof will show that, in addition, we have $2^{j/2} \Sigma_\phi D_{2^j}^{-1} s \to 0$ as $j \to -\infty$, where the convergence takes place in $L^2(\mathbb{Z})$ whenever $\phi(\cdot - n)$ is merely a Bessel sequence.

Proof. Since every Riesz basis is a Bessel sequence the sampling–reconstruction operator P_0 is bounded. The same holds for the dilated versions P_j with the same bound.

Therefore, it is enough to show the above theorem for s in some dense set. Indeed, the usual approximation argument works: for arbitrary r we write $r = s + \eta$, where the theorem holds for s and η is small in norm. Since the P_j are uniformly bounded we have $\limsup_{j \to -\infty} \|P_j r\| \leq \|P_0\| \, \|\eta\|$, which may be made arbitrary small.

Therefore, suppose s is continuous and compactly supported, say with support contained in $[-c, +c]$, and consider the sampling values

$$\alpha_j(k) = 2^{j/2} \Sigma_\phi D_{2^j} s(k) = 2^{-j/2} \int_{|t| \leq c2^j} dt\, s(2^{-j}t)\phi(t-k)$$

$$= 2^{-j/2} \int_{|t-k| \leq c2^j} dt\, s(2^{-j}t)\phi(t).$$

By the Cauchy–Schwarz inequality we can estimate

$$\|\alpha_j\|^2_{L^2(\mathbb{Z})} \leq \|s\|^2_{L^2(\mathbb{R})} \sum_{k \in \mathbb{Z}} \int_{|t-k| \leq 2^j} dt\, |\phi(t)|^2$$

$$= \|s\|^2_{L^2(\mathbb{R})} \int_{|t| \leq 2^j} dt \sum_{k \in \mathbb{Z}} |\phi(t+k)|^2.$$

Since the sum is a locally absolutely integrable function it follows that in the $\lim j \to -\infty$ the sequence α_j tends to 0 in $L^2(\mathbb{Z})$. The theorem follows since the reconstruction operator with respect to a Riesz basis is continuous. □

To obtain an approximation of the identity we have to assume a little more, however.

Theorem 6.0.2

Let $\rho \in L^2(\mathbb{R})$ be such that $\hat{\rho}$ is continuous in $\omega = 0$ and satisfies at $|\hat{\rho}(0)| = 1$. Suppose, further, that $\sum_{k \in \mathbb{Z}} |\hat{\rho}(\omega + 2k\pi)|^2$ is equal almost everywhere to a function that is continuous at $\omega = 0$, where it takes on the value 1. The same should hold for ϕ. Then we have

$$\lim_{j \to \infty} P_j s = s$$

for all $s \in L^2(\mathbb{R})$ and the convergence takes place in $L^2(\mathbb{R})$.

Remark. The proof will show that in addition we have $\|2^{j/2} \Sigma_\rho D_{2^j} s\|^2_{L^2(\mathbb{Z})} \to \|s\|^2_{L^2(\mathbb{R})}$ as $j \to \infty$ whenever $\phi(\cdot - n)$ is merely a Bessel sequence that satisfies the properties stated in the theorem.

Remark. The smoothness hypothesis on $\hat{\rho}$ of the theorem is satisfied in particular if ρ is compactly supported. Indeed, in this case the sampling sequence $\tilde{\rho} * \rho(k)$ is of finite length and its Fourier transform is precisely $\sum_{k \in \mathbb{Z}} |\hat{\rho}(\omega + 2k\pi)|^2$, which is therefore a smooth function.

Proof. Again, it is enough to show the convergence $P_j s \to s$ for s in a dense subset of $L^2(\mathbb{R})$. Therefore, suppose that $\hat{s}$ is of compact support and bounded. By an overall dilation, and if necessary changing j to some $j + N$, we may assume that the support is contained in $[-\pi, +\pi]$. The Poisson summation formula now shows that $s_j = P_j s$ reads in Fourier space

$$\hat{s}_j(\omega) = \hat{\phi}(\omega 2^{-j}) \sum_{k \in \mathbb{Z}} \bar{\hat{\rho}}(2^{-j}\omega + 2\pi k)\hat{s}(\omega + 2\pi 2^j k). \tag{6.0.1}$$

The main contribution to the sum of (6.0.1) comes from $k = 0$. Since $\hat{\phi}$ and $\hat{\rho}$ are—after modifications on a set of measure zero—continuous in 0 we have $\hat{\phi}(2^{-j}\omega)\bar{\hat{\rho}}(2^{-j}\omega)\hat{s}(\omega) \to \hat{s}(\omega)$, where the convergence holds in $L^2(\mathbb{R})$ thanks to dominated convergence.

We now estimate the remaining contribution. Because the supports of the functions under the sum are disjoint we have that its norm equals

$$\int_{-\infty}^{+\infty} d\omega \, |\hat{\phi}(\omega 2^{-j})|^2 \sum_{k \in \mathbb{Z} - \{0\}} |\hat{\rho}(2^{-j}\omega + 2\pi k)|^2 |\hat{s}(\omega + 2\pi 2^j k)|^2 .$$

Because $\hat{s}$ is supported by $[-\pi, \pi]$ and because it is bounded, this is estimated by

$$O(1) \sum_{k \in \mathbb{Z} - \{0\}} \int_{-\pi + 2^j k\pi}^{+\pi + 2^j k\pi} d\omega \, |\hat{\phi}(\omega 2^{-j})|^2 |\hat{\rho}(2^{-j}\omega + 2\pi k)|^2$$

$$= O(1) 2^j \int_{-\pi 2^{-j}\pi}^{+\pi 2^{-j}\pi} d\omega \sum_{k \in \mathbb{Z} - \{0\}} |\hat{\phi}(\omega + 2\pi k)|^2 \, |\hat{\rho}(\omega)|^2$$

The sum is by hypothesis almost everywhere equal to a function that is continuous at the origin where it takes on the value 0. Therefore, the mean value given by the integral tends to 0 as $j \to \infty$ and we are done. □

7 Multi-resolution analysis of $L^2(\mathbb{Z})$

In this section we will show how to define a multi-resolution analysis of the sequence space $L^2(\mathbb{Z})$. This is of great importance since in any practical application of the wavelet transform only sampled data are available. One way to obtain a decomposition of $L^2(\mathbb{Z})$ into wavelets is through the sampling operator. Indeed, let $\eta \in L^2(\mathbb{R})$ be a scaling function that generates a multi-resolution analysis over $\mathbb{R}$ with wavelet ρ. Then the sequences

$$\psi_{k,j} = \psi_j(\cdot - 2^j k), \qquad \psi_j = 2^{j/2}\Sigma_\eta D_{2^j}\eta, \qquad j \in \mathbb{N}_0,$$

are an orthonormal base of $L^2(\mathbb{Z})$, as follows from the unitarity of the dilation $2^{j/2}D_{2^j}$ and Lemma 2.3.1.

But there is also an intrinsic way defining a multi-resolution analysis of $L^2(\mathbb{Z})$, as we will now see. It is essentially based on the QMF-systems.

We first need a definition of a dilation or scale-changing operator that acts on sequences in $L^2(\mathbb{Z})$. We would like to satisfy similar properties to those of dilation over $\mathbb{R}$. Let us call $D^{\mathbb{Z}}$ for the moment the dilation operator that we want to construct. It should be the sequence equivalent of a dilation by 2. Therefore, we require the following commutation relation with the shift operator $S\colon s(n) \mapsto s(n-1)$:

$$D^{\mathbb{Z}} S = S^2 D^{\mathbb{Z}} \tag{7.0.1}$$

This is in complete analogy with the commutation relation of translations and dilations acting on functions over the real line. From this it is clear that it is enough to know $D^{\mathbb{Z}}\delta$ to know $D^{\mathbb{Z}}$ for all sequences. Indeed, for any other sequence $s \in L^2(\mathbb{Z})$ we would obtain by superposition

$$D^{\mathbb{Z}} s(n) = \sum_{k \in \mathbb{Z}} \phi(n-2k)s(k), \tag{7.0.2}$$

where $\phi = D^{\mathbb{Z}}\delta$, and vice versa every such dilation operator satisfies (7.0.1). We therefore write $D^{\mathbb{Z}}_\phi$ to stress the dependency of the dilation over $\mathbb{Z}$ on the filter ϕ.

Consider the adjoint of the down-sampling operator $\Sigma^{\mathbb{Z}}_2$:

$$(\Sigma^{\mathbb{Z}}_2)^*\colon L^2(\mathbb{Z}) \mapsto L^2(\mathbb{Z}). \qquad s(n) \mapsto \begin{cases} s(n/2) & \text{if } n = 0 \bmod 2, \\ 0 & \text{otherwise.} \end{cases}$$

This is the dilation with wholes (see Figure 7.1). In Fourier space this operator reads, thanks to the Poisson summation formula,

$$(\Sigma^{\mathbb{Z}}_2)^*\colon \hat{s}(\omega) \mapsto \hat{s}(2\omega).$$

Then the dilation by 2 for sequences can also be written in the following concise form:

$$D^{\mathbb{Z}}_{2,\phi} s = \phi * ((\Sigma^{\mathbb{Z}}_2)^* s),$$

which reads in Fourier space

$$D^{\mathbb{Z}}_{2,\phi}\colon \hat{s}(\omega) \mapsto \hat{\phi}(\omega)\cdot\hat{s}(2\omega). \tag{7.0.3}$$

Therefore, the dilation that we have constructed consists of blowing up the sequence s by putting 0 at every odd position. These wholes are then

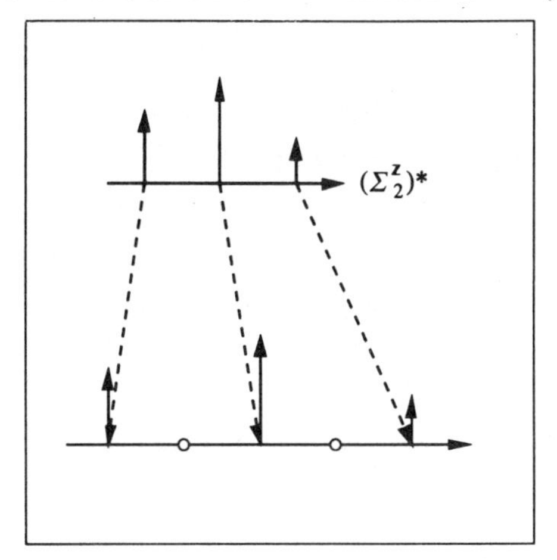

Fig. 7.1 The dilation with wholes.

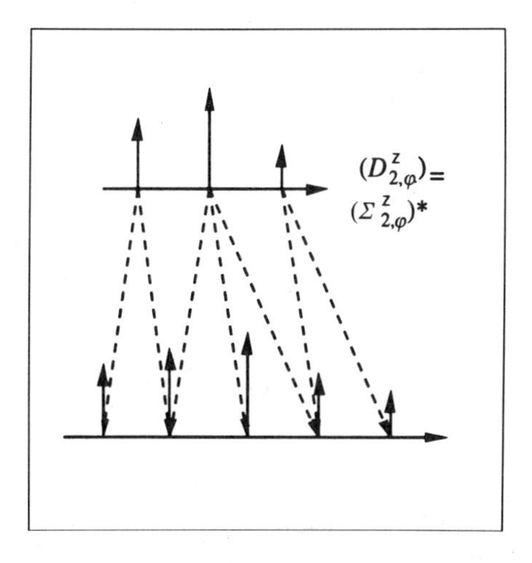

Fig. 7.2 The dilation operator for sequences.

filled up by the smoothing filter ϕ. Clearly, the dilation operator that we have constructed depends on the filter ϕ. See Figure 7.2 for an illustration. Observe also that this dilation operator is nothing other than the adjoint of the imperfect sampling operator $\Sigma_{\phi,2}^{\mathbb{Z}} s = \Sigma_2^{\mathbb{Z}}(\tilde{\phi} * s)$:

$$\langle s \mid D_{2,\phi}^{\mathbb{Z}} r \rangle_{\mathbb{Z}} = \langle \Sigma_{\phi,2}^{\mathbb{Z}} s \mid r \rangle_{\mathbb{Z}}. \tag{7.0.4}$$

There is still another argument for a dilation operator of the form of (7.0.2). Consider a sampling space V with translation invariant orthonormal base

$\eta(\cdot - n)$, $n \in \mathbb{Z}$. Let us suppose that V is scaling invariant in the sense that

$$s \in V \Rightarrow s(\cdot/2) \in V,$$

as is the case if V comes from a multi-resolution analysis. Since the imperfect sampling operator associated to η:

$$\Sigma_\eta\colon V \to L^2(\mathbb{Z}), \qquad \Sigma_\eta s(n) = \langle \eta(\cdot - n) \mid s \rangle_{\mathbb{R}}$$

is an isomorphism we may compute the dilation D_2 in sequence space. By straightforward computation we see that this is achieved through the following commutative diagram:

$$D^{\mathbb{Z}}_{2,\phi}\Sigma_\eta = \Sigma_\eta D_2, \qquad \begin{array}{ccc} V & \xrightarrow{\Sigma_\eta} & L^2(\mathbb{Z}) \\ \downarrow{\scriptstyle D_2} & & \downarrow{\scriptstyle D^{\mathbb{Z}}_{2,\phi}} \\ V & \xrightarrow{\Sigma_\eta} & L^2(\mathbb{Z}) \end{array}, \qquad \phi = \Sigma_\eta D_2 \eta. \tag{7.0.5}$$

Therefore, the dilation over the real line for functions in V is given by the action of $D^{\mathbb{Z}}_{2,\phi}$ on their (imperfect) sample sequences.

We are now ready for the definition of a multi-resolution analysis over $\mathbb{Z}$.

Definition 7.0.1
A descending chain of spaces

$$L^2(\mathbb{Z}) = V_0 \supset V_1 \supset \cdots \supset V_j \cdots$$

is called a multi-resolution analysis of $L^2(\mathbb{Z})$ with interpolation filter $\phi \in L^2(\mathbb{Z})$ iff

(i) $\bigcap_{j=0}^{\infty} V_j = \{0\}$,

(ii) $s \in V_j \Rightarrow D^{\mathbb{Z}}_{2,\phi} s \in V_{j+1}$ *and this defines an isomorphism for* $j = 0$.

Because of

$$D^{\mathbb{Z}}_{2,\phi} S = S^2 D^{\mathbb{Z}}_{2,\phi},$$

it follows that $\{\phi(\cdot - 2k)\}$ is a Riesz basis for the sampling space V_0 since it is the isomorphic image of the orthonormal basis $\{\delta(\cdot - k)\}$. For exactly the same reason it follows by iteration that $\{\phi_j(\cdot - 2^j k)\}$ is a Riesz basis for V_j with

$$\phi_j = (D^{\mathbb{Z}}_{2,\phi})^j \delta.$$

In Fourier space this reads as follows from (7.0.3):

$$\hat{\phi}_j(\omega) = \prod_{k=0}^{j-1} \hat{\phi}(2^k \omega).$$

Actually, ϕ may be chosen in such a way that $D^{\mathbb{Z}}_{2,\phi}$ is an isometry and, accordingly, ϕ_j generates an orthonormal base of V_j.

Theorem 7.0.2
If we replace ϕ by ρ given by

$$\hat{\rho}(\omega) = m(\omega)\hat{\phi}(\omega), \qquad m(\omega) = \sqrt{2}/\sqrt{|\hat{\phi}(\omega)|^2 + |\hat{\phi}(\omega + \pi)|^2},$$

then the multi-resolution analyses generated by ϕ and ρ are the same.

Proof. By Theorem 2.1.5 the translates by 2 of ρ generate an orthonormal base of V_1. The dilation $D^{\mathbb{Z}}_{2,\rho}$ is therefore an isometry and thus the translates by 2^j of

$$\rho_j = (D^{\mathbb{Z}}_{2,\rho})^j\delta$$

generate a translation invariant orthonormal base of some space V_j'. We have to show that $V_j' = V_j$. Now by construction of ρ it follows that

$$\hat{\rho}_j(\omega) = m_j(\omega)\hat{\phi}_j(\omega), \qquad m_j(\omega) = \prod_{k=0}^{j-1} m(2^k\omega),$$

and since $m \sim 1$ it follows that $m_j \sim 1$, too. In addition, m_j is $2\pi 2^{-j-1}$-periodic. Therefore, $V_j = V_j'$. □

Therefore, we may assume that in addition ϕ satisfies the QMF-condition

$$|\hat{\phi}(\omega)|^2 + |\hat{\phi}(\omega + \pi)|^2 = 2.$$

In this case the orthonormal basis $\{\delta(\cdot - n)\}$ of $L^2(\mathbb{Z})$ is now mapped onto the orthonormal basis $\{\phi(\cdot - 2k)\}$ of the sampling space V_1. Hence $D^{\mathbb{Z}}_{2,\phi}$ is an isometry.

In the following subsection we will insert a short digression about isometries in general. It helps to understand the relation between wavelets over $\mathbb{Z}$ and the QMF.

7.1 Isometries and the shift operator

An isometric operator U acting in a Hilbert space $\mathscr{H}$ is characterized by the fact that it preserves the scalar product and hence the norm.

$$\langle Us \mid Ur \rangle = \langle s \mid r \rangle.$$

It follows at once that an isometry is injective. Its range is a closed subspace of $\mathscr{H}$ and U is unitary iff this subspace is all of $\mathscr{H}$. In the case where the image is a true subspace of $\mathscr{H}$ we have a descending sequence of Hilbert spaces and U acts via shifting from one space to the next larger V one. More precisely, we have the following theorem. See, for example, Hoffmann (1962) for a proof.

Theorem 7.1.1
Let U be an isometry with $V \subset \mathcal{H}$ its range. Suppose that $W = \mathcal{H} \ominus V$ is different from $\{0\}$. Then the successive images $V_{n+1} = UV_n$ with $V_0 = V$ are a descending chain of spaces

$$V_{n+1} \subset V_n.$$

Their intersection is a closed invariant space

$$K = \bigcap_{n=1}^{\infty} V_n, \qquad UK = K.$$

If K is different from 0 then the restriction of U to K is unitary. By taking orthogonal complements

$$W_n = V_n \ominus V_{n+1},$$

we have the following decomposition of $\mathcal{H}$:

$$\mathcal{H} = K \oplus \bigoplus_{n=0}^{\infty} W_n$$

and U acts by shifting

$$s \in W_k \Leftrightarrow Us \in W_{k+1}.$$

Therefore, the action of U is on the orthogonal complement of K in $\mathcal{H}$ conjugate to a shift operator. More precisely, consider the space of semi-infinite sequences v whose entries v_k are elements in some Hilbert space $\mathcal{K}$ and for which the following expression is finite:

$$\|v\|^2 = \sum_{k \in \mathbb{N}} \|v_k\|_{\mathcal{K}}^2 < \infty.$$

This is a norm that makes this space a Hilbert space on its own. It will be denoted by $L^2(\mathbb{N}_0, \mathcal{K})$. Now let U be an isometry with K as in Theorem 7.1.1. We may then identify $s \in K^{\perp}$ in an obvious way with a sequence in $L^2(\mathbb{N}_0, W)$, and U and U^* are then acting as a shift

$$U: [v_0, v_1, v_2, \ldots] \mapsto [0, v_0, v_1, v_2, \ldots], \qquad U^*: [v_0, v_1, v_2, \ldots] \mapsto [v_1, v_2, \ldots].$$

7.2 QMF and multi-resolution analysis over $\mathbb{Z}$

As we have seen, to every multi-resolution analysis there is associated a QMF that generates a translation invariant orthonormal basis of V_0. In this section we want to discuss when a QMF-filter gives rise to a multi-resolution analysis. To be more precise let $\phi \in L^2(\mathbb{Z})$ be a QMF. Then the dilation operator for sequences is an isometric operator in $L^2(\mathbb{Z})$, defined, as usual by

$$D_{2,\phi}^{\mathbb{Z}}: s \mapsto \phi * (\Sigma_2^{\mathbb{Z}})^* s, \qquad \hat{s}(\omega) \mapsto \hat{\phi}(\omega)\hat{s}(2\omega).$$

We now define the spaces V_j as

$$V_j = (D^{\mathbb{Z}}_{2,\phi})^j L^2(\mathbb{Z}).$$

These spaces are now trivially sampling spaces and the only condition that remains to be checked is that

$$\bigcap_{j=0}^{\infty} V_j = 0.$$

In other words, we are looking for conditions under which the operator $D^{\mathbb{Z}}_{2,\phi}$ is conjugate to a shift. Equivalently, we may ask whether the iterations of the adjoint operator tend strongly to 0:

$$(D^{\mathbb{Z}}_{2,\phi}{}^*)^n s \to 0 \quad (n \to \infty), \qquad s \in L^2(\mathbb{Z}).$$

Theorem 7.2.1
Suppose $\hat{\phi} \in L^2(\mathbb{T})$ is the Fourier transform of a QMF,

$$|\hat{\phi}(\omega)|^2 + |\hat{\phi}(\omega + \pi)|^2 = 2,$$

that satisfies in addition at

$$|\hat{\phi}(\omega)| \neq 1$$

on a set of positive Lebesgue measure. Then V_j is a multi-resolution analysis of $L^2(\mathbb{Z})$.

Therefore, essentially all QMFs give rise to a multi-resolution analysis over $\mathbb{Z}$.

Proof.[6] Consider the following operator:

$$B\colon L^2(\mathbb{T}) \to L^2(\mathbb{T}), \qquad \kappa(\omega) \mapsto (\hat{\phi}(\omega/2)\kappa(\omega/2) + \hat{\phi}(\omega/2 + \pi)\kappa(\omega/2 + \pi))/2.$$

It is the adjoint of $F^{\mathbb{Z}} D^{\mathbb{Z}}_{2,\phi} (F^{\mathbb{Z}})^*$. We want to show that

$$\lim_{n \to \infty} \|B^n s\| = 0, \qquad s \in L^2(\mathbb{T}).$$

From the Riesz representation theorem we may write

$$\|B^n s\|^2 = \sup_r |\langle r \mid B^n s\rangle| = \sup_r |\langle (B^*)^n r \mid s\rangle|,$$

where the sup is over all $r \in L^2(\mathbb{T})$ with $\|r\| = 1$. Thanks to $F^{\mathbb{Z}} D^{\mathbb{Z}}_{2,\phi} (F^{\mathbb{Z}})^*\colon r(\omega) \mapsto \hat{\phi}(\omega) r(2\omega)$, this last expression reads explicitly

$$\left| \int_{-\pi}^{+\pi} d\omega \prod_{j=0}^{n-1} \bar{\hat{\phi}}(2^j\omega) \bar{r}(2^n\omega) s(\omega) \right|$$

[6] A. Cohen, private communication.

and its square may be estimated by the Cauchy–Schwarz inequality to be

$$\leqslant \int_{-\pi}^{+\pi} d\omega \prod_{j=0}^{n-1} |\hat{\phi}(2^j\omega)|^2 \, |s(\omega)|^2 \, \|r\|^2. \tag{7.2.1}$$

Let us call the product ρ_n. We now introduce the baker's map in $[-\pi, +\pi]$ by means of

$$T\colon [-\pi, +\pi) \to [-\pi, +\pi), \qquad \omega \mapsto 2\omega \bmod[-\pi, +\pi).$$

It is an ergodic, measure preserving endomorphism. Then the logarithm of the product can be written as

$$\frac{1}{n}\log(\rho_n(\omega)) = \frac{1}{n}\sum_{j=0}^{n-1} u(T^j\omega), \qquad u(\omega) = 2\log(\hat{\phi}(\omega)).$$

Suppose now that $u \in L^1(\mathbb{T})$. Recall that by the Birkhoff ergodic theorem, the average on the right-hand side converges almost everywhere to the mean value

$$\frac{1}{\pi}\int_{-\pi}^{+\pi} d\omega \log(|\hat{\phi}(\omega)|).$$

We now show that this mean value is actually negative, from which it follows that the ρ_n tend to 0 almost everywhere. Since $\hat{\phi}$ satisfies the QMF-condition the above mean value equals

$$\frac{1}{\pi}\int_{0}^{\pi} d\omega \,[\log(|\hat{\phi}(\omega)|^2) + \log(2 - |\hat{\phi}(\omega)|^2)].$$

Since $\log(x) + \log(2 - x) < 0$ for $0 < x \leq 2$ and $x \neq 1$, where this expression is 0, it follows that, if $|\hat{\phi}(\omega)| \neq 1$ on a set of positive Lebesgue measure, the integral is <0. Therefore, the product ρ_n tends to 0 almost everywhere. By Egorov's theorem (e.g. Torchinsky 1988) it converges almost everywhere uniformly. More precisely for every $\epsilon > 0$ there is a subset $I \subset [0, 2\pi)$ of measure $2\pi - \epsilon$ on which ρ_n converges uniformly. If we set $s = s_1 + s_2$, where s_1 vanishes on the complement of I, then $\|s_2\|$ may be chosen to tend to 0 with ϵ. By what we have seen, $B^n s_1 \to 0$ as $n \to \infty$ and since B is the adjoint of an isometry we have $\|B^n s_2\| \leq \|s_2\|$, which becomes arbitrary small with ϵ, and we are done for $\log \hat{\phi} \in L^1(\mathbb{T})$.

In the remaining case we may replace $|\hat{\phi}|$ in (7.2.1) by $|\hat{\phi}| + \eta$ with η a non-negative function chosen such that $\log(|\hat{\phi}| + \eta) \in L^1(\mathbb{T})$ with still negative mean value $\int_{-\pi}^{+\pi} d\omega \log(|\hat{\phi}(\omega)| + \eta(\omega))$. □

7.3 Wavelets over $\mathbb{Z}$

We now return to the construction of wavelets over $\mathbb{Z}$ out of the multi-resolution V_k. As we have seen, we may assume that ϕ satisfies the QMF-condition which makes the dilation $D^{\mathbb{Z}}_{2,\phi}$ an isometry in $L^2(\mathbb{Z})$. By property (i) and the preceding lemma we know that $L^2(\mathbb{Z})$ splits into a direct sum

$$L^2(\mathbb{Z}) = \bigoplus_{k=1}^{\infty} W_k,$$

where W_k is the orthogonal complement of V_{k+1} in V_k. The dilation $D^{\mathbb{Z}}_{2,\phi}$ acts by shifting. As orthogonal complements the W_j are sampling spaces, too, with decimation factor 2^j (Theorem 3.0.1). Therefore, each W_j has an orthonormal base of the form $\{\psi_{k,j} = \psi_j(\cdot - 2^j k)\}$ with ψ_j given by

$$\psi_j(n) = (D^{\mathbb{Z}}_{2,\phi})^{j-1}\psi, \qquad \psi(n) = (-1)^n \bar{\phi}(1-n),$$

and any sequence s can be decomposed into wavelets according to

$$s = \sum_{j=0}^{\infty} \sum_{n \in \mathbb{Z}} \langle \psi_{k,j} \mid s \rangle_{\mathbb{Z}} \psi_{k,j}.$$

8 QMF and multi-resolution analysis

As we have seen, every multi-resolution analysis over $\mathbb{R}$ gives rise via sampling to a QMF pair. More precisely, if ϕ, $\psi \in L^2(\mathbb{R})$ are its scaling function and wavelet then $\alpha = \sqrt{2}\Sigma_\phi D_2 \phi$ and $\beta = \sqrt{2}\Sigma_\phi D_2 \psi$ are a QMF pair in $L^2(\mathbb{Z})$. We now consider two inverse problems. The first is to what extent a multi-resolution analysis over $\mathbb{R}$ may be recovered from its QMF. The second question is what are the QMF that result from a multi-resolution analysis.

We now attack the first problem. Since the sampling operator on V_0 may be inverted it follows that ϕ satisfies the following scaling equation:

$$\phi(t/2)/2 = \sum_{k \in \mathbb{Z}} \alpha(k)\phi(t-k), \qquad \alpha = \Sigma_\phi D_2 \phi. \tag{8.0.1}$$

And the orthonormality of the translates of ϕ implies that α satisfies the following QMF-like condition:

$$|\hat{\alpha}(\omega)|^2 + |\hat{\alpha}(\omega + \pi)|^2 = 1.$$

In Fourier space the scaling equation for ϕ reads

$$\hat{\phi}(2\omega) = m(\omega)\hat{\phi}(\omega), \qquad m(\omega) = \hat{\alpha}(\omega).$$

Therefore, upon iteration we obtain (formally)

$$\hat{\phi}(\omega) = \hat{\phi}(0) \prod_{j=1}^{\infty} m(2^{-j}\omega). \tag{8.0.2}$$

In many cases this product actually makes sense. Suppose, for instance, that m comes from a multi-resolution analysis over $\mathbb{R}$ with generating function $\phi \in L^1(\mathbb{R}) \cap L^2(\mathbb{R})$. For each j we have

$$\hat{\phi}(\omega) = \prod_{k=1}^{j} m(2^{-k}\omega)\hat{\phi}(2^{-j}\omega). \tag{8.0.3}$$

Since $\hat{\phi}$ is continuous it follows that the infinite product converges pointwise and uniformly on compact sets. It follows in particular from the infinite product of (8.0.2) that $\hat{\phi}(0) \neq 0$, since otherwise by (8.0.3) that would imply $\hat{\phi}(\omega) = 0$ for all ω.

But there is also a way of recovering a continuous $\phi \in L^1(\mathbb{R}) \cap L^2(\mathbb{R})$ from m (or α) without passing through the Fourier space representation. For this note that the imperfect sampling sequence of the dilated scaling function may be computed by iteration of the dilation operators for sequences

$$\Sigma^{\mathbb{Z}}_{\phi,2}(D_2)^j\phi = (D^{\mathbb{Z}}_{2,\alpha})^j\delta.$$

The left-hand side equals $2^{-j}\langle\phi(\cdot - k) \mid \phi(2^{-j}\cdot)\rangle$, which is essentially the weighted mean value of ϕ around the dyadic point $k2^{-j}$. Now $\int \phi = c \neq 0$, as we have seen, and it follows, thanks to the approximation of the identity lemma, that pointwise at a dyadic point $t = k2^{-j}$ we have

$$\lim_{j\to\infty} 2^j(D^{\mathbb{Z}}_{2,\alpha})^j\delta(k) = \lim_{j\to\infty} 2^j(D^{\mathbb{Z}}_{2,\alpha})^j\delta(t2^j) = c\phi(2^{-j}k) = c\phi(t).$$

Therefore, we obtain ϕ at every dyadic point and by continuity of ϕ we have it everywhere. Note also that if the hypothesis of continuity is dropped nothing can be said anymore since the dyadic points are of measure zero.

As we have seen, a multi-resolution analysis may be recovered from its QMF under fairly general assumptions. We now attack the second problem: what are the QMF that define, via (8.0.2), a scaling function of a multi-resolution analysis, and what properties of the wavelet can be read from the QMF. We will not give an exhaustive discussion of this rich field of research. As a general reference for this and related questions we refer to Daubechies (1992).

We first show that the product in (8.0.2) is well defined in many cases.

Theorem 8.0.1

Let $m \in L^\infty(\mathbb{T})$ be continuous at 0 and regular such that for some neighbourhood I of 0 we have

$$\sum_{k=1}^{\infty} |m(0) - m(\omega 2^{-k})| \leq c, \qquad \omega \in I, \tag{8.0.4}$$

with some $c > 0$. Then consider the limit $N \to \infty$ in the following expression:

$$\prod_{j=1}^{N} m(\omega 2^{-j}).$$

If $|m(0)| < 1$ or $m(0) = 1$ it converges uniformly on compact sets and in the topology of $S'(\mathbb{R})$. In the first case this limit is 0. If $|m(0)| > 1$ or $m(0) \neq 1$, the limit does not exist either pointwise or in $S'(\mathbb{R})$.

Remark. The regularity requirement for m is automatically satisfied if m satisfies the following Hölder condition:

$$|m(0) - m(\omega)| \leq O(\omega^{\alpha}), \qquad \alpha > 0.$$

Proof. First consider $|m(0)| \leq 1$. For any compact set K we can find j sufficiently large that $2^{-j}K$ is in the set I where (8.0.4) holds. Writing

$$\prod_{k=1}^{N} m(2^{-k}\omega) = \prod_{k=1}^{j-1} m(2^{-k}\omega) \prod_{k=j}^{N} m(2^{-k}\omega),$$

we see that the product converges uniformly on compact sets for $|m(0)| < 1$ or $m(\omega) = 1$, whereas in all other cases it diverges. Next, we want to show that for all N we have

$$\left|\prod_{k=0}^{N} m(2^{-k}\omega)\right| \leq 0(1)(1 + |\omega|^2)^{n/2} \tag{8.0.5}$$

for some $n \in N$. Then the pointwise convergence in connection with the dominated convergence theorem implies the convergence in $S'(\mathbb{R})$, as stated in the theorem.

First note that, thanks to the uniform convergence on compact sets, we have

$$\left|\prod_{k=0}^{N} m(2^{-k}\omega)\right| \leq 0(1), \qquad |\omega| \leq 1,\ N \geq 1. \tag{8.0.6}$$

For $|\omega| > 1$ we may pick $j \in N$ such that $2^{j-1} \leq |\omega| < 2^{j}$. As before, we may write

$$\left|\prod_{k=0}^{N} m(2^{-k}\omega)\right| \leq \left|\prod_{k=0}^{j-1} m(2^{-k}\omega)\right| \left|\prod_{k=j}^{N} m(2^{-k}\omega)\right|.$$

The last product is uniformly bounded by 0(1) given in (8.0.6), whereas the first product may be estimated by $\|m\|_{\infty}^{j}$. This shows that (8.0.5) holds and the theorem is proved for $|m(0)| \leq 1$. In all other cases consider $m(\omega)/m(0)$ and we are back to the former situation. □

Now that we have settled the existence of the infinite product we are concerned with some properties of the limit distribution itself. First we show that in the case where m satisifes the QMF-condition then the infinite product is actually a function in $L^2(\mathbb{R})$ (e.g. Mallat 1989).

Theorem 8.0.2
Let $m \in L^\infty(\mathbb{T})$ satisfy at

$$|m(\omega)|^2 + |m(\omega + \pi)|^2 = 1.$$

Suppose that

$$\lim_{N \to \infty} \prod_{k=1}^{N} m(2^{-k}\omega) = \eta(\omega)$$

exists for almost all ω. Then $\eta \in L^2(\mathbb{R})$.

Proof. Consider the following operator associated to m:

$$B: L^2(\mathbb{R}) \to L^2(\mathbb{R}), \qquad s(\omega) \mapsto m(\omega/2)s(\omega/2).$$

Starting from the characteristic function of $[-\pi/2, +\pi/2]$ we define a sequence

$$\hat{\phi}_n = B^n \chi_{[-\pi/2, +\pi/2]}, \qquad \hat{\phi}_n(\omega) = \prod_{j=1}^{n} m(2^{-j}\omega)\chi_{[-\pi 2^j, +\pi 2^j]}(\omega).$$

This sequence converges pointwise towards the limit function given by the infinite product. In addition we have, as we will now show,

$$\|\hat{\phi}_n\|_2^2 = 2\pi. \tag{8.0.7}$$

Fatou's lemma (Lemma 7.0.3. in Chapter 1) allows us to conclude. We still have to show the conservation of the norms. The condition on m implies that $m = \hat{\alpha}$ with some sequence $\alpha \in L^2(\mathbb{Z})$. The approximations $\hat{\phi}_n$ are the Fourier transforms of some function $\phi_n \in L^2(\mathbb{R})$. They are the band-limited approximations given by

$$2^{-n/2}\phi_n(2^{-n}t) = \sum_{k \in \mathbb{Z}} \sigma_n(k) \operatorname{sinc}(t - k).$$

The expansion coefficients satisfy at $\hat{\sigma}_n(\omega) = 2^{n/2} \prod_{j=0}^{n-1} m(2^j\omega)$ and thus

$$\sigma_{n+1} = D_{2,\alpha}^{\mathbb{Z}} \sigma_n, \qquad \sigma_0 = \delta,$$

as can be seen from the Fourier space picture of the dilation in $L^2(\mathbb{Z})$ (see 7.0.3). The conservation of energy of (8.0.7) now follows from the isometric property of $D_{2,\alpha}^{\mathbb{Z}}$—$\sqrt{2}\alpha$ is a QMF—and the orthonormality of the translates of the sinc function. □

Until now we have shown that the limit function of the infinite product is in $L^2(\mathbb{R})$. In addition, each approximation ϕ_n of the previous proof satisfies at

$$\langle \phi_n(\cdot - k) | \phi_n \rangle = \delta(k) \Leftrightarrow \sum_{k \in \mathbb{Z}} |\hat{\phi}_n(\omega + 2\pi k)|^2 = 1, \text{ a.e.}$$

But this does not imply that the same equation holds for the limit function ϕ, too. Indeed, we have not shown so far that the convergence $\phi_n \to \phi$ takes place in $L^2(\mathbb{R})$ and in fact, as the following counterexample shows, it does not do so in general.

Let $m(\omega) = (1 + e^{3i\omega})/2$ or equivalently $\alpha = (\delta + \delta(\cdot - 3))/2$. This is clearly a QMF; however the limit function obtained via the infinite product is $\chi_{[0,3]}/3$. The following theorem due to A. Cohen (Cohen 1990) now settles the case of $L^2(\mathbb{R})$ convergence.

We first need a definition.

Definition 8.0.3
A set I is called congruent modulo $[-\pi, +\pi]$ iff it is of measure 2π and iff for all $\omega \in [-\pi, \pi]$ at least one of the numbers $2\pi k + \omega$, $k \in \mathbb{Z}$, lies in I.

Alternatively, I is congruent modulo $[-\pi, +\pi]$ iff the following two conditions hold:

$$\left.\begin{aligned} &\text{(i) } \Pi\chi_I(\omega) = \sum_{k\in\mathbb{Z}} \chi_I(\omega + 2\pi k) \geq 1 \quad \text{for all } \omega, \\ &\text{(ii) } \Pi\chi_I(\omega) = \sum_{k\in\mathbb{Z}} \chi_I(\omega + 2\pi k) = 1 \quad \text{for almost all } \omega, \end{aligned}\right\} \tag{8.0.8}$$

where χ_I is the characteristic function of I.

Theorem 8.0.4
Let m be a 2π-periodic function in $L^\infty(\mathbb{T})$, Hölder continuous at $\omega = 0$, $|1 - m(\omega)| \leq O(\omega^\epsilon)$, $\epsilon > 0$, that satisfies at

$$|m(\omega)|^2 + |m(\omega + \pi)|^2 = 1, \qquad m(0) = 1.$$

Suppose in addition that there is a compact set I, congruent modulo $2\pi\mathbb{Z}$ with $[-\pi, +\pi]$, with 0 in its interior such that

$$m(\omega) \neq 0 \quad \textit{for all } \omega \in 2^{-j}I, \quad j = 1, 2, \ldots. \tag{8.0.9}$$

Then the infinite product

$$\hat{\phi}(\omega) = \prod_{k=0}^{\infty} m(2^{-k}\omega)$$

converges almost everywhere to a non-zero function $\hat{\phi}$ that satisfies at

$$\sum_{k\in\mathbb{Z}} |\hat{\phi}(\omega + 2\pi k)|^2 = 1 \Leftrightarrow \tilde{\phi} * \phi(k) = \delta(k)$$

for almost all ω, or, what, amounts to the same, $\langle \phi(\cdot - k) \mid \phi \rangle = \delta(k)$. Vice versa, for any $m \in L^\infty(\mathbb{T})$, continuous at the origin and $m(0) = 1$, such that the infinite product exists for all ω and defines a continuous function $\hat{\phi}$ for which this last equation holds for all ω, there is a compact set I of the above type.

Remark. Since I is a compact set there is a j_0 such that $2^{-j_0}I$ is close enough to 0 to be certain that $m(\omega 2^{-j}) \neq 0$ for all $\omega \in I$ and $j \geq j_0$. Therefore, only a finite number of terms have to be checked.

Remark. In the case where m is trigonometric polynomial the existence of such a set I clearly is sufficient for the infinite product to converge to a function whose integer translates are an orthonormal set.

Remark. The condition is automatically satisfied if m does not vanish in $[-\pi/2, +\pi/2]$. This is precisely Mallat's condition (Mallat 1989).

Proof. We first show the necessity. Suppose that $\hat{\phi}$ as given by the infinite product is continuous and satisfies for all ω at

$$\sum_{k \in \mathbb{Z}} |\hat{\phi}(\omega + 2\pi k)|^2 = 1.$$

It follows that for each $\omega \in [-\pi, \pi]$ there is at least one $k \in \mathbb{Z}$ such that $\hat{\phi}(\omega + 2\pi k) \neq 0$. Since $\hat{\phi}$ is continuous this holds with the same k for a whole interval containing ω. Since $[-\pi, +\pi]$ is compact a finite number of such intervals is enough to cover $[-\pi, +\pi]$. Note that, in particular, some interval may be chosen to contain 0 in its interior since $\hat{\phi}(0) = 1$. Let $\{I_m\}$ be the family of intervals and k_m the respective integral numbers. Then

$$I = \bigcup_m (I_m + 2\pi k_m)$$

is congruent to $[-\pi, +\pi]$ if we choose the I_m appropriately. We still have to show (8.0.9). With j_0 at our disposal we may write

$$\hat{\phi}(\omega) = \prod_{j=1}^{j_0} m(2^{-j}\omega)\hat{\phi}(2^{-j_0}\omega)$$

The set I is compact and therefore if j_0 is large enough then the last term $\hat{\phi}(2^{-j_0}\omega)$ is never 0 for $\omega \in I$. By construction, the left-hand side does not vanish on I either and the same must hold for the finite product. But none of these terms to be 0 is precisely the condition of the theorem.

We now show the sufficiency part. Consider again the operator B, defined already:

$$B: r(\omega) \mapsto m(\omega/2)r(\omega/2).$$

Let us define the following family of approximations starting from the characteristic function of $I/2$:

$$\hat{\phi}_n = B^n \chi_{I/2}, \qquad \hat{\phi}_n(\omega) = \prod_{k=1}^{n} m(\omega 2^{-k})\chi_I(\omega 2^{-n}).$$

These functions converge pointwise towards the limit function ϕ since 0 is in the interior of I. We claim that with some $c > 0$ we have

$$|\hat{\phi}_n(\omega)| \leq c|\hat{\phi}(\omega)|. \tag{8.0.10}$$

The latter being in $L^2(\mathbb{R})$, we may evoke the dominated convergence theorem to conclude for the convergence of $\hat{\phi}_n$ towards $\hat{\phi}$ in $L^2(\mathbb{R})$. As we will show, for each n we have

$$\phi_n * \tilde{\phi}_n(k) = \langle \phi_n(\cdot - k) \mid \phi_n \rangle = \delta(k). \tag{8.0.11}$$

Because the convergence of ϕ_n takes place in $L^2(\mathbb{R})$ we may go to the limit, which proves the theorem.

We now show (8.0.10). Note that with some constant $c > 0$ we have

$$\chi_I(\omega) \leq c|\hat{\phi}(\omega)|. \tag{8.0.12}$$

where χ_I is the characteristic function of I. Indeed, with j_0 at our disposal we may write

$$\hat{\phi}(\omega) = \prod_{k=1}^{j_0} m(2^{-k}\omega) \prod_{k=j_0+1}^{\infty} m(2^{-k}\omega).$$

Since I is compact we may chose j_0 large such that for all $\omega \in I$, $2^{-j_0}\omega$ is close enough to 0 such that the infinite product on the left is $\geqslant c_1 > 0$. By hypothesis on I the finite product is larger in absolute value than some $c_2 > 0$. This shows (8.0.12).

Now from this it follows that for all n we have

$$|\hat{\phi}_n(\omega)| = \prod_{k=1}^{n} |m(2^{-k}\omega)| \chi_I(2^{-n}\omega)$$

$$\leq c \prod_{k=1}^{n} |m(2^{-k}\omega)|\, |\hat{\phi}(2^{-n}\omega)| = |\hat{\phi}(\omega)|.$$

We now show (8.0.11). Let η be such that $\hat{\eta} = \chi_I$. Consider the space that is spanned by the integer translates of η. Because of (8.0.8) it is a perfect sampling space with η as generator of the translation invariant orthonormal basis. As before, we have

$$\phi_n = \sum_{k \in \mathbb{Z}} \sigma_n(k) \eta_n(\cdot - k2^{-n}), \qquad \eta_n(t) = 2^{n/2}\eta(2^n t),$$

and the expansion coefficients satisfy the following recursion ($\hat{\alpha} = \sqrt{2}m$):

$$\sigma_{n+1} = D^{\mathbb{Z}}_{2,\alpha}\sigma_n, \qquad \sigma_0 = \delta.$$

Since $D^{\mathbb{Z}}_{2,\alpha}$ with respect to the QMF α is an isometry it follows that the integer translates $\{\sigma_n(\cdot - k2^n)\}$ are an orthonormal set of sequences. Since the $\eta_n(\cdot - k2^{-n})$ are an orthonormal family it follows that the $\phi_n(\cdot - k)$ are an orthonormal family, too, and we are done. □

8.1 Compact support

The following theorem shows that the trigonometric polynomials give rise through the infinite product to compactly supported distributions.

Theorem 8.1.1
Let $m(\omega) = \sum_{k=N_1}^{N_2} \beta(k)\, e^{ik\omega}$ *be a trigonometric polynomial. Then if* $m(0) = 1$ *the infinite product*

$$\prod_{j=1}^{\infty} m(2^{-j}\omega)$$

is the Fourier transform of a distribution in $S'(\mathbb{R})$ *with support contained in* $[N_1, N_2]$.

Proof. As we have already shown, the product converges in $S'(\mathbb{R})$. Let $\hat{\alpha} = m$. This is a modulated δ-comb with support in $[N_1/2, N/2]$. The successive product corresponds to

$$\alpha * D_{1/2}\alpha * D_{1/4}\alpha \cdots.$$

It converges in $S'(\mathbb{R})$ since the Fourier transform is continuous in $S'(\mathbb{R})$. For a finite number of convolutions the support of the resulting distribution is contained in $[N_1(1/2 + 1/4 + \cdots)/2, N_2(1/2 + 1/2 + \cdots)/2]$ and we may pass to the limit. □

8.2 An easy regularity estimate

Now that we have shown which QMFs give rise to a multi-resolution analysis over $\mathbb{R}$ we are now concerned with the question of the regularity of the so-obtained scaling function and wavelet.

We measure regularity here by estimating the rate of growth of the Fourier transform. For this we split the QMF m into two parts $m(\omega) = m_1(\omega)m_2(\omega)$. For the first part we control the infinite product, and for the second part we have an estimation of its growth. Thanks to the formula[7]

$$\prod_{k=1}^{\infty} m_1(\omega 2^{-k}) = e^{i\omega}\,\mathrm{sinc}(\omega), \qquad m_1(\omega) = (1 + e^{i\omega})/2,$$

[7] This formula follows easily from the two-scale equation

$$\chi(t/2) = \chi(t) + \chi(t-1),$$

valid for the characteristic function of $[0, 1]$.

good candidates for m_1 are the powers of $(1 + e^{i\omega})/2$ and we write

$$m(\omega) = \left(\frac{1 + e^{i\omega}}{2}\right)^n m_2(\omega).$$

Then ϕ as given by the infinite product can be written as

$$\phi = \rho * \mu, \quad \hat{\rho}(\omega) = e^{in\omega} \operatorname{sinc}^n(\omega). \quad \hat{\mu}(\omega) = \prod_{k=1}^{\infty} m_2(2^{-k}\omega).$$

As we know from Section (1.6), ρ is a compactly supported n-times differentiable function. The convolution with μ destroys some of the regularity, depending on the growth of its Fourier transform. As we will soon see, $\hat{\mu}$ is of at most polynomial growth. It therefore acts in terms of regularity like a fractional derivative.

We then have the following estimate of the growth of infinite products.

Lemma 8.2.1
Suppose $0 \leq m(\omega) \leq c$ *is a smooth function with* $m(0) = 1$. *Then*

$$\prod_{j=1}^{\infty} m(2^{-j}\omega) \leq O(1)(1 + |\omega|^2)^{\gamma/2}, \qquad \gamma = -\log c/\log 2.$$

Proof. With k at our disposal we may write

$$\prod_{j=1}^{\infty} m(2^{-j}\omega) = \prod_{j=1}^{k} m(2^{-j}\omega) \prod_{j=k+1}^{\infty} m(2^{-j}\omega).$$

Given ω we choose k as small as possible to have $|\omega|2^{-k} \leq \pi$. In the first product we estimate $m \leq c$ and hence the product is estimated by c^k. The remaining product equals

$$\prod_{j=1}^{\infty} m(2^{-j}\xi), \qquad \xi = 2^{-j}\omega.$$

Now $\xi \in [-\pi, +\pi]$ and since the infinite product converges uniformly on compacts sets it may be estimated by $O(1)$. □

Remark. By applying this result to $m(\omega)m(2\omega)\cdots m(2^n\omega)$ instead of m the result can be optimized (e.g. Volkmer 1992).

Altogether we have shown the following theorem due to Daubechies (1988).

Theorem 8.2.2
Let $m \in C^1(\mathbb{T})$ *satisfy at* $m(0) = 1$.

Suppose that the following factorization holds:

$$m(\omega) = \left(\frac{1 + e^{i\omega}}{2}\right)^n \eta(\omega),$$

with $|\eta| \leq c$ *and some* $n \in \mathbb{N}$. *Then* ϕ *given in Fourier space by*

$$\hat{\phi}(\omega) = \prod_{k=1}^{\infty} m(2^{-j}\omega)$$

satisfies

$$|\hat{\phi}(\omega)| \leq O(|\omega|^{-\gamma}), \qquad |\omega| \to \infty,$$

where $\gamma = n - \log c/\log 2$.

9 The dyadic interpolation spaces

As a first application of the preceding theorems, we will now construct a family of perfect sampling spaces that will eventually lead to the construction of compactly supported wavelets. The dyadic interpolation spaces have been used by Deslaurier and Dubuc in order to construct some self-similar fractals (Deslaurier and Dubuc 1987). In the context of wavelet analysis they have been revealed as useful in some fast algorithms for computing the continuous wavelet transform efficiently (Holschneider *et al.* 1988 and 1989). We will come back to this later on, in Section 16.

To understand the idea of the construction let s be a function over the real line of which only the values at the integral points are known. We now want to reconstruct s approximately. If s is continuous one might guess that a good approximation to s at the mid-points might be given by interpolation, that is, we set (in the case of linear interpolation)

$$s(n + 1/2) = (s(n) + s(n + 1))/2.$$

However, to be more general we may allow more general interpolation coefficients and we write as 'best guess'

$$s(n + 1/2) = \sum_{k \in \mathbb{Z}} \alpha(n - k) s(k),$$

with some interpolation filters α. Clearly, we can continue in this way by successively adding the midpoints using the previous generation and interpolating it. This defines s at all dyadic points, that is, at all points of the form $k2^{-j}$, where $k, j \in \mathbb{Z}$. These points are dense and if s is continuous it is therefore uniquely determined by this construction.

Another starting point for the discussion of such spaces is whether there is a perfect sampling space V over the real line that is scaling invariant in the sense that $D_{1/2}V \subset V$. In this case the dilation operator for functions in V may be computed in sequence space, where this time the perfect sampling operator may be used. Let h be the unique function in V that satisfies at $h(k) = \delta(k)$. Then we have

$$\Sigma D_2 = D_{2,\gamma}^{\mathbb{Z}} \Sigma, \qquad \begin{array}{ccc} V & \xrightarrow{\Sigma} & L^2(\mathbb{Z}) \\ \downarrow D_2 & & \downarrow D_{2,\gamma}^{\mathbb{Z}} \\ V & \xrightarrow{\Sigma} & L^2(\mathbb{Z}) \end{array}, \qquad \gamma = \Sigma D_2 h.$$

Observe the following relation between the interpolation filter γ and the interpolation coefficients α as follows from $h(k) = \delta(k)$:

$$2\gamma = \delta + S(\Sigma_2^{\mathbb{Z}})^* \alpha, \qquad 2\gamma(k) = \begin{cases} 1 & \text{for } k = 0, \\ 0 & \text{for } k \text{ even, } k \neq 0, \\ \alpha([k+1]/2) & \text{for } k \text{ odd.} \end{cases}$$

Suppose we are given such a dilation invariant sampling space, then h may be recovered from the interpolation coefficients α or γ by applying the interpolation procedure that we described previously to the sequence δ. These successive interpolations can be conveniently rephrased in terms of distributions by considering the following sequence of measures:

$$h_{j+1} = \gamma_j * h_j, \qquad \gamma_{j+1} = \gamma_j(2\cdot),$$

where

$$\gamma_0 = \delta/2 + \sum_{k \in \mathbb{Z}} \alpha(k)\delta(\cdot - k - 1/2)/2, \qquad h_0 = \delta.$$

Note the particular shape of the interpolator γ_0: it is only supported by 0 and the half-odd positions. This is because the interpolation procedure is hierarhical in the sense that only new values are added and no old values are destroyed.

In Fourier space this can be rewritten as

$$\hat{h}_j(\omega) = \prod_{k=0}^{j} m(2^{-k}\omega),$$

$$m(\omega) = \hat{\gamma}_0(\omega) = 1 + \sum_{k \in \mathbb{Z}} \alpha(k)\, e^{2i\omega(k+1/2)}.$$

We can therefore use the theorems proved in the previous section to analyse such infinite products.

9.1 The Lagrange interpolation spaces

Obvious examples of such dyadic interpolation spaces are the band-limited functions and the splines of odd order. However, in both cases the interpolation filters are of infinite length, which makes them difficult to use in numerical applications. For the rest of this section we will therefore focus on the special case, where the interpolation is done with the help of a Lagrange polynomial of odd order N. In this case the filter coefficients α read

$$\alpha(k + (1 - N)/2) = \prod_{n=0, n \neq k}^{N} \frac{N/2 - n}{k - n}, \qquad k = -N/2 + 1/2, N/2 - 1/2.$$

We have listed the first four interpolators in the following table:

$$N = 1: \frac{1}{2}, \frac{1}{2}$$

$$N = 3: -\frac{1}{16}, \frac{9}{16}, \frac{9}{16}, \frac{-1}{16}$$

$$N = 5: \frac{3}{256}, -\frac{25}{256}, \frac{75}{256}, \frac{75}{256} - \frac{25}{256}, \frac{3}{256}$$

$$N = 7: -\frac{5}{2048}, \frac{49}{2048}, -\frac{245}{2048}, \frac{1225}{2048}, \frac{1225}{2048}, -\frac{245}{2048}, \frac{49}{2048}, -\frac{5}{2048}.$$

Since the interpolation filter is now compactly supported it follows that h itself has compact support. In Figures 9.1 and 9.2 we have sketched h for the third-order Lagrange interpolation and its first derivative.

Theorem 9.1.1

The Lagrange interpolation spaces of odd order are perfect sampling spaces.

Proof. We first show that $\hat{\gamma}$ is non-negative with the only exception being $\omega = \pi$.

Theorem 9.1.2

Let γ be the coefficients of a Lagrange interpolation of odd order N. Then we have

$$\hat{\gamma}(\omega) = 1 - c_N \int_0^{\omega} d\xi \, \sin^N(\xi), \qquad c_N^{-1} = \int_0^{\pi} d\xi \, \sin^N(\xi).$$

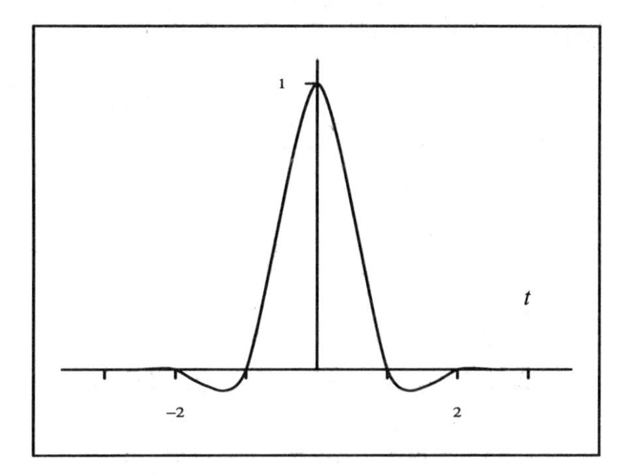

Fig. 9.1 The cubic interpolator.

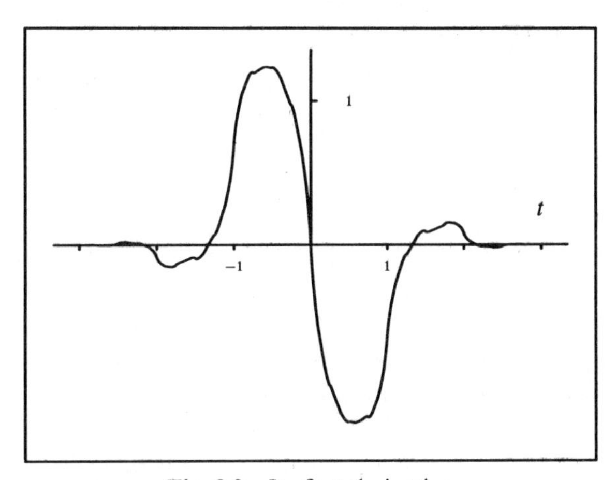

Fig. 9.2 Its first derivative.

Proof.[8] First observe that the dyadic interpolation with the help of a Lagrange polynomial preserves the polynomials up to degree N. More precisely, suppose P is a polynomial of degree not exceeding N, then the interpolation procedure applied to the sampling sequence $P(k)$ gives back P everywhere. If we specify $P(t) = t^n$, $0 \leq n \leq N$, we obtain in particular

$$D_{2,\gamma}^{\mathbb{Z}} \Sigma P = D_2 P = 2^{-1-n} \Sigma P.$$

[8] C. Miccelli, private communication.

In Fourier space this reads in the sense of distributions:

$$\delta^{(n)}(\omega) = \hat{\gamma}(\omega)\delta^{(n)}(\omega), \qquad 0 \le n \le N,$$

where $\delta^{(n)}$ is the nth derivative of the delta distribution. This implies that

$$\hat{\gamma}(\omega) = 1 + O(\omega^N), \qquad (\omega \to 0).$$

Now γ is symmetric and thus $\hat{\gamma}(\omega) = \hat{\gamma}(-\omega)$. Therefore, we even have, since N is odd,

$$\hat{\gamma}(\omega) = 1 + O(\omega^{N+1}), \qquad (\omega \to 0).$$

Now $2\gamma(2k) = \delta(k)$ and we therefore have, by the Poisson summation formula,

$$\hat{\gamma}(\omega) + \hat{\gamma}(\omega + \pi) = 1.$$

Since $\hat{\gamma}(0) = 1$ we have the following behaviour around π

$$\gamma(\pi + \omega) = O(\omega^{N+1}).$$

The derivative of γ thus has a zero of order N at 0 and at π. In addition, it is a trigonometric polynomial of degree $2N$. Therefore, we necessarily have

$$\partial\hat{\gamma}(\omega) = c \sin^N(\omega),$$

with some $c \in \mathbb{C}$, and the equation above follows by integration. □

In particular, we have that $\hat{\gamma}$ is non-negative and it does not vanish on $[-\pi/2, +\pi/2]$. Therefore, we can use Theorem 8.0.4 with $m = \sqrt{\hat{\gamma}}$ to conclude that ϕ given by the infinite product of m is the generator of a translation invariant orthonormal set. Now $h = \tilde{\phi} * \phi$ since $\hat{h}(\omega) = |\hat{m}(\omega)|^2$. This shows that h is continuous and satisfies at $h(k) = \tilde{\phi} * \phi(k) = \delta(k)$. □

It comes now as no surprise that, actually, the Lagrange interpolation spaces give rise to a multi-resolution analysis. Let V_0 be the linear span of $h(\cdot - k)$. By dilation we obtain a scale of such spaces upon setting $V_j = D_{2^{-j}}V_0$.

Theorem 9.1.3
The spaces V_j are a multi-resolution analysis over $\mathbb{R}$.

Proof. The inclusion property

$$V_j \subset V_{j+1}$$

follows from the hierarchical nature of the construction. Condition (iii) follows from the previous theorem and the fact that all perfect sampling spaces have a translation invariant Riesz basis. We will have to check for completeness and uniqueness.

The uniqueness follows from Theorem 6.0.1.

To see the completeness we show that the functions s of regularity C^1 with compact support may be approximated in $L^\infty(\mathbb{R})$ by functions in V_j upon sampling at the dyadic points and reconstructing with a scaled version of h:

$$s_j = \sum_{k \in \mathbb{Z}} s(2^{-j}k)h(2^j t - k).$$

Since the support of the approximations is contained in a fixed compact set, it follows from uniform convergence that the convergence is in $L^2(\mathbb{R})$, too. Completeness then follows from a standard density argument. Note that from the Poisson summation formula we have

$$\Pi h = \sum_{k \in \mathbb{Z}} h(\cdot - k) = 1,$$

since $\hat{h}(2\pi k) = \delta(k)$, as follows from the infinite product. Therefore,

$$s(t) - s_j(t) = \sum_{k \in \mathbb{Z}} (s(2^{-j}k) - s_j(t))h(2^j t - k).$$

Now, thanks to the regularity of s, around the dyadic points we have

$$|s(2^{-j}k) - s_j(t)| \leq O(1)2^{-j}|2^{-j}k - t| \quad \text{for } t \text{ in the support of } h(2^j t - k).$$

We therefore have for all t

$$|s(t) - s_j(t)| \leq O(1)2^{-j} \sum_{k \in \mathbb{Z}} |2^{-j}k - t|\,|h(2^j t - k)|.$$

The sum is just another constant that is not dependent on j and thus

$$|s(t) - s_j(t)| \leq O(1)2^{-j}$$

and we have convergence in $L^\infty(\mathbb{R})$. □

Now the associated wavelets and scaling functions can be obtained by the standard methods of Section 4. Note, however, that none of them is compactly supported. However, they are exponentially localized as follows from the fact that the Fourier multiplier that maps h into ϕ is analytic in some strip.

10 Compactly supported wavelets

Until now we have constructed families of orthonormal wavelets that have been exponentially localized. In the special case of the multi-resolution analysis generated by the Lagrange interpolation spaces we have found a compactly supported elementary generating function h, $h(k) = \delta(k)$, whose translates from a Riesz basis but are not orthonormal.

We now show how to construct a compactly supported orthonormal wavelet basis from the Lagrange interpolation spaces. The relation will be that the auto-correlation function $\tilde{\phi} * \phi$ has precisely the dyadic interpolation property with respect to some Lagrange interpolator. That is, in some sense the generating function of the multi-resolution analysis with orthonormal wavelets corresponds to the square root of the basic function h of the dyadic interpolation space.

We now come to the details. If the generating function ϕ of some multi-resolution analysis is compactly supported it follows that the associated QMF as given by $\sqrt{2}\Sigma_\phi D_2 \phi$ has only a finite number of non-zero elements. Vice versa, Theorem 8.1.1 proves that in this case the infinite product is the Fourier transform of a distribution with compact support. If, in addition, the polynomial satisfies the condition of Theorem 8.0.4, then we have generated an orthonormal basis of compactly supported wavelets. The task is therefore to find trigonometric polynomials m that satisfy at

$$|m(\omega)|^2 + |m(\omega + \pi)|^2 = 1 \tag{10.0.1}$$

and the condition of Theorem 8.0.4. In Section 13 we show how to obtain all polynomials solutions of (10.0.1).

Together with m its squared value is again a trigonometric polynomial. So $r(\omega) = |m(\omega)|^2$ has to satisfy at

$$r(\omega) + r(\omega + \pi) = 1, \qquad r(\omega) \geq 0. \tag{10.0.2}$$

As the following theorem shows, given a positive polynomial, we may take the square root in the space of polynomials. If r satifies (10.0.2), this square root is then a polynomial that satisfies (10.0.1). The square root of polynomials is found with the help of the following lemma due to Riesz.

Lemma 10.0.1

Let $v(\omega) = \sum_{k=-N}^{N} \alpha(k)\, e^{i\omega k}$ be a real-valued trigonometric polynomial that satisfies $v(\omega) \geq 0$. Then there is a trigonometric polynomial of the form $w(\omega) = \sum_{k=0}^{N} \beta(k)\, e^{i\omega k}$, such that $|w(\omega)|^2 = v(\omega)$. In addition, if the coefficients $\alpha(k)$ are real valued, the coefficients $\beta(k)$ may be chosen real valued, too.

Proof. The proof is adapted from Pólya and Szegö (1971). We include it here since the construction is widely used nowadays in the design of wavelets. Let us consider the polynomial

$$G = \sum_{k=0}^{2N} \alpha(k - N) z^k.$$

Then we have

$$v(\omega) = e^{-iN\omega} G(e^{i\omega}). \tag{10.0.3}$$

Since v is real we have the Hermitian symmetry $\alpha(-k) = \bar{\alpha}(k)$. Therefore, it follows that on the unit circle $z = e^{i\omega}$ we have $z^{2n}\bar{G}(1/z) = G(z)$, where $\bar{G}$ stands for the polynomial in z with coefficients $\bar{\alpha}(k)$. Thus they are identical in the whole complex plane. In particular, it follows that if $z \neq 0$ is a zero of G then $1/\bar{z}$ is a zero, too. The same is true if multiplicities are counted, as can be seen by looking at the derivative. Therefore, there are three classes of zero. Either $|\zeta_n| = 1$, or $|\xi_k| \notin \{0, 1\}$, in which case $1/\overline{\xi_k}$ is another zero, and finally $z = 0$. Accordingly, the polynomial G factorizes:

$$G(z) = cz^{N_0} \prod_{n=0}^{N_1 - 1} (z - \zeta_n) \prod_{k=0}^{N_2 - 1} (z - \xi_k)(z - 1/\overline{\xi_k}),$$

where $N_0 + N_1 + 2N_2 = N$. We claim that N_1 is even. Indeed, $G(\zeta_k) = 0$ iff $v(\omega_k) = 0$ with $\zeta_k = e^{i\omega_k}$. Since $v(\omega) \geq 0$ the order of the zero at ω_k must be even. Taking the derivative of (10.0.3) at ω_k we see that this must also be true for the order of ζ_k as zero of G. Therefore, N_1 is even, and we end up with

$$v(\omega) = |\omega(\omega)|^2 = c' \left| \prod_{n=0}^{N_1/2 - 1} (e^{i\omega} - \zeta_n) \sum_{k=0}^{N_2 - 1} (e^{i\omega} - \xi_k) \right|^2. \qquad \square$$

We still have to solve (10.0.1) in order to construct the orthonormal wavelets.

There is an immediate solution to equation (10.0.1) given by the Lagrange interpolations filters of the previous section. More precisely, consider the numbers c_k that are defined via

$$c_k \int_0^\pi d\omega \sin^{2k+1}(\omega) = 1.$$

Then set

$$r_k(\omega) = 1 - c_k \int_0^\omega d\omega' \sin^{2k+1}(\omega').$$

Clearly, r_k is a trigonometric polynomial with $r_k(\omega) + r_k(\omega + \pi) \equiv 1$, and $0 \leq r_k(\omega) \leq 1$. In addition, we have $r_k(\omega) > 0$ for $\omega \in (-\pi, +\pi)$. We now set

$$\hat{\phi}(\omega) = \prod_{k=1}^{\infty} m(2^{-k}\omega),$$

where m is a trigonometric polynomial that satisfies at $|m|^2 = r_k$. By Theorem 8.0.4 the translates $\phi(\cdot - k)$ are an orthonormal family of compactly supported scaling functions. Using the results of Section 6 we see that they generate a multi-resolution analysis of $L^2(\mathbb{R})$. By the usual construction we therefore obtain the wavelets, which in turn are also compactly supported.

We thus have shown the following theorem due to Daubechies (1987).

Theorem 10.0.2
There is an infinite family of compactly supported wavelet bases.

11 Wavelet frames

Suppose now that we are given a trigonometric polynomial that satisfies at

$$|m(\omega)|^2 + |m(\omega+\pi)|^2 = 1, \qquad m(0) = 1,$$

but for which the condition of Theorem 8.0.4 does not hold. The infinite product

$$\hat{\phi}(\omega) = \prod_{j=1}^{\infty} m(2^{-j}\omega)$$

still defines the Fourier transform of a square integrable function. In addition, it satisfies the scaling equation

$$\phi(t/2)/2 = \sum_{k\in\mathbb{Z}} \alpha(k)\phi(t-k), \qquad \hat{\alpha} = m. \tag{11.0.1}$$

However, its integer translates are no longer an orthonormal family. We may still define

$$\psi = \sum_{k\in\mathbb{Z}} \beta(k)\phi(\cdot - k), \qquad \beta(k) = (-1)^k\bar{\alpha}(1-k), \quad \hat{\alpha} = m, \tag{11.0.2}$$

as we did in the orthonormal case. We then have the following theorem due to Lawton (1991).

Theorem 11.0.1
The family $\psi_{j,k} = 2^{j/2}\psi(2^j\cdot - k)$ *is a tight frame, that is, we have for every* $s \in L^2(\mathbb{R})$

$$s = \sum_{k\in\mathbb{Z}} \sum_{j\in\mathbb{Z}} \langle \psi_{j,k} \mid s\rangle \psi_{j,k} \tag{11.0.3}$$

and the sum converges unconditionally.

Proof. It is clearly enough to show that

$$\sum_{k\in\mathbb{Z}} \sum_{j\in\mathbb{Z}} |\langle \psi_{j,k} \mid s\rangle|^2 = \|s\|_2^2.$$

Indeed, this would show that $\{\psi_{j,k}\}$ is a Bessel family. The sampling operator associated to it is an isometry from $L^2(\mathbb{R})$ to a closed subspace of $L^2(\mathbb{Z})$. It may be inverted by its adjoint, and hence (11.0.3) holds. Consider now the operators

$$P_j = 2^{j/2}\Sigma_\phi D_{2^j}, \qquad Q_j = 2^{j/2}\Sigma_\psi D_{2^j}.$$

Both operators are bounded from $L^2(\mathbb{R})$ to $L^2(\mathbb{Z})$ since both functions ϕ and ψ are in $L^2(\mathbb{R})$ and of compact support. To show the theorem it is therefore enough to show that

$$\sum_{j\in\mathbb{Z}} \|Q_j s\|^2_{L^2(\mathbb{Z})} = \|s\|^2_2.$$

We claim that we have

$$\|P_j s\|^2_{L^2(\mathbb{Z})} + \|Q_j s\|^2_{L^2(\mathbb{Z})} = \|P_{j+1} s\|^2_{L^2(\mathbb{Z})}. \tag{11.0.4}$$

This then allows us to write

$$\sum_{j=-N}^{N} \|Q_j s\|^2_{L^2(\mathbb{Z})} = \|P_{N+1} s\|^2_{L^2(\mathbb{Z})} - \|P_{-N} s\|^2_{L^2(\mathbb{Z})}.$$

The first term goes to $\|s\|^2_{L^2(\mathbb{R})}$ and the second goes to 0, as follows from the remarks after Theorems 6.0.1 and 6.0.2, and the theorem is proved.

We still have to show (11.0.4). Because of the scaling equation (11.0.1) and the construction of ψ out of ϕ we may write

$$\Sigma_\phi(D_2)^* = (D^{\mathbb{Z}}_{2,\alpha})^*\Sigma_\phi, \qquad \Sigma_\psi(D_2)^* = (D^{\mathbb{Z}}_{2,\beta})^*\Sigma_\phi,$$

where $D^{\mathbb{Z}}_{2,\alpha}$ and its adjoint are the dilation in $L^2(\mathbb{Z})$, as defined in Section 7, and α, β are given by (11.0.2). Now $(\sqrt{2}\alpha, \sqrt{2}\beta)$ is a QMF-system. Hence, it preserves the energy under down-sampling, namely,

$$\|(D^{\mathbb{Z}}_{2,\alpha})^* v\|^2_{L^2(\mathbb{Z})} + \|(D^{\mathbb{Z}}_{2,\beta})^* v\|^2_{L^2(\mathbb{Z})} = 2\|v\|^2_{L^2(\mathbb{Z})}.$$

Therefore, in particular,

$$\begin{aligned}\|P_{-1} s\|^2_{L^2(\mathbb{Z})} + \|Q_{-1} s\|^2_{L^2(\mathbb{Z})} &= 2^{-1}\|\Sigma_\phi(D_2)^* s\|^2_{L^2(\mathbb{Z})} + 2^{-1}\|\Sigma_\psi(D_2)^* s\|^2_{L^2(\mathbb{Z})}\\ &= 2^{-1}\|(D^{\mathbb{Z}}_{2,\alpha})^*\Sigma_\phi s\|^2_{L^2(\mathbb{Z})} + 2^{-1}\|(D^{\mathbb{Z}}_{2,\beta})^*\Sigma_\phi s\|^2_{L^2(\mathbb{Z})}\\ &= \|P_0 s\|^2_{L^2(\mathbb{Z})}.\end{aligned}$$

The other j of (11.0.4) follow by dilation co-variance. □

12 Bi-orthogonal expansions

As we have seen, in the continuous case the analysing wavelet and the reconstruction wavelet need not be the same. In the discrete case we have encountered this phenomenon in form of bi-frames. Here we want to construct some bi-orthogonal Riesz basis of wavelets. This brings a new freedom to the analysis.

Since each W_j is a sampling space we may find a bi-orthogonal system, too, as we saw in Section 1.4. Let g and h be two functions in W_0 that satisfy at

$$\bar{\hat{g}}(\omega)\hat{h}(\omega) = |\hat{\psi}(\omega)|^2.$$

It follows from the results of Section 1.4 that $\{g(\cdot - k)\}$ and $\{h(\cdot - k)\}$ are a bi-orthogonal system for W_0. It follows by dilation that we have the following expansion of an arbitrary function $s \in L^2(\mathbb{R})$:

$$s = \sum_{j \in \mathbb{Z}} \sum_{k \in \mathbb{Z}} \langle g_{j,k} \mid s \rangle_{\mathbb{R}} h_{j,k}, \qquad g_{j,k} = 2^{j/2} g(2^j \cdot - k), \quad h_{j,k} = 2^{j/2} h(2^j \cdot - k),$$

and $\{g_{j,k}\}$ and $\{h_{j,k}\}$ are a bi-orthogonal system for $L^2(\mathbb{R})$:

$$\langle g_{j,k} \mid h_{j,k} \rangle = \delta_{j,j'} \delta_{k,k'}.$$

We now want to dispense with the restriction that g and h are in the same sampling space. In the case of a multi-resolution analysis over $\mathbb{R}$ it was useful to consider the case of a multi-resolution analysis over $\mathbb{Z}$ in parallel since the sampling operator allowed us to go from V_0 to $L^2(\mathbb{Z})$. We will pursue the same strategy here in the case of bi-orthogonal expansions.

12.1 Bi-orthogonal expansions of $L^2(\mathbb{Z})$

Since, by definition, the translates by $2\mathbb{Z}$ of a QMF pair α, β are an orthonormal basis of $L^2(\mathbb{Z})$ we have the following decomposition of the identity into a sum of two projection operators:

$$(\Sigma^{\mathbb{Z}}_{\alpha,2})^* \Sigma^{\mathbb{Z}}_{\alpha,2} + (\Sigma^{\mathbb{Z}}_{\beta,2})^* \Sigma^{\mathbb{Z}}_{\beta,2} = \mathbb{1}.$$

We are now looking for sequences α, α', β, $\beta' \in L^2(\mathbb{Z})$ such that, again, we have the possibility of down-sampling and reconstruction through up-sampling

$$(\Sigma^{\mathbb{Z}}_{\alpha',2})^* \Sigma^{\mathbb{Z}}_{\alpha,2} + (\Sigma^{\mathbb{Z}}_{\beta',2})^* \Sigma^{\mathbb{Z}}_{\beta,2} = \mathbb{1}. \tag{12.1.1}$$

In order to give a precise meaning to the operators on the left-hand side we assume that the set of $2\mathbb{Z}$ translates of all sequences generate Riesz bases for the spaces they span. In Fourier space this operator acts as follows:

$$\begin{aligned}\hat{s}(\omega) \mapsto & \tfrac{1}{2}(\hat{\alpha}'(\omega)\bar{\hat{\alpha}}(\omega) + \hat{\beta}'(\omega)\bar{\hat{\beta}}(\omega))\hat{s}(\omega) \\ & + \tfrac{1}{2}(\hat{\alpha}'(\omega)\bar{\hat{\alpha}}(\omega + \pi) + \hat{\beta}'(\omega)\bar{\hat{\beta}}(\omega + \pi))\hat{s}(\omega + \pi),\end{aligned}$$

or, in vector form,

$$\begin{bmatrix} \hat{s}(\omega) \\ \hat{s}(\omega+\pi) \end{bmatrix} \mapsto \frac{1}{2} \begin{bmatrix} \bar{\hat{\alpha}}(\omega) & \bar{\hat{\beta}}(\omega) \\ \bar{\hat{\alpha}}(\omega+\pi) & \bar{\hat{\beta}}(\omega+\pi) \end{bmatrix} \begin{bmatrix} \hat{\alpha}'(\omega) & \hat{\alpha}'(\omega+\pi) \\ \hat{\beta}'(\omega) & \hat{\beta}'(\omega+\pi) \end{bmatrix} \begin{bmatrix} \hat{s}(\omega) \\ \hat{s}(\omega+\pi) \end{bmatrix}.$$

Now perfect reconstruction is equivalent to the matrix product being the identity for a.a. ω. Upon commuting the matrices we have, equivalently,

$$\bar{\hat{\alpha}}(\omega)\hat{\alpha}'(\omega) + \bar{\hat{\alpha}}(\omega+\pi)\hat{\alpha}'(\omega+\pi) \equiv \bar{\hat{\beta}}(\omega)\hat{\beta}'(\omega) + \bar{\hat{\beta}}(\omega+\pi)\hat{\beta}'(\omega+\pi) \equiv 2,$$

$$\bar{\hat{\alpha}}(\omega)\hat{\beta}'(\omega) + \bar{\hat{\alpha}}(\omega+\pi)\hat{\beta}'(\omega+\pi) \equiv \bar{\hat{\beta}}(\omega)\hat{\alpha}'(\omega) + \bar{\hat{\beta}}(\omega+\pi)\hat{\alpha}'(\omega+\pi) \equiv 0,$$

or, equivalently, thanks to the Poisson summation formula, we have shown that perfect sampling–reconstruction of (12.1.1) holds iff

$$\langle \alpha(\cdot - 2k) \mid \alpha' \rangle = \langle \beta(\cdot - 2k) \mid \beta' \rangle = \delta(k),$$

$$\langle \alpha(\cdot - 2k) \mid \beta' \rangle = \langle \beta(\cdot - 2k) \mid \alpha' \rangle = 0.$$

Consider now the linear spaces that are spanned by the translates by 2 of α, α', β, and β', which we call V, V', W, and W' respectively. Then the above conditions imply that V' is orthogonal to W and V is orthogonal to W'. To put it differently, if we write $P + Q$ for the sum of (12.1.1) then this condition means precisely that P is a projector on V' and Q is a projector on W'. In addition, the range of the one is in the null-space of the other:

$$P^2 = P, \qquad Q^2 = Q, \qquad PQ = QP = 0.$$

Note that, in general, P and Q are no longer orthogonal projectors since α and α' do not necessarily span the same space.

Now suppose we are given α and α' whose translates by 2 generate a Riesz basis for their spans. Can we find β and β' such that (12.1.1) holds? An immediate solution to this is given by setting

$$\left.\begin{aligned} \beta(k) &= (-1)^k \bar{\alpha}'(1-k) \Leftrightarrow \hat{\beta}(\omega) = e^{-i\omega}\, \bar{\hat{\alpha}}'(\omega+\pi), \\ \beta'(k) &= (-1)^k \bar{\alpha}(1-k) \Leftrightarrow \hat{\beta}'(\omega) = e^{-i\omega}\, \bar{\hat{\alpha}}(\omega+\pi). \end{aligned}\right\} \qquad (12.1.2)$$

Obviously, β and β' are again a system of bi-orthogonal Riesz bases.

To summarize we have shown the following theorem.

Theorem 12.1.3

Suppose a projection operator in $L^2(\mathbb{Z})$ has the following form:

$$P\colon s \mapsto \sum_{k \in \mathbb{Z}} \langle \alpha(\cdot - 2k) \mid s \rangle \alpha'(\cdot - 2k),$$

where we suppose that the translates by 2 of $\alpha, \alpha' \in L^2(\mathbb{Z})$ generate a Riesz basis for their linear spans. Then $Q = \mathbb{1} - P$ has the same form with α and α'

replaced by β, β' *given by* (12.1.2). *These sequences again generate Riesz bases for the span of their translates by* $2\mathbb{Z}$.

12.2 Bi-orthogonal expansions in $L^2(\mathbb{R})$

Suppose we are given two multi-resolution analyses V_j, V'_j over $\mathbb{R}$ with associated generating functions ϕ and ϕ' respectively. We do not assume for the moment that their translates generate an orthonormal basis of their span. We only require them to be a Riesz basis. Now, instead of taking orthogonal projections into V_j, we consider the projection along V_j. More precisely, we set,

$$P_j = D_{2^j}^{-1} P_0 D_{2^j}, \qquad P_0 s = \sum_{k \in \mathbb{Z}} \langle \phi'(\cdot - k) \mid s \rangle \phi(\cdot - k). \qquad (12.2.1)$$

In order that this is actually a projection ($P_j^2 = P_j$) a necessary and sufficient condition is

$$\langle \phi'(\cdot - k) \mid \phi \rangle = \delta(k) \quad \Leftrightarrow \quad \sum_{k \in \mathbb{Z}} \overline{\hat{\phi}(\omega + 2\pi k) \hat{\phi}'(\omega + 2\pi k)} \equiv 1, \quad (12.2.2)$$

as can be seen by straightforward computation of P_j^2. This will be assumed to hold from now on. Note that, because of the inclusion property of the V_j and V'_j, we have

$$P_j P_{j'} = P_{\min\{j, j'\}}. \qquad (12.2.3)$$

As before, we now take the difference between two successive approximations and set

$$Q_j = P_{j+1} - P_j.$$

This is again a projector because of (12.2.3), as direct computation shows. We now want to show that, again, it is a sampling–reconstruction operator of the form

$$Q_j = D_{2^j}^{-1} Q_0 D_{2^j}, \qquad Q_0 s = \sum_{k \in \mathbb{Z}} \langle \psi'(\cdot - k) \mid s \rangle \psi(\cdot - k).$$

By dilation co-variance it is enough to show this for $Q_{-1} = P_0 - P_{-1}$. Thus we go into sequence space by setting

$$K = \Sigma_{\phi'} Q_{-1} (\Sigma_\phi)^* = \mathbb{1} - B,$$

with

$$B = \Sigma_{\phi'} P_{-1} (\Sigma_\phi)^*,$$

since by condition (12.2.2) we have $\Sigma_{\phi'} P_0 (\Sigma_\phi)^* = \mathbb{1}_{L^2(\mathbb{Z})}$. By direct computation we see that B is a sampling–reconstruction operator in $L^2(\mathbb{Z})$:

$$B = (\Sigma_{2,\alpha}^{\mathbb{Z}})^* \Sigma_{2,\alpha'}^{\mathbb{Z}}, \qquad \alpha = \sqrt{2} \Sigma_{\phi'} D_2 \phi, \quad \alpha' = \sqrt{2} \Sigma_\phi D_2 \phi'.$$

The orthogonality condition of (12.2.2) now reads in sequence space as follows:

$$\langle \alpha(\cdot - 2k) \mid \alpha' \rangle = \delta(k) \Leftrightarrow \bar{\hat{\alpha}}(\omega)\alpha'(\omega) + \bar{\hat{\alpha}}(\omega + \pi)\hat{\alpha}'(\omega + \pi) = 2.$$

By Theorem 12.1.3 it follows that $K = \mathbb{1} - B$ is again a sampling–reconstruction operator with respect to the sequences

$$\beta(k) = (-1)^k \overline{\alpha'}(1-k), \qquad \beta'(k) = (-1)^k \bar{\alpha}(1-k).$$

In addition, the translates by 2 of β and β' generate a Riesz basis for their span. This is preserved if we go back over the real line and thus we have shown that Q_{-1} and hence, by dilation invariance, all Q_j are sampling–reconstruction operators of the form of (12.2.4), with ψ and $\psi' \in L^2(\mathbb{R})$ given by

$$\sqrt{2}D_2\psi = (\Sigma_{\phi'})^*\beta, \qquad \sqrt{2}D_2\psi' = (\Sigma_{\phi})^*\beta'. \tag{12.2.4}$$

The bi-orthogonality of the β and β' implies that

$$\langle \psi(\cdot - k) \mid \psi' \rangle = \delta(k).$$

By looking at $P_{N+1} - P_{-N} = \sum_{j=-N}^{N} Q_j$ we have thus shown the following theorem.

Theorem 12.2.1

Let V_j and V'_j be two multi-resolution analyses over $\mathbb{R}$. Suppose, further, that the projectors (12.2.1) *satisfy at*

$$P_j \to \mathbb{1}, \quad (j \to \infty), \qquad P_j \to 0, \quad (j \to -\infty),$$

where both limits hold strongly. Then the following limit holds for all $s \in L^2(\mathbb{R})$:

$$s = \lim_{N \to \infty} \sum_{j=-N}^{N} \sum_{k \in \mathbb{Z}} 2^j \langle \psi(2^j \cdot - k) \mid s \rangle \psi'(2^j \cdot - k),$$

where ψ and ψ' are given by (12.2.4) *and the limit holds in $L^2(\mathbb{R})$.*

So far we have obtained a non-orthogonal expansion into wavelets of any functions $s \in L^2(\mathbb{R})$. The question, however, is whether the above sum converges unconditionally, or, equivalently, is the set of dilates and translates of ψ (respectively ψ') a Bessel family of $L^2(\mathbb{R})$. For this it is enough to show that

$$\sum_{j \in \mathbb{Z}} \sum_{k \in \mathbb{Z}} |\langle \psi_{j,k} \mid s \rangle|^2 \leq c\|s\|^2, \quad \text{and} \quad \sum_{j \in \mathbb{Z}} \sum_{k \in \mathbb{Z}} |\langle \psi'_{j,k} \mid s \rangle|^2 \leq c\|s\|^2, \tag{12.2.5}$$

with some constant c not depending on $s \in L^2(\mathbb{R})$. For fixed j this is obvious since the translates of ψ are actually a Riesz basis. Concerning the sum over j a sufficient condition for this has been given in Theorem 13.0.1 in Chapter 2.

It is interesting to observe that for ψ and ψ', as constructed above, condition (12.2.5) implies that the family $\psi_{j,k}$ *and* $\psi'_{j,k}$ are a bi-orthogonal system of Riesz bases and we have

$$\langle \psi_{j,k} \mid \psi'_{j,k} \rangle = \delta(k-k')\delta(j-j'). \tag{12.2.6}$$

To see this let W_j be the space generated by the $\{\psi_{j,k} : k \in \mathbb{Z}\}$ and W'_j the analogue for ψ'. As already noted, in the sequence case we have that W_0 is orthogonal to V'_0 and hence to all W'_j with $j < 0$. For the same reason we have that W'_0 is orthogonal to all W_j, $j < 0$. Hence, W_0 is orthogonal to all W'_j except for $j = 0$. By dilation co-variance (12.2.6) follows.

13 QMF and loop groups

In this section we want to show that the QMFs have a group structure related to loop groups of unitary operators. In particular, we will show how to parameterize all FIR-QMFs since they correspond to polynomial loops. The following section is mainly based on Gröchenig (1987) and Holschneider and Pinkall (1993).

Recall that α, $\beta \in L^2(\mathbb{Z})$ is a QMF-system iff the set of $2\mathbb{Z}$-translated functions satisfies at (in this section we use T_k to denote the shift and the translation operator)

$$\{T_k\alpha : k \in 2\mathbb{Z}\} \cup \{T_k\beta : k \in 2\mathbb{Z}\} \quad \text{is an o.n.b. of } L^2(\mathbb{Z}).$$

13.1 The group of unitary operators with $[U, T_2] = 0$

We now want to show that the set of QMF-systems has a natural group structure that is isomorphic to a specific subgroup of unitary operators acting in the Hilbert space $L^2(\mathbb{Z})$. We will denote the group of all unitary operators acting in $L^2(\mathbb{Z})$ by $\mathcal{U}(L^2(\mathbb{Z}))$. Let U be a unitary operator acting on sequences in $L^2(\mathbb{Z})$. Suppose, further, that it commutes with the translations by 2:

$$[T_2, U] = T_2U - UT_2 = 0 \quad \text{or, equivalently, } T_2UT_2^* = U. \tag{13.1.1}$$

Because of the second equation this class of operators will be denoted by $\mathrm{Fix}(T_2, \mathcal{U}(L(\mathbb{Z})))$. Now let α, $\beta \in L^2(\mathbb{Z})$ be a QMF-system. Then the image of α and β under any unitary U satisfying (13.1.1) is again clearly a QMF-system. Indeed, as unitary images they are again an orthonormal base and in addition we have for $i, i' \in \{0, 1\}$ and $k, k' \in \mathbb{Z}$, $\phi_0 = \alpha$, $\phi_1 = \beta$,

$$\begin{aligned}\langle T_{2k}U\phi_i \mid T_{2k'}U\phi_{i'} \rangle &= \langle UT_{2k}\phi_i \mid UT_{2k'}\phi_{i'} \rangle \\ &= \langle T_{2k}\phi_i \mid T_{2k'}\phi_{i'} \rangle = \delta(i-i')\delta(k-k').\end{aligned}$$

Now, clearly, the sequences $\alpha = \delta$, $\beta = T_1\delta$ are a QMF-system. We claim that any other QMF-system can be obtained as an image of δ and $T_1\delta$ under some unitary operator U that commutes with the $2\mathbb{Z}$-translates. To see this let α, β be a QMF-system. Clearly, any operator that commutes with the translations with respect to $2\mathbb{Z}$ is completely determined by its image on one single fundamental domain $\mathbb{Z}/2\mathbb{Z} = \{0, 1\}$. Therefore, if we have

$$U\delta = \alpha, \qquad UT_1\delta = \beta, \tag{13.1.2}$$

then there is at most one such unitary U acting on all of $L^2(\mathbb{Z})$ and satisfying $[T_2, U] = 0$. There actually is one and it is given by

$$U: s \mapsto (\Pi_0 s) * \alpha + (\Pi_1 s) * (T_1^* \beta), \tag{13.1.3}$$

with $\Pi_0 = (\Sigma_2^{\mathbb{Z}})^* \Sigma_2^{\mathbb{Z}}$, the projection on the even positions, and $\Pi_1 = T_1 \Pi_0 T_1^*$, the projection on the odd positions.

This operator obviously commutes with T_2 and we only have to check its unitarity. But since it sends the orthonormal base $\{T_{2k}\delta_0, T_{2k}\delta_1\}$ into the orthonormal base $\{T_{2k}\alpha, T_{2k}\beta\}$, we are done. Therefore, we have proved the following theorem:

Theorem 13.1.1

There is an explicit bijection between the group $Fix(T_2, \mathcal{U}(L^2(\mathbb{Z})))$ and the QMF-systems α, β. This relation is explicitly given by (13.1.2) *and* (13.1.3).

This shows that the QMFs have a natural group structure. In particular, one can produce new ones from old ones by multiplication of their associated unitary operators.

13.2 The Fourier space picture

Since we have already established the equivalence between QMF-systems and unitary operators U that commute with T_2 let us now exploit this relation by writing the action of the unitary operator U in Fourier space. From (13.1.3) it follows that we have

$$U: \hat{s}(\omega) \mapsto \tfrac{1}{2}(\hat{s}(\omega) + \hat{s}(\omega + \pi))\hat{\alpha}(\omega) + \tfrac{1}{2}(\hat{s}(\omega) - \hat{s}(\omega + \pi))\hat{\beta}(\omega)\, e^{i\omega}.$$

But now the action of U on $\hat{s}(\omega + \pi)$ looks similar and therefore if we introduce the vector

$$\begin{bmatrix} \hat{s}(\omega) \\ \hat{s}(\omega + \pi) \end{bmatrix}$$

we may write

$$U: \begin{bmatrix} \hat{s}(\omega) \\ \hat{s}(\omega + \pi) \end{bmatrix} \mapsto \frac{1}{2} \begin{bmatrix} \hat{\alpha}(\omega) + \hat{\beta}(\omega)\, e^{i\omega} & \hat{\alpha}(\omega) - \hat{\beta}(\omega)\, e^{i\omega} \\ \hat{\alpha}(\omega + \pi) + \hat{\beta}(\omega + \pi)\, e^{i\omega} & \hat{\alpha}(\omega + \pi) - \hat{\beta}(\omega + \pi)\, e^{i\omega} \end{bmatrix} \times \begin{bmatrix} \hat{s}(\omega) \\ \hat{s}(\omega + \pi) \end{bmatrix},$$

or, equivalently,

$$U\colon \begin{bmatrix} \hat{s}(\omega) \\ \hat{s}(\omega+\pi) \end{bmatrix} \mapsto \frac{1}{2}\begin{bmatrix} \hat{\alpha}(\omega) & \hat{\beta}(\omega) \\ \hat{\alpha}(\omega+\pi) & \hat{\beta}(\omega+\pi) \end{bmatrix}\begin{bmatrix} 1 & 1 \\ e^{-i\omega} & -e^{-i\omega} \end{bmatrix}\begin{bmatrix} \hat{s}(\omega) \\ \hat{s}(\omega+\pi) \end{bmatrix}.$$

This is a product of two unitary matrices and hence is unitary itself. This is a matrix-valued function $\eta\colon \mathbb{T} \to U(2)$ that[9] satisfies at

$$\eta(\omega+\pi) = \begin{bmatrix} 0 & 1 \\ 1 & 0 \end{bmatrix}\eta(\omega)\begin{bmatrix} 0 & 1 \\ 1 & 0 \end{bmatrix}.$$

The space of such mappings form a group under pointwise multiplication. Vice versa, suppose we are given such a map η. Upon setting

$$U\colon \begin{bmatrix} \hat{s}(\omega) \\ \hat{s}(\omega+\pi) \end{bmatrix} \mapsto \eta(\omega)\begin{bmatrix} \hat{s}(\omega) \\ \hat{s}(\omega+\pi) \end{bmatrix}, \tag{13.2.1}$$

we obtain an operator acting on $\hat{s}(\omega)$ and the above condition ensures that this operator is well defined, as can be seen upon replacing ω with $\omega + \pi$. It is clear that this operator is unitary since $\eta(\omega)$ is unitary for each ω. The translation operator T_2 becomes a multiple of the identity

$$T_2\colon \begin{bmatrix} \hat{s}(\omega) \\ \hat{s}(\omega+\pi) \end{bmatrix} \mapsto \begin{bmatrix} e^{-2i\omega} & 0 \\ 0 & e^{-2i\omega} \end{bmatrix}\begin{bmatrix} \hat{s}(\omega) \\ \hat{s}(\omega+\pi) \end{bmatrix}$$

and therefore U commutes with T_2. We have therefore shown the following theorem.

Theorem 13.2.1

There is a bijection of the subgroup of all unitary operators acting in $L^2(\mathbb{Z})$ with $[T_2, U] = 0$ and the group of measurable mappings $\eta\colon \mathbb{T} \to U(2)$ satisfying (almost everywhere)

$$\eta(\omega+\pi) = \begin{bmatrix} 0 & 1 \\ 1 & 0 \end{bmatrix}\eta(\omega)\begin{bmatrix} 0 & 1 \\ 1 & 0 \end{bmatrix}.$$

The correspondence is given by (13.2.1).

As we know already, QMF is equivalent to the fact that the matrix

$$\frac{1}{\sqrt{2}}\begin{bmatrix} \hat{\alpha}(\omega) & \hat{\alpha}(\omega+\pi) \\ \hat{\beta}(\omega) & \hat{\beta}(\omega+\pi) \end{bmatrix} \tag{13.2.2}$$

is unitary for all ω. We call this matrix the QMF-matrix associated with (α, β). The set of such matrices coincides with the matrix-valued function

[9] We denote by $U(N)$ the group of unitary $N \times N$ matrices.

$\chi : \mathbb{T} \to U(2)$ satisfying

$$\chi(\omega + \pi) = \chi(\omega)\begin{bmatrix} 0 & 1 \\ 1 & 0 \end{bmatrix}. \tag{13.2.3}$$

This shows in particular that the QMF-matrices are not a group. However, they are a coset as we will now see.

13.3 QMF and loop groups

A map from the torus $\mathbb{T}$ to the space of unitary matrices $U(2)$ is called a closed loop in $U(2)$. The space of all loops, $\mathrm{Map}(\mathbb{T} \to U(2))$, is a group under pointwise multiplication (e.g. Pressley and Segal 1976).

If we identify $\mathbb{T}$ with $\mathbb{R}/2\pi\mathbb{Z}$ then the loop conditions reads

$$\xi(\omega) \in U(2), \qquad \xi(\omega + 2\pi) = \xi(\omega).$$

The set of functions $\xi : \mathbb{T} \to U(2)$ such that $T_\pi \xi(\omega) = \xi(\omega + \pi) = \xi(\omega)$ forms a (non-normal) subgroup, which we denote by $\mathrm{Fix}(T_\pi, \mathrm{Map}(\mathbb{T} \to U(2)))$. This group may be identified with $\mathrm{Map}(\mathbb{T} \to U(2))$ itself upon replacing $\xi(\omega)$ by $\xi(\omega/2)$. Next consider the QMF-matrices of (13.2.2), or, what amounts to the same, a loop $\chi(\omega)$ satisfying (13.2.3). The set of such loops does not form a group. However, this set of loops is a coset; to be more precise let $\chi_1(\omega)$ and $\chi_2(\omega)$ be two such loops. Then $\xi(\omega) = \chi_1(\omega)\chi_2^*(\omega)$ satisfies at

$$\xi(\omega + \pi) = \chi_1(\omega + \pi)\chi_2^*(\omega + \pi) = \chi_1(\omega)\chi_2^*(\omega) = \xi(\omega),$$

and is therefore an element of $\mathrm{Fix}(T_\pi, \mathrm{Map}(\mathbb{T} \to U(2)))$. In addition, all loops in this subgroup can be obtained in this way since for

$$\xi \in \mathrm{Fix}(T_\pi, \mathrm{Map}(\mathbb{T} \to U(2)))$$

we may set

$$\chi_2(\omega) = \xi^*(\omega)\chi_1(\omega),$$

and this loop satisfies at (13.2.3) and it gives back ξ by setting $\xi(\omega) = \chi_1(\omega)\chi_2^*(\omega)$. To summarize, we have proved the following theorem.

Theorem 13.3.1

The set of QMF-loops of (13.2.2) *is a right coset of* $\mathit{Fix}(T_\pi, \mathit{Map}(\mathbb{T} \to U(2)))$. *Therefore, let* $\chi(\omega)$ *be any QMF-matrix* (13.2.2). *Then all others can be obtained in a unique way as*

$$\eta(\omega) = \rho(\omega)\chi(\omega) \quad \textit{where } \rho \textit{ runs through } \mathit{Fix}(T_\pi, \mathit{Map}(\mathbb{T} \to U(2))).$$

Let us now look at the other combinations, that is, consider the set of loops $\xi(\omega) = \chi_1^*(\omega)\chi_2(\omega)$. Note that this set is not just the conjugate set of the previous one since $\chi^*(\omega)$ no longer satisfies (13.2.3). Rather, we have

$$\xi(\omega + \pi) = \begin{bmatrix} 0 & 1 \\ 1 & 0 \end{bmatrix} \xi(\omega) \begin{bmatrix} 0 & 1 \\ 1 & 0 \end{bmatrix}.$$

This is again a (non-normal) subgroup of all loops. Its elements are called twisted loops. If we introduce the operator σ acting on the loopgroup via conjugation:

$$\sigma : \rho(\omega) \mapsto \begin{bmatrix} 0 & 1 \\ 1 & 0 \end{bmatrix} \rho(\omega) \begin{bmatrix} 0 & 1 \\ 1 & 0 \end{bmatrix},$$

then we can write $\mathrm{Fix}(\sigma T_\pi, \mathrm{Map}(\mathbb{T} \to U(2)))$ for the group of twisted loops. Again, we have that the set of QMFs is a coset which this time is a left-coset.

Theorem 13.3.2
The set of QMF-loops of (13.2.2) *is a left coset of* $\mathit{Fix}(\sigma T_\pi, \mathit{Map}(\mathbb{T} \to U(2)))$. *Therefore, let* $\chi(\omega)$ *be any QMF-matrix* (13.2.2). *Then all other QMF-matrices can be obtained in a unique way as*

$$\eta(\omega) = \chi(\omega)\rho(\omega) \quad \textit{where } \rho \textit{ runs through } \mathit{Fix}(\sigma T_\pi, \mathit{Map}(\mathbb{T} \to U(2))).$$

13.4 Some subclasses of QMF

We now consider various subgroups of the whole loop group and their associated QMF-systems. The correspondence is made in the spirit of the preceding theorems, that is, the set of QMF matrices will be identified with a right coset of a given subgroup of all loops. The following classes of loops are clearly subgroups and we leave the proofs for the reader.

(a) The smooth loops. Here smoothness means that every matrix element of $\xi(\omega)$ is a function in $C^\infty(\mathbb{T})$. The space of all smooth loops, $\mathrm{Map}(\mathbb{T} \to U(2)\colon C^\infty)$, is a Lie group under pointwise multiplication (e.g. Pressley and Segal 1976). As we know already, these loops correspond to QMF-systems α, β, whose Fourier transforms are in $C^\infty(\mathbb{T})$, or, what amounts to the same, whose sequences are arbitrarily well localized in the sense that

$$\sup_{n \in \mathbb{Z}} (1 + |n|)^m |\alpha_i(n)| < \infty, \quad \text{for all } m > 0, \quad i \in \{0, 1\}.$$

(b) The polynomial loops. This is the class of loops for which the matrix elements of $\xi(\omega)$ are trigonometric polynomials (i.e. Laurent polynomials in $z = e^{i\omega}$). We will denote this set by $\mathrm{Map}(\mathbb{T} \to U(2)\colon \mathrm{poly})$. It corresponds to the set of QMF-systems with finite impulse response (i.e. their convolution with a delta sequence has only a finite number of non-vanishing terms).

(c) Analytic loops. The class of loops for which each entry has an analytic extension into some strip $\{|\Im z| \leq c\}$ is a subgroup, too. It corresponds to QMFs that are exponentially localized, as was shown in Theorem 1.7.1.

(d) The vanishing moment loops. Consider the subset of smooth loops that satisfy

$$\xi(\omega) = \mathbb{1} + o(\omega^n) \quad (\omega \to 0).$$

This is again a normal subgroup of Map($\mathbb{T} \to U(2)$: C^∞). It corresponds to the class of highly localized QMF-systems such that

$$\frac{1}{\sqrt{2}}\begin{bmatrix} \hat{\alpha}(\omega) & \hat{\alpha}(\omega+\pi) \\ \hat{\beta}(\omega) & \hat{\beta}(\omega+\pi) \end{bmatrix} = \mathbb{1} + o(\omega^n) \quad (\omega \to 0),$$

which implies that

$$\hat{\alpha}(\omega) = \sqrt{2} + o(\omega^n), \qquad \hat{\beta}(\omega) = o(\omega^n) \quad (\omega \to 0).$$

The second condition is equivalent to the fact that the first n moments of β vanish:

$$\sum_{m \in \mathbb{N}} m^p \beta(m) = 0, \qquad p = 0, 1, \ldots, n.$$

13.5 The factorization problem

Since the polynomial loops correspond to QMFs of finite length they are of particular importance in signal processing. We will now give a technique for generating all finite impulse response QMF-systems.

Given that the polynomial loops are an infinite-dimensional Lie group, the question arises as to whether or not there is a small family of loops that generate all polynomial loops. Equivalently, the problem is how to generate, in an economic way, all finite impulse response QMFs. Such a factorization exists, as can be seen by general theorems on loop groups (e.g. in Pressley and Segal 1976). Instead of reproducing the demonstration given there we present a different, more geometric construction.

As we have seen we may equally well consider the group of unitary operators that commute with the translates with respect to the sublattice $2\mathbb{Z}$.

Consider the fundamental domain $J = \{0, 1\}$ of $2\mathbb{Z}$ in $\mathbb{Z}$. Recall that a fundamental domain is a set $J \subset \mathbb{Z}$ whose translates with respect to $2\mathbb{Z}$ are a mutually disjoint cover of $\mathbb{Z}$. Suppose that U is a unitary operator acting in $L^2(\mathbb{Z})$ that commutes with T_2. Clearly, by translation invariance such an operator is completely determined by its action on the fundamental domain J. We now pick a unitary 2×2 matrix u in $U(2)$. It acts in a natural way in the two-dimensional vector space $L^2(J)$ which we consider in a natural way as a subspace of $L^2(\mathbb{Z})$. Now consider the well-defined operator

U_u, with $[U_u, T_2] = 0$, whose restriction to the fundamental domain J is given by u. Since this operator preserves the L^2-norm in each of the translated fundamental domains we find that U_u is unitary. We have therefore found a family of unitary operators that commute with the translations by 2. This family is indexed by points in the four-dimensional manifold $U(2)$.

This family, together with the translation operators T_1, actually generates all finite impulse response QMFs.

Theorem 13.5.1

Let $U: L^2(\mathbb{Z}) \to L^2(\mathbb{Z})$ be a unitary operator with $[U, T_2] = 0$. Suppose in addition that the associated QMF $\alpha = U\delta_0$ and $\beta = U\delta_1$ is of finite impulse response with length not larger than 2N (i.e. each sequence has at most 2N non-zero coefficients). Then U can be factorized as follows:

$$U = T_p U_{u_{2N}} T_1 U_{u_{2N-1}} \cdots T_1 U_{u_1},$$

with some $p \in \mathbb{Z}$. If all coefficients of α and β are real valued then the u_n can be chosen in $O(2)$.

Proof. Let $\alpha = U\delta_0$ and $\beta = U\delta_1$ be the associated QMF-system. By an overall translation we may suppose that the support of α and β is contained in $[0, 2N - 1]$. By the QMF-property we have

$$\langle \alpha \mid T_{2(N-1)}\alpha \rangle = \langle \beta \mid T_{2(N-1)}\beta \rangle = \langle \alpha \mid T_{2(N-1)}\beta \rangle = \langle \beta \mid T_{2(N-1)}\alpha \rangle = 0,$$

unless $N = 1$. This shows that among the four vectors

$$e_0^l = [\alpha(0), \alpha(1)],\ e_0^r = [\alpha(2N-2), \alpha(2N-1)],$$

$$e_1^l = [\beta(0), \beta(1)],\ e_1^r = [\beta(2N-2), \beta(2N-1)],$$

the 'r' and 'l' are orthogonal. We can thus find $u_N \in SU(2)$ such that

$$\langle u_N e_0^l \mid [1, 0] \rangle = \langle u_N e_1^l \mid [1, 0] \rangle = \langle u_N e_0^r \mid [0, 1] \rangle = \langle u_N e_1^l \mid [0, 1] \rangle = 0.$$

Therefore, $U_{u_N}\alpha$ and $U_{u_N}\beta$ are supported by a subset of $\{1, 2N - 2\}$. We may thus repeat the procedure with $T_1^* U_{u_N}\alpha$ and $T_1^* U_{u_N}\beta$, and so on, until $N = 1$. Here we are left with a two-dimensional vector space with two orthogonal vectors. We may again find a $u_1 \in U(2)$ that maps these vectors to $[1, 0]$ and $[0, 1]$. Upon reverting all operations we find the desired factorization. The case where all sequences are real valued follows in the same way. □

Example 13.5.2

We chose as the first fundamental domain the set $J_1 = \{0, 1\}$. Since we are only interested in real-valued wavelets we pick an element g in $SO(2)$. This can be—up to a parity that we dispense with for the sake of simplicity—

parameterized by one angle γ:

$$g = g(\gamma) = \begin{bmatrix} \cos(\gamma) & \sin(\gamma) \\ -\sin(\gamma) & \cos(\gamma) \end{bmatrix}.$$

Now the first family of unitary operators is defined by mixing the points in the fundamental domain J_1 by one of these matrices and extending this by translation by 2 to all of $\mathbb{Z}$. Thus, applied to δ_0 and δ_1, we obtain a one-parameter family of QMF, namely,

$$\alpha_0^1 = \cos(\gamma_1)\delta_0 - \sin(\gamma_1)\delta_1, \qquad \alpha_1^1 = \sin(\gamma_1)\delta_0 + \cos(\gamma_1)\delta_1.$$

Starting from this pair of QMFs we produce more by picking a new fundamental domain, say $J_2 = \{-1, 0\}$, and mixing the values in J_2 and all its congruent versions by another orthogonal matrix given by a different angle γ_2. We thus obtain

$$\begin{aligned} \alpha_0^2 = {} & \cos(\gamma_1)\sin(\gamma_2)\delta_{-1} + \cos(\gamma_1)\cos(\gamma_2)\delta_0 \\ & - \cos(\gamma_2)\sin(\gamma_1)\delta_1 + \sin(\gamma_1)\sin(\gamma_2)\delta_2 \end{aligned}$$

and

$$\begin{aligned} \alpha_1^2 = {} & \sin(\gamma_1)\sin(\gamma_2)\delta_{-1} + \cos(\gamma_2)\sin(\gamma_1)\delta_0 \\ & + \cos(\gamma_1)\cos(\gamma_2)\delta_1 - \cos(\gamma_1)\sin(\gamma_2)\delta_2. \end{aligned}$$

We may now use the previous fundamental domain again but with a third parameter to obtain

$$\begin{aligned} \alpha_0^3 = {} & \cos(\gamma_1)\sin(\gamma_2)\sin(\gamma_3)\delta_{-2} \\ & + \cos(\gamma_1)\cos(\gamma_3)\sin(\gamma_2)\delta_{-1} \\ & + (\cos(\gamma_1)\cos(\gamma_2)\cos(\gamma_3) - \cos(\gamma_2)\sin(\gamma_1)\sin(\gamma_3))\delta_0 \\ & - (\cos(\gamma_2)\cos(\gamma_3)\sin(\gamma_1) - \cos(\gamma_1)\cos(\gamma_2)\sin(\gamma_3))\delta_1 \\ & + \cos(\gamma_3)\sin(\gamma_1)\sin(\gamma_2)\delta_2 \\ & - \sin(\gamma_1)\sin(\gamma_2)\sin(\gamma_3)\delta_3 \end{aligned}$$

and

$$\begin{aligned} \alpha_1^3 = {} & \sin(\gamma_1)\sin(\gamma_2)\sin(\gamma_3)\delta_{-2} \\ & + \cos(\gamma_3)\sin(\gamma_1)\sin(\gamma_2)\delta_{-1} \\ & + (\cos(\gamma_2)\cos(\gamma_3)\sin(\gamma_1) + \cos(\gamma_1)\cos(\gamma_2)\sin(\gamma_3))\delta_0 \\ & + (\cos(\gamma_1)\cos(\gamma_2)\cos(\gamma_3) - \cos(\gamma_2)\sin(\gamma_1)\sin(\gamma_3))\delta_1 \\ & - \cos(\gamma_1)\cos(\gamma_3)\sin(\gamma_2)\delta_2 \\ & + \cos(\gamma_1)\sin(\gamma_2)\sin(\gamma_3)\delta_3. \end{aligned}$$

We have thus obtained a three-parameter family of QMFs.

14 Multi-resolution analysis over $\mathbb{T}$

In this section we wish to construct a multi-resolution analysis of functions over the torus $\mathbb{T}$. As in the continuous case we can use the periodization operator to obtain a family of orthonormal wavelets over the circle from a wavelet family over the real line. Indeed, as we saw in Section 2.5, the periodization of a sampling space over $\mathbb{R}$ yields a sampling space over $\mathbb{T}$ provided the periodization length is an integral multiple of the sampling distance. In addition, mutually orthogonal sampling spaces over $\mathbb{R}$ are mapped into mutually orthogonal sampling spaces over $\mathbb{T}$. Therefore, a family of orthogonal wavelets over $\mathbb{T}$ is obtained by periodizing the wavelets over $\mathbb{R}$, that is, if $\psi \in L^1(\mathbb{R}) \cap L^2(\mathbb{R})$ is the generator of an orthonormal wavelet base of $L^2(\mathbb{R})$, then

$$\left.\begin{aligned} \psi_{k,j} &= \psi_j(\cdot - 2\pi k 2^{-j}),\\ \psi_j &= \sqrt{2\pi 2^{-j}}\,\Pi D_{2\pi 2^{-j}}\psi = \sqrt{\frac{2^j}{2\pi}} \sum_{n\in\mathbb{Z}} \psi(2^j \cdot /2\pi + 2^j n) \end{aligned}\right\} \tag{14.0.1}$$

is an orthonormal family. As we will see below, their linear span is $L^2(\mathbb{T}) \ominus \{1\}$: the constant functions cannot be expanded into wavelets since these are of 0-mean.

Apart from this direct construction by means of the periodization operator there is also an intrinsic way of defining a multi-resolution analysis over $\mathbb{T}$ which will be presented in the sequel.

Since multi-resolution analysis is based on a hierarchy of spaces that are deduced from each other by a re-scaling procedure—over the real line this was just dilation by 2—we first have to construct the dilation over the circle. An immediate candidate for dilation by 1/2 might be

$$D^{\mathbb{T}}_{1/2}\colon L^2(\mathbb{T}) \to L^2(\mathbb{T}), \qquad s(t) \mapsto s(2t).$$

A good candidate for dilation by 2 for functions over the circle is now given by the adjoint mapping

$$D^{\mathbb{T}}_2\colon L^2(\mathbb{T}) \to L^2(\mathbb{T}), \qquad s(t) \mapsto \frac{1}{2}\left\{s\left(\frac{t}{2}\right) + s\left(\frac{t+\pi}{2}\right)\right\}.$$

This is nothing other than the periodization operator for periodic functions. Note the following relation with the dilatations over the real line which shows that $D_2^{\mathbb{T}}$ is actually the appropriate candidate:

$$D_2^{\mathbb{T}}\Pi = \Pi D_2, \qquad \begin{array}{ccc} L^1(\mathbb{R}) & \xrightarrow{\Pi} & L^1(\mathbb{T}) \\ \downarrow{\scriptstyle D_2} & & \downarrow{\scriptstyle D_2^{\mathbb{T}}} \\ L^1(\mathbb{R}) & \xrightarrow{\Pi} & L^1(\mathbb{T}) \end{array} \tag{14.0.2}$$

There is still another argument for using $D_2^{\mathbb{T}}$ as we have done. By direct computation using the Poisson summation formula we obtain the Fourier space representation of dilation by 2 over the circle in the Fourier space:

$$D_2^{\mathbb{T}}: \hat{s}(n) \mapsto \hat{s}(2n).$$

This means that $D_2^{\mathbb{T}}$ throws away all the odd frequencies. In addition we have the following commutation law:

$$D_2^{\mathbb{T}} T_\alpha = T_{2\alpha} D_2^{\mathbb{T}}, \qquad \alpha \in \{2\pi k 2^{-j}: j, k \in \mathbb{N}_0\}.$$

By looking at the action on δ_α with α a dyadic point we see that $D_2^{\mathbb{T}}$ is the only reasonable linear operator that satisfies this commutation law.

Now we are ready for the following definition.

Definition 14.0.1

An ascending chain of spaces

$$\{1\} = V_0 \subset \cdots \subset V_j \subset V_{j+1} \subset \cdots \subset L^2(\mathbb{T}), \qquad j \in \mathbb{N}_0,$$

is called a multi-resolution analysis of $L^2(\mathbb{T})$ *iff*

(i) $\bigcup_{j \in \mathbb{N}} V_j$ *is dense in* $L^2(\mathbb{T})$,

(ii) $s \in V_{j+1} \Rightarrow D_2^{\mathbb{T}} s \in V_j$,

(iii) V_j *is a* 2^j*-dimensional sampling space.*

Some comments are in order. Let P_j again be the projector on V_j. Condition (i) shows again that arbitrary functions $s \in L^2(\mathbb{R})$ can be approximated by the projections $P_j s$.

Condition (ii) is the analogue for dilation by 2 in the former situation of a multi-resolution analysis over the real line. It states that $D_2^{\mathbb{T}}$ shifts functions from V_j to V_{j-1}. It is intuitively clear that application of $D_2^{\mathbb{T}}$ to some function s averages out small scale features of s. This shows again that the more small scale details the V_j contain, the higher the exponent j. Thus V_j again contains all details up to scale 2^{-j}.

Since the dilation over $\mathbb{T}$ that we use to go from one scale to another leaves the space of constant functions invariant, it follows from (ii) that either all V_j contain the constant functions or none of them do. The condition $V_0 = \{1\}$ is therefore pure convention.

Condition (iii) states, as we saw in Section 1 that V_j has a translation invariant orthonormal base. That is, for every j there is—up to phase factor in Fourier space—a uniquely defined function ϕ_j such that the collection $\phi_{n,j} = \phi_j(t - n2^{-j})$ with $n = 0, \ldots, 2^j - 1$ is an orthonormal base for V_j.

As before, we develop $L^2(\mathbb{T})$ according to

$$L^2(\mathbb{T}) = \{1\} \oplus \overline{\bigoplus_{j \in \mathbb{N}_0} W_j}, \qquad V_{j+1} = V_j \oplus W_j.$$

Note that the space of constant functions must be added separately since the orthogonal complements W_j cannot contain any constants.

From Section 3.3 it follows that the complement spaces W_j are sampling spaces, too. If $\{\phi_j(\cdot - 2\pi k 2^{-j})\}$ is an orthonormal base for V_j then $\{\psi_j(\cdot - 2\pi k 2^{-j})\}$ is one for W_j, where

$$\psi_j = \sum_{k=0}^{2^j - 1} \beta(k)\phi_j(\cdot - 2\pi k 2^{-j}), \qquad \beta(k) = (-1)^n \bar{\alpha}(k-1), \quad \alpha = \Sigma^{\mathbb{T}}_{2^j, \phi_j} \phi_{j-1}.$$

We have therefore proved the following theorem.

Theorem 14.0.2

For every multi-resolution analysis over $\mathbb{T}$ *there is a family of wavelets* $\psi_{n,j}(t) = \psi_j(t - 2\pi n 2^{-j})$ *which constitute an orthonormal basis of* $L^2(\mathbb{T}) \ominus \{1\}$. *Any function in* $L^2(\mathbb{T})$ *may be decomposed according to*

$$s = \sum_{j=0}^{\infty} \sum_{n=0}^{2^j - 1} \langle \psi_{n,j} \mid s \rangle_{\mathbb{T}} \psi_{n,j} + \int_{\mathbb{T}} s.$$

The family of periodized wavelets (14.0.1) is an orthonormal base for $L^2(\mathbb{T}) \ominus \{1\}$. Indeed, we have the following theorem.

Theorem 14.0.3

Let ϕ *be the generating function of a multi-resolution analysis* U_j *over* $\mathbb{R}$ *with wavelet* ψ. *Suppose that* $\phi \in L^1(\mathbb{R}) \cap L^2(\mathbb{R})$. *Then let* $V_j \subset L^2(\mathbb{T})$ *consist of all periodized functions* $\sum_{k \in \mathbb{Z}} s(\cdot/2\pi - k)$ *with* $s \in U_j$ *and where* s *is a finite superposition of translates* $\phi(2^j \cdot + k)$, $k \in \mathbb{Z}$. *Then* V_j *is a multi-resolution analysis over* $\mathbb{T}$. *The scaling functions and wavelets are given by*

$$\phi_{k,j} = \phi_j(\cdot - 2\pi k 2^{-j}), \quad \phi_j = \sqrt{2\pi 2^{-j}}\, \Pi D_{2\pi 2^{-j}} \phi = \sqrt{\frac{2^j}{2\pi}} \sum_{n \in \mathbb{Z}} \phi(2^j \cdot / 2\pi + 2^j n),$$

$$\psi_{k,j} = \psi_j(\cdot - 2\pi k 2^{-j}), \quad \psi_j = \sqrt{2\pi 2^{-j}}\, \Pi D_{2\pi 2^{-j}} \psi = \sqrt{\frac{2^j}{2\pi}} \sum_{n \in \mathbb{Z}} \psi(2^j \cdot / 2\pi + 2^j n).$$

Proof. The inclusion property is obvious. Condition (ii) and (iii) have been shown in Section 2.5 and 2.6. Only the completeness (i) remains to be shown.

For this we have to show that

$$\lim_{j\to\infty} P_j s = s, \qquad s \in L^2(\mathbb{T}),$$

where the limit holds in $L^2(\mathbb{T})$. We first study this limit if $s = e_k = e^{ik\omega}$. If we show that the above limit holds for all pure frequencies we are done. Indeed, as the projection operator, P_j is bounded. Now any $s \in L^2(\mathbb{T})$ may be written as $\rho + \eta$ where ρ is a trigonometric polynomial and $\|\eta\|_2$ may be arbitrarily small. Since $P_j\rho \to \rho$ we have

$$\limsup_{j\to\infty} \|P_j s - s\|_2 \le \limsup_{j\to\infty}(\|P_j\eta - \eta\|_2 + \|P_j\rho - \rho\|_2) \le 2\|\eta\|_2$$

and because the right-hand side may become arbitrarily small we are done.

By direct computation using the Poisson summation formula we obtain

$$P_j\colon e_k \mapsto \sum_{n\in\mathbb{Z}} \bar{\hat{\phi}}(2\pi k 2^{-j})\hat{\phi}(\pi k 2^{-j} + 2\pi n)\, e_{k+2^j n}.$$

Now the term with $n = 0$ reads

$$|\hat{\phi}(2\pi k 2^{-j})|^2 e_k \to e_k, \qquad (j \to \infty)$$

since $\hat{\phi}(0) = 1$ and $\hat{\phi}$ is continuous. On the other hand,

$$\|P_j e_k\|_2^2 = 2\pi \sum_{n\in\mathbb{Z}} |\hat{\phi}(2\pi k 2^{-j})|^2\, |\hat{\phi}(2\pi k 2^{-j} + 2\pi n)|^2 = 2\pi|\hat{\phi}(2\pi k 2^{-j})|^2,$$

which tends to $2\pi = \|e_k\|_2^2$ and thus the theorem is proved. □

15 Multi-resolution analysis over $\mathbb{Z}/2^M\mathbb{Z}$

We now end with the multi-resolution analysis of functions over the discretized circle $\mathbb{Z}/N\mathbb{Z}$ in the case where N is a power of 2. This is probably the most common situation in practical applications, at least when it concerns the numerical computation of wavelet transforms. As in the circle case we may obtain a wavelet decomposition of $\mathbb{Z}/2^M\mathbb{Z}$ by periodizing a multi-resolution analysis over $\mathbb{Z}$. Alternatively, one may sample a wavelet basis over $\mathbb{T}$ with the help of the (imperfect) sampling operator to obtain a family of orthogonal periodic sequences.

Recall that the Fourier transform $\hat{s}$ of a sequence of $s \in L^2(\mathbb{Z}/N\mathbb{Z})$ is nothing other than the evaluation of the associated trigonometric polynomial at the roots of unity

$$\hat{s}(n) = F^{\mathbb{Z}/N\mathbb{Z}} s(n) = \sum_{k=0}^{N-1} s(k)\, e^{-2i\pi nk/N}.$$

Clearly, sequences in $\mathbb{Z}/N\mathbb{Z}$ will be identified with periodic sequences over $\mathbb{Z}$ without writing this identification explicitly.

We first construct a dilation by 2 acting on sequences in $\mathbb{Z}/N\mathbb{Z}$. Taking our inspiration from the circle case and the integer case we define the dilation by 2 over $\mathbb{Z}/N\mathbb{Z}$ as

$$D_{2,\phi}^{\mathbb{Z}/N\mathbb{Z}}: s \mapsto D_{2,\phi}^{\mathbb{Z}} \Pi_2^{\mathbb{Z}/N\mathbb{Z}} s,$$

where ϕ is some sequence over $\mathbb{Z}/N\mathbb{Z}$. We refer to Section 2.7 and 7 for the notations. Clearly, we must have $N = 0 \bmod 2$ in order to have a well-defined dilation by 2 over the circle. More explicitly, we have

$$D_{2,\phi}^{\mathbb{Z}/N\mathbb{Z}} s(n) = \frac{1}{2} \sum_{k=0}^{N-1} \phi(n - 2k)(s(k) + s(k + N/2)).$$

In Fourier space the dilation by 2 acts as

$$D_{2,\phi}^{\mathbb{Z}/N\mathbb{Z}}: \hat{s}(n) \mapsto \hat{\phi}(n)\hat{s}(2n), \tag{15.0.1}$$

which is in remarkable formal coincidence with the circle case.

By direct computation we can verify that the following diagram is commutative:

$$\begin{array}{ccc} L^1(\mathbb{Z}) & \xrightarrow{\Pi_N^{\mathbb{Z}}} & L^1(\mathbb{Z}/N\mathbb{Z}) \\ \big\downarrow D_\eta^{\mathbb{Z}} & & \big\downarrow D_{2,\phi}^{\mathbb{Z}/N\mathbb{Z}} \\ L^1(\mathbb{Z}) & \xrightarrow{\Pi_N^{\mathbb{Z}}} & L^1(\mathbb{Z}/N\mathbb{Z}) \end{array}$$

and the relation between η and ϕ is given by

$$\phi = \Pi_N^{\mathbb{Z}} \eta.$$

This is clearly the analogue of relation (14.0.2).

We obtain an additional relation to the dilation in sequence space in the following way. Let V be some equidistant sampling space with some translation invariant orthonormal base $\eta(\cdot - k/N)$, $k = 0, \ldots, N - 1$. We recall that the imperfect sampling operator was defined by

$$\Sigma_\eta^{\mathbb{T}}: L^2(\mathbb{T}) \mapsto L^2(\mathbb{Z}/N\mathbb{Z}), \qquad s(k) \mapsto \langle \eta(\cdot - k/N) \mid s \rangle_{\mathbb{T}}.$$

If now $D_2^{\mathbb{T}} V \subset V$, as would be the case for any V taken from a multi-resolution analysis over the circle, then the following diagram is commutative:

$$\begin{array}{ccc} V & \xrightarrow{\Sigma_N^{\mathbb{T}}} & L^2(\mathbb{Z}/N\mathbb{Z}) \\ \big\downarrow D_2^{\mathbb{T}} & & \big\downarrow D_{2,\phi}^{\mathbb{Z}/N\mathbb{Z}} \\ V & \xrightarrow{\Sigma_\eta^{\mathbb{T}}} & L^2(\mathbb{Z}/N\mathbb{Z}) \end{array}$$

and the relation between η and ϕ is given by

$$\phi = \Sigma_\eta D_2^{\mathbb{T}} \eta.$$

The proof is straightforward and is left to the reader.

We now give the following definition.

Definition 15.0.1
A sequence of spaces ($N = 2^M$):

$$\{1\} = V_0 \subset \cdots \subset V_j \subset \cdots \subset V_M = L^2(\mathbb{Z}/N\mathbb{Z}), \qquad V_j \neq V_{j+1},$$

is called a multi-resolution analysis over $\mathbb{Z}/2^M\mathbb{Z}$ *with respect to the interpolation filter* ϕ *iff*

$$D_{2,\phi}^{\mathbb{Z}/N\mathbb{Z}} V_j \subset V_{j-1}.$$

In this highly discrete case only one single condition is left and we obtain a complete characterization of all possible multi-resolution analyses. It turns out that the space of multi-resolution analysis is a 2^{M-1}-dimensional solid torus.

Theorem 15.0.2
Let $\phi \in L^2(\mathbb{Z}/N\mathbb{Z})$, $N = 2^M$, *satisfy at*

$$|\hat{\phi}(n)|^2 + |\hat{\phi}(n + N/2)|^2 = 2. \tag{15.0.2}$$

Then ϕ *generates a multi-resolution analysis and all multi-resolution analyses can be obtained in this way. Two such functions* ϕ *and* ρ *generate the same multi-resolution analysis iff*

$$\hat{\phi}(n) = m(n)/\hat{\rho}(n) \quad \textit{where } m(n + N/2) = m(n),\ |m(n)| = 1$$

Proof. From (15.0.1) it follows that the dimension of V_j is at most 2^j. Since all V_j are different it follows that each has dimension 2^j. In particular, it follows that V_{M-1} has dimension $N/2$. It follows that V_{M-1} is generated by the translates $\phi(\cdot - 2k)$, $k = 0, N/2 - 1$ since this is the image of $\delta(\cdot - k)$. For dimensional reasons this collection must be a basis. For the same reason V_j is generated by the translates $\{\phi_j(\cdot - k2^{N-j})\}$ with

$$\phi_j = (D_{2,\phi}^{\mathbb{Z}/N\mathbb{Z}})^{N-j+1}\phi.$$

For dimensional reasons they must again be a basis for V_j. In Fourier space we have the following by iteration of (15.0.1):

$$\hat{\phi}_j(k) = \prod_{n=1}^{j} \hat{\phi}(2^n k).$$

We now want to show that ϕ may be supposed to satisfy the QMF-condition

$$|\hat{\phi}(n)|^2 + |\hat{\phi}(n + N/2)|^2 = 2.$$

Since the $2\mathbb{Z}$-translates of ϕ are a basis they certainly satisfy this equation with '=' replaced by '$\sim$'. Therefore, we can find a translation invariant orthonormal basis of V_{N-1} generated by some ρ given by $\hat{\rho} = m\hat{\phi}$. It follows

that

$$\phi_j(n) = \gamma(n)\hat{\rho}_j(n), \qquad \gamma(n) = \prod_{k=1}^{j} m(n2^k).$$

Since $\gamma \sim 1$ and since γ is 2^j periodic it follows that the space spanned by $\{\rho_j(\cdot - k2^j)\}$ is again V_j. We may therefore supppose that the QMF-condition holds for generating function ϕ.

In this case we actually obtain a multi-resolution analysis. Indeed, in this case the dilation $D_{2,\phi}^{\mathbb{Z}/N\mathbb{Z}}$ maps a translation invariant orthonormal base into another (having half the number of elements of course). More precisely, let $s(n - k2^j)$, $k = 0, \ldots, 2^{N-j} - 1$ be an orthonormal set. Consider the set $r(n - k2^{j+1})$, $k = 0, \ldots, 2^{N-j-1}$, with $r = D_{2,\phi}^{\mathbb{Z}/N\mathbb{Z}}s$. In Fourier space we have, because of the orthogonality of the shifted versions of s,

$$\sum_{k=0}^{2^j-1} |\hat{s}(n + k2^{N-j})|^2 \equiv 2^j$$

and we have to verify that this implies

$$\sum_{k=0}^{2^{j+1}-1} |\hat{r}(n + k2^{N-j-1})|^2 \equiv 2^{j+1}.$$

Replacing $\hat{r}$ by $\hat{r}(n) = \hat{\phi}(n)\hat{s}(2n)$ we obtain that this sum equals

$$\sum_{k=0}^{2^{j+1}-1} |\hat{\phi}(n + k2^{N-j-1})|^2\, |\hat{s}(2n + k2^{N-j})|^2.$$

Splitting the sum into two sums $\sum_{k=0}^{2^j-1} + \sum_{k=2^j}^{2^{j+1}-1}$ and using the periodicity of $\hat{s}$ we obtain

$$\sum_{k=0}^{2^j-1} |\hat{\phi}(n + k2^{N-j-1})|^2\, |\hat{s}(2n + k2^{N-j})|^2$$
$$+ \sum_{k=0}^{2^j-1} |\hat{\phi}(n + 2^{N-1} + k2^{N-j-1})|^2\, |\hat{s}(2n + k2^{N-j})|^2,$$

and hence, because of (15.0.2), this equals

$$(|\hat{\phi}(n + k2^{N-j-1})|^2 + |\hat{\phi}(n + 2^{N-1} + k2^{N-j-1})|^2) \sum_{k=0}^{2^j-1} |\hat{s}(2n + k2^{N-j})|^2 \equiv 2^{j+1},$$

and the proof is finished. □

Let us again take orthonormal complements

$$V_{j+1} = W_j \oplus V_j.$$

Again, W_j is a sampling space and, accordingly, an explicit translation invariant orthonormal base given by $\psi_j(n - k2^{N-j})$, $k = 0, \ldots, 2^j$, with

$$\psi_j = (D_{2,\phi}^{\mathbb{Z}/N\mathbb{Z}})^{N-j}\psi, \qquad \psi(k) = (-1)^k\phi(1-k).$$

Accordingly, we have the following decomposition of an arbitrary 2^M-periodic sequence into wavelets:

$$s(n) = \sum_{j=0}^{M-1} \sum_{k=0}^{2^j-1} \langle \psi_{n,j} \mid s \rangle_{\mathbb{Z}/N\mathbb{Z}} \psi_{n,j} + 2^{-N} \sum_{k=0}^{N-1} s(k).$$

Note that, as in the circle case, the constant sequences must be added separately. The decomposition of a sequence over s into its wavelet coefficients is the analogue of the decomposition

$$L^2(\mathbb{Z}/2^M\mathbb{Z}) - \{1\} = \bigoplus_{j=1}^{N} L^2(\mathbb{Z}/2^{M-j}\mathbb{Z}).$$

A multi-resolution over $\mathbb{Z}/N\mathbb{Z}$ is obtained by sampling a multi-resolution analysis over the torus in the obvious way. In the same way we may periodize a multi-resolution analysis over $\mathbb{Z}$. We leave the details to the reader.

16 Computing the discrete wavelet transform

We now come to the computation of the wavelet transform with the help of fast algorithms based on the hierarchic structure of the wavelet analysis over a dyadic grid. Essentially, they are obtained from the commutative diagrams that describe how to compute the dilation operator in sequence space.

We start by describing the general schema of these algorithms. For the sake of notational simplicity we only consider orthogonal expansion schemes. The bi-orthogonal schemes follow easily.

Let V_j be a multi-resolution analysis over the real line with associated wavelet spaces W_j. The orthogonal projectors on V_j and W_j will be called P_j and Q_j respectively. Clearly, by construction we have the following identity:

$$P_{j+1} = P_j \oplus Q_j.$$

Suppose s is one of these approximation spaces, say $s \in V_0$. Then we may decompose s without loss of information into a sequence of functions according to

$$s_{\text{low}}^{(j+1)} = P_{-j-1} s_{\text{low}}^{(j)}, \qquad s_{\text{high}}^{(j+1)} = Q_{-j-1} s_{\text{low}}^{(j)},$$

with initial value $s_{\text{low}}^{(0)} = s$. At each step we split the 'low frequency' part $s_{\text{low}}^{(j)}$ into a part with still lower frequencies and a difference part $s_{\text{high}}^{(j)}$ containing the high frequencies that have been dropped by passing from j to $j+1$. Schematically we have

$$\begin{array}{ccccccccc}
s & \xrightarrow{P_{-1}} & s_{\text{low}}^{(1)} & \xrightarrow{P_{-2}} & s_{\text{low}}^{(2)} & \cdots \xrightarrow{P_{-N+1}} & s_{\text{low}}^{(N-1)} & \xrightarrow{P_{-N}} & s_{\text{low}}^{(N)} \\
\downarrow Q_{-1} & & \downarrow Q_{-2} & & \downarrow Q_{-3} \cdots & & \downarrow Q_{-N} & & \\
s_{\text{high}}^{(1)} & & s_{\text{high}}^{(2)} & & s_{\text{high}}^{(3)} \cdots & & s_{\text{high}}^{(N)} & &
\end{array}$$

and reconstruction is obtained by simple addition

$$s^{(j-1)} = s^{(j)} + s^{(j)}_{\text{high}},$$

with initial value $s^{(N)} = s^{(N)}_{\text{low}}$. Again, schematically we have

$$\begin{array}{ccccccccccc} s^{(N)}_{\text{low}} & \xrightarrow{\mathbb{1}} & \oplus & \xrightarrow{\mathbb{1}} & \oplus & \xrightarrow{\mathbb{1}} & \cdots & \xrightarrow{\mathbb{1}} & \oplus & \xrightarrow{\mathbb{1}} & s \\ & & \downarrow{\scriptstyle \mathbb{1}} & & \downarrow{\scriptstyle \mathbb{1}} & & & & \downarrow{\scriptstyle \mathbb{1}} & & \\ & & s^{(N)}_{\text{high}} & & s^{(N-1)}_{\text{high}} & & & & s^{(0)}_{\text{high}} & & \end{array}$$

Unfortunately, we have to change our projection operator in the decomposition at each step. This is due to the fact that the different functions $s^{(j)}_{\text{low}}$ are on different scales. On the other hand, each space V_{-j} is nothing other than a dilated version of V_0. Therefore, the projection operators P_{-j} and Q_{-j} may be computed in V_0 by re-scaling

$$P_{-j} = D_{2^j} P_0 D_{1/2^j}, \qquad Q_{-j} = D_{2^j} P_0 D_{1/2^j}.$$

Therefore, if instead we compute the re-scaled functions $r^{(j)}_{\text{low}} = 2^{-j/2} D_{1/2^j} s^{(j)}_{\text{low}}$ and $r^{(j)}_{\text{high}} = 2^{-j/2} D_{1/2^j} s^{(j)}_{\text{high}}$ we can do this without leaving V_0, and therefore we only need one fixed pair of operators. More precisely, let

$$A = \sqrt{1/2} P_0 D_{1/2}, \qquad B = \sqrt{1/2} Q_0 D_{1/2}. \tag{16.0.1}$$

Then the sequence $r^{(j)}_{\text{low}}$ and $r^{(j)}_{\text{high}}$ can be obtained via the following recursion:

$$r^{(j+1)}_{\text{low}} = A r^{(j)}_{\text{low}}, \qquad r^{(j+1)}_{\text{high}} = B r^{(j)}_{\text{low}},$$

with initial value $r^{(0)}_{\text{low}} = s$. In diagrams we have

$$\begin{array}{cccccccccc} s & \xrightarrow{A} & r^{(1)}_{\text{low}} & \xrightarrow{A} & r^{(2)}_{\text{low}} & \cdots & \xrightarrow{A} & r^{(N-1)}_{\text{low}} & \xrightarrow{A} & r^{(N)}_{\text{low}} \\ \downarrow{\scriptstyle B} & & \downarrow{\scriptstyle B} & & \downarrow{\scriptstyle B} & \cdots & & \downarrow{\scriptstyle B} & & \\ r^{(1)}_{\text{high}} & & r^{(2)}_{\text{high}} & & r^{(3)}_{\text{high}} & \cdots & & r^{(N)}_{\text{high}} & & \end{array} \tag{16.0.2}$$

The reconstruction is then obtained with the help of the adjoint operators

$$A^* = \sqrt{2} D_2 P_0, \qquad B^* = \sqrt{2} D_2 Q_0.$$

We have $s = r^{(0)}$, where $r^{(j)}$ is obtained by the following iteration:

$$r^{(j-1)} = A^* r^{(j)} + B^* r^{(j)}_{\text{high}},$$

with initial value $r^{(N)} = r^{(N)}_{\text{low}}$. Graphically, we have

$$\begin{array}{ccccccccccc} r^{(N)}_{\text{low}} & \xrightarrow{A^*} & \oplus & \xrightarrow{A^*} & \oplus & \xrightarrow{A^*} & \cdots & \xrightarrow{A^*} & \oplus & \longrightarrow & s \\ & & \uparrow{\scriptstyle B^*} & & \uparrow{\scriptstyle B^*} & & & & \uparrow{\scriptstyle B^*} & & \\ & & r^{(N)}_{\text{high}} & & r^{(N-1)}_{\text{high}} & & & & r^{(0)}_{\text{high}} & & \end{array}$$

Note that the last two decomposition and reconstruction schemes are based on the fact that on V_0 we have

$$\mathbb{1} = A^*A + B^*B,$$

and vice versa, for all operators acting in some Hilbert space that satisfy the equation above, we obtain a decomposition and reconstruction schema as above. Indeed, by successive insertion of the identity operator between A^* and A, we obtain the sated analysis–reconstruction schema

$$\mathbb{1} = A^*A + B^*B = A^*A^*AA + A^*B^*BA + B^*B.$$

We now wish to apply this framework to the computation of the wavelet transform over the real line.

16.1 Filterbanks over $\mathbb{Z}$

In the special case where A and B are acting in sequence space $L^2(\mathbb{Z})$ as the dilation by 2 and its adjoint, we speak of a filterbank. That is, we define for any sequence $v \in L^2(\mathbb{Z})$ two new sequences via

$$\begin{aligned} v_{\text{low}} &= Av = (D^{\mathbb{Z}}_{2,\phi})^* v = \Sigma^{\mathbb{Z}}_2(\tilde{\phi} * v), \\ v_{\text{high}} &= Bv = (D^{\mathbb{Z}}_{2,\psi})^* v = \Sigma^{\mathbb{Z}}_2(\tilde{\psi} * v). \end{aligned}$$

As we know we have perfect reconstruction

$$A^*A + B^*B = \mathbb{1}$$

iff (ϕ, ψ) is a QMF-system.

The iteration of the above system in the sense that we have presented in (16.0.2) is called a *filterbank*. Graphically, we can write

$$\begin{array}{ccccccccc} v & \to \boxed{\tilde{\phi}\downarrow 2} \to & v^{(1)}_{\text{low}} & \to \boxed{\tilde{\phi}\downarrow 2} \to & v^{(2)}_{\text{low}} & \cdots \to \boxed{\tilde{\phi}\downarrow 2} \to & v^{(N-1)}_{\text{low}} & \to \boxed{\tilde{\phi}\downarrow 2} \to & v^{(N)}_{\text{low}} \\ \downarrow & & \downarrow & & \downarrow & & \downarrow & & \\ \boxed{\tilde{\psi}\downarrow 2} & & \boxed{\tilde{\psi}\downarrow 2} & & \boxed{\tilde{\psi}\downarrow 2} & & \boxed{\tilde{\psi}\downarrow 2} & & \\ \downarrow & & \downarrow & & \downarrow & & \downarrow & & \\ v^{(1)}_{\text{high}} & & v^{(2)}_{\text{high}} & & v^{(3)}_{\text{high}} & & v^{(N)}_{\text{high}} & & \end{array}$$

where we have written $\to \boxed{\tilde{\phi}\downarrow 2} \to$ for the operation that consists of filtering the sequence with a filter $\tilde{\phi}$ and dropping every second point. Reconstruction from the sequences $v^{(j)}_{\text{low(high)}}$ is obtained via

$$\begin{array}{cccccccc} v^{(N)}_{\text{low}} & \to \boxed{\uparrow 2\,\phi} \to & \oplus & \to \boxed{\uparrow 2\,\phi} \to & \oplus & \to \boxed{\uparrow 2\,\phi} \to \cdots \to \boxed{\uparrow 2\,\phi} \to & \oplus & \to s \\ & & \uparrow & & \uparrow & & \downarrow & \\ & & \boxed{\uparrow 2\,\psi} & & \boxed{\uparrow 2\,\psi} & & \boxed{\uparrow 2\,\psi} & \\ & & \uparrow & & \uparrow & & \uparrow & \\ & & v^{(N)}_{\text{high}} & & v^{(N-1)}_{\text{high}} & & v^{(0)}_{\text{high}} & \end{array}$$

Here, $\rightarrow \boxed{\uparrow 2\phi} \rightarrow$ is acting as the insertion of a zero between every second sample, with a subsequent smoothing.

Some words about computational complexity. A natural quantity for measuring the complexity of an algorithm might be the number of elementary operations, $u \mapsto au + b$, that are necessary to compute the result. Since our algorithm transforms an infinite sequence, into a new set of infinite sequences, we should, rather, consider the number of such operations per unit input. Let $|\phi|$ denote the number of non-zero elements of the sequence ϕ. Then a convolution of a very long sequence v with the filter ϕ computed directly takes $|\phi|$ elementary operations per unit input of v. Since in a filterbank the frequency is divided by 2 at every iteration step, we obtain

$$\#\text{operations}/[\text{sample}] \leq (|\phi| + |\psi|)(1 + 1/2 + 1/4 + \cdots) \leq 2(|\phi| + |\psi|).$$

16.2 Computing the orthonormal wavelet transform over a dyadic grid

Consider a multi-resolution analysis V_j over the real line with scaling function η and wavelet ρ. Suppose, again, that $s \in V_0$. We now wish to compute the orthogonal wavelet transform, that is, the set of numbers

$$\langle \rho_{k,j} \mid s \rangle_{\mathbb{R}}, \quad \text{with } \rho_{k,j}(t) = 2^{-j/2}\rho(2^{-j}t - k), \quad k \in \mathbb{Z}, j \in \mathbb{N}_0.$$

In order to treat this problem numerically we have to sample s in a suitable way, since only sequences of numbers may be treated by computers. Here we will use in the first step the imperfect sampling operator associated with the generating function η:

$$\Sigma_\eta\colon V_0 \rightarrow L^2(\mathbb{Z}), \qquad s \mapsto \langle \eta(\cdot - n) \mid s \rangle_{\mathbb{R}},$$

This mapping is an isometry. The operators A and B as given by (16.0.1) can now be computed in the sequence space. Indeed, the adjoint A^* of A as the operator on V_0 is nothing other than dilation by 2. By (7.0.5) we may write the dilation over the real line with the help of $D^{\mathbb{Z}}_{2,\phi}$ and, therefore, the operator $A = \sqrt{1/2}\,P_0 D_{1/2}$ is given in sequence space by the adjoint of (7.0.4) of dilation by 2 in sequence space with the sampled dilated scaling function as the interpolation filter. In the same way we obtain the operator B as the adjoint of dilation by 2 over the sequence space with the interpolation filter given by the sampled dilated wavelet. This shows that the wavelet transform of a function $s \in V_0$ can be computed with the help of a filterbank, where the discrete filters ϕ and ψ of the filterbank are given by

$$\left.\begin{aligned} \phi(k) &= \sqrt{2}\,\Sigma_\eta D_2 \eta(k) = \frac{1}{\sqrt{2}} \int_{-\infty}^{+\infty} dt\, \bar{\eta}(t - k)\eta(t/2), \\ \psi(k) &= \sqrt{2}\,\Sigma_\eta D_2 \rho(k). \end{aligned}\right\} \tag{16.2.1}$$

The kth output of the jth iteration of the filterbank gives the wavelet coefficient at position $k2^j$, and scale 2^j:

$$v_{\text{high}}^{(j)}(k) = \langle \rho_{j,k} \mid s \rangle_{\mathbb{R}} = 2^{j/2} \mathscr{W}_\rho s(k2^j, 2^j).$$

16.3 More general wavelet

How can we use the fast algorithm above to compute the wavelet transform with respect to more general wavelets than just the orthogonal ones, and over a denser set than just the dyadic grid? Clearly, we have to suppose again that we are in a sampling space—say V_0—of a multi-resolution analysis over $\mathbb{R}$. If not, one has to take a suitable projection. Then the first problem is easily solved. Suppose that $g \in V_0$ and $s \in V_0$. We wish to compute in a first step the wavelet transform of s with respect to g over a dyadic grid, that is, the set

$$\{2^{j/2} \mathscr{W}_g s(k2^j, 2^j) \colon k \in \mathbb{Z}, j \in \mathbb{N}_0\}.$$

Let η be a generating function of the multi-resolution analysis V_j. Then $g \in V_0$ can be expanded as

$$g = \sum_{k \in \mathbb{Z}} \gamma(k) \eta(\cdot - k), \qquad \gamma = \Sigma_\eta g.$$

Therefore, the wavelet transform $v^{(j)}(k) = 2^{j/2} \mathscr{W}_g s(k2^j, 2^j)$ sampled over a dyadic grid of a function $s \in V_0$ with respect to a wavelet $g \in V_0$ can then be computed via ($v = \Sigma_\eta s$, $\phi = \Sigma_\eta D_2 \eta$, $\gamma = \Sigma_\eta g$):

$$\begin{array}{ccccccccccc}
v & \rightarrow \boxed{\tilde{\phi}\downarrow 2} \rightarrow & v_{\text{low}}^{(1)} & \rightarrow \boxed{\tilde{\phi}\downarrow 2} \rightarrow & v_{\text{low}}^{(2)} & \cdots \rightarrow \boxed{\tilde{\phi}\downarrow 2} \rightarrow & v_{\text{low}}^{(N-1)} & \rightarrow \boxed{\tilde{\phi}\downarrow 2} \rightarrow & v_{\text{low}}^{(N)} \\
\downarrow & & \downarrow & & \downarrow & & \downarrow & & \\
\boxed{\tilde{\gamma}} & & \boxed{\tilde{\gamma}} & & \boxed{\tilde{\gamma}} & \cdots & \boxed{\tilde{\gamma}} & & \\
\downarrow & & \downarrow & & \downarrow & & \downarrow & & \\
v^{(1)} & & v^{(2)} & & v^{(3)} & \cdots & v^{(N)} & &
\end{array} \tag{16.3.1}$$

Here we have written $\begin{array}{c}\downarrow\\ \boxed{\tilde{\gamma}}\\ \downarrow\end{array}$ for the convolution of a sequence with the discrete filter $\tilde{\gamma}$. The complexity in this case is clearly given by

$$\#\text{operations}/[\text{sample}] \leq 2(|\phi| + |\gamma|).$$

16.4 Denser grids

We now come to the second problem, that is, the computation of wavelet transforms over a denser grid, than just the dyadic one. That is, we wish to compute the set of numbers

$$\{\mathscr{W}_g s(k, 2^j) \colon k \in \mathbb{Z}, j \in \mathbb{N}_0\}.$$

Clearly, we assume again that g and s are in the approximation space V_0 of some multi-resolution analysis over $\mathbb{R}$. There are at least two ways of computing the wavelet transform over a denser grid. The first consists of interpolating the sampling values over a dyadic grid by using the fact that each voice is in a perfect sampling space having the dyadic interpolation property. The other method which we will consider now is to use a slightly different scheme of computation right from the beginning.

Since g, $s \in V_0$, and V_0 is a sampling space the convolution can now be computed in sequence space via

$$\Sigma(\tilde{r} * s) = \Sigma_{\tilde{\eta}} \tilde{r} * \Sigma_\eta s = \widetilde{\Sigma_\eta r} * \Sigma_\eta s \tag{16.4.1}$$

We may therefore use the commutative diagram (7.0.5) to write

$$\begin{aligned} \mathscr{W}_g s(k, 2^j) &= \Sigma((D_2)^j \tilde{g} * s)(k) \\ &= \widetilde{(\Sigma_\eta (D_2)^j g)} * \Sigma_\eta s(k) \\ &= \widetilde{((D^{\mathbb{Z}}_{\phi, 2})^j \Sigma_\eta g)} * \Sigma_\eta s(k) \qquad \text{with } \phi = \Sigma_\eta D_2 \eta. \end{aligned}$$

This expression can be simplified considerably thanks to the following considerations. To simplify the notation we write H for the adjoint of the down-sampling operator $\Sigma_2^{\mathbb{Z}}$. So H is the dilation of a sequence by inserting a 0 between each element. This is called 'dilation with wholes'. In Fourier space we have, by iteration of (7.0.3) for $r \in L^2(\mathbb{Z})$

$$(D^{\mathbb{Z}}_{\phi, 2})^j \colon \hat{r}(\omega) \mapsto \hat{\phi}(\omega)\hat{\phi}(2\omega)\hat{\phi}(2^2\omega) \cdots \hat{\phi}(2^{j-1}\omega)\hat{r}(2^j\omega).$$

And hence, since $H \colon \hat{s}(\omega) \mapsto \hat{s}(2\omega)$,

$$\widetilde{((D^{\mathbb{Z}}_{\phi, 2})^j \gamma)} * v = H^j \tilde{\gamma} * H^{j-1} \tilde{\phi} * \cdots * \tilde{\phi} * v.$$

Therefore, $2^{j/2} \mathscr{W}_g s(k, 2^j) = v^{(j)}_{\text{low}}(k)$ if we apply the following recursion:

$$v^{(j+1)}_{\text{low}} = H^j \tilde{\phi} * v^{(j)}_{\text{low}}, \qquad v^{(j+1)}_{\text{high}} = H^j \tilde{\gamma} * v^{(j)}_{\text{low}},$$

with initial value $v = v_{\text{low}}^{(0)} = \Sigma_\eta s$ and $\gamma = \Sigma_\eta g$. Graphically, we have

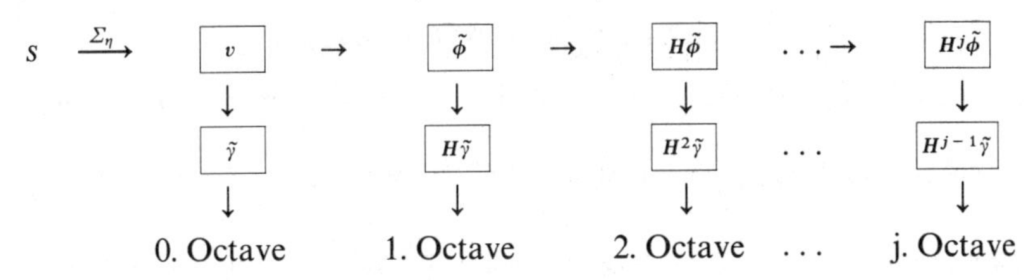

In this case, the complexity grows linearly in the number of voices (=octaves), that we compute. Indeed, dilation with wholes does not change the effective length of filter, since only zeros are added. Therefore,

$$\#\text{additions}/[\text{sample}] = j_{\max}(|\phi| + |\gamma|).$$

A naïve algorithm using direct convolution with the dilated sampled wavelet gives an exponential growth in the number of voices.

16.5 Interpolation of the voices

It may be useful to stock only the wavelet coefficients with respect to an orthonormal wavelet ψ over a dyadic grid. However, to obtain diagrams that can be interpreted it might be useful to interpolate the values in an appropriate way. Clearly, the sampling operator over the dyadic grid is invertible on the image of the wavelet transform. An explicit inversion is given by the superposition of reproducing kernels. However, the computational amount is quite considerable. For the interpolation of the voices with dilation parameters $a = 2^j$, however, a trick helps.

The convolution $\tilde{g} * s$ of two functions g, s in V_0 is in general[10] no longer in V_0 and neither is it any other of the voices $\mathcal{W}_g s(\cdot, 2^j)$. We therefore have a closer look at the space $\tilde{V}_0 * V_0$ consisting of all functions $\tilde{s} * r$ with $s, r \in V_0$ and $\tilde{s}(t) = \bar{s}(-t)$ as usual. First observe that it is a perfect sampling space. Indeed, since V_0 is generated by the translates $\eta(t-n)$ with $n \in \mathbb{Z}$, the space $\tilde{V}_0 * V_0$ is generated by the integer translates $\kappa(t-n)$ with

$$\kappa = \tilde{\eta} * \eta, \qquad \hat{\kappa}(\omega) = |\hat{\eta}(\omega)|^2.$$

From the Fourier representation of κ and the fact that the translates of η are an orthonormal set, we see that $\kappa(t-n)$ is an elementary generating function of $\tilde{V}_0 * V_0$ in the sense that

$$\kappa(n) = \delta(n).$$

[10] Essentially, the only exceptions are the band-limited functions.

In addition, it is scaling invariant in the sense that

$$s \in \tilde{V}_0 * V_0 \Rightarrow s(\cdot/2) \in \tilde{V}_0 * V_0.$$

Therefore, $s \in \tilde{V}_0 * V_0$ has the dyadic interpolation property and ($k = 1 \bmod 2$)

$$s(k2^{-j}) = \sum_{n \in \mathbb{Z}} \mu(n) s([k-1]2^{-j} + n2^{1-j}), \qquad j \geq 0,$$

where $\mu(n) = \kappa(n - 1/2)$. Using the sampling relation for the convolution (16.4.1) and the definition of ϕ in (16.2.1) we obtain

$$v(n) = h(n/2) = 2\Sigma D_2 h(n) = 2\Sigma(D_2\tilde{\eta} * D_2\eta)(n) = 2\tilde{\phi} * \phi(n),$$

and μ is related to v by

$$v = \delta_0 + SH\mu,$$

as follows from $h(n) = \delta_0(n)$.

Now every voice $\mathcal{W}_g s(\cdot, 2^j)$ is in a scaled version of $\tilde{V}_0 * V_0$ and we therefore obtain the missing points through dyadic interpolation: ($k = 1 \bmod 2$)

$$\mathcal{W}_g s(k2^{m-1}, 2^j) = \sum_{n \in \mathbb{Z}} \mu(n) \mathcal{W}_g s([k-1]2^{m-1} - n2^m, 2^j), \qquad m = j, j-1, \ldots.$$

In practice, one could also use the iterations of the up-sampling operator $\boxed{\uparrow v}$ in the following way. If $v_{\text{high}}^{(j)}$ is the output of algorithm (16.3.1) the densely sampled voices can be obtained through

$$v_{\text{high}}^{(j)} \quad \rightarrow \quad \overbrace{\boxed{\uparrow v}\,\boxed{\uparrow v} \cdots \boxed{\uparrow v}}^{j\text{-times}} \quad \rightarrow \quad \mathcal{W}_g s(\cdot, 2^j).$$

Remark. In the case of the Daubechies wavelets the interpolation coefficients are given by the Lagrange interpolation polynomials.

16.6 The 'à trous' algorithm

This algorithm computes the wavelet transform over a dense grid. Until now we have used the imperfect sampling operator to come from a function in some approximation space to a sequence which could now be treated numerically. However, it might often be more convenient to use the perfect sampling operator

$$\Sigma\colon V_0 \rightarrow L^2(\mathbb{Z}), \qquad s \mapsto s(k).$$

Therefore, one reason to use the 'à trous' scheme might be that it is much more difficult to compute the orthogonal projections of a given function into the approximation space V_0 than just to sample it at all integers, say. Another reason to use the 'à trou' algorithm might be that it may happen that the filters involved in the computation have only finite length, whereas the filters in the orthonormal case are infinite. Indeed, suppose we take V_0 to be given by one of the Lagrange dyadic interpolation spaces of Section 8.2. The associated scaling functions and wavelets are not of compact support. However, the cardinal spline $h \in V$ with $h(k) = \delta(k)$ is compactly supported.

The general setting of the 'à trou' is as follows. Given a dilation invariant perfect sampling space V_0, compute the numbers

$$\mathscr{W}_g s(k, 2^j), \qquad k \in \mathbb{Z}, j \in \mathbb{N}_0,$$

efficiently for $g, s \in V_0$. As we have seen in V_0 the following diagram allows us to compute the dilation in sequence space from the perfect sampling sequence. In this case, dilation by 2 in sequence space reads

$$\Sigma D_2 = D^{\mathbb{Z}}_{2,v}\Sigma, \qquad \begin{array}{ccc} V_0 & \xrightarrow{\Sigma} & L^2(\mathbb{Z}) \\ \downarrow{\scriptstyle D_2} & & \downarrow{\scriptstyle D^{\mathbb{Z}}_{2,v}} \\ V_0 & \xrightarrow{\Sigma} & L^2(\mathbb{Z}) \end{array}$$

with

$$v(k) = \Sigma D_2 h(k) = h(k/2)/2.$$

As before, the convolution of two functions $s \in V_0$, $r \in \tilde{V}_0$ can again be computed in sequence space:

$$\Sigma(r * s) = \Sigma r * F * \Sigma s,$$

with F a so-called pre-integrator

$$F(k) = \int_{-\infty}^{+\infty} dt\, h(t-k)h(t).$$

For compact supported h, the pre-integrator is of finite length.

To compute the wavelet transform of $s \in V_0$ with respect to $g \in V_0$ we may write, as before, $(\gamma = \Sigma g)$

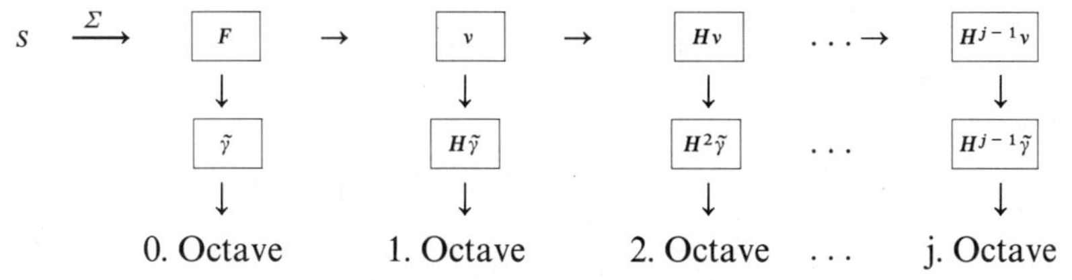

$s \xrightarrow{\Sigma}$	F	$\rightarrow$	v	$\rightarrow$	Hv	$\ldots \rightarrow$	$H^{j-1}v$
	$\downarrow$		$\downarrow$		$\downarrow$		$\downarrow$
	$\tilde{\gamma}$		$H\tilde{\gamma}$		$H^2\tilde{\gamma}$	$\ldots$	$H^{j-1}\tilde{\gamma}$
	$\downarrow$		$\downarrow$		$\downarrow$		$\downarrow$
	0. Octave		1. Octave		2. Octave	$\ldots$	j. Octave

For the computational complexity we again obtain a linear grow in the number of voices:

$$\#\text{additions}/[\text{sample}] = |F| + j_{\max}(|v| - 1 + |\gamma|).$$

16.7 Computation over $\mathbb{Z}/2^N\mathbb{Z}$

Here the schema for the computation of the associated wavelet coefficients is obtained by periodizing the previous one. Therefore, one could use the previous algorithm where the input sequence is now 2^N-periodic. But one may also reduce the dimension of the sequence spaces at each step, gaining computation time. That is, let V_j be a multi-resolution analysis over $\mathbb{Z}/2^N\mathbb{Z}$ with scaling function ϕ and wavelet ψ. Then the discrete wavelet transform is obtained via

$$v_{\text{low}}^{(j+1)} = \Sigma_2^{\mathbb{Z}/N_j\mathbb{Z}}(\tilde{\phi}^{(j)} * v_{\text{low}}^{(j)})(n),$$

$$v_{\text{high}}^{(j+1)} = \Sigma_2^{\mathbb{Z}/N_j\mathbb{Z}}\tilde{\psi}^{(j)} * v_{\text{low}}^{(j)},$$

$$\phi^{(j+1)} = \Pi_2^{\mathbb{Z}/N_j\mathbb{Z}}\phi^{(j)},$$

$$\psi^{(j+1)} = \Pi_2^{\mathbb{Z}/N_j\mathbb{Z}}\psi^{(j)},$$

$$N_{j+1} = N_j/2.$$

With initial values $v_{\text{low}}^{(0)} = v$, $\phi^{(0)} = \phi$, $\psi^{(0)} = \psi$, and $N_0 = 2^N$ we have (cf. Section 15),

$$\langle \psi_{k,j} \mid s \rangle = v_{\text{high}}^{(j)}(k).$$

The reconstruction is obtained through the adjoint operators. In Figure 16.1 we show the flow-graph of the analysis and the reconstruction associated to the simplest discrete wavelet $\phi(0) = \phi(1) = 1$ and $\psi(0) = -1, \psi(1) = 1$.

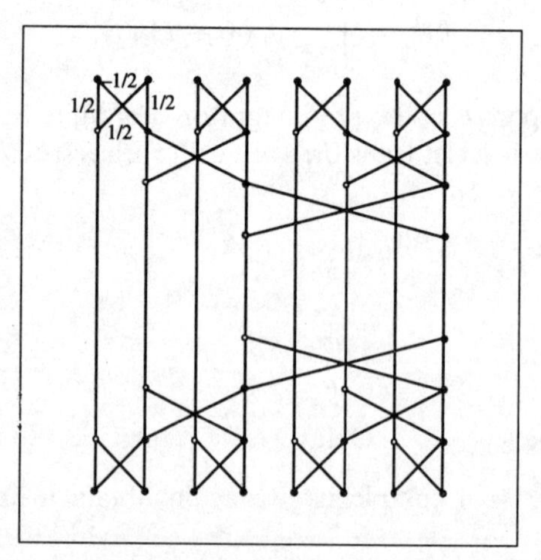

Fig. 16.1 The analysis and the reconstruction scheme for $\mathbb{Z}/8\mathbb{Z}$ with respect to the discrete Haar wavelet.

Appendix

17 Computing over $\mathbb{R}$ by using data over $\mathbb{Z}/N\mathbb{Z}$

In many applications one would like to compute with functions over $\mathbb{R}$ by using only a finite number of sampling values. Here we present a family of spaces over $\mathbb{R}$ that are determined by their sampling values at $t = 0, 1, \ldots, N-1$. We therefore end this chapter with a remark concerning the computation of the Fourier transform of a function over $\mathbb{R}$ through the discrete Fourier transform that can be implemented very efficiently on computers.

Combining the Poisson summation formulas that we have encountered, we have the following commutative diagram:

$$\begin{array}{ccccc}
(\mathbb{R}) & \xrightarrow{\Pi} & (\mathbb{T}) & \xrightarrow{\Sigma_N^{\mathbb{T}}} & (\mathbb{Z}/N\mathbb{Z}) \\
\downarrow F & & \downarrow F^{\mathbb{T}} & & \downarrow \frac{2\pi}{N} F^{\mathbb{Z}/N\mathbb{Z}} \\
(\mathbb{R}) & \xrightarrow{\Sigma} & (\mathbb{Z}) & \xrightarrow{\Pi_N^{\mathbb{Z}}} & (\mathbb{Z}/N\mathbb{Z}) \\
\downarrow F & & \downarrow F^{\mathbb{Z}} & & \downarrow F^{\mathbb{Z}/N\mathbb{Z}} \\
(\mathbb{R}) & \xrightarrow{\Pi} & (\mathbb{T}) & \xrightarrow{\Sigma_N^{\mathbb{T}}} & (\mathbb{Z}/N\mathbb{Z})
\end{array} \tag{17.0.1}$$

This shows that the only way to come from functions over $\mathbb{R}$ to functions over $\mathbb{Z}/N\mathbb{Z}$ in this diagram is through the successive application of the sampling and periodization operators. Let $v \in L^1(\mathbb{Z})$, $h \in L^2(\mathbb{R})$, $\hat{h} \in L^1(\mathbb{R}) \cap L^2(\mathbb{R})$ be such that

$$\sum_{m \in \mathbb{Z}} v(n + mN) \equiv 1, \qquad h(n) = \delta_{n,0},$$

or, equivalently,

$$F^{\mathbb{Z}} v\left(\frac{2\pi n}{N}\right) = \frac{N}{2\pi} \delta_{n,0}, \qquad \sum_{n \in \mathbb{Z}} \hat{h}(\omega + 2\pi n) \equiv 1.$$

Consider the space $V \subset L^2(\mathbb{R})$ consisting of those functions that can be written as

$$s(t) = \sum_{n \in \mathbb{Z}} s^{\#}(n) v(n) h(t - n),$$

with $s^{\#}$ an N-periodic sequence. As a consequence of all previous considerations the expansion coefficients are uniquely determined by

$$s \in V \Rightarrow s^{\#} = \Pi_N^{\mathbb{Z}} \Sigma s.$$

By the commutation diagram of (17.0.1) we can compute the Fourier transform of s with the help of the discrete Fourier transform of $s^{\#}$ via

$$\hat{s}(\omega) = \hat{h}(\omega) \sum_{n \in \mathbb{Z}/N\mathbb{Z}} (F^{\mathbb{Z}/N\mathbb{Z}} s^{\#})(n) r\left(\omega - \frac{2\pi n}{N}\right),$$

where $r(\omega) = F^{\mathbb{Z}} v(\omega)$.

4

FRACTAL ANALYSIS AND WAVELET TRANSFORMS

This chapter is devoted to a short introduction to fractals in general and their analysis through wavelet transforms.

We may distinguish at least three aspects in fractals

(i) an underlying dynamical system,
(ii) global self-similarity,
(iii) local self-similarity.

Whereas for the first two points we do not know yet how to use the wavelet technique as an analysis tool in general, the third point is quite well understood now. In particular, we have strong theorems about the analysis of local regularity through wavelet transforms. We therefore present how the wavelet analysis works as a mathematical microscope to reveal the small scale features of the analysed function. We will apply the wavelet technique to Brownian motion, where we will re-demonstrate Lévi's law, to the Weierstrass function, and to the Riemann–Weierstrass function. Further, we give a short overview on fractal dimensions and how they may be estimated through wavelet transforms, at least for a special class of highly self-similar fractal measures.

1 Self-similarity and the re-normalization group

We do not give a precise definition of what we call a fractal. Roughly speaking, a fractal is an object—for us a function or distribution—that has structure at all length scales. Therefore, if we look at our object under a microscope, more and more details will appear. Thus these objects cannot be modelled by a smooth function. A smooth function looks at small scale essentially like a constant function. Very often, fractals display a behaviour known as self-similarity. This means that while looking at smaller and smaller scales we obtain the same features again and again. To be more precise, consider the local function s_{loc} that describes the fluctuations of s around some point τ_0:

$$s_{\text{loc}}(t) = s(\tau_0 + t) - s(\tau_0).$$

Then local self-similarity would mean that for some re-scaling factor λ we have in some sense

$$s_{\mathrm{loc}}(\lambda t) \simeq \lambda^{\alpha} s_{\mathrm{loc}}(t). \tag{1.0.1}$$

The only functions that satisfy this scaling invariance for all λ are the homogeneous functions. This motivates the following definition.

Definition 1.0.1
A function s satisfies the exact scaling condition (ESC) at τ_0 if

$$s_{\mathrm{loc}}(t) = c_{-}|t|_{-}^{\alpha} + c_{+}|t|_{+}^{\alpha} + o(t^{\alpha}), \qquad (t \to 0).$$

The exponent α is called the local scaling exponent at τ_0.

If, instead equation (1.0.1) does not hold for all re-scalings λ but only for the discrete set of scalings $\lambda = \gamma^n$, $n = 0, 1, \ldots$, we have another class of functions, called the periodic scaling class.

Definition 1.0.2
A function s satisfies the periodic scaling condition (PSC) at τ_0 if

$$s_{\mathrm{loc}}(t) = c_{-}(t)|t|_{-}^{\alpha} + c_{+}(t)|t|_{+}^{\alpha} + o(t^{\alpha}), \qquad (t \to 0),$$

where the $c_{\pm}$ satisfy the discrete scale-invariance

$$c_{\pm}(\lambda t) = c_{\pm}(t)$$

with some constant $\lambda \geq 1$.

Remark. The number α is called the *local scaling exponent*, or *local fractal dimension.* It is actually a dimension because it relates the t scale to the s-scale, as, for instance, the 'volume-scale' is related to the 'length-scale' by a third power, which is the dimension of a volume.

Remark. We note that the exact scaling condition may be seen as a special case of the discrete scaling condition. In turn, we may identify the periodic scaling condition with superpositions of functions satisfying the exact scaling condition, but with the complex scaling exponent $\alpha + i\beta$.

Clearly, the functions that satisfy PSC are locally self-similar in the sense described above. The greater part of this chapter will be devoted to giving a sufficient criterion in wavelet space to show that the analysed function actually satisfies ESC or PSC. The principal term has already been treated in Section 29 of Chapter 1. The remainder, however, is much more difficult to treat, as we will see.

Sometimes the concepts of ESC or PSC are too strong and will be replaced by the following weaker condition. We have defined the local function s_{loc} as the fluctuation of s around its value at a given point. However, to be more general, we consider the fluctuations of s around any local polynomial approximation of s, that is, if there is a polynomial P_n of order n such that

$$s(\tau_0 + t) = P_n(t) + o(t^n), \qquad (t \to 0),$$

then we set, for the highest such n

$$s_{\text{loc}}(t) = s(\tau_0 + t) - P_n(t).$$

Therefore, to look at the fluctuations of a function around some local polynomial approximation amounts to considering the function modulo some polynomials. As we saw in Section 24 of Chapter 1 this can be done by taking s as a distribution in $S_0'(\mathbb{R})$.

To look at the fluctuations of s under a microscope can now be formalized as re-scaling the t-coordinate by $t \mapsto \lambda t$ with some $0 < \lambda < 1$ and re-scaling the fluctuation scale of s in such a way that this process eventually becomes stabilized in a non-trivial way. In more mathematical terms this means that we can find a sequence c_n such that the sequence

$$s_{n+1}(t) = c_n s_n(\lambda t), \qquad s_0(t) = s(t),$$

satisfies

$$\lim_{n \to \infty} s_n = s^* \neq 0,$$

where the limit holds in $S_0'(\mathbb{R})$. The function s^* is called the *local re-normalization* of s. By translation we can define the re-normalization around every point in the obvious way.

Example 1.0.3

Clearly, the functions satisfying ESC or PSC have a local re-normalization. Let ϕ be a smooth function with compact support. For $0 < \alpha, \beta < 1$, set

$$f(t) = \sum_{n \in \mathbb{Z}} \phi\left(\frac{t - \beta^n}{\alpha^n}\right).$$

It is an easy exercise to see that f has a local re-normalization, namely,

$$(\alpha/\beta)^m f(\beta^m t) \to \sum_{n \in \mathbb{Z}} \alpha^n \delta(t - \beta^n).$$

This shows that the limit may actually be a distribution, even though we started from a function.

Usually, the re-scaling constants c_n are fixed by taking some function $g \in S_0(\mathbb{R})$—supposed to 'measure', in some sense, the fluctuation scale—and then by requiring that (in the sense of distributions)

$$s_n(g) = \int s_n \cdot g = 1, \qquad \text{for all } n \in \mathbb{N}_0. \tag{1.0.2}$$

Let $\Lambda_g \subset S_0'(\mathbb{R})$ be the set of vectors η for which $\eta(D_\lambda g) \neq 0$ for all $\lambda > 0$. We then define the re-normalization map as follows:

$$\mathscr{R}_\lambda : \Lambda_g \to \Lambda_g, \qquad s \mapsto cs(\lambda t),$$

where c is fixed by (1.0.2), that is, $c = c(s) = (\int dts(\lambda t)g(t))^{-1}$. The re-normalization map is a non-linear map that satisfies the semigroup property

$$\mathscr{R}_\lambda \circ \mathscr{R}_\gamma = \mathscr{R}_{\lambda \cdot \gamma}.$$

To have a local re-normalization means that s is in the attracting domain of some fixed point s^* of some re-normalization map

$$\mathscr{R}_\lambda^n s \to s^*, \qquad (n \to \infty), \quad \text{and} \quad \mathscr{R}_\lambda s^* = s^*.$$

The local fractal dimension is again defined as the relation between the t-scale and the fluctuation scale. That is, we set

$$\alpha = \alpha(t_0) = -\lim_{n \to \infty} \frac{\log c_n}{\log \lambda}.$$

The limit actually exists since c_n converges if s has a local re-normalization.

1.1 Re-normalization in wavelet-space

We now want to compute the re-normalization transformations in wavelet space. Suppose s has a local re-normalization at $t_0 = 0$ with scale-fixing functional g. It is clear from the co-variance of the wavelet transform under dilation that the re-normalization procedure reads, in wavelet ($\mathscr{W} = \mathscr{W}_g s$),

$$\mathscr{R}_\lambda : \mathscr{W}_n(b, a) \mapsto c\mathscr{W}_n(\lambda b, \lambda a),$$

where the scale is fixed by

$$\mathscr{W}_n(0, 1) = 1.$$

Theorem 1.1.1

A function or distribution $s \in S_0'(\mathbb{R})$ has a local re-normalization iff for some wavelet $g \in S_0(\mathbb{R})$ its wavelet transform $\mathscr{W} = \mathscr{W}_g s$ satisfies at:

(i) $\lim_{n \to \infty} \mathscr{W}_n(b, a) = \mathscr{W}^*(b, a) \not\equiv 0$

pointwise for every $(b, a) \in \mathbb{H}$ and

(ii) $|\mathscr{W}_n(b, a)| \leq O(1)(1 + |b|)^m(a + a^{-1})^m$

uniformly in n for some $m \geq 0$.

Proof. We first show the sufficiency. Let $r \in S_0(\mathbb{R})$. As we saw in Section 24 of Chapter 1, we may write the action of $s_n = c_n s(\lambda^n)$ on any function

$r \in S_0(\mathbb{R})$ as an absolutely convergent integral over the half-plane:

$$s_n(r) = \int_0^{\infty} \frac{da}{a} \int_{-\infty}^{+\infty} db\, \mathscr{W}_n(b, a) \mathscr{W}_g r(b, a).$$

By hypothesis this function is uniformly majorized by an integrable function and we may evoke the dominated convergence theorem to write

$$s_n(r) \to \int_0^{\infty} \frac{da}{a} \int_{-\infty}^{+\infty} db\, \mathscr{W}^*(b, a) \mathscr{W}_g r(b, a), \qquad (n \to \infty).$$

This shows that s has a local re-normalization in $S_0'(\mathbb{R})$.

Now the condition is necessary. Indeed, if s has a local re-normalization, the pointwise convergence of the $\mathscr{W}_g s_n(b, a)$ is trivial, and since as a convergent sequence it is bounded, there is some polynomially bounded function over the half-plane that majorizes all the $c_n \mathscr{W}_g s(\lambda^n b, \lambda^n a)$. □

2 The order of magnitude of wavelet coefficients

As we saw in Section 12 of Chapter 1, the rapid decrease of the Fourier transform of the analysed function at infinity and the rapid decrease of the wavelet at zero frequency is mirrored in a rapid decrease of the wavelet coefficients at small scale. In terms of the time representation, this means that a high global regularity of the analysed function and many vanishing moments of the wavelet imply a rapid decrease of the wavelet coefficients. This relation will be clarified in this section. In particular, we will show that a local regularity of s implies a local decrease of the wavelet coefficients.

In order to measure the regularity of a function s in a point t_0 we estimate how well it can be approximated locally by a polynomial P_n of degree n. That is, we suppose that we may write

$$s(t_0 + t) = P_n(t) + s_{\text{loc}}(t), \qquad (2.0.1)$$

where the remainder becomes small compared to the principal term of P_n, that is, $s_{\text{loc}}(t) = o(t^n)$ as $t \to 0$.

Example 2.0.1

If s is continuous in t_0 we may write $s(t) = s(t_0) + o(1)$ and thus $P_0(t) = s(t_0)$. If s is differentiable at t_0, then $s(t_0 + t) = s(t_0) + ts'(t_0) + o(t)$, and therefore we may choose $P_1(t) = s(t_0) + ts'(t_0)$.

In general, the faster $s_{\text{loc}}(t)$ goes to 0, together with t, the higher is the regularity of s at t_0. Therefore, we introduce a hierarchy of functions with which we can compare the remainder s_{loc}.

Definition 2.0.2

Let s be a function that satisfies (2.0.1) *with some* $n \in \mathbb{N}_0$. *We say that s is of regularity* Λ^α *at* t_0, $n < \alpha \leq n+1$, *iff*

$$s_{\mathrm{loc}}(t) = O(t^\alpha) \qquad (t \to 0).$$

We say that s is of regularity λ^α, $n \leq \alpha < n+1$, *if*

$$s_{\mathrm{loc}}(t) = o(t^\alpha) \qquad (t \to 0).$$

We say that s is of regularity $\Lambda_{\log}^{\alpha,\beta}$, $n < \alpha \leq n+1$, $\beta > 0$, *if*

$$s_{\mathrm{loc}}(t) = O(t^\alpha \log^\beta t) \qquad (t \to 0).$$

We say that s is of regularity $\lambda_{\log}^{\alpha,\beta}$, $n < \alpha \leq n+1$, $\beta > 0$, *if*

$$s_{\mathrm{loc}}(t) = o(t^\alpha \log^\beta t) \qquad (t \to 0).$$

We say that s is in $\Lambda^\alpha(\mathbb{R})$, $\lambda^\alpha(\mathbb{R})$, $\Lambda_{\log}^{\alpha,\beta}(\mathbb{R})$, $\lambda_{\log}^{\alpha,\beta}(\mathbb{R})$ *if* $|s(t)| \leq c(1+|t|^\alpha)$ *and the respective local estimations hold uniformly in* t_0.

We have the following scale of local regularities ($\alpha < \alpha'$, $\beta < \beta'$, $\gamma > 0$):

$$\Lambda_{\log}^{\alpha,\beta} \Leftarrow \Lambda_{\log}^{\alpha,\beta'} \Leftarrow \Lambda^\alpha \Leftarrow \lambda_{\log}^{\alpha,\beta} \Leftarrow \lambda_{\log}^{\alpha,\beta'} \Leftarrow \lambda^\alpha \Leftarrow \Lambda_{\log}^{\alpha',\gamma}.$$

The local regularity of the analysed function implies a decrease of the wavelet transform at small scales whenever the analysing wavelet is localized and has a sufficient number of vanishing moments. This decrease is of the same type as the decrease of s_{loc}, the difference of s, and its local polynomial approximation P_n.

Theorem 2.0.3

Let s be a polynomial bounded function

$$|s(t)| \leq c(1+|t|^\alpha).$$

If s is of regularity Λ^α *at some* τ_0, *then the wavelet transform* $\mathcal{W}_g s$ *of s with respect to the wavelet g satisfies*

$$\mathcal{W}_g s(\tau_0 + b, a) = O(a^\alpha + |b|^\alpha) \qquad (b, a \to 0).$$

If s is of regularity λ^α, *then*

$$\mathcal{W}_g s(\tau_0 + b, a) = o(a^\alpha + |b|^\alpha) \qquad (b, a \to 0).$$

In both cases the wavelet should satisfy $g \in L^1(\mathbb{R})$ *and* $t^\alpha g \in L^1(\mathbb{R})$. *If s is of regularity* $\Lambda_{\log}^{\alpha,\beta}$, *then*

$$\mathcal{W}_g s(\tau_0 + b, a) = O(a^\alpha \log^\beta a + |b|^\alpha \log^\beta |b|) \qquad (b, a \to 0).$$

If s is of regularity $\lambda_{\log}^{\alpha,\beta}$, *then*

$$\mathcal{W}_g s(\tau_0 + b, a) = o(a^\alpha \log^\beta a + |b|^\alpha \log^\beta |b|) \qquad (b, a \to 0).$$

In both cases, the wavelet should be localized such that $g \in L^1(\mathbb{R})$ *and* $t^\alpha \log^\beta tg \in L^1(\mathbb{R})$. *In addition, in all four cases the first n moments of g should vanish:*

$$\int_{-\infty}^{+\infty} dt\, t^m g(t) = 0, \qquad \textit{for } m = 0, 1, \ldots, n,$$

where n is the only integer that satisfies $\alpha \leq n < \alpha + 1$.

This theorem shows that, for example, the local Hölder regularity of degree α is mirrored by a decrease of order a^α along every straight line in the half-plane passing through the point τ_0, that is, for fixed $(b, a) \in \mathbb{H}$ we have

$$\mathcal{W}_g s(\tau_0 + \lambda a, \lambda b) = O(\lambda^\alpha), \qquad (\lambda \to 0).$$

We will prove the theorem for an even larger class of local regularities.

Definition 2.0.4

A non-negative, non-decreasing function r over $\mathbb{R}_+$ *is called submultiplicative if there is a constant* $c > 0$ *such that*

$$r(tu) \leq c r(t) r(u)$$

for all $t, u \in \mathbb{R}_+$. *A function s over* $\mathbb{R}$ *is called submultiplicative if* $s(t)$ *and* $s(-t)$, $t \geq 0$, *are submultiplicative.*

Note that the submultiplicativity means precisely that $t \mapsto \log(r(e^t))$ is subadditive.

We remark that for all regularity classes we have encountered so far, the local function s_{loc} may be majorized by a submultiplicative function. Since the wavelet is operating modulo polynomials, it only sees the local fluctuation s_{loc}, but not the polynomial part.

Theorem 2.0.5

Let r be a submultiplicative function and let s satisfy at

$$s(\tau_0 + t) = P_n(t) + O(r(|t|)), \qquad (t \in \mathbb{R}),$$

with some polynomial P_n *of degree n. Suppose the analysing wavelet satisfies at*

(i) $g \in L^1(\mathbb{R})$, $r \cdot g \in L^1(\mathbb{R})$,

(ii) $\displaystyle\int_{-\infty}^{+\infty} dt\, g(t) t^m = 0$, $m \in \{0, 1, \ldots, n\}$.

Then

$$\mathcal{W}_g s(\tau_0 + b, a) = O(r(b) + r(a)),$$

uniformly in $\mathbb{H}$. *If in addition s satisfies at*

$$s(\tau_0 + t) = P_n(t) + o(r(|t|)), \qquad (t \to 0),$$

then we have, in addition,

$$\mathcal{W}_g s(\tau_0 + b, a) = o(r(b) + r(a)), \qquad (b, a) \to 0.$$

Proof. The argument is well known in the Λ^α case and may be merely translated to the general submultiplicative situation. Since under the conditions stated in the theorem the wavelet does not see the polynomial part, the theorem follows from the following two lemmata.

Lemma 2.0.6

Let $s(t) = s(-t)$ *be an even, submultiplicative function. Then the wavelet transform of s satisfies uniformly in* $(b, a) \in \mathbb{H}$

$$\mathcal{W}_g s(b, a) = O(s(b) + s(a)).$$

The analysing wavelet should satisfy $g \in L^1(\mathbb{R})$ *and* $sg \in L^1(\mathbb{R})$.

Proof. First note that by symmetry and monotonicity we have

$$s(t + u) = s(|t + u|) \leq s(|t| + |u|).$$

Now either $2|t| > |t| + |u|$ or $2|u| > |t| + |u|$. Therefore, by montonicity and submultiplicativity,

$$s(t + u) \leq s(2|t|) + s(2|u|) \leq c(s(|t|) + s(|u|)).$$

Therefore, we may estimate

$$\begin{aligned} |\mathcal{W}_g s(b, a)| &= \left| \int_{-\infty}^{+\infty} dt \, \frac{1}{a} \bar{g}\left(\frac{t - b}{a}\right) s(t) \right| = \left| \int_{-\infty}^{+\infty} dt \; \bar{g}(t) s(at + b) \right| \\ &\leq \int_{-\infty}^{+\infty} dt \; |g(t)| (s(a|t|) + s(|b|)) \\ &\leq O(1) \|sg\|_{L^1(\mathbb{R})} s(a) + O(1) \|g\|_{L^1(\mathbb{R})} s(b) \end{aligned}$$

and the lemma is proved. □

The case o is treated by the following lemma.

Lemma 2.0.7

Let r be a non-negative, even submultiplicative, function, and let s be a function that satisfies

(i) $|s(t)| \leq cr(t)$ *for all t*,
(ii) $s(t) = o(r(t))$ $(t \to 0)$,

then

$$\mathscr{W}_g s(b, a) = o(r(b) + r(a)) \qquad (b, a \to 0).$$

The wavelet should again satisfy $g \in L^1(\mathbb{R})$, $rg \in L^1(\mathbb{R})$.

Proof. For every $\epsilon > 0$ we can find an $\eta > 0$ such that $|s(t)| < \epsilon|r(t)|$ for $|t| < \eta$. Splitting the integral into two parts we may write

$$|\mathscr{W}_g s(b, a)| \leq \int_{-\infty}^{+\infty} dt\, |\bar{g}(t)|\, |s(at + b)|$$
$$= \left\{ \int_{|at+b|<\eta} + \int_{|at+b|\geq\eta} \right\} dt\, |\bar{g}(t)|\, |s(at + b)| = X_1 + X_2.$$

In the first term we have, by hypothesis on η and submultiplicativity of r,

$$X_1 \leq \epsilon O(1) \int_{|at+b|<\eta} dt\, |\bar{g}(t)|(r(a)r(t) + r(b))$$
$$\leq \epsilon O(1)\|rg\|_{L^1(\mathbb{R})} r(a) + \epsilon O(1)\|g\|_{L^1(\mathbb{R})} r(b) \leq \epsilon O(r(a) + r(b)).$$

The second term may be estimated using the global estimation of s:

$$X_2 \leq O(1) \int_{|at+b|\geq\eta} dt\, |\bar{g}(t)|(r(a)r(t) + r(b))$$
$$\leq O(r(b) + r(a)) \int_{|at+b|\geq\eta} dt\, |g(t)|(r(t) + 1).$$

Now $|g(t)|(r(t) + 1)$ is integrable, and therefore for a and b small enough the integral is smaller than ϵ, since it runs over a smaller and smaller neighbourhood of infinity, and thus

$$X_2 < \epsilon O(r(b) + r(a)).$$

Since ϵ was arbitrary the lemma is proved. □

This also proves Theorem 2.0.5. □

2.1 Inverse theorems for global regularity

Here we will prove some inverse theorems concerning the proof of global regularity through the wavelet coefficients. The general setting will be as follows. Suppose we have a position-scale representation of a distribution s:

$$s(t) = \int_0^\infty \frac{da}{a} \int_{-\infty}^{+\infty} db\, \mathscr{T}(b, a) \frac{1}{a} h\left(\frac{t-b}{a}\right)$$

with some polynomial bounded scale-position coefficients

$$|\mathscr{T}(b, a)| \leq c(1 + |b|)^n (a + 1/a)^n.$$

We may then split s into two parts:

$$\begin{aligned} s(t) &= \int_0^\infty \frac{da}{a} \int_{-\infty}^{+\infty} db\, \mathscr{T}(b, a) \frac{1}{a} h\left(\frac{t-b}{a}\right) \\ &= \left\{\int_0^1 \frac{da}{a} + \int_0^\infty \frac{da}{a}\right\} \int_{-\infty}^{+\infty} db\, \mathscr{T}(b, a) \frac{1}{a} h\left(\frac{t-b}{a}\right) \\ &= s_{\text{small}}(t) + s_{\text{large}}(t). \end{aligned}$$

Now the large-scale reconstruction s_{large} is a smooth, polynomially bounded function. Therefore, the local behaviour of s is only determined by the small scale behaviour of $\mathscr{T}$. In the following theorem we show that uniform Hölder regularity and similar regularity may be completely characterized by the decay of the modulus of the scale–space coefficients as the scale becomes small.

Theorem 2.1.1

Let $\mathscr{T}$ be some scale–space coefficients of s. Suppose that for large a, $\mathscr{T}$ is rapidly decreasing. Then in the limit $a \to 0$ we obtain the following classification:

$$|\mathscr{T}(b, a)| \leq ca^\alpha \Rightarrow s \in \Lambda^\alpha(\mathbb{R}),$$

$$|\mathscr{T}(b, a)| = o(a^\alpha) \Rightarrow s \in \lambda^\alpha(\mathbb{R}),$$

$$|\mathscr{T}(b, a)| \leq ca^\alpha \log^\beta a \Rightarrow s \in \Lambda_{\log}^{\alpha,\beta}(\mathbb{R}),$$

$$|\mathscr{T}(b, a)| = o(a^\alpha \log^\beta a) \Rightarrow s \in \lambda_{\log}^{\alpha,\beta}(\mathbb{R}).$$

The reconstruction wavelet should be compactly supported and $[\alpha] + 1$ *times*[1] *continuously differentiable.*

Together with Theorem 2.0.3, this shows that the uniform regularity with non-integral regularity exponent may be completely characterized through the small scale behaviour of the absolute value of the wavelet coefficients. The proof is again a corollary of the two more general lemmata that show that actually this kind of uniform local regularity analysis applies to submultiplicative regularities in general.

[1] We denote by $[t]$ the biggest integer $\leq t$.

Theorem 2.1.2

Let r be a non-negative, monotonic even submultiplicative function that satisfies, for some $n \in \mathbb{N}_0$,

$$\text{(i)} \quad \int_0^1 \frac{dt}{t^{1+n}} r(t) < \infty, \text{ and } r(t) = o(t^n) \qquad (t \to 0)$$

$$\text{(ii)} \quad \int_1^\infty \frac{dt}{t^{2+n}} r(t) < \infty, \text{ and } r(t) = o(t^{n+1}) \qquad (t \to \infty).$$

Let $\mathcal{T}$ *be some scale–space coefficients of some function s. Suppose* $\mathcal{T}(\cdot, a) = 0$, *for* $a > 1$. *Then if*

$$|\mathcal{T}(b, a)| \leq cr(a)$$

it follows that s is n-times continuously differentiable and its derivative satisfies uniformly at

$$|\partial^n s(t + u) - \partial^n(t)| \leq O(r(u)/u^n).$$

If the scale–space coefficients satisfy in addition at

$$|\mathcal{T}(b, a)| = o(r(a)) \qquad (a \to 0), \text{ uniformly in } b,$$

then we have uniformly in t

$$|\partial^n s(t + u) - \partial^n s(t)| = o(r(u)/u^n), \qquad (u \to 0).$$

In both cases the reconstruction wavelet is supposed to be compactly supported having $n + 1$ *continuous derivatives.*

Condition (i) ensures a certain decay at small scales. Condition (ii) ensures that this decay is not too fast. Indeed, because of the submultiplicativity of r we have $r(t) \geq c/r(1/t)$. Now by (ii) we have $t^{-1-n}r(t) \to 0$ as $t \to \infty$. Therefore, we have $t^{-n-1}r(t) \to \infty$ as $t \to 0$, or what amounts to the same,

$$t^{1+n} = o(r(t)), \qquad (t \to 0).$$

Thus in some sense conditions (i) and (ii) ensure that the local fluctuation is 'bounded away' from the polynomial behaviour, which would be $\sim t^n$ and $\sim t^{n+1}$.

Before we come to the proof we recall the schema of finite differences. Let Δ be the following operator:

$$\Delta: s(t) \mapsto t^{-1}(s(t) - s(0)), \text{ and set } \tilde{\Delta}^n s(t_0, t) = \tilde{\Delta}^n T_{-t_0} s(t).$$

Then we say that the nth difference-quotient of s exists at $t = 0$ (and by translation at t_0) iff

$$\Delta^m s(t) = \tilde{\Delta}^m s(0, t)$$

converges to some finite number as $t \to 0$ for $m = 1, \ldots, n$. The nth difference quotient exists at t_0 iff there is a polynomial of degree n such that

$$s(t_0 + t) = P_n(t) + r(t), \qquad (t \to 0)$$

with some $r(t) = o(t^n)$, or equivalently

$$|\tilde{\Delta}^n s(t_0, t) - \tilde{\Delta}^n s(t_0, 0)| \leq O(r(t)/t^n).$$

If this holds uniformly in t_0, then $\tilde{\Delta}^n s(t_0, t)$ may be replaced by $\partial^n s(t_0 + t)$. For later use we note the following commutation relations:

$$\tilde{\Delta}^n(T_b s)(t_0, t) = \tilde{\Delta} s(t_0 - b, t), \qquad \Delta^n D_a = a^{-n} D_a \Delta^n.$$

In particular, we have for $h_{b,a} = T_b D_a h$

$$\tilde{\Delta}^n h_{b,a}(t_0, t) = \frac{1}{a^n}\, \tilde{\Delta}^n h\left(\frac{t_0 - b}{a}, \frac{t}{a}\right).$$

Therefore, if h is n times continuously differentiable and compactly supported we have in particular for all $t \in \mathbb{R}$

$$\|\tilde{\Delta}^n h_{b,a}(\cdot, t)\|_{L^1(\mathbb{R})} \leq c a^{-n}, \tag{2.1.1}$$

with some $c > 0$.

Proof. By translation invariance it is enough to analyse s around 0. By an overall re-scaling we may suppose that the support of h is contained in $[-1/2, +1/2]$. We only need to consider the case $t > 0$, since the case $t < 0$ is analogous. By hypothesis on s we may write pointwise $(0 < t \leq 1)$

$$\begin{aligned} s(t) &= \left\{ \int_0^t \frac{da}{a} + \int_t^1 \frac{da}{a} \right\} \int_{-\infty}^{+\infty} db \frac{1}{a} h\left(\frac{t - b}{a}\right) \mathscr{T}(b, a) \\ &= X_1(t) + X_2(t). \end{aligned}$$

We now prove the O part.

X_1. Using the decrease of the wavelet coefficients $\mathscr{T} = O(r(a))$ at small scales we may write

$$\begin{aligned} |X_2| &\leq \int_0^t \frac{da}{a} \int_{-\infty}^{+\infty} db \left| \frac{1}{a} h\left(\frac{t - b}{a}\right) \right| |\mathscr{T}(b, a)| \\ &= O(1) \|h\|_{L^1(\mathbb{R})} \int_0^t \frac{da}{a} r(a) \\ &= O(r(t)). \end{aligned}$$

Because of condition (i) we have $r(t) = o(t^n)$ as $t \to 0$ and thus $X_1(t)$ has a finite differential quotient of order n at 0. In the last equation we have used the fact that because of the submultiplicativity of r we have

$$\int_0^t \frac{da}{a} r(a) = \int_0^1 \frac{da}{a} r(at) \leq O(r(t)) \int_0^1 \frac{da}{a} r(a) = O(r(t)). \qquad (2.1.2)$$

$\boldsymbol{X_2}$. Since h is $n+1$ times continuously differentiable we may write for $h_{b,a} = T_b D_a h$

$$\Delta^n h_{b,a}(t) - \Delta^n h_{b,a}(0) = t\Delta^{n+1} h_{b,a}(t).$$

In particular, from (2.1.1) we have

$$\int_{-\infty}^{+\infty} db \, |\Delta^n h_{b,a}(t) - \Delta^n h_{b,a}(0)| \leq O(ta^{-n-1}).$$

Thus we may write

$$\begin{aligned} |\Delta^n X_2(t) - \Delta^n X_2(0)| &\leq \int_t^1 \frac{da}{a} \int_{-\infty}^{+\infty} db \, |\Delta^n h_{b,a}(t) - \Delta^n h_{b,a}(0)| \, |\mathcal{T}(b, a)| \\ &\leq O(t) \int_t^1 \frac{da}{a} a^{-n-1} r(a) \\ &\leq O(r(t)/t^n). \end{aligned}$$

In the last equation we have again used the submultiplicativity of r via

$$\begin{aligned} t \int_t^1 \frac{da}{a^{2+n}} r(a) &= \int_1^{1/t} \frac{da}{a^{2+n}} r(at) t^{-n} \\ &\leq O(r(t)/t^n) \int_1^{1/t} \frac{da}{a^{2+n}} r(a) = O(r(t)/t^n). \qquad (2.1.3) \end{aligned}$$

The last estimate holds since, by hypothesis on r, the integral is convergent.

Altogether by translation invariance this shows that $\tilde{\Delta}^n s(t_0, t)$ satisfies the stated estimates uniformly in t_0.

To prove the o part of the lemma, we remark that it is enough to replace estimations (2.1.2) and (2.1.3) by their o version. Suppose, therefore, that v is a non-negative function that satisfies

$$v(t) = o(r(t)) \qquad (t \to 0).$$

Then for every $\epsilon > 0$ we can find an η, $0 < \eta < 1$, such that $|v(t)| < \epsilon r(t)$ whenever $0 \leq t \leq \eta$. Therefore, for t small enough,

$$\int_0^t \frac{da}{a} v(a) \leq \epsilon \int_0^t \frac{da}{a} r(a) \leq \epsilon r(t),$$

proving because ϵ was arbitrarily chosen, the o analog of (2.1.2).

For the analogue of (2.1.3) we obtain, for fixed ϵ and t small enough, with the same η as previously,

$$t\int_0^t \frac{da}{a^{2+n}}\, v(a) = t\left\{\int_t^\eta \int_\eta^1\right\} \frac{da}{a^{2+n}}\, v(a) = \sigma_1(t) + \sigma_2(t).$$

Now by submultiplicativity of r we have, as before,

$$\sigma_2(t) \leq O(t) \int_\eta^1 \frac{da}{a^{2+n}}\, v(a) \leq O(r(t)/t^n) \int_{\eta/t}^{1/t} \frac{da}{a^{2+n}}\, r(a).$$

By condition (ii), the value of the integral tends to 0 as $t \to \infty$ and thus $\sigma_2(t) = o(r(t)/t^n)$.

By hypothesis (ii) on r we have for t small enough, as in (2.1.3),

$$\sigma_1(t) \leq t \int_t^\eta \frac{da}{a^{2+n}}\, v(a) \leq \epsilon t \int_t^1 \frac{da}{a^{2+n}}\, r(a) \leq \epsilon O(r(t)/t^n).$$

Since ϵ was arbitrary the theorem follows. □

2.2 The class of Zygmund

Until now we have had to exclude the case $\Lambda^\alpha(\mathbb{R})$ with $\alpha \in \mathbb{N}$. We will now show that the appropriate global regularity class of functions that can be analysed by the wavelet transform in the case of integer exponents is the class of Zygmund.

Definition 2.2.1

A function s is in the class of Zygmund—written $\Lambda^*(\mathbb{R})$—if it is continuous and if it satisfies

$$|s(t_0 + t) + s(t_0 - t) - 2s(t_0)| \leq O(t) \qquad (t \to 0) \tag{2.2.1}$$

uniformly in t_0. We say that s is in $\lambda^*(\mathbb{R})$ if the same estimate holds with O replaced by o.

Functions in $\Lambda^*(\mathbb{R})$ do not have cusps, as can be seen from the identity

$$s(t_0 + t) + s(t_0 - t) - 2s(t_0) = (s(t_0 + t) - s(t_0)) - (s(t_0) - s(t_0 - t)).$$

Therefore, for example, the function $|t| \log|t|$ is not in $\Lambda^*(\mathbb{R})$. However, the function $t \log|t|$ is in the class of Zygmund. We now show how to characterize these regularity classes in wavelet space.

Theorem 2.2.2

Let s be in the class of Zygmund $\Lambda^*(\mathbb{R})$. Then

$$\mathscr{W}_g s(b, a) = O(a), \qquad (a \to 0).$$

If $s \in \lambda^*(\mathbb{R})$ then the same estimate holds with O replaced by o. We suppose that g is in $S_0(\mathbb{R})$.

Remark. The assumption on g is much too strong and is for technical convenience only. In the proof we will see how it can be relaxed.

Proof. First, suppose that g is even:

$$g(t) = g(-t).$$

Then we have by symmetry and $\int_{-\infty}^{+\infty} dt g(t) = 0$ as usual

$$\mathscr{W}_g s(b, a) = \int_{-\infty}^{+\infty} dt \frac{1}{a} \bar{g}\left(\frac{t}{a}\right) s(t + b)$$

$$\frac{1}{2} \int_{-\infty}^{+\infty} dt \frac{1}{a} \bar{g}\left(\frac{t}{a}\right) \{s(t + b) + s(-t + b) - 2s(b)\}.$$

Using the estimate (2.2.1) and the fact that s is bounded we may estimate the brases by $O(t)$ and the theorem follows. Suppose that h is not symmetric. The wavelet transform with respect to h is obtained from the one with respect to g by the action of the cross kernel (see Chapter 1, Section 20.1). This function is sufficiently localized and thus it does not change the decrease of the wavelet coefficients at small scale (see e.g. Chapter 4, Section 3).

The proof of the o part is analogous. □

But also the inverse theorem holds.

Theorem 2.2.3
Let s be a function with position-scale coefficients $\mathscr{T}$ supported by $a \leq 1$ that satisfy

$$\mathscr{T}(b, a) = O(a)$$

uniformly in b. Then $s \in \Lambda^*(\mathbb{R})$. If the same estimate holds with O replaced by o, then $s \in \lambda^*(\mathbb{R})$. The reconstruction wavelet should be two times continuously differentiable and compactly supported.

Proof. By translation invariance it is enough to estimate

$$s(t) + s(-t) - 2s(0).$$

To each term corresponds an integral over some influence cone of h. The integrals over the region $0 < a \leq t$ may be estimated as previously. In the only new term X_3, say, we may write, using the regularity of h,

$$h\left(\frac{t-b}{a}\right)+h\left(\frac{t-b}{a}\right)-2h\left(-\frac{b}{a}\right)=t^2\frac{1}{a^2}h''\left(\frac{\tilde{t}-b}{a}\right)$$

with some $\tilde{t}\in[0,t]$. We therefore end up with

$$X_3=\int_t^1\frac{da}{a}\int_{-\infty}^{+\infty}db\,\frac{1}{a}\left\{h\left(\frac{t-b}{a}\right)+h\left(\frac{t-b}{a}\right)-2h\left(-\frac{b}{a}\right)\right\}\mathcal{T}(b,a)$$

$$=O(t^2)\|h''\|_{L^1(\mathbb{R})}\int_t^1\frac{da}{a}a^{-1}=O(t).$$

The proof of the o part is similar. □

By what we have shown so far this implies, via the characterization of the Hölder spaces through wavelet transforms, the following for the regularity of the Zygmund class. We have

$$\Lambda^1(\mathbb{R})\Rightarrow\Lambda^*(\mathbb{R})\Rightarrow\Lambda^\alpha(\mathbb{R})$$

for $0<\alpha<1$. But, as follows easily from the proof of Theorem 2.1.2, we even have a little more precise information: a function in $\Lambda^*(\mathbb{R})$ satisfies

$$|s(t_0+t)-s(t_0)|=O(t\log t)\qquad(t\to 0).$$

Indeed, the logarithmic correction comes from the term X_2 in the proof of Theorem 2.1.3 (see also Corollary 2.3.3 below.)

2.3 Inverse theorems for local regularity

As we have seen, a local Hölder regularity implies a local decrease of the wavelet coefficients. On the other hand, as we just have proved, an overall decrease of the scale–space coefficients proves a global regularity of the reconstructed function. However, to prove pointwise local regularity through scale–space coefficients, we must suppose some stronger conditions.

Theorem 2.3.1

Let s be a scale–space representation $\mathcal{T}$ that satisfies at small scale $a<1$ for some $\gamma>0$:

(i) $\mathcal{T}(b,a)=O(a^\gamma)$ *uniformly in b,*
(ii) $\mathcal{T}(\tau_0+b,a)=O(a^\alpha)+O(b^\alpha\log^{-1}b)$ $(b,a\to 0)$.

At large scale we suppose that $\mathcal{T}$ is rapidly decaying. Then s has at τ_0 a local *regularity exponent α in the sense that*

$$|s(\tau_0 + u) - P_n(u)| = O(u^\alpha) \qquad (u \to 0),$$

with some polynomial of order n, $\alpha - 1 < n < \alpha$.

Again, this theorem holds for submultiplicative remainders in general.

Theorem 2.3.2
Let r be a non-negative, submultiplicative function that satisfies, for some $n \in \mathbb{N}_0$,

(i) $\displaystyle\int_0^1 \frac{dt}{t^{n+1}} r(t) < \infty$, *and* $r(t) = o(t^n) \qquad (t \to 0)$

(ii) $\displaystyle\int_1^\infty \frac{dt}{t^{n+2}} r(t) < \infty$, *and* $r(t) = o(t^{1+n}) \qquad (t \to \infty)$.

Let s have scale–space coefficients $\mathcal{T}$ supported by $a \leq 1$ that satisfy, with some $\gamma > 0$,

(iii) $\mathcal{T}(b, a) = O(a^\gamma)$ *uniformly in b,*
(iv) $\mathcal{T}(\tau_0 + b, a) = O(r(a)) + O(r(b)/\log r(b))$ $(b, a \to 0)$.

Then there is a polynomial P_n of degree n such that

$$s(\tau_0 + t) = P_n + O(r(t)), \qquad (t \to 0).$$

If we have, in addition,

(iv′) $\mathcal{T}(\tau_0 + b, a) = o(r(a)) + o(r(b)/\log r(b))$ $(b, a \to 0)$,

then

$$s(\tau_0 + t) = P_n(t) + o(r(t)), \qquad (t \to 0).$$

The reconstruction wavelet should be $n + 1$ times continuously differentiable and compactly supported.

Proof. We may suppose that $\gamma < 1$. Then with the help of the function $\eta(t) = r(t)^{2/\gamma}$ we may split the integral into several parts:

$$s(t) = \int_0^{\eta(t)} \frac{da}{a} \int_{-\infty}^{+\infty} db \frac{1}{a} h\left(\frac{t-b}{a}\right) \mathcal{T}(b, a) \qquad (X_1)$$

$$+ \int_{\eta(t)}^{t} \frac{da}{a} \int_{-\infty}^{+\infty} db \frac{1}{a} h\left(\frac{t-b}{a}\right) \mathcal{T}(b, a) \qquad (X_2)$$

$$+ \int_t^1 \frac{da}{a} \int_{-\infty}^{+\infty} db \frac{1}{a} h\left(\frac{t-b}{a}\right) \mathcal{T}(b, a). \qquad (X_3)$$

The contributions of these terms will be estimated independently.

X_1. Using the global Hölder regularity of s we may estimate $\mathcal{T} = O(a^\gamma)$, which leads us to

$$|X_1| \leq \int_0^{\eta(t)} \frac{da}{a} \int_{-\infty}^{+\infty} db \left|\frac{1}{a} h\left(\frac{t-b}{a}\right)\right| |\mathcal{T}(b, a)|$$

$$= O(1)\|h\|_{L^1(\mathbb{R})} \int_0^{\eta(t)} \frac{da}{a} a^\gamma.$$

By the choice of η we obtain $X_1 = O(r(t))$.

X_2. Since h is compactly supported we have under the integral $a \leq O(t)$ and $b \leq O(t)$. Therefore, we may estimate

$$|\mathcal{T}(b, a)| = O(r(a)) + O(r(t)/\log r(t)).$$

The contribution of the first term may be estimated as the term X_1 in the proof of Theorem 2.1.2 to be of order $O(r(t))$, and thus we end up with

$$|X_2| \leq \int_{\eta(t)}^{t} \frac{da}{a} \int_{-\infty}^{+\infty} db \left|\frac{1}{a} h\left(\frac{t-b}{a}\right)\right| |\mathcal{T}(b, a)|$$

$$= O(1) \frac{r(t)}{|\log r(t)|} \int_{\eta(t)}^{t} \frac{da}{a} \int_{-\infty}^{+\infty} db \left|\frac{1}{a} h\left(\frac{t-b}{a}\right)\right| + O(r(t))$$

$$= \|h\|_{L^1(\mathbb{R})} O(1) \frac{r(t)}{|\log r(t)|} \log\left(\frac{t}{\eta(t)}\right) + O(r(t))$$

$$= O(r(t)),$$

by choice of η and since, by hypothesis (ii), on r we have $\log(t) = O(\log(r(t))$.

X_3. This term may be treated in the same way as X_2 in the proof of Theorem 2.1.2, and the 'O'-part of the theorem is done.

The 'o' part follows by considerations as before. □

Corollary 2.3.3
If $\mathcal{T}$ satisfies (with $0 < \alpha \leq 1$)

$$\text{(i) } |\mathcal{T}(b, a)| \leq O(a^\gamma) \quad \textit{with some } \gamma > 0,$$

$$\text{(ii) } |\mathcal{T}(\tau_0 + b, a)| \leq O(b^\alpha + a^\alpha) \qquad (b, a \to 0),$$

then $|s(\tau_0 + t) - s(\tau_0)| = O(t^\alpha \log(t)), (t \to 0)$.

Proof. The logarithm is exactly the contribution of the second term in the previous demonstration. □

2.4 Pointwise differentiability and wavelet analysis

We now wish to study pointwise differentiability with the help of wavelet transforms. As we saw in Section 2.0, the wavelet transform of a polynomial bounded function s that is differentiable in τ_0 satisfies

$$\mathscr{T}(\tau_0 + b, a) = o(|b| + a).$$

The analysing wavelet should satisfy $g \in L^1(\mathbb{R})$, $tg \in L^1(\mathbb{R})$, and $\int g = \int tg = 0$. To generalize the setting we will once more consider fluctuations around polynomial approximations:

$$s(\tau_0 + t) = P_n(t) + o(t^n), \qquad (t \to 0),$$

or, what amounts to the same,

$$\Delta^n(t) = o(1), \qquad (t \to 0),$$

where $\Delta^{k+1}(t) = t^{-1}(\Delta^k(t) - \Delta^k(0))$ and $\Delta^o(t) = s(\tau_0 + t)$. In this case we say that the nth differential quotient of s exists in τ_0. In wavelet space this regularity implies a decrease of

$$\mathscr{W}_g s(b, a) = o(a^n + |b|^n), \qquad (b, a \to 0),$$

whenever the wavelet is sufficiently localized and its first n moments vanish (see Theorem 2.0.5).

Now we will prove an inverse theorem that relates the local decrease of the scale–space coefficients of the function to the differentiability of the function itself. We will see that it is not possible to prove the full inversion of the preceding, but a slightly stronger hypothesis on the wavelet side is needed to prove the differentiability of a function through its scale–space coefficients.

Theorem 2.4.1

If the scale–space coefficients $\mathscr{T}(b, a)$, $\mathscr{T} = 0$ for $a > 1$, of some function s satisfy

(i) $\mathscr{T}(b, a) = O(a^\gamma)$ *uniformly in* b,
(ii) $\mathscr{T}(\tau_0 + b, a) = O(r(a) + r(b))$,

with an arbitrary $\gamma > 0$ and with a non-negative, monotonic function $r = r(|a|)$ satisfying the condition of Dini,

$$\int_0^1 \frac{da}{a^{n+2}} r(a) < \infty,$$

then the nth differential quotient of s exists in τ_0. Furthermore, the condition on r is optimal. The reconstruction wavelet h should be $n + 1$ times continuously differentiable and compactly supported.

Proof. As usual, we suppose that, $\tau_0 = 0$, $t > 0$ and $\gamma < 1$. As in the proof of Theorem 2.3.1 we only need to estimate the small scale integral. Again, with the help of the function $\eta(t) = r(t)^{2/\gamma}$, we may split the reconstruction integral into several parts,

$$s(t) = \int_0^{\eta(t)} \frac{da}{a} \int_{-\infty}^{+\infty} db \frac{1}{a} h\left(\frac{t-b}{a}\right) \mathscr{T}(b, a) \qquad (X_1)$$

$$+ \int_{\eta(t)}^{t} \frac{da}{a} \int_{-\infty}^{+\infty} db \frac{1}{a} h\left(\frac{t-b}{a}\right) \mathscr{T}(b, a) \qquad (X_2)$$

$$+ \int_{t}^{1} \frac{da}{a} \int_{-\infty}^{+\infty} db \frac{1}{a} h\left(\frac{t-b}{a}\right) \mathscr{T}(b, a), \qquad (X_3)$$

which will be estimated independently.

As in the proof of Theorem 2.3.2 we have $X_1(t)$, $X_2(t) = O(r(t)) = o(t^n)$. Hence the nth differential quotient exists at $t = 0$ and we only need to consider $\Delta^n X_3$. As for X_2 in the proof of Theorem 2.1.2 we have, for $|\epsilon| \leq t$,

$$|\Delta^n X_3(t) - \Delta^n X_3(0)| \leq O(t) \int_t^1 \frac{da}{a^{2+n}} r(a)$$

and we have to show that this expression becomes arbitrarily small as $t \to 0$. Clearly, $\lim_{a \mapsto 0} a^{-n-1} r(a) = 0$ and therefore, given $\rho > 0$, we can find $\delta > 0$ such that $a^{-n-1} r(a) < \rho$ for $a < \delta$. Thus

$$t \int_t^1 \frac{da}{a} a^{-n-1} r(a) \leq t\rho \int_t^{\delta} \frac{da}{a} + t \int_{\delta}^1 \frac{da}{a} a^{-n} r(a).$$

The second term goes to 0 with t and the first term can be made arbitrarily small since ρ was arbitrary. Therefore, the first part of the theorem is proved.

That the condition on r is actually the weakest possible can be seen as follows. We limit ourselves to $n = 1$; the general case is similar. Consider a function g in the Schwarz-class whose Fourier transform is supported by $[1, 2]$ such that the functions $2^{j/2} g(2^j t)$, $j = 1, 2, \ldots$ are an orthonormal set. In addition, we suppose that $g'(0) = 1$. Then let

$$s(t) = \sum_{j=1}^{\infty} 2^{-j} \varrho(2^{-j}) g(2^j t)$$

with a continuous, positive, monotonic, bounded function ϱ. The wavelet transform of s with respect to g satisfies

$$\mathscr{W}_g s(0, 2^{-j}) = 2^{-j} \varrho(2^{-j}), \qquad j = 1, 2, \ldots.$$

Now s is differentiable for $t \neq 0$ and therefore $\Delta(t) = t^{-1}(s(t) - s(0))$ can be written as

$$\Delta(t) = \sum_{j=1}^{\infty} \varrho(2^{-j})h'(2^j\tau),$$

with some $\tau = o(1)$ for $t \to 0$. Now ϱ satisfies a condition of *Dini* if and only if

$$\sum_{j=1}^{\infty} \varrho(2^{-j}) < \infty$$

and therefore this same condition implies the differentiability of s at $t = 0$.

On the other hand, suppose that s is differentiable at $t = 0$. By an overall dilation we may suppose that $g'(t) \sim 1$ for $t \in [-1, 1]$. Then

$$\Delta(t) \sim \sum_{2^{-j} \geq \tau} \varrho(2^{-j}) + \sum_{2^{-j} < \tau} \varrho(2^{-j})g'(2^j\tau) = X_1 + X_2.$$

Since ϱ is uniformly bounded and since we have

$$\sum_{2^{-j} < \tau} |g'(2^j\tau)| \leq \sup_{\xi \in [1/2, 1]} \sum_{j=1}^{\infty} |g'(2^j\xi)| < \infty,$$

it follows that X_2 stays bounded as $t \to 0$. The same must then be true for X_1 since $\Delta(t)$ stays bounded. This shows that ϱ satisfies a condition of *Dini* and so the wavelet transform of s satisfies the condition of Theorem 2.4.1 if and only if s is differentiable in 0. □

We may even use the wavelet transform to compute the value of the derivative, if it exists. The following theorem is a generalization of a theorem due to Fatou.

Theorem 2.4.2

Let s be a periodic function or a measure. Suppose that at τ_0 the nth derivative $\partial_t s(\tau_0)$ exists.[2] *Let $g \in L^1(\mathbb{R})$ and $t^n g \in L^1(\mathbb{R})$ be an analysing wavelet that satisfies*

(i) $\displaystyle\int_{-\infty}^{+\infty} dt\, t^k g(t) = 0$, *for* $k = 0, \ldots, n-1$,

(ii) $\displaystyle\int_{-\infty}^{+\infty} dt\, t^n g(t) = n! \Leftrightarrow (i\partial)^n \hat{g}(0) = 2\pi n!$,

then

$$\lim_{a \to 0} a^{-n} \mathcal{W}_g s(\tau_0, a) = \partial_t^n s(\tau_0).$$

[2] By this we mean that the finite difference quotients have a limit.

Proof. Writing $s(t) = P_{n-1}(t) + \partial_t^n s(\tau_0)t^n/n! + o(t^n)$ we obtain

$$\mathcal{W}_g s(b, a) = \int_{-\infty}^{+\infty} dt \, \frac{1}{a} \bar{g}\left(\frac{t-b}{a}\right)(\partial_t^n s(\tau_0)t^n/n! + o(t^n))$$

$$= a^n \partial_t s(\tau_0) + o(a^n),$$

and the theorem follows. □

2.5 The class W^α

As we have seen, global regularity can be characterized by a certain decrease of the modulus of the wavelet coefficients at small scale. Local regularity in the sense of fluctuations around a polynomial approximation could not be completely characterized: a little more is needed on the wavelet side. However, it is possible to define classes of local 'wavelet regularity' that are characterized through the modulus of the wavelet coefficients. Let r be a monotonic, non-negative, submultiplicative function that satisfies some global estimate of the form

$$r(t) \leq O(1 + t^2)^{\gamma/2}$$

for some $\gamma > 0$.

Definition 2.5.1
We then say that a function (or distribution) s is of local W^r regularity at τ_0 iff for some admissible wavelet $g \in S_0(\mathbb{R})$ we have

$$|\mathcal{W}_g s(\tau_0 + b, a)| \leq O(r(b) + r(a)), \qquad (b, a \to 0).$$

This is well defined since we have the following theorem.

Theorem 2.5.2
The definition does not depend on the analysing wavelet: if s is of regularity W^r with respect to some admissible $g \in S_0(\mathbb{R})$ it is of the same regularity with respect to any other $h \in S_0(\mathbb{R})$.

Proof. Indeed, the wavelet transform of s with respect to $h \in S_0(\mathbb{R})$ is obtained through convolution over the half-plane with $\Pi = c_g^{-1}\mathcal{W}_h g$, which is a highly localized function. Therefore, we may exchange the integrals in the following expression and obtain

$$\int_0^\infty \frac{da'}{a'} \int_{-\infty}^{+\infty} db' \, \frac{1}{a'} \Pi\left(\frac{b-b'}{a'}, \frac{a}{a'}\right) r(b') = G_a * r(b),$$

with $G_a(t) = G(t/a)/a$ and

$$G(t) = \int_0^\infty \frac{da'}{a'} \Pi(t/a', 1/a').$$

This function is again in $S_0(\mathbb{R})$ and thus by Lemma 2.0.6 this first expression is estimated by $O(r(b) + r(a))$.

In the same way we have

$$\int_0^\infty \frac{da'}{a'} \int_{-\infty}^{+\infty} db' \frac{1}{a'} \Pi\left(\frac{b-b'}{a'}, \frac{a}{a'}\right) r(a') = \int_0^\infty \frac{da'}{a'} H(a/a') r(a'),$$

where

$$H(a) = \int_{-\infty}^{+\infty} db\, \Pi(b, a).$$

This function is highly localized in the sense that

$$|H(a)| \leq O((a + 1/a)^{-\alpha})$$

for all $\alpha > 0$. We therefore have

$$\begin{aligned} \int_0^\infty \frac{da'}{a'} |H(a/a')| r(a') &= \int_0^\infty \frac{da'}{a'} H(1/a') r(aa') \\ &\leq O(r(a)) \int_0^\infty \frac{da'}{a'} H(1/a') r(a'), \end{aligned}$$

which is again $O(r(a))$ because the integral is convergent. □

3 Asymptotic behaviour at small scales

In this section we give some asymptotic expansion of the small scale behaviour of the wavelet coefficients. In particularly, we will give a sufficient condition that s satisfies locally the exact scaling of Definition 1.0.2. Asymptotic expansions may be useful because in general the wavelet transform at a given point in the half-plane depends on the analysed function as well as on the analysing wavelet, and in general it is difficult to say what is responsible for what. However, in the asymptotic limit of small length scales, as we shall see, we can clearly distinguish the influence of the analysed function and that of the analysing wavelet.

We recall the definition of an asymptotic series expansion of a function s. Let r_n, $n = 0, 1, \ldots$ be a family of functions that satisfy

$$r_{n+1}(t) = o(r_n(t) \qquad (t \to 0).$$

For instance, the family $r_n(t) = t^n$ will work. Then we say s has an asymptotic expansion of order N, in terms of r_n, if there are coefficients c_n such that we have

$$s(t_0 + t) = s(t_0) + \sum_{n=1}^{N} c_n r_n(t) + o(r_N(t)) \qquad (t \to 0),$$

and we write

$$s(t_0 + t) \simeq s(t_0) + \sum_{n=1}^{N} c_n r_n(t) \qquad (t \to 0).$$

We write

$$s(t_0 + t) \simeq s(t_0) + \sum_{n=1}^{\infty} c_n r_n(t) \qquad (t \to 0)$$

if N may be chosen arbitrarily large. Note that the infinite series may diverge everywhere except for $t = 0$. If a function s has an asymptotic expansion (finite or infinite) the expansion coefficients c_n are uniquely determined by s.

As an immediate application of the theorems of the preceding section and the explicit formula of Chapter 1, Section 29 we find that the following theorem holds.

Theorem 3.0.1

Let s be a polynomial bounded function. Suppose that s has the following infinite asymptotic expansion around t_0:

$$s(t_0 + t) \simeq s(t_0) + \sum_{n=1}^{\infty} c_{+,n}|t|_+^{\alpha_n} + c_{-,n}|t|_-^{\alpha_n} + \sum_{n=1}^{\infty} \gamma_n t^n,$$

with $\alpha_n \notin \mathbb{N}_0$, and $\Re\alpha_n$ monotonic growing in n and not constant as $n \to \infty$ and some arbitrary constants γ_n. Then the wavelet transform satisfies the following asymptotic expansion:

$$\mathcal{W}_g s(b, a) = \sum_{n=1}^{\infty} (c_{+,n} + c_{-,n}\, e^{i\pi\alpha_n}) \mathrm{U}(\alpha_n, b/a) a^{\alpha_n},$$

with

$$\mathrm{U}(\alpha, u) = \int_0^{\infty} dt\, t^{\alpha} \bar{g}(t - u).$$

Vice-versa, suppose this expansion holds. If $s \in \Lambda^{\epsilon}(\mathbb{R})$ for some $\epsilon > 0$ it follows that s has the above local asymptotic expansion. The wavelet g is supposed to be progressive and highly time frequency localized, $g \in S_+(\mathbb{R})$.

Thus the constants α_n, $c_{+,n}$, $c_{-,n}$, and β_n may be determined from the asymptotic behaviour of the wavelet transform. The polynomial behaviour γ_n, however, is invisible in wavelet space.

Let us look at the principal part of the expansion ($\alpha = \alpha_1$):

$$\mathscr{W}_g s(t_0 + b, a) \simeq a^\alpha U_\alpha(b/a),$$

$$U_\alpha(u) = (c_+ - c_- \, e^{i\pi\alpha}) \int_0^\infty dt \, t^\alpha \bar{g}(t - u)$$

$$= -\frac{i\Gamma(\alpha + 1)}{2\pi}(c_+ \, e^{-i\pi\alpha/2} - c_- \, e^{i\pi\alpha/2}) \int_0^\infty d\omega \, \omega^{-\alpha - 1} \bar{\hat{g}}(\omega) \, e^{i\omega u}.$$

It corresponds to the local cusp

$$s(t_0 + t) \simeq s(t_0) + P_n(t) + c_+|t|_+^\alpha + c_-|t|_-^\alpha,$$

where P is some polynomial.

First suppose that α is real valued. Then if $\alpha \notin \mathbb{N}_0$ or if $\alpha \in \mathbb{N}_0$ but $c_- \neq (-1)^\alpha c_+$ the constants α, c_-, and c_+ can always be recovered from the asymptotic form of the wavelet coefficients. Upon a translation we may suppose that $t_0 = 0$. We then simply choose a line $b/a = c$ in the half-plane, passing through the location of the singularity such that for all α we have

$$\int_0^\infty dt \, t^\alpha \bar{g}(t - c) \neq 0.$$

We say in this case that the zoom $b/a = c$ is allowed. For many wavelets every zoom will work; in the general case, however, it should be checked. To fix the ideas we suppose that $b/a = 0$ is an allowed zoom. We thus have at small scale

$$\mathscr{W}_g s(0, a) \simeq c \, e^{i\phi} \, a^\alpha \qquad (a \to 0), \quad c \geq 0.$$

If either c_+ or c_- is different from 0 then $c \neq 0$ and we obtain, in a double logarithmic representation presenting $\log(|\mathscr{W}_g s|)$ as a function of $\log a$ along this line asymptotically, a straight line whose slope reveals the exponent α as sketched in Figure 4.1.

The small scale behaviour of the phase along this line is related to the phase of the complex number $c_+ \, e^{+i\alpha\pi/2} - c_- \, e^{-i\alpha\pi/2}$. Thus it determines the relative size of the constants c_+. Therefore, it gives the qualitative aspect of the local singularity, that is, up to re-scaling of the singularity. For simplicity let us suppose that $\hat{g}$ is real valued and positive. Then

$$\arg(c_+ \, e^{+i\alpha\pi/2} - c_- \, e^{-i\alpha\pi/2}) = \phi - \pi/2.$$

In Figure 4.2 we have shown different local geometries for various phase values. The constant c determines the size of c_+, c_- via

$$|c_+ \, e^{+i\alpha\pi/2} - c_- \, e^{-i\alpha\pi/2}| = c.$$

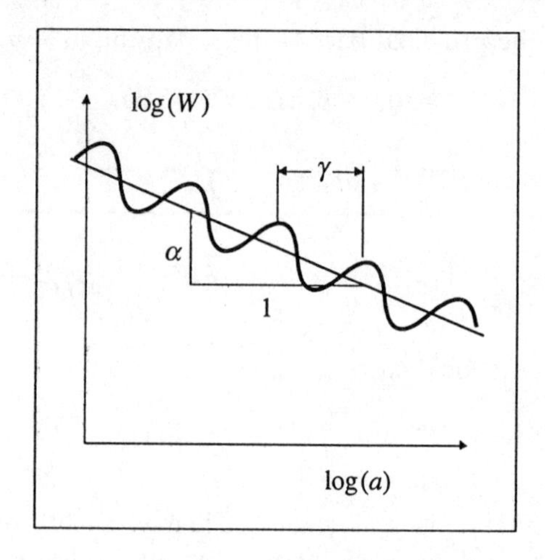

Fig. 4.1 The asymptotic behaviour of the modulus of an exact scaling singularity in a log–log representation. The slope gives the scaling exponent.

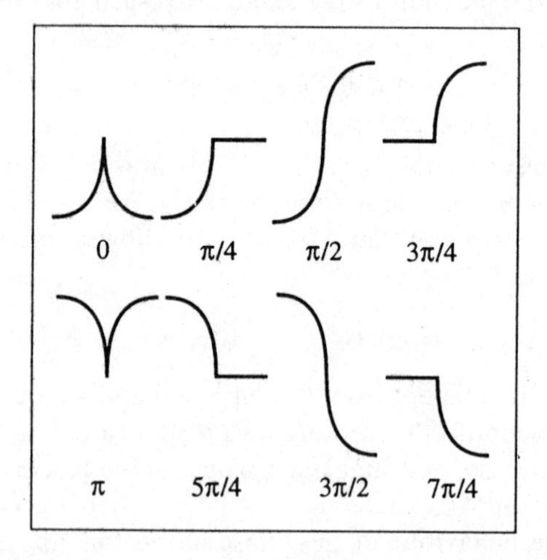

Fig. 4.2 The small scale value of the phase gives the local geometry of a singularity.

In the case where the exponent α is complex valued the modulus will show oscillations that are related to the imaginary part of α (see Fig. 4.1).

4 The Brownian motion

As a first application of the regularity analysis through wavelet transforms we analyse the regularity of a typical trajectory of a Brownian motion.

In particular, we wish to re-demonstrate Lévi's theorem, which states that with probability 1 it is of regularity $\Lambda_{\log}^{1/2, 1/2}$. Consider a modulated delta-comb

$$\gamma_\lambda(t) = \sum_{n=-\infty}^{+\infty} c_\lambda(n)\delta(t - \lambda n), \tag{4.0.1}$$

where the amplitudes c_λ are independent, real-valued random variables. Each of them is distributed according to a Gaussian law with mean value $\mu = 0$ and variance $\sigma_\lambda^2 > 0$:

$$\text{Proba}\{c_\lambda \leq t\} = \int_{-\infty}^{t} d\xi \, \frac{1}{\sqrt{2\pi\sigma}} e^{-\xi^2/2\sigma_\lambda} \, d\xi.$$

Consider a rapidly decaying continuous function. The action of the random measure of (4.0.1) on s it itself a random variable

$$\gamma_\lambda(s) = \sum_{n=-\infty}^{+\infty} s(\lambda n).$$

Because the sum of Gaussian random variables with mean values μ_1 and μ_2 and variances σ_1^2 and σ_2^2 is again a Gaussian random variable with mean value $\mu = \mu_1 + \mu_2$ and variance $\sigma^2 = \sigma_1^2 + \sigma_2^2$, we have that $\mu_\lambda(s)$ is distributed with a Gaussian law with $\mu = 0$ and variance

$$\sigma^2 = \sigma_\lambda^2 \sum_{n=-\infty}^{+\infty} |s(\lambda n)|^2.$$

Therefore, if we chose $\sigma_\lambda = \sqrt{\lambda}$, the distribution of the random variable γ_λ will tend as $\lambda \to 0$ to a Gaussian distribution with

$$\mu = 0, \qquad \sigma^2 = \int_{-\infty}^{+\infty} dt \, |s(t)|^2.$$

The limit random measure $\gamma_\lambda \to W$ is known as the Wiener measure, or white noise. Consider now the primitive of the measures γ_λ:

$$F_\gamma(t) = \int_0^t \gamma_\lambda(du) = \sum_{0 < \lambda n \leq t} c_\lambda(n).$$

This function can be interpreted as a random walk with independent random increments at all points λn, $n \in \mathbb{Z}$. The limit random walk $\lambda \to 0$ exists and is known as Brownian motion:

$$B(t) = \int_0^t W(dt).$$

Now consider wavelet coefficients of the Wiener measure with respect to some orthonormal wavelet $g \in S_0(\mathbb{R})$:

$$\mathcal{W}_{j,k} = \mathcal{W}_g(k2^{-j}, 2^{-j}) = \int_{-\infty}^{+\infty} W(dt) 2^j g(2^j t - k).$$

These numbers are again random variables. As we have seen already, in Section 21 of Chapter 1, the correlation between two values of wavelet transform is given by the reproducing kernel. Therefore, in the case of orthonormal wavelets these random variables are independent. Each of them has a Gaussian distribution with

$$\mu_{j,k} = 0, \qquad \sigma_{j,k}^2 = 2^j.$$

The absolute value $|\mathcal{W}_{j,k}|$ has a χ distribution with density

$$\text{Proba}\{|\mathcal{W}_{j,k}| \le t\} = \sqrt{\frac{2}{\pi\sigma_{j,k}}} \int_0^t du\, e^{-u^2/2\sigma_{j,k}} = \sqrt{\frac{2}{\pi}} \int_0^{2^{-j/2}t} du\, e^{-u^2/2}.$$

We claim that from this it follows that, given $c > -\log 2$ and given any interval $I \subset \mathbb{R}$, we can find with probability 1 an integer j_0 such that

$$|\mathcal{W}_{j,k}| \le \sqrt{c|j|} 2^{j/2} \tag{4.0.2}$$

for all indices j, k such that $2^{-j}k \in I$ and $j \ge j_0$. Translated into continuous wavelet transform language this implies that almost surely we have

$$\mathcal{W}_g W(b, a) = O(a^{-1/2} \log^{1/2}(a)), \qquad (a \to 0).$$

Thus for Brownian motion—the primitive of W is essentially obtained through multiplication by a in wavelet space (compare Section 26.1 of Chapter 1)—we have then proved the following theorem.

Theorem 4.0.1

Brownian motion is, with probability 1, *in the class* $\Lambda_{\log}^{1/2, 1/2}$, *namely*

$$\textit{Proba}\left\{\limsup_{u \to +0} \frac{|B(t+u) - B(t)|}{\sqrt{u \log(1/u)}} < \infty \textit{ holds for all } t \in [0, 1]\right\} = 1.$$

Proof. It remains to show that the assertion of (4.0.2) holds with probability 1 for j large enough and $k2^{-j}$ in some interval. By dilation co-variance it is enough to consider the interval $[0, 1)$. Here we need the famous 0, 1 law of probability theory. Roughly, it states the following: let $x_1, x_2, \ldots$ be an infinite family of independent random variables. Suppose Λ is an event that depends only on the infinite tail, namely, on $x_n, x_{n+1}, \ldots$ for n arbitrarily large. Then either Λ occurs with probability 0 or with probability 1.

We start by noting that the random variables $|\mathcal{W}_{j,k}|$ with $2^{-j}k \in [0, 1)$, $j \le 0$, may be enumerated in such a way that the small scale behaviour is in the infinite tail of the sequence. Consider now the event that all numbers $|\mathcal{W}_{j,k}|$ satisfy the inequality above for all j larger than some j_0. This event clearly depends only on the infinite tail of the ordered sequence of random variables. Its probability is therefore either 0 or 1. Let us compute the probability that $j_0 = 1$. This probability is a lower bound for the former probability. It is given by the infinite product

$$\text{Proba}\{|\mathcal{W}_{j,k}| \leq \sqrt{cj}\,2^{j/2}\} = \prod_{j=1}^{\infty}\left\{\sqrt{\frac{2}{\pi}}\int_0^{\sqrt{cj}} du\, e^{-u^2/2}\right\}^{2^j}.$$

Here we have used the fact that for a given **j** the 2^j random variables $|\mathcal{W}_{j,k}|$ with $2^{-j}k \in [0, 1)$ have the same probability law. From the asymptotic form of the integral,

$$\int_t^{\infty} du\, e^{-u^2/2} = O\left(\frac{e^{-t^2}}{t}\right) \qquad (t \to \infty),$$

we see that the infinite product is convergent. Indeed, convergence of the product is assured by the absolute convergence of the series:

$$\sum_{j=1}^{\infty} \frac{1}{\sqrt{j}} 2^j e^{-cj} = \sum_{j=1}^{\infty} \frac{1}{\sqrt{j}} e^{-(c+\log 2)|j|}, \qquad c > -\log 2.$$

Therefore, for all $c > \log(1/2)$ we can almost surely find some j_0 such that the required estimation holds for all scales smaller than 2^{-j_0}. □

5 The Weierstrass non-differentiable function

In the year 1872, Weierstrass introduced his famous function (see Figure 5.1)

$$\sigma(t) = \sum_{n=1}^{\infty} \alpha^n \cos(\beta^n t), \qquad 0 < \alpha < 1.$$

He showed that this function is continuous but nowhere differentiable whenever the product $\alpha \cdot \beta$ exceeds a certain value. Later, Hardy showed that $\alpha \cdot \beta > 1$ is all that is needed. We now wish to re-prove Hardy's result with the help of the wavelet transform. It will become a two-line proof.

By computation in Fourier space we obtain

$$\mathcal{W}_g \sigma(b, a) = \sum_{n=1}^{\infty} \alpha^n \hat{g}(\beta^n a)\, e^{i\beta^n b}.$$

We may choose $g \in S_+(\mathbb{R})$ in such a way that $\operatorname{supp} \hat{g} \subset [1, \beta]$. Then the different frequencies de-couple and we obtain

$$|\mathcal{W}_g \sigma(b, a)| = \sum_{n=1}^{\infty} \alpha^n |\hat{g}(\beta^n a)|.$$

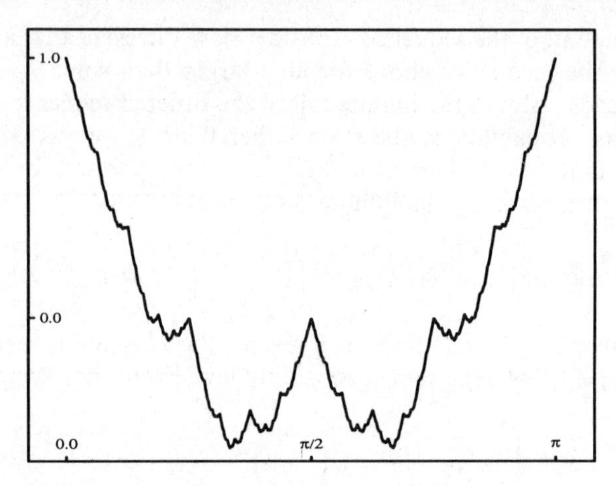

Fig. 5.1 Weierstrass function for $\beta = 2$ and $\alpha = 1/2$.

Choosing $a_m = \beta^{-m}$ we see that this function does not decay with $o(a)$, thereby showing the non-differentiability of σ. Instead, we have,

$$\mathscr{W}_g\sigma(b, a) = O(a^{\log\beta/\log\alpha})$$

and this is the best possible estimation. Therefore, $\sigma \in \Lambda^{\log\beta/\log\alpha}(\mathbb{R})$ but not in $\lambda^{\log\beta/\log\alpha}(\mathbb{R})$. In particular, σ is nowhere differentiable.

We may even choose $\alpha \cdot \beta = 1$. In this case we find that $\sigma \in \Lambda^*(\mathbb{R})$ but not in $\lambda^*(\mathbb{R})$, and whence σ is still not differentiable.

Remark. Clearly, the same argument applies to other lacunary Fourier series, that is, for sums of the kind

$$\sum_{k=1}^{\infty} \gamma_k \cos(\lambda_k t),$$

with $\lambda_k/\lambda_{k+1} > 1 + \epsilon$, $\epsilon > 0$, for all k.

6 The Riemann–Weierstrass function

The history of non-differentiable functions even goes back to Riemann, who is reported to have proposed the function

$$W(t) = \sum_{n=1}^{\infty} n^{-2} \sin(\pi n^2 t),$$

which is less lacunar than the Weierstrass function, as a continuous but nowhere differentiable function. However, neither he nor Weirstrass succeeded in proving this. However, Hardy later showed that the function is not differentiable at any irrational point and at some specific rationals. Later, Gerver and Queffelec solved the problem of the differentiability of W completely. In particular, they showed that W is differentiable at any rational point P/Q with $P = Q = = 1 \bmod 2$.

We now generalize the setting by looking at the following family of functions:

$$W_\beta(t) = \frac{2}{\pi^\beta} \sum_{n=1}^{\infty} n^{-2\beta} e^{i\pi n^2 t}, \qquad \text{for } \beta > 1/2. \tag{6.0.1}$$

Look at Figures 6.1 and 6.2, where we have shown the real and imaginary parts of W_1.

In this section we will apply the results of the previous sections to the analysis of the fractal Riemann–Weierstrass function. We will show that this fractal function has a countable infinity of points where it is asymptotically self-similar in the ESC sense due to local cusps. We will compute the asymptotic behaviour of these cusps completely. In addition, we will show that there is another countable infinite set of points where this function is differentiable for $\beta > 3/4$. And finally, on a third set, we will show that the Riemann–Weierstrass function has a local re-normalization. We will essentially follow Holschneider (1988) and Holschneider and Tchamitchian (1991).

By direct computation we can verify that the following scale–space representation for W_β holds:

$$W_\beta(t) = \int_0^\infty \frac{da}{a} \int_{-\infty}^{+\infty} db \, \frac{1}{a} h\left(\frac{t-b}{a}\right) \mathscr{T}(b + ia), \tag{6.0.2}$$

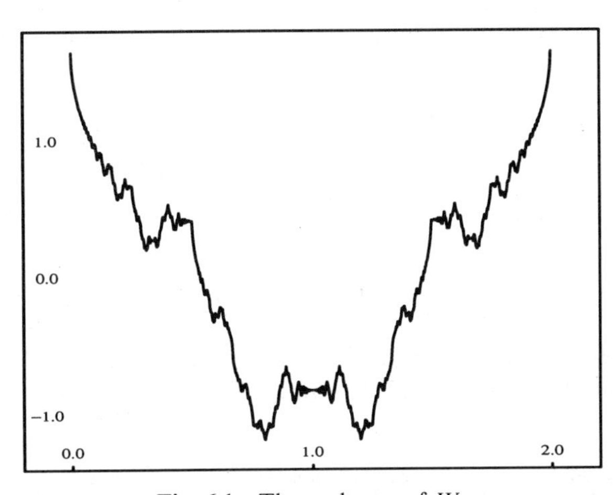

Fig. 6.1 The real part of W_1.

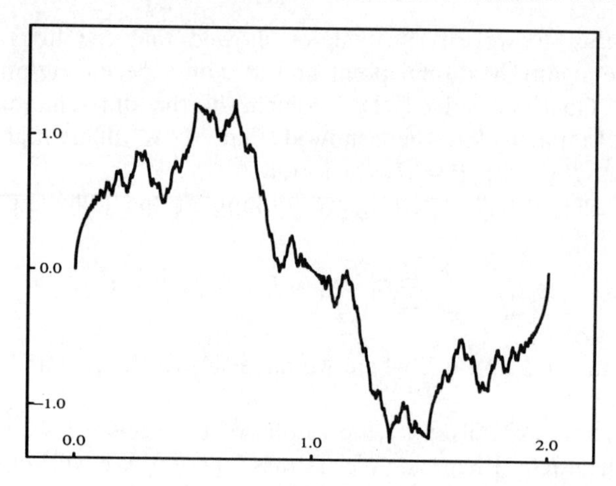

Fig. 6.2 The imaginary part of W_1.

with $\mathscr{T}(\tau) = (\Im\tau)^\beta \vartheta(\tau)$, $\tau = b + ia$, $\Im\tau = a$ and

$$\vartheta(\tau) = 1 + 2 \sum_{n=1}^{\infty} e^{i\pi n^2 \tau}, \qquad \text{for } \Im\tau > 0, \tag{6.0.3}$$

a *Jacobi* theta function and h, any integrable function, that satisfies

$$\int_0^\infty \frac{da}{a} \hat{h}(a) a^\beta e^{-a} = 1 \quad \text{and} \quad \hat{h}(0) = 0. \tag{6.0.4}$$

Thus the analysis of the local behaviour of W_β is transformed into the investigation of the behaviour of an analytic function near the boundary of its domain of analyticity.

The fact that analytic functions appear as scale–space representations of functions is not surprising if we consider the wavelet

$$\phi(t) = \frac{\Gamma(\beta + 1)}{(1 - it)^{\beta+1}}, \qquad \text{for fixed } \beta > 0.$$

Looking at the family of its dilates and translates we see that we can write[3]

$$\frac{1}{a} \bar{\phi}\left(\frac{t - b}{a}\right) = (\Im\tau)^\beta \Gamma(\beta + 1) \left(\frac{i}{\tau - t}\right)^{\beta+1}, \qquad \tau = b + ia.$$

Thus a wavelet analysis of a bounded function with respect to these wavelets yields—up to the pre-factor $(\Im\tau)^\beta$—an analytic function over the complex

[3] In all formulas involving powers of complex numbers we use the convention that the argument of a complex number is continuous, with the exception of the negative real axis.

upper half-plane. Condition (6.0.4) says that h is a reconstruction wavelet for ϕ.

The wavelet transform of W_β with respect to the wavelet ϕ is given by

$$\mathscr{W}[\phi, W_\beta](b, a) = \mathscr{W}_\phi W_\beta(\tau) = (\Im\tau)^\beta(\vartheta(\tau) - 1). \tag{6.0.5}$$

Note, however, that the analysing wavelet does not satisfy the hypothesis of Theorem 2.4.1. Therefore, the apparent decrease of at most $O(a^\beta)$ does not tell us anything about the local differentiability properties of W_β. However, since it is known that (e.g. Mumford 1983)

$$\vartheta(\tau) = O((\Im\tau)^{-1/2}), \tag{6.0.6}$$

it follows from Theorem 2.3.1 as a first result that W_β satisfies uniformly a Hölder condition with regularity exponent $\beta - 1/2$ for $\beta > 1/2$ and $2\beta \notin 2\mathbb{N}$. In the latter case we have that W_β—or some derivative of it—is in the class of Zygmund $\Lambda^*(\mathbb{R})$.

We now come to the analysis of ϑ near the real axis. An efficient tool is the modular group acting on the upper half-plane. Let us introduce the following transformations of the complex plane:

$$g\colon \tau \mapsto \frac{a + bz}{c + dz}, \qquad \text{with } a, b, c, d \in \mathbb{R},\ ad - bc = 1.$$

These transformations form the modular group[4] G. Each element is a meromorphic map that leaves invariant the upper and lower half-planes, the real line, and the rational numbers. We write $g_1 g_2$ for the composition map $\tau \mapsto g_1(g_2(\tau))$.

Consider the two elements of the modular group:

$$K\colon \tau \to \tau + 1 \quad \text{and} \quad U\colon \tau \to -\frac{1}{\tau}.$$

The first transformation is nothing other than the translation, whereas the second is the inversion. These two transformations generate G, that is, every element $g \in G$ can be written as

$$g = K^{n_N} U K^{n_{N-1}} U \cdots K^{n_1}.$$

However, this decomposition is not unique. The geometry of these transformations becomes clearer if one considers the circles and the straight lines lying at least partially in the half-plane. Every circle or straight line is mapped again on a circle or a straight line.

For later reference note that the action of K and U on the real line looks,

[4] For simplicity we do not distinguish between the group and its faithful representation.

in the continuous fraction representation,

$$t \leftrightarrow \alpha_0 + \cfrac{1}{a_1 + \cfrac{1}{a_2 + \cdots}} \leftrightarrow [a_0, a_1, a_2, \ldots]$$

as follows:

$$\left.\begin{aligned} K&: [\alpha_0, \alpha_1, \alpha_2, \ldots] \to [\alpha_0 + 2, \alpha_1, \alpha_2, \ldots], \\ U&: [\alpha_0, \alpha_1, \alpha_2, \ldots] \to [0, -\alpha_0, -\alpha_1, -\alpha_2, \ldots]. \end{aligned}\right\} \tag{6.0.7}$$

This means that U acts essentially as a shift in the continuous fraction expansion.

The important point is that the Jacobi theta function is co-variant under a subgroup of the modular group.

Lemma 6.0.1
The following transformation formulas for ϑ are valid:

$$\vartheta(K^2\tau) = \vartheta(\tau), \qquad \vartheta(U\tau) = \sqrt{-i\tau}\,\vartheta(\tau). \tag{6.0.8}$$

Proof. The first is completely clear of the Fourier series of ϑ, and merely states that ϑ is a function over the half-cylinder. The second follows from the Poisson summation formula over $\mathbb{R}$:

$$\sum_{n \in \mathbb{Z}} s(n) = \sum_{n \in \mathbb{Z}} \hat{s}(2\pi n)$$

applied to the Fourier transform pair ($\Im z > 0$)

$$s(t) = e^{i\pi z t^2} \leftrightarrow \hat{s}(\omega) = \sqrt{\frac{1}{-iz}}\, e^{-i\omega^2/4\pi z}. \qquad \square$$

The subgroup generated by K^2 and U is called the *theta* group G_ϑ. The transformation of ϑ under any element $g \in G_\vartheta$ can be determined from

$$\vartheta(g(\tau)) = f_g(\tau)\vartheta(\tau),$$

where the multiplier f_g is determined by

$$\left.\begin{aligned} &\text{(i) } g_1, g_2 \in G_\vartheta \Rightarrow f_{g_1 g_2}(\tau) = f_{g_1}(g_2(\tau)) g_2(\tau), \\ &\text{(ii) } f_{K^2}(\tau) = 1, \quad f_U(\tau) = \sqrt{-i\tau}. \end{aligned}\right\} \tag{6.0.9}$$

The first condition (i) is similar to the chain rule of derivation. We therefore conclude, since $f_{K^2}(z) = \partial_z K^2(z)$ and $f_U(z) = \zeta(\partial_z U(z))^{-1/4}$ with some $\zeta^8 = 1$,

that

$$f_g(z) = \zeta(\partial_z g(z))^{-1/4}, \quad \text{with some } \zeta^8 = 1. \tag{6.0.10}$$

Now consider two subsets S_0, S_1, $S_0 \cup S_1 = \mathbb{Q}$ of the rational numbers $\mathbb{Q}$:

$$S_0 = \left\{\frac{2n+1}{2m} : n, m \in \mathbb{Z}\right\} \cup \left\{\frac{2m}{2n+1} : n, m \in \mathbb{Z}\right\},$$

$$S_1 = \left\{\frac{2n+1}{2m+1} : n, m \in \mathbb{Z}\right\}.$$

It turns out that the first set is the orbit of 0 under G_ϑ, whereas the second set is the orbit of 1. The local behaviour of the Riemann–Weierstrass function at these two sets will be treated separately.

But before we go any further we show some numerical results on the Jacobi theta function. In Figures 6.3 and 6.4 we have plotted the modulus $|\vartheta|$ as a function of the position parameter $b = \Re z$. The scale $a = \Im z$ is fixed at 10^{-2} and 10^{-6}. As the scale becomes small, the diagrams seem to become stabilized when $|\vartheta|$ is rescaled with a re-scaling factor depending only on the scale $\Im z$. More precisely, let us look at the following family of functions

$$s_\lambda(t) = \lambda^{-\alpha}|\vartheta(\lambda + it)|.$$

For every λ the exponent α is fixed via

$$s_\lambda(0) = 1.$$

What is observed numerically is that the limit $\lambda \to 0$ exists at least at some points. These points must correspond to the strongest singularity, since,

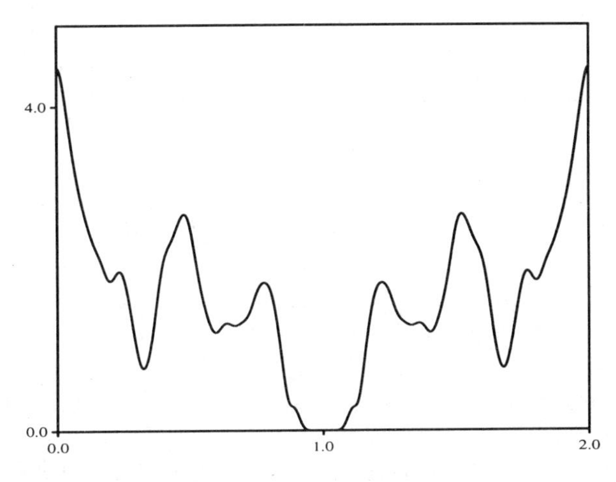

Fig. 6.3 The modulus of the theta function at scale $a = 10^{-2}$.

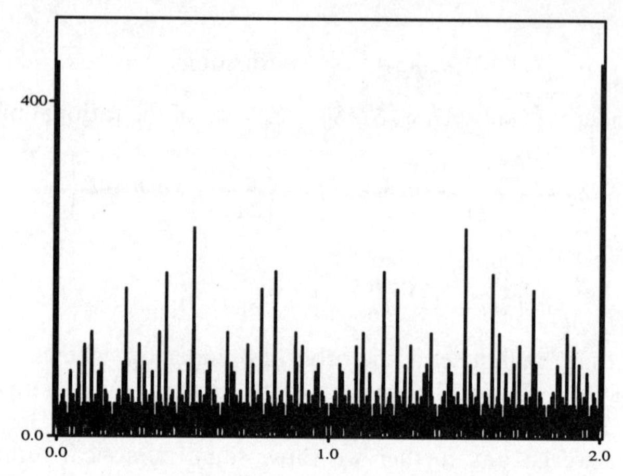

Fig. 6.4 The re-normalized theta function at scale $a = 10^{-6}$.

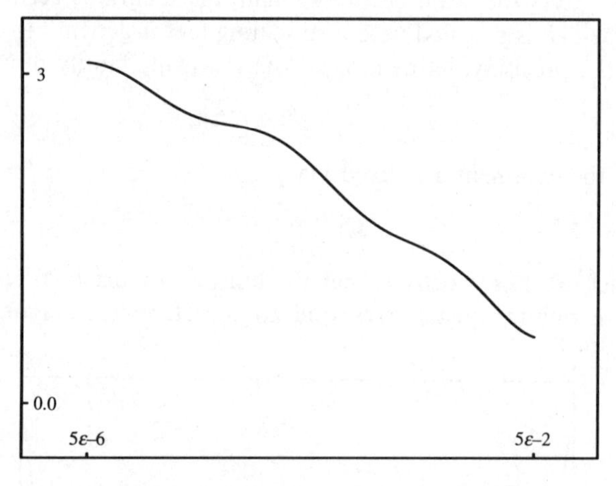

Fig. 6.5 The scaling of ϑ at $b = 1 + \sqrt{3}$.

otherwise, the re-normalization procedure would explode somewhere. All points, at which the limit exists, have the same local scaling exponent. The amplitude of the re-normalized theta function is connected to the constants $c_{\pm}$ that describe the local behaviour of the cusp to the left and to the right.

In Figure 6.5 we have plotted log a against $\log|\vartheta(b^*, a)|$ with $b^* = 1 + \sqrt{3}$. The oscillations around the slope indicate that W might be in a local periodic scaling class. We will not prove this but, however, it turns out that locally at this point—and many others—the theta function has a local, discrete re-normalization.

6.1 The orbit of 0

It will turn out that the points in the orbit of 0 correspond to those positions at which the global re-normalization s_λ has a limit. Indeed, the local behaviour of the scale–space representation of (6.0.2) is determined by the following theorem.

Theorem 6.1.1
At any point $t \neq \infty$ in the orbit of 0 the scale–space coefficients look like

$$(\Im\tau)^\beta \vartheta(t+\tau) = c_t (\Im\tau)^\beta \sqrt{\frac{i}{\tau}} + O(\tau^{2\beta - 1/2}) \qquad (\tau \to 0) \qquad (6.1.1)$$

for all $\beta > 1/2$. The constant c_t satisfies the following transformation formulas:

$$c_{K^2 t} = c_t, \qquad c_{Ut} = \sqrt{-\frac{i}{t}}\, c_t = e^{-i\pi\,\mathrm{sign}(t)/2} \frac{1}{\sqrt{|t|}},$$

which, together with $c_0 = 1$, determines c_t along the orbit of under G_ϑ.

The proof of this theorem is an immediate consequence of the following two lemmata.

Lemma 6.1.2
Theorem 6.1.1 *holds at $t = 0$.*

Proof. From the series expansion ((6.0.3)) of ϑ we can easily read the following estimation:

$$\vartheta(\tau) = 1 + O(e^{-\Im\tau}) \qquad (\Im\tau \to \infty). \qquad (6.1.2)$$

From the transformation behaviour (6.0.8) of ϑ we can write

$$\vartheta(\tau) = \sqrt{\frac{i}{\tau}}\, \vartheta\left(-\frac{1}{\tau}\right).$$

Using the estimate (6.1.2) and the global estimate (6.0.6) with τ replaced by $-1/\tau$ we can read the following behaviour at 0:

$$\vartheta(\tau) = \sqrt{\frac{i}{\tau}} + \rho(\tau)$$

and the remainder can be estimated by

$$\rho(b+ia) = \begin{cases} O((b^2+a^2)^{-1/4}\, e^{-a/(b^2+a^2)}), & \text{for } \dfrac{a}{b^2+a^2} > 1, \\ O(a^{-1/2}(b^2+a^2)^{1/4}), & \text{uniformly.} \end{cases}$$

From this it is not difficult to see that

$$a^{\beta}\rho(b+ia) = O((b^2+a^2)^{\beta-1/4}$$

and this shows that (6.1.1) holds at $t=0$. □

It remains to show that everything is well behaved under the action of the theta group. Therefore, it is enough to check on the generators of G_{ϑ}. Since this is of course true for K^2 the theorem will be proved by the following lemma.

Lemma 6.1.3
Suppose that at some $t \in \mathbb{R}$, $t \neq 0$,

$$\vartheta(t+\tau) = c_t\sqrt{\frac{i}{\tau}} + \rho(\tau)$$

holds with ρ satisfying estimate (6.1.1). *Then at $t_0 = Ut = -1/t$ we have*

$$\vartheta(t_0+\tau) = \sqrt{-\frac{i}{\tau}}\, c_t\sqrt{\frac{i}{\tau}} + \tilde{\rho}(\tau),$$

where $\tilde{\rho}$ again satisfies the estimate (6.1.1).

Proof. By direct computation using the transformation behaviour (6.0.8) of ϑ we can write

$$\vartheta\left(-\frac{1}{t}+\tau\right)\cdot\sqrt{\frac{it}{t\tau-1}}\,\vartheta\left(t+\frac{t^2\tau}{1-t\tau}\right),$$

and by hypothesis on the behaviour of ϑ in t:

$$\vartheta\left(-\frac{1}{t}+\tau\right) = c_t\sqrt{\frac{-i}{t\tau}} + \rho\left(\frac{t^2\tau}{1-t\tau}\right).$$

Since $1/(1-t\tau) \simeq 1$ for $\tau \to 0$ the remainder again satisfies the estimate of (6.1.1). Therefore, the lemma is proved and so is Theorem 6.1.1. □

We now interpret these results in terms of W_{β}. By the theorems of Section 3 this means that W_{β} is in the exact scaling class at all points of the orbit of 0 and thus we have the following theorem.

Theorem 6.1.4
At any finite point t in the orbit of 0 the function of Riemann and Weierstrass has local cusps of the following explicit form ($\gamma = \beta - 1/2$):

$$W_\beta(t+u) = c_t^- |u|_-^\gamma + c_t^+ |u|_+^\gamma + \rho(u),$$

where $|u|_\pm = |u \mp |u||/2$. *The remainder* ρ *is differentiable in* 0 *for* $\beta > 3/4$, *whereas for* $1/2 < \beta \leq 3/4$ *it is of local regularity* $\Lambda_{\log}^{2\beta - 1/2, 1}$. *The two complex constants* $c_t^\pm$ *determining the local behaviour of* W_β *to the left (to the right) of* t *obey the transformation equations:*

$$c_{K^2 t}^\pm = c_t^\pm, \qquad c_{Ut}^\pm = \sqrt{-\frac{i}{t}}\, c_t^\pm,$$

which, together with

$$c_0^\pm = -\frac{2}{\sqrt{\pi}} \cdot \frac{e^{\pm i\gamma\pi/2}}{\sin(\pi\gamma)\Gamma(\gamma)},$$

determine $c_t^\pm$ *along the orbit of* 0 *under* G_ϑ.

Corollary 6.1.5
Riemann's classical function[5] *has a dense set of points where it is differentiable to the right (left) but not to the left (right) and the value of the respective derivative is* $-i\pi$.

See Figures 6.6 and 6.7 for an illustration.

Proof. Since $\beta = 1$ we have $\arg c_0^{+(-)} = \pi/4$. The action of K^2 does not change the argument, whereas U turns the phase by $\pm\pi/4$ depending on whether $t > 0$ or $t < 0$. Therefore, we can, by successive application of K^{2n} and U, construct, an infinite and dense set of points $\{t\}$ for which the argument of the constant $c_t^{+(-)}$ equals 0 or π, and therefore $\Im c_t^{+(-)}$, vanishes. Since the function of Riemann is the imaginary part of W_1, the differentiability to the right (to the left) follows from the differentiability of the remainder in the previous theorem. The value of the derivative follows from Theorem 4.4.2 and the behaviour at small scale of the wavelet transform (6.0.5), since we may replace ϕ by $-2i\Im\phi$ because W_1 is progressive. □

6.2 The orbit of 1

Along the orbit of 1 under G_ϑ the local regularity of W_β is characterized by the following theorem.

[5] We mean here the series of (6.0.1) for $\beta = 1$.

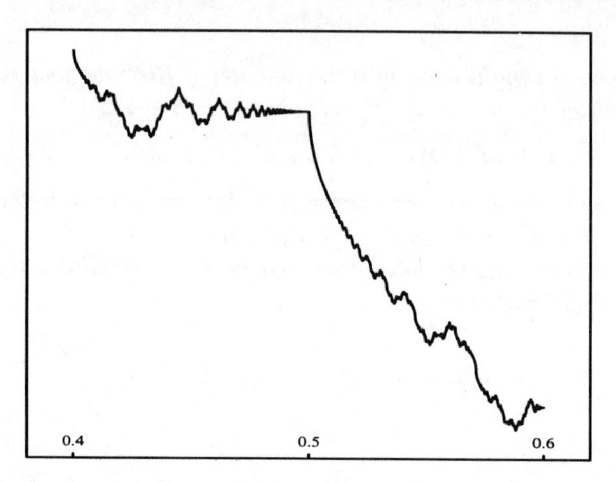

Fig. 6.6 The local cusps at $t = 1/2$ for the real part of W_1. Note that the one-sided derivative exists.

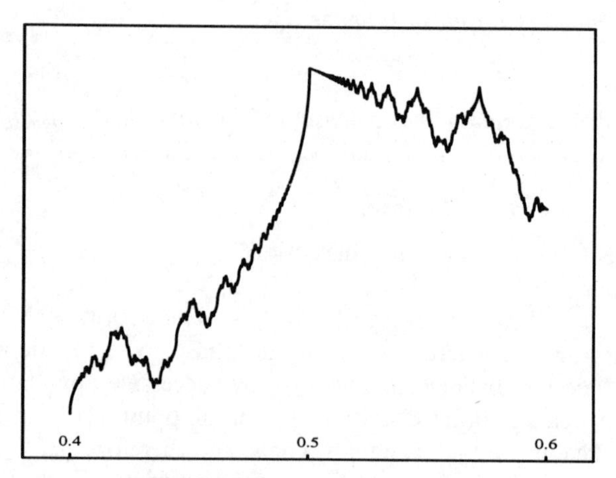

Fig. 6.7 The imaginary part of W_1.

Theorem 6.2.1

At any point of the orbit of 1 *under the group* G_ϑ *the Riemann–Weierstrass function is differentiable for* $\beta > 3/4$. *For* $\beta = 1$, *the value of the derivative is equal to* $-i\pi$ *at all these points. For* $1/2 < \beta \leq 3/4$ *it is of regularity* $\Lambda_{\log}^{2\beta-1/2,1}$.

This shows that the classical function of Riemann and Weierstrass has derivative $-\pi$ at these points whereas the associated cosine series has derivative 0. Look at Figures 6.8 and 6.9 for a sketch.

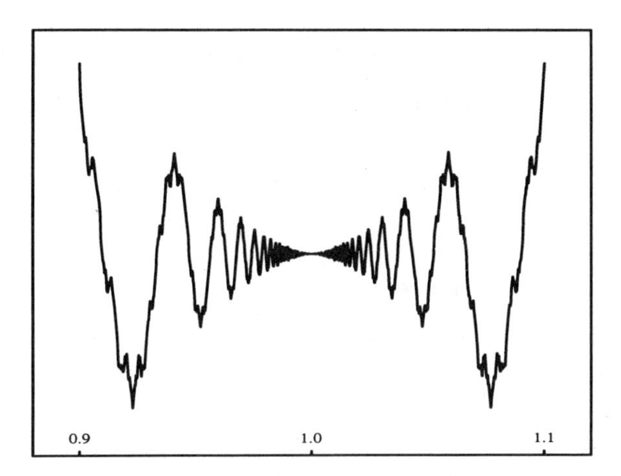

Fig. 6.8 The local behaviour at $t = 1$ for the real part of W_1. Note that the derivative exists.

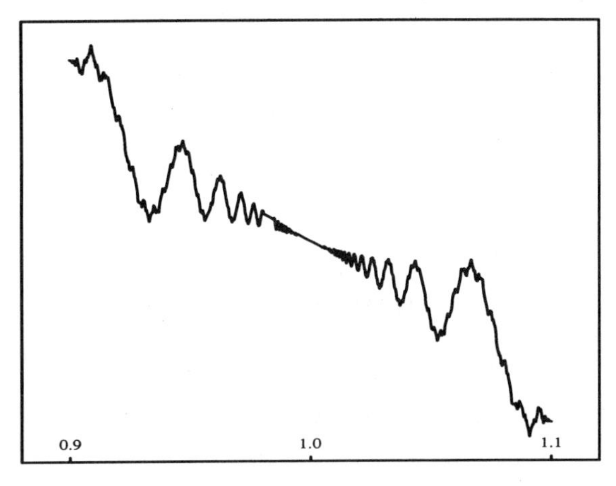

Fig. 6.9 The imaginary part of W_1.

Proof. Denote by $\Theta(\tau) = \vartheta(1 + \tau)$ the translate of ϑ. Then we may write

$$\begin{aligned}\Theta(\tau) &= \sum_{j=-\infty}^{\infty} (-1)^j e^{i\pi j^2 \tau} \\ &= \sum_{j \text{ even}} e^{i\pi j^2 \tau} - \sum_{j \text{ odd}} e^{i\pi j^2 \tau} \\ &= 2 \sum_{j \text{ even}} e^{i\pi j^2 \tau} - \vartheta(\tau) \\ &= 2\vartheta(4\tau) - \vartheta(\tau).\end{aligned}$$

Therefore, the local behaviour of ϑ in the points of the orbit of 0 implies that Θ satisfies the same estimate as the remainder in (6.1.1) at every point of the orbit of 1. The theorem then follows from Theorem 2.4.1 and Theorem 2.3.1.

We still have to compute the value of the derivative. It follows from Theorem 2.4.2 and (6.0.5).

6.3 The non-degenerated fixed points

We now come to the third set of points. Consider an element g of the theta group G_ϑ that has a real-valued, non-degenerate, fixed point t^*, that is, we have

$$g(t^*) = t^*, \qquad |\partial_z g(t^*)| \neq 1.$$

Without loss of generality we may assume that $|\partial_z g(t_0)| < 1$, otherwise we could consider the inverse mapping. Then there is an attracting domain $B \ni t^*$ such that for every $z \in B$ the sequence of iterates $z_{n+1} = g(z_n)$, $z_0 = z$, converges to t_0:

$$z \in B \Rightarrow \lim_{n \to \infty} z_n = t^*.$$

Now according to the transformation rule (6.0.9) the corresponding sequence of $\vartheta_n = \vartheta(z_n)$ satisfies

$$\vartheta_n = f_g(z_{n-1}) f_g(z_{n-2}), \ldots, f_g(z_0)\vartheta_0.$$

Both sequences become, asymptotically, a geometric progression:

$$z_n / z_{n+1} \to \partial_z g(t^*), \qquad \vartheta_n / \vartheta_{n+1} \to f_g(t^*),$$

and therefore we might expect scaling behaviour of the following type:

$$\vartheta_n \sim z_n^{-1/4 + i\gamma},$$

where $\gamma \bmod 2\pi = \Im \log f_g(t_0) / \log \partial_z g(t_0)$. Note that we have used relation (6.0.10). In the following theorem we will make this argument rigorous by showing that ϑ has a local re-normalization. But note here that the fixed points t^* have a continuous fraction expansion that eventually becomes periodic:

$$t^* \leftrightarrow [\alpha_0, \alpha_1, \ldots, \alpha_n, \alpha_{n+1}, \ldots, \alpha_{n+N}, \alpha_n + 1, \ldots],$$

as can be seen from the expansion of (6.0.7).

Theorem 6.3.1

The theta function ϑ has a pointwise re-normalization around every non-

degenerate fixed point, in the sense that the following limit:

$$\vartheta_n(t^* + \tau) = \lambda^{-\alpha n}\vartheta(t^* + \lambda^n \tau) \to \vartheta^*(\tau), \qquad (n \to \infty)$$

holds uniformly for τ in any compact subset of the half-plane.

Proof. Let $\lambda = \partial_z g(t^*)$. As we have seen, λ is real valued. Since $|\lambda| \neq 1$ we may suppose that $0 < \lambda < 1$, if not we take g^{-1}, or g^{-2} instead of g. Let $\alpha \in \mathbb{C}$ satisfy at

$$\lambda^\alpha = f_g(t^*).$$

Note that $\Re\alpha = -1/4$. By the transformation properties given above we may write

$$\begin{aligned}\vartheta_n(t^* + \tau) &= \lambda^{-\alpha n}\vartheta(t^* + \lambda^n\tau) \\ &= \lambda^{-\alpha n} f_{g^n}(g^{-n}(t^* + \lambda^n\tau))\vartheta(g^{-n}(t^* + \lambda^n\tau)).\end{aligned}$$

In suitable coordinates, the transformation g having a non-degenerate fixed point at $t^* \in \mathbb{R}$ is conjugated to a dilation by a dilation factor $\partial_g(t^*)$ around t^* and the conjugation map is regular at t^*. Therefore, we have

$$g^{-n}(t^* + \lambda^n\tau) \to t^* + \tau$$

and the limit is uniform in any compact set of $\mathbb{H}$. On the other hand, by the co-chain property we can write

$$f_{g^n}(\tau) = \prod_{j=0}^{n-1} f_g(g^j(\tau)).$$

Now since g is conjugated to a dilation with dilation factor $0 < \lambda < 1$ the iterates $g^n z$ tend asymptotically in a geometric progression towards t^*. Thus for any compact subset of the half-plane there is a constant $c > 0$ and another constant q with $0 < q < 1$ such that

$$|g^j(t^* + \tau) - t^*| \leq cq^n.$$

By construction of α we have $\lambda^{-\alpha} f_g(t^*) = 1$ and because f_g is regular in a neighbourhood of t^* we find that the infinite product

$$\lim_{n\to\infty} \lambda^{-\alpha n} f_{g^n}(\tau) = \prod_{j=0}^{\infty} \lambda^{-\alpha} f_g(g^j(\tau))$$

converges uniformly in every compact of $\mathbb{H}$ and thus the same type of convergence holds for ϑ_n.

6.4 The irrational points

For any irrational point t, Hardy and Littlewood have shown (Hardy and Littlewood 1916) that there is a positive constant c such that

$$a_n^{1/4}|\Im\vartheta(t + ia_n)| > c \quad \text{and} \quad a_n^{1/4}|\Re\vartheta(t + ia_n)| > c$$

for a sequence of positive real numbers $a_n \to 0$. This implies, via Theorem 2.0.5, that neither W_β nor its real or imaginary part can be Lipschitz continuous and hence they are not differentiable at any irrational point for any $\beta \in (1/2, 5/4)$ since the wavelet transform (6.0.5) is only decreasing as $O(a^{1-\epsilon})$, $\epsilon > 0$, at these points.

We remark here that this proof essentially follows the reasoning of Hardy, who previously proved the non-differentiability of W_β at these points. He used the angular derivative of the *Poisson* integral for functions on the circle, which in this case was the theta function ϑ.

7 The baker's map

Until now, we have used the wavelet transform only for the analysis of 'static' objects. As a first simple example of how to use wavelets to analyse dynamic behaviour, consider the following dynamical system:

$$\phi\colon [0,1) \mapsto \begin{cases} 2t & \text{for } 0 \le t \le 1/2 \\ 2t - 1 & \text{for } 1/2 < t < 1 \end{cases} = 2t \bmod 1.$$

Consider some mass distribution μ or $\mu(t)\,dt$ over the state space $[0,1]$. Under the action of the dynamic ϕ,[6] the mass distribution evolves according to

$$\mu(t) \mapsto U\mu(t) = (1/2)\{\mu(t/2) + \mu(t/2 + 1/2)\}.$$

The action of U is actually dilation by 2 over the circle (see Section 14 of Chapter 3). During the iteration the original distribution will become more and more 'mixed up' and will finally tend to an equilibrium state. Here this final state is given by the constant functions, which are the only fixed points of U in the space of continuous functions. We will now discuss in which way the speed of this transition towards equilibrium depends on the local regularity of the original mass distribution.

Theorem 7.0.1

Let $s \in \Lambda^\alpha(\mathbb{T})$ with $\alpha > 0$. Then

$$\|(\Pi_1 - \mathbb{1})U^n s\|_{L^2(\mathbb{T})} \le c2^{-\alpha n},$$

where Π_1 is the orthogonal projector on the constant functions.

---

[6] By this we mean the so-called *Frobenius* operator. In general it reads

$$\mu(t)\,dt \mapsto U\mu(t)\,dt = \sum_{\phi(u)=t} \frac{\mu(u)\,dt}{|\phi'(u)|}$$

if the dynamic is differentiable. If μ is a probability measure then $U\mu(t)\,dt$ gives the probability of distribution t after one iteration of the dynamic if in the original state the points were distributed according to μ.

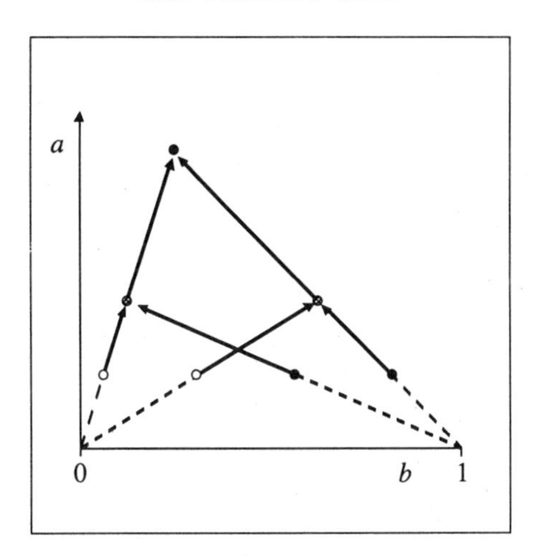

Fig. 7.1 The baker's map shifts small scale energy towards the large scales.

This shows that the higher the regularity of s, the faster is equilibrium approached.

Proof. The analysis tool is the continuous wavelet transform over the circle as presented in Section 10 of Chapter 2. First observe that in wavelet space the operator U can be written

$$U\colon \mathcal{W}_g^{\mathbb{T}} s(b, a) \mapsto \frac{1}{2}\left\{ \mathcal{W}_g^{\mathbb{T}} s\left(\frac{b}{2}, \frac{a}{2}\right) + \mathcal{W}_g^{\mathbb{T}} s\left(\frac{b+1}{2}, \frac{a}{2}\right)\right\}. \tag{7.0.1}$$

Indeed, from the commutative diagram of (14.0.2 of Chapter 3) the dilation over the torus is a periodized dilation over the real line. On the other hand, the wavelets over the circle are periodized wavelets (equation 10.0.1, Chapter 2). In Figure 7.1 we have shown the geometry of equation (7.0.1). We observe that the dynamic shifts the small scale features towards the large scales.

On the other hand, the wavelet transform decays arbitrarily fast at large scale, as follows from Theorem 11.1.3 in Chapter 2. In particular, as we have seen, we may find a wavelet $g \in S_0(\mathbb{R})$ such that $\mathcal{W}_g r(b, a) = 0$ for $a > 1$ whenever $r \in L^1(\mathbb{T})$. From the regularity of s we have $|\mathcal{W}_g s(b, a)| \leq c a^{\alpha}$. By iteration of (7.0.1) we may estimate

$$|\mathcal{W}_g U^n r(b, a)| \leq c 2^{-\alpha n} a^{\alpha},$$

and the theorem follows from the unitarity of the wavelet transform and the fact that the wavelet operates modulo the constant functions. □

8 A family of dynamical systems and fractal measures

Our knowledge about the analysis of a dynamical system through wavelet transforms is in general still very poor. Only for a special class of dynamical systems, known as IFS (i.e. iterated function system) have we some idea of the possible relations. We will not present the general theory of dynamical systems. Rather we stick to concrete examples. A dynamical system consists in the first place of some metric space Ω—the state space—and some mapping $\phi: \Omega \mapsto \Omega$—the (discrete) dynamic. If the system was in the state $x_0 \in \Omega$ at 'time' $t = 0$, then the sequence

$$x_0 \xrightarrow{\phi} x_1 \xrightarrow{\phi} x_2 \cdots$$

describes the evolution of the system under the dynamic ϕ, and $\phi^n(x_0)$ is the state of the system at time $t = n$. We now wish to present the prototype of a dynamical system that we will work with in this chapter .

Suppose we are given a finite collection of closed intervals I_n, $n = 0, N - 1$, that satisfy

(i) $|I_n \cap I_m| = 0$ for $n \neq m$,
(ii) $I_n \subset [0, 1]$,
(iii) $0 \in I_0$, $1 \in I_{N-1}$.

We write $|I|$ for the Lebesgue measure of I. The first condition states that two different intervals have at most one point in common, in which case it lies on the border. If no two different intervals intersect, there are necessarily gaps between the intervals and we call this an IFS with gaps. Conditions (ii) and (iii) essentially mean, that $[0, 1]$ is the smallest interval that contains all I_n.

For each of the intervals I_n there are exactly two affine maps, whose restriction to I_n maps I_n onto $[0, 1]$. We choose one of them for each interval and we call it A_n. It can be written in the following form:

$$A_n: \mathbb{R} \mapsto \mathbb{R}, \qquad t \mapsto \frac{t - t_n^*}{l_n}.$$

The N numbers l_n are called the length scales of the IFS. They satisfy $|l_n| = |I_n|$. The IFS dynamical system is now defined by

$$\phi: [0, 1] \cup \{\infty\} \mapsto [0, 1] \cup \{\infty\}, \qquad t \mapsto \begin{cases} A_n t & \text{for } t \in I_n, \\ \infty & \text{for } t \text{ not in any } I_m. \end{cases}$$

For the sake of simplicity we will sometimes limit ourselves to the case of two intervals. The general case obviously follows from that. Therefore, henceforth we are given two intervals, a left one I_L, and a right one I_R, and

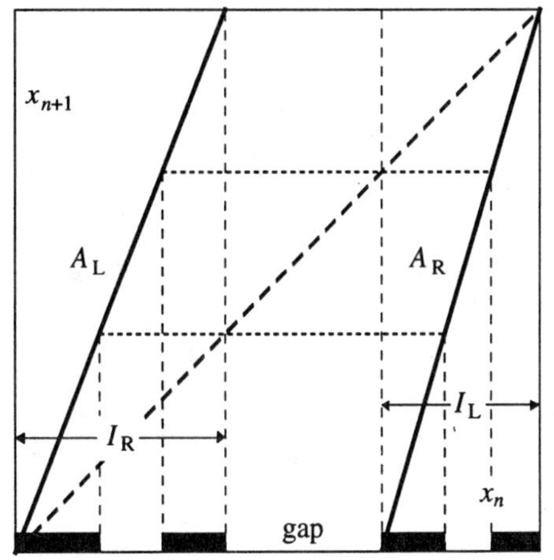

Fig. 8.1 An example of a dynamical IFS system.

two length scales, $l_L = |I_L|$ and $l_R = |I_R|$, together with two affine maps A_L and A_R. In addition, we suppose, unless otherwise stated, that $I_L \cup I_R$ is a true subset of [0, 1], that is, we suppose that there is a gap between I_L and I_R.

Let us pick a point in $t_0 \in [0, 1]$ and let us look at the orbit of t_0 under the dynamic, that is, we look at the sequence of iterates (see Figure 8.1)

$$t_{n+1} = \phi(t_n).$$

It may happen that the sequence ends in the point ∞ and stays there:

$$\exists n: t_n = \infty.$$

When we iterate this system, more and more points will leave the interval [0, 1] because they are mapped to ∞, which is the only attracting fixed point. See Figure 8.2 for the first three iterates.

However, there is a set of points that will never leave the interval. This set will be called Λ_∞:

$$\Lambda_\infty = \bigcap_{n=0}^{\infty} \phi^{-n}([0, 1]).$$

We write $\phi^{-1}I$ for the set of points t, such that $\phi(t) \in I$. The set of these points, Λ_∞, is a Cantor set that is invariant under the dynamic

$$\phi\Lambda_\infty \subset \Lambda_\infty, \qquad \phi^{-1}\Lambda_\infty \subset \Lambda_\infty.$$

Every point in Λ_∞ can be coded by an infinite sequence of LR symbols,[7] as we have sketched in Figure 8.2. This sequence of symbols is sometimes

[7] In the general case by N symbols.

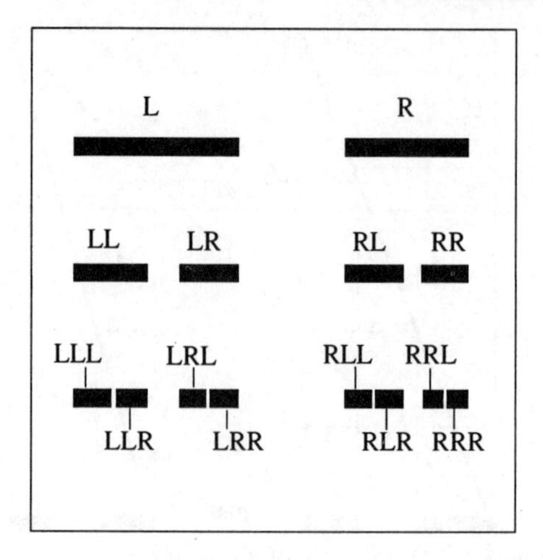

Fig. 8.2 Upon iterating, more and more points leave the interval [0, 1]. At the end only a Cantor set will remain.

called a 'kneading sequence'. If we introduce the sequence space $\{L, R\}^{\mathbb{N}}$ consisting of all semi-infinite sequences that can be made from the two symbols, say L for 'left' and R for 'right':

$$\{L, R\}^{\mathbb{N}} = \{[S(0), S(1), \ldots]: S(n) \in \{L, R\}, \text{ for } n \in \mathbb{N}_0\},$$

then the last statement means, that there is a surjective[8] map

$$p: \{L, R\}^{\mathbb{N}} \to \Lambda_\infty.$$

Let us introduce the left-shift operator of the sequences

$$\Sigma: \{L, R\}^{\mathbb{N}} \mapsto \{L, R\}^{\mathbb{N}}, \qquad [S(0), S(1), S(2) \cdots] \mapsto [S(1), S(2), S(3) \cdots].$$

It can be seen from Figure 8.2 that, on the invariant set Λ_∞, the action of the dynamic is given by a shift in symbol space, that is, the following diagram is commutative:

$$\Sigma p = p\phi, \qquad \begin{array}{ccc} \{L, R\}^{\mathbb{N}} & \xrightarrow{\Sigma} & \{L, R\}^{\mathbb{N}} \\ \downarrow p & & \downarrow p \\ \Lambda_\infty & \xrightarrow{\phi} & \Lambda_\infty \end{array}$$

This description of the dynamic in symbol space is sometimes called the symbolic dynamic.

[8] In general, this map is not injective. However, for an IFS with gaps, p is bijective.

Example 8.0.1
The baker's map:

$$\phi: [0, 1[\mapsto [0, 1[, \qquad t \mapsto 2t \bmod 2.$$

Here the whole interval $[0, 1]$ is the invariant set, and the infinite sequence of two symbols is nothing other than a sequence of binary digits

$$t \leftrightarrow 1001011000 \cdots.$$

Note that here the map p is not bijective, since the sequence of digits is not unique; for example, $0111 \cdots = 1000 \cdots$. This is due to the fact that there is no gap between the two intervals.

Let us look at the trajectory $t_{n+1} = \phi(t_n)$ of a point $t_0 \in \Lambda_\infty$. As we have seen, it lies completely on Λ_∞, and the dynamic corresponds to a shift in symbol space. Therefore, we can now distinguish two qualitatively different types of behaviour.

(i) The sequence t_n may end up in a *limit cycle*, that is,

$$\exists n, \exists N: m \geq n \Rightarrow t_{m+N} = t_m.$$

The smallest such N is called the period of the limit cycle. Let t^* be an element of such a limit cycle. Since after N iterations of ϕ we come back to t^*, we see that every point in a limit cycle can be written as the fixed point of ϕ^N, that is, $\phi^N(t^*) = t^*$. Vice versa, every fixed point of ϕ^N lies in a limit cycle of period N of ϕ. Clearly, in sequence space the limit circles may be identified with the sequences that eventually become periodic.

(ii) In all other cases, if the kneading sequence is aperiodic, the sequence of iterates describes a pseudo-random movement over the invariant set Λ_∞. These trajectories are called chaotic.

8.1 Self-similar fractal measure

We now wish to construct a probability measure that is supported by the invariant set Λ_∞ and that is somehow in close connection with the dynamical system ϕ. We therefore choose some numbers p_k that satisfy

$$0 \leq p_k \leq 1, \qquad \sum_{k=0}^{N-1} p_k = 1.$$

These quantities are called the *a priori* probabilities. The action in measure space that is induced by ϕ goes as follows. We re-scale and translate the support of the measure μ in such a way that it is now supported by one of the intervals I_n. At the same time we re-scale the overall probability by p_n. This is repeated for all intervals I_k and the new measure is the superposition of the N measures obtained in this way. We have thus constructed the following operator:

$$U: \mu(t) \mapsto \sum_{k=0}^{N-1} \frac{p_k}{l_k} \mu\left(\frac{t - t_k^*}{l_k}\right) dt = \sum_{k=0}^{N-1} \alpha(k) \mu\left(\frac{t - t_k^*}{l_k}\right). \quad (8.1.1)$$

In Figure 8.3 we have sketched the evolution of the density for the first three iterations of the dynamic when we start with the Lebesgue measure as the initial condition.

Remark. Because we use the inverse dynamic for the construction of this evolution in measure space it is called the *backward iteration* of the IFS.

8.2 The evolution in wavelet space

Let us look at the evolution in wavelet space. By elementary computation using the co-variance of the wavelet transform, we obtain the following evolution over the half-plane:

$$U: \mathscr{W}(b, a) \mapsto \sum_{k=0}^{N-1} \frac{p_k}{l_k} \mathscr{W}\left(\frac{b - t_k^*}{l_k}, \frac{a}{l_k}\right),$$

with $\mathscr{W} = \mathscr{W}_g \mu$. This time the evolution produces small scale features, as can be seen in Figure 8.4. The dynamic shifts the energy from the large scales to the smaller scales.

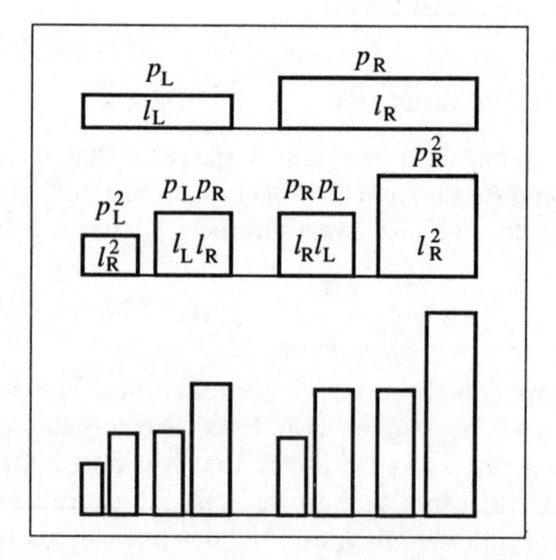

Fig. 8.3 The first three iterates in measure space. The evolution of the density.

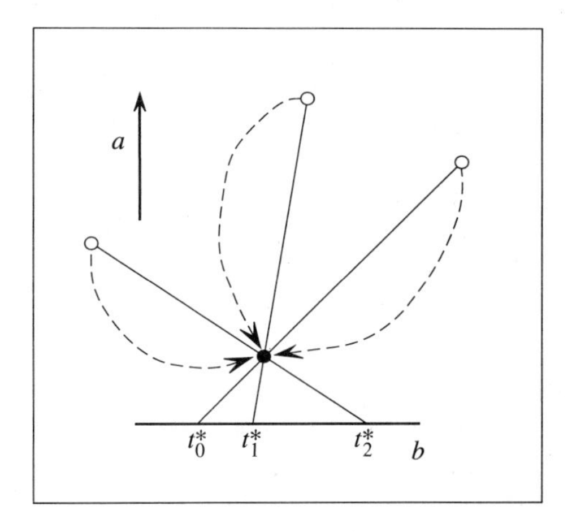

Fig. 8.4 The geometry of the scale transition.

8.3 Some fractal measures

As we will now show, this sequence of backward iterated measures that we have constructed converges towards a limit measure μ_∞.

Theorem 8.3.1
Let $s \in L^1(\mathbb{R})$, compactly supported with $\int s = 1$. Then

$$\lim_{n \to \infty} U^n s = \mu_\infty \neq 0$$

exists and is unique. The convergence takes place in $S'(\mathbb{R})$ and the limit μ_∞ is a positive measure.

Actually, as the proof shows, we have for any g with compact support and regularity Λ^α, $\alpha > 0$

$$\lim_{n \to \infty} \int g \cdot U^n s = \int g \cdot \mu_\infty .$$

Proof. Let $r = Us - s$. Then we may write

$$U^M s = \sum_{k=0}^{M-1} U^k r + s.$$

Therefore, if we show that for all compactly supported $g \in \Lambda^\alpha(\mathbb{R})$, $\alpha > 0$,

we have

$$\sum_{n=0}^{\infty} \left| \int \bar{g} \cdot U^n r \right| < \infty, \tag{8.3.1}$$

it follows that $\int \bar{g} U^n s$ is a Cauchy sequence and hence we have convergence in $S'(\mathbb{R})$. Indeed, passing to the modulus we have, for $n > m$,

$$\left| \int \bar{g} \cdot (U^n s - U^m s) \right| \leq \sum_{k=m}^{n-1} \left| \int \bar{g} \cdot U^k r \right|,$$

which tends to 0 as $m \to \infty$.

We now wish to show that (8.3.1) holds. First, note that the iterates U^n of U satisfy the same kind of equation as U but with *a priori* probabilities $p_{k,n}$ and length scales $l_{k,n}$ obtained from the p_k and l_k by taking products of n terms. The positions will be denoted by $t_{k,n}$. By co-variance of the wavelet transform (note that g is not necessarily admissible, but this does not matter in that context) we have

$$\int \bar{g} \cdot U^n r = \mathscr{W}_g U^n r(0, 1) = \sum_k \frac{p_{k,n}}{l_{k,n}} \mathscr{W}_g r(-t_{k,n}/l_{k,n}, 1/l_{k,n}). \tag{8.3.2}$$

Now $\int Us = \int s$ and thus $\int r = 0$. Hence, for any compactly supported $g \in \Lambda^\alpha(\mathbb{R})$, we have

$$|\mathscr{W}_g r(b, a)| = \frac{1}{a} \left| \mathscr{W}_r g\left(\frac{-b}{a}, \frac{1}{a}\right) \right| \leq O(a^{-1-\alpha}) \qquad (a \to \infty).$$

The last estimate follows from Theorem 2.0.3. Thus the sum in (8.3.2) may be estimated by

$$\left| \int \bar{g} \cdot U^n r \right| \leq c \sum_k p_{k,n} l_{k,n}^{\alpha n},$$

with some c not depending on n. Since $|l_k| \leq \lambda < 1$ we have $|l_{k,n}| \leq \lambda^n$ and since the probabilities always satisfy at $\sum_k p_{k,n} = 1$ we may estimate

$$\sum_{n=0}^{\infty} \left| \int \bar{g} \cdot U^n r \right| \leq O(1) \sum_{n=0}^{\infty} \lambda^{\alpha n} < \infty,$$

and we are done for the convergence.

In particular, we have seen that for $r \in L^1(\mathbb{R})$ with compact support and $\int r = 0$ we have $U^n r \to 0$. This clearly shows the uniqueness since for s and $s^\#$ with $\int s = \int s^\# = 1$ we have $\int (s^\# - s) = 0$.

To see that the limit is a measure it is enough to observe that each $U^n s$ is a positive measure if s was chosen to be a non-negative function. □

The limit measure is obviously supported by the invariant Cantor set

$$\mu_\infty(I) = \mu_\infty(I \cap M_\infty).$$

In addition, by construction, it is invariant under the backward iteration of the dynamic

$$\mu_\infty(I) = \mu_\infty(\phi^{-1} I),$$

where $\mu(I)$ stands for the measure of the interval I. It is a typical example of a *multi-scale fractal measure*. These measures satisfy the following multi-scale equation:

$$\mu_\infty(t) = \sum_{k=0}^{N-1} \alpha(k)\mu_\infty([t - t_k^*]/l_k),$$

which holds in the sense of distributions. It is in formal analogy with the two-scale equations that are satisfied by the orthonormal wavelets of equation (8.0.1) in Chapter 3.

By co-variance the wavelet transform of such a multi-scale fractal measure satisfies the following global symmetry:

$$\mathcal{W}_g\mu_\infty(b, a) = \sum_{k=0}^{N-1} \frac{p_k}{l_k} \mathcal{W}_g\mu_\infty\left(\frac{b - t_k^*}{l_k}, \frac{a}{l_k}\right).$$

This coupling relation between the large and the small scales explains the whole global self-similarity of Figure 8.5, showing the wavelet transform of triadic Cantor set with balanced measure.

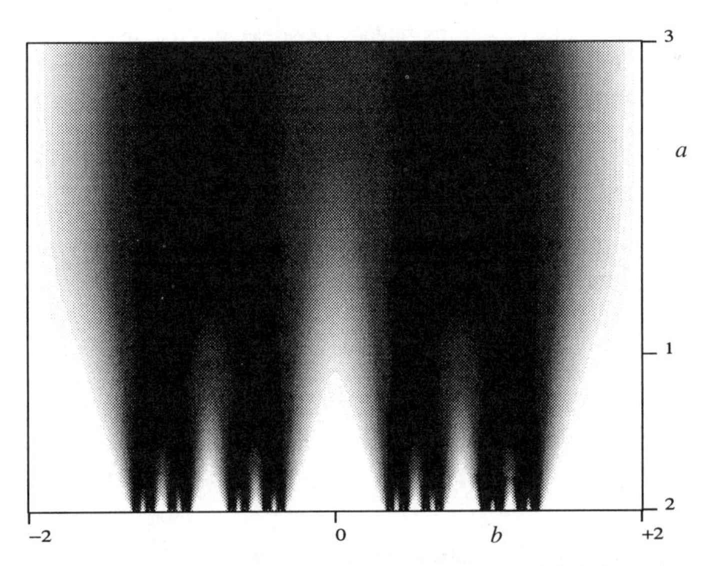

Fig. 8.5 The wavelet transform of a triadic Cantor set with balanced measure.

9 Fractal dimensions

Historically the first dimension that was used to describe the dimension of a set was the Hausdorff dimension, whose definition will now be presented briefly.

Consider a set[9] $S \subset [0, 1]$. We call a set of intervals $K = \{I_k\}$ a cover of S if

$$\bigcup_k I_k \supset S.$$

A cover is called an ϵ-cover of S, $\epsilon > 0$, if every interval $I_k \in K$ satisfies $|I_k| \leq \epsilon$. The set of all ϵ-covers of S will be called $K_\epsilon(S)$. For every cover of S we define a partition function

$$F(K, q) = \sum_{I_k \in K} |I_k|^q.$$

Let us call $H(\epsilon, q)$[10] the infimum of F over all ϵ-covers

$$H(\epsilon, q) = \inf_{K \in K_\epsilon(S)} F(K, q), \tag{9.0.1}$$

and finally let us look at the limit as the length scale of the cover tends to 0:

$$H(q) = \lim_{\epsilon \to \infty} H(\epsilon, q).$$

The limit actually exists because $K_\epsilon \subset K_{\epsilon'}$ for $\epsilon \leq \epsilon'$, showing that $H(\epsilon, q)$ is monotonic growing as $\epsilon \to 0$ for fixed q because the 'inf' in (9.0.1) is taken over smaller and smaller sets. It can easily be shown that there is a well-defined number d_H, called the *Hausdorff dimension* of S, such that

$$q < d_H \Rightarrow H(q) = \infty,$$

$$q > d_H \Rightarrow H(q) = 0.$$

Unfortunately, the Hausdorff dimension is not very easy to handle. In particular, in numerical experiments, where all estimations have to be done with a finite resolution, the Hausdorff dimension is very difficult to estimate.

9.1 Capacity

Therefore, one usually uses the so-called fractal dimension or capacity. Let S again be a bounded set of points, say $S \subset [0, 1]$. Let $N(\epsilon)$ be the number

[9] Clearly, the definition can be extended to any metric space.
[10] This quantity is also called the outer Hausdorff measure of S.

of intervals of length ϵ that are at least necessary to cover the set S. At small scale, $\epsilon \to 0$, scaling behaviour can in general be observed:

$$N(\epsilon) \sim \epsilon^{-d_C}.$$

The exponent d_C is called the fractal dimension or capacity of S. More precisely, it is defined as

$$d_C = -\liminf_{\epsilon \to 0} \frac{\log N(\epsilon)}{\log \epsilon}. \tag{9.1.1}$$

In numerical experiments, however, this definition is still simplified. One covers the real line with intervals of length ϵ, and counts how many boxes contain some points of the set S. This number is taken as an approximation of $N(\epsilon)$. This is the so-called box-counting technique. In a double-logarithmic plot of N and ϵ, a straight line—or oscillations around this line—can be observed, the slope of which gives an estimation of the capacity dimension of the set S.

For the invariant set of an IFS dynamical system with gaps, which we constructed at the beginning, the capacity can be computed directly.

Theorem 9.1.1

For the nvariant set Λ_∞ of an IFS with gaps, d_C is the unique solution of

$$\sum_{n=0}^{n-1} l_n^{d_C} = 1.$$

The numbers l_k are the n length scales of the IFS.

Proof. Let $N(\epsilon, S)$ be the lowest number of intervals of length ϵ that are necessary to cover S. Clearly, this number is invariant under translation of S and co-variant under dilations

$$N(\epsilon, aS + b) = N(\epsilon/a, S).$$

Now consider two sets, S_1 and S_2, having a positive distance

$$\Delta = \inf_{t \in S_1} \inf_{u \in S_2} |t - u| > 0.$$

By construction, the optimal cover for $S_1 \cup S_2$ with intervals of length $\epsilon < \Delta/2$ is obtained by optimizing the cover of S_1 and S_2 separately and we have

$$N(\epsilon, S_1 \cup S_2) = N(\epsilon, S_1) + N(\epsilon, S_2).$$

By the construction of the Cantor set Λ_∞ we have

$$\Lambda_\infty = \bigcup_k A_k^{-1} \Lambda_\infty.$$

The sets $A_k^{-1}\Lambda_\infty$ have positive distance, since we consider an IFS with gaps. Therefore, for epsilon small enough, we obtain the following equation for $N(\epsilon) = N(\epsilon, \Lambda_\infty)$:

$$N(\epsilon) = \sum_{k=0}^{n-1} N(\epsilon/l_k). \tag{9.1.2}$$

A heuristic argument might consist of replacing $N(\epsilon)$ by its asymptotic form ϵ^{-d_C}, yielding

$$\epsilon^{-d_C} \sim \epsilon^{-d_C} \sum_{k=0}^{n-1} l_k^{d_C}.$$

Thus the equation must hold, as stated in the theorem. This heuristic argument is actually true, as we will now show.

We start by rephrasing a limit expression such as (9.1.1) in the language of critical exponents.

Lemma 9.1.2

Let $s(t) > 0$ *be a positive function defined for* $t > 0$. *Then we have*

$$\liminf_{t\to 0} \frac{\log s(t)}{\log t} = \sup\{\gamma \in \mathbb{R}: s(t) = O(t^\gamma), (t \to 0)\},$$

$$\limsup_{t\to 0} \frac{\log s(t)}{\log t} = \inf\{\gamma \in \mathbb{R}: t^\gamma = O(s(t)), (t \to 0)\},$$

$$\liminf_{t\to \infty} \frac{\log s(t)}{\log t} = \sup\{\gamma \in \mathbb{R}: t^\gamma = O(s(t)), (t \to \infty)\},$$

$$\liminf_{t\to \infty} \frac{\log s(t)}{\log t} = \inf\{\gamma \in \mathbb{R}: s(t) = O(t^\gamma), (t \to \infty)\}.$$

Proof. (Although this is well known we give the proof anyway.) Consider the first equation. Call the left-hand side α and the right-hand side β. For any $\gamma < \beta$ there is a constant $c > 0$ such that for t small enough we have

$$s(t) \leq ct^\gamma.$$

Therefore, for $0 < t < 1$ we have

$$\frac{\log s(t)}{\log t} \geq \frac{c}{\log t} + \gamma,$$

and thus upon passing to the limit $t \to 0$ we obtain $\alpha \geq \gamma$, and since $\gamma < \beta$ was arbitrary we have

$$\alpha \geq \beta.$$

On the other hand, for every $\gamma > \beta$ there is a sequence $t_n \to 0$ such that

$$t_n^{-\gamma} s(t_n) \to \infty, \qquad (n \to \infty).$$

Therefore, $(\log s(t_n) - \gamma \log t_n) \to \infty$ and thus, in particular, if n is large enough

$$\log s(t_n) > \gamma \log t_n.$$

Therefore, we have

$$\alpha = \liminf_{t \to 0} \frac{\log s(t)}{\log t} \leq \liminf_{t_n \to 0} \frac{\log s(t_n)}{\log t_n} \leq \gamma.$$

Since $\gamma > \beta$ was arbitrary we have $\alpha \leq \beta$, and we are done. The other formulas are proved in the same way, and we leave them to the reader. □

We now exploit the scaling relation (9.1.2).

Lemma 9.1.3

Let $s(t)$ be a positive function defined for $t \in (0, 1]$ that is bounded away from 0 and infinity,

$$\forall t \in (0, 1], \qquad 0 < \inf_{u \in [t,1]} s(u) \leq \sup_{u \in [t,1]} s(u) < \infty,$$

and that satisfies pointwise at

$$s(t) = \sum_{n=1}^{N} \alpha_n s(t/\beta_n), \qquad t \in (0, \min\{\beta_n\}],$$

with some constants $\alpha_n > 0$, $\beta_n \in (0, 1)$. Then

$$\limsup_{t \to 0} \frac{\log s(t)}{\log t} = \liminf_{t \to 0} \frac{\log s(t)}{\log t} = \delta$$

is the unique solution of

$$\sum_{n=1}^{N} \alpha_n \beta_n^{-\delta} = 1.$$

Proof. We first wish to show that

$$\sum_{n=1}^{N} \alpha_n < 1 \Rightarrow \lim_{t \to 0} s(t) = 0,$$

$$\sum_{n=1}^{N} \alpha_n > 1 \Rightarrow \lim_{t \to 0} s(t) = \infty.$$

The lemma then follows by considering $t^{-\gamma} s(t)$ that satisfies the same type of equation as s but where α_n is replaced by $\sum_{n=1}^{N} \alpha_n \beta_n^{-\gamma}$.

Now consider

$$s^+(t) = \min\left\{\sup_{u\in[t,1]} s(u),\ \sup_{u\in(0,t]} s(u)\right\},$$

$$s^-(t) = \max\left\{\inf_{u\in[t,1]} s(u),\ \inf_{u\in(0,t]} s(u)\right\}.$$

These are the smallest (largest) monotonic functions that majorize (minorize) s for t small enough; that is, any other monotone function γ that majorizes (minorizes) s on $(0, \tau)$ with $\tau < 1$ satisfies $\gamma(t) \geq s^+(t)$ (respectively $\gamma(t) \geq s^-(t)$) for t small enough. For $t \in (0, 1]$ we have

$$0 < s^-(t) \leq s^+(t) < \infty,$$

since suppose $s^+(t) = \infty$, say, then $\sup_{u\in[t,1]} s(u) = \infty$, which contradicts our hypothesis on s. We may estimate for $0 < t \leq \min\{\beta_n\}$:

$$s(t) = \sum_{n=1}^{N} \alpha_n s(t/\beta_n) \leq \sum_{n=1}^{N} \alpha_n s^+(t/\beta_n),$$

$$s(t) = \sum_{n=1}^{N} \alpha_n s(t/\beta_n) \geq \sum_{n=1}^{N} \alpha_n s^-(t/\beta_n).$$

The right-hand sides are again monotonic functions that majorize (minorize) s and thus since $s^\pm$ is optimal we have for t small enough,

$$s^+(t) \leq \sum_{n=1}^{N} \alpha_n s^+(t/\beta_n),$$

$$s^-(t) \geq \sum_{n=1}^{N} \alpha_n s^-(t/\beta_n).$$

By monotonicity we may write

$$s^+(t) \leq \left(\sum_{n=1}^{N} \alpha_n\right) s^+(t/\beta^+),$$

$$s^-(t) \leq \left(\sum_{n=1}^{N} \alpha_n\right) s^+(t/\beta^-),$$

where $\beta^\pm$ are given by either $\min\{\beta_n\}$ or $\max\{\beta_n\}$, respectively, depending on whether $s^\pm$ are non-decreasing or non-growing. Since $\beta^\pm \in (0, 1)$ we are done. □

This shows the Theorem 9.1.1. □

The same arguments applied to the partition function (9.0.1) prove that the Hausdorf dimension satisfies the same equation.

9.2 The generalized fractal dimensions

In order to take into account the probabilistic nature of fractals—as, for example, the invariant fractal measure of the IFS that we have constructed—generalized fractal dimensions have been introduced (e.g. Hentschel and Procaccia 1983). Consider a probability measure μ that is supported by some set $S \subset [0, 1]$, say. Let K be a cover of S with intervals I_k of length $|I_k| = \epsilon$. Denote by p_k the probability of every interval I_k:

$$p_k = \mu(I_k).$$

We define a partition function associated to the cover K via

$$G(K, \epsilon, q) = \sum_{I_k \in K} \mu(I_k)^q = \sum_{I_k \in K} p_k^q, \qquad q \in \mathbb{R},$$

where we use the convention that $0^q = 0$, for $q \in \mathbb{R}$. For $q \in \mathbb{N}_0$ this function is nothing other than the q-moment[11] of the distribution of the set of numbers $\{p_k\}$. Now we optimize this partition function by setting

$$G(\epsilon, q) = \inf G(K, \epsilon, q),$$

where the inf runs over all covers of S with intervals of length ϵ. At small scale $\epsilon \to 0$ scaling behaviour can be observed:

$$G(\epsilon, q) \sim \epsilon^{\tau(q)}$$

and we define

$$\tau(q) = \liminf_{\epsilon \to 0} \frac{\log G(\epsilon, q)}{\log \epsilon}.$$

The generalized fractal dimensions are then defined by

$$d_q = \frac{\tau(q)}{q - 1}.$$

In practice the definition is again simplified using a box-counting technique. One works with a fixed partition of $\mathbb{R}$ into intervals of length ϵ and one looks at the scaling behaviour of G in a log–log plot. Note, however, that this procedure is, at least for $q < 0$, numerically unstable because low probability boxes contribute strongly in the partition function. A second disadvantage is of mathematical nature. The 'box-counting' dimensions are *a priori* not translation invariant.

[11] We recall that the moments of a measure ν are defined as $\int \nu(dt) t^q$.

$q = 0$
Note that for $q = 0$ we have

$$G(K, \epsilon, q) = \text{Number of intervals in } K$$

and therefore the generalized dimension d is nothing other than the capacity of S, the support of the measure

$$d_0 = d_C(S).$$

$q = 1$
For $q \to 1$ we obtain

$$\lim_{q \to 1} \frac{1}{q-1} \log \sum_{k=0}^{n-1} p_k^q = \log \sum_{k=0}^{n-1} p_k \log p_k.$$

Therefore, the dimension d_1 is sometimes called the *information* dimension, or *entropy* dimension, because it measures the scaling behaviour of the entropy $\sum_{k=0}^{n-1} p_k \log p_k$ of the distribution of probabilities.[12]

$q = 2, 3, \ldots$
For $q = 2$ the dimensions may be interpreted as correlation dimensions. Let us pick two points, t_1, t_2, randomly according to the probability measure μ and let us consider the probability that these two points are closer than ϵ, that is, we consider the correlation

$$C(\epsilon) = \text{Proba}\{|t_1 - t_2| \leq \epsilon\}.$$

Then $C(\epsilon)$ scales with $\epsilon^{d_{\text{Corr}}}$, where d_{Corr} is the so-called *correlation* dimension. Heuristically, one could argue that the two points whose distance is smaller than ϵ are in some interval of length ϵ. The probability that both points are in the interval I_k is p_k^2. Therefore, one might argue that $C(\epsilon) \sim \sum p_k^2$, showing that $d_{\text{Corr}} = d_2$.

For $q = 3, 4, \ldots$ the dimensions can be interpreted in the same way as correlation dimensions of q points.

Again for the IFS everything may be computed explicitly.

Theorem 9.2.1
The generalized fractal dimensions of the invariant measure of an IFS with length scale l_k *and a priori probabilities* p_k, $k = 0, \ldots, n-1$ *satisfy*

$$\sum_{k=0}^{n-1} \frac{p_k^q}{l_k^\tau} \equiv 1,$$

and $d_q = \tau(q)/(q-1)$ *as usual.*

---

[12] Note, however, that this is only a heuristic argument. We did not prove that the limit $\epsilon \to 0$ and $q \to 1$ could be exchanged.

Proof. By a self-similarity argument that is completely analogous to the one used in the proof of Theorem 9.1.1 we obtain the following recursion relation:

$$G(\epsilon, q) = \sum_{k=0}^{n-1} p_k^q G(\epsilon/l_k, q).$$

We may now apply Lemma 9.1.3 to show that the exponents τ have to satisfy the equation stated in the theorem. □

9.3 Fractal dimensions and wavelet transforms

We now wish to give a definition of fractal dimensions with the help of the wavelet transform. This definition will apply not only to measures, but to arbitrary functions, or distributions. We will show that in the case of the invariant IFS measures the wavelet dimensions coincide with the generalized fractal dimensions.

From now on we are only interested in local properties; therefore we suppose that all analysed distributions $\eta \in S'(\mathbb{R})$ are well behaved at infinity. More precisely, we require that for all $s \in S(\mathbb{R})$ we have $s * \eta \in S(\mathbb{R})$. This condition will be assumed throughout. It is of a purely technical nature and may be relaxed considerably from case to case. In particular, note that the Fourier transform of such a distribution is actually a smooth function (however it is not localized). In addition, we assume that $\eta \neq 0$.

For every scale a we look at the mean-q-energy at scale a:

$$G_g(a, q) = \|\mathcal{W}_g\eta(\cdot, a)\|_q^q = \int_{-\infty}^{+\infty} db |\mathcal{W}_g\eta(b, a)|^q \quad \text{with } q \geq 1. \tag{9.3.1}$$

Note that for $q = 2$ this is actually an energy since the wavelet transform is an isometry. At small scale a scaling behaviour of the form $G(a, q) \sim a^{\kappa(q)}$ can be observed—at least for affine self-similar measures as, for example, the triadic Cantor set with Bernoulli measure (e.g. Holschneider 1988)—giving rise to the definition of the fractal dimensions $\kappa(q)$. However, we will use a slightly modified definition in view of an application in Section 10. We set

$$\Gamma_g(a, q) = \int_a^1 \frac{d\alpha}{\alpha} G_g(\alpha, q).$$

For every $q \geq 1$ this is a monotonic function of a. Therefore, the limit $a \to 0$ exists, but it may not be infinite. This will always be the case if η is singular enough. In the opposite case when this limit is finite we subtract the constant $\int_0^1 (d\alpha/\alpha) G(\alpha, q)$ and we set instead,

$$\Gamma_g(a, q) = \int_0^a \frac{d\alpha}{\alpha} G_g(\alpha, q).$$

To summarize, we have for a small enough

$$\Gamma_g(a, q) = \min\left\{\int_a^1 \frac{d\alpha}{\alpha} G_g(\alpha, q), \int_0^a \frac{d\alpha}{\alpha} G_g(\alpha, q)\right\}.$$

Note that $G_g(a, q) \sim a^{\kappa(q)}$ implies $\Gamma_g(a, q) \sim a^{\kappa(q)}$ unless $\kappa(q) = 0$. The generalized dimensions $\kappa^+(q)$ and $\kappa^-(q)$ are now defined as follows:

$$\kappa^+(q) = \limsup_{a \to 0} \frac{\log \Gamma_g(a, q)}{\log a}, \qquad \kappa^-(q) = \liminf_{a \to 0} \frac{\log \Gamma_g(a, q)}{\log a}.$$

In the case where $\eta = 0$ we set $\kappa^{\pm}(q) = -\infty$. We will refer to these numbers as the upper and lower q-wavelet dimensions. Note that in this form the dimensions are defined for any distribution, in particular for measures and functions.

The fractal dimensions $\kappa^{\pm}(q)$ are actually well defined, as is shown by the following theorem.

Theorem 9.3.1

The dimensions $\kappa^{\pm}(q)$ do not depend on the analysing wavelet $g \in S_+(\mathbb{R})$ for $q \geq 1$, provided $g \neq 0$ and $|\kappa^{\pm}(q)| < \infty$.

Proof. We start by modifying the definition of the partition function Γ_g a little. Here we have to distinguish two cases according to whether

$$\lim_{a \to 0} \int_a^1 \frac{d\alpha}{\alpha} G_g(\alpha, q)$$

is infinite or not. Consider, first, the case where it is infinite. Note that we may suppose, for all $m > 0$,

$$\hat{\eta}(\omega) \leq (\omega^m) \qquad (\omega \to 0). \tag{9.3.2}$$

Indeed, the fractal dimensions of η and $\eta + s$ with $s \in S(\mathbb{R})$ are the same because of the fast decay of $\mathcal{W}_g s$ at small scale $a \to 0$. Therefore, if we choose some $\psi \in S(\mathbb{R})$ satisfying for all m at $\psi(\omega) = 1 + O(\omega^m)$ as $\omega \to 0$, we may replace η by $\eta - \psi * \eta$ without modifying the dimensions. Now this later distribution satisfies at (9.3.2). Therefore, the wavelet transform decays fast as $a \to \infty$ and thus we may replace $\int_a^1$ by $\int_a^\infty$ in the definition of Γ_g, that is, we may consider

$$\Gamma_g(a, q) = \int_a^\infty \frac{d\alpha}{\alpha} \|\mathcal{W}_g \eta(\cdot, \alpha)\|_q^q.$$

We now attack the estimates. Note that from equation (20.1.1) in Chapter 1 it follows that with

$$K_{a',a}(b) = \frac{1}{a'} \Pi_{g \to h}\left(\frac{b}{a'}, \frac{a}{a'}\right)$$

the passage from $\mathscr{W}_g\eta$ to $\mathscr{W}_h\eta$ reads

$$\mathscr{W}_h\eta(\cdot, a) = \int_0^\infty \frac{da'}{a} K_{a',a} * \mathscr{W}_g\eta(\cdot, a').$$

Using Minkowky and Hölder's inequalities we may now write

$$\begin{aligned}\|\mathscr{W}_h\eta(\cdot, a)\|_q^q &\leq \left\{\int_0^\infty \frac{da'}{a'} \|K_{a',a} * \mathscr{W}_g\eta(\cdot, a')\|_q\right\}^q \\ &\leq \left\{\int_0^\infty \frac{da'}{a'} \|K_{a',a}\|_1 \, \|\mathscr{W}_g\eta(\cdot, a')\|_q\right\}^q.\end{aligned}$$

Now we have

$$\|K_{a',a}\|_1 = \int_{-\infty}^{+\infty} db \frac{1}{a'} \left|\Pi_{g \to h}\left(\frac{b}{a'}, \frac{a}{a'}\right)\right| = H(a/a'),$$

with

$$H(a) = \int_{-\infty}^{+\infty} db \, |\Pi_{g \to h}(b, a)|.$$

This is a non-negative function that is rapidly decaying as $a + 1/a$ becomes large. Using Jensen's inequality, we now obtain

$$\|\mathscr{W}_h\eta(\cdot, a)\|_q^q \leq \left\{\int_0^\infty \frac{da'}{a'} H(a/a')\right\}^{q-1} \int_0^\infty \frac{da'}{a'} H(a/a') \|\mathscr{W}_g\eta(\cdot, a')\|_q^q.$$

By the high localization of H the first integral is a finite constant and thus

$$\begin{aligned}\Gamma_h(\epsilon, q) &\leq O(1) \int_\epsilon^\infty \frac{da}{a} \int_0^\infty \frac{da'}{a'} H(a/a') \|\mathscr{W}_g\eta(\cdot, a')\|_q^q \\ &= O(1) \int_0^\infty \frac{da'}{a'} H(1/a') \int_\epsilon^\infty \frac{da}{a} \|\mathscr{W}_g\eta(\cdot, aa')\|_q^q \\ &= O(1) \int_0^\infty \frac{da'}{a'} H(1/a') \int_{\epsilon a'}^\infty \frac{da}{a} \|\mathscr{W}_g\eta(\cdot, a)\|_q^q.\end{aligned}$$

Thus we have

$$\Gamma_h(\epsilon, q) \leq O(1) \int_0^\infty \frac{da}{a} H(\epsilon/a) \Gamma_g(a, q).$$

The same type of relation holds for g and h exchanged provided $h \neq 0$. Therefore, the theorem will be now an immediate consequence of the following lemma, which deals with such mean values.

Lemma 9.3.2

Let $s(t)$, $t > 0$, be a non-negative, monotonic (i.e. either non-decreasing or non-increasing) function of at most polynomial growth near 0,

$$s(t) \leq O(t^{-m}), \qquad (t \to 0), \quad \textit{for some } m > 0.$$

Near ∞ we assume that it is bounded:

$$s(t) \leq O(1), \qquad t \geq 1.$$

Further, let *$H(t)$ be a non-negative function, not identically* 0, *that is arbitrarily well polynomially localized, namely,*

$$H(t) = O((t + 1/t)^{-n}), \quad \textit{for all } n > 0.$$

Then for the mean values

$$r(t) = \int_0^\infty \frac{da}{a} H(t/a) s(a),$$

we have

$$\alpha^- = \liminf_{t \to 0} \frac{\log r(t)}{\log t} = \liminf_{t \to 0} \frac{\log s(t)}{\log t},$$

$$\alpha^+ = \limsup_{t \to 0} \frac{\log r(t)}{\log t} = \limsup_{t \to 0} \frac{\log s(t)}{\log t}.$$

Proof. Throughout this demonstration c will be some positive number not always the same.

We pick ϵ, $0 < \epsilon < 1$, and keep it fixed. For $0 < t < 1$ we split the integral defining $r(t)$ into four parts:

$$r(t) = \left\{ \int_0^{t^{1+\epsilon}} + \int_{t^{1+\epsilon}}^{t^{1-\epsilon}} + \int_{t^{1-\epsilon}}^1 + \int_1^\infty \right\} \frac{da}{a} H(t/a) s(a) = X_1 + X_2 + X_3 + X_4.$$

In the last term we may estimate that $s(t) = O(1)$ and thus

$$X_4 \leq O(1) \int_{1/t}^\infty \frac{da}{a} H(1/a).$$

Since $H(t)$ is arbitrarily well polynomially localized it follows that $X_4 = O(t^n)$ for all $n > 0$ as $t \to 0$.

In X_1 and X_3 we may estimate that $s(t) \leq O(t^{-m})$ and thus

$$X_1 \leq O(1) t^{-m} \int_0^{t^\epsilon} \frac{da}{a} H(1/a) a^{-m},$$

$$X_3 \leq O(1) t^{-m} \int_{t^{-\epsilon}}^\infty \frac{da}{a} H(1/a) a^{-m}.$$

Since H is arbitrarily well polynomially localized the integrals are rapidly decaying and thus, again, $X_1, X_3 = O(t^n)$ for all $n > 0$ in the limit $t \to 0$.

It remains the middle term X_2 which is the main contribution. Since $s(t)$ is monotonically growing as $t \to 0$ we may estimate $s(a) \geq s(t^{1-\epsilon})$ or $s(a) \geq s(t^{1+\epsilon})$ for $t^{1+\epsilon} \leq a \leq t^{1-\epsilon}$ depending on whether s is non-growing or non-decreasing. Therefore, we end up with

$$X_2 \geq cs(t^{1\pm\epsilon}) \int_{t^\epsilon}^{t^{-\epsilon}} \frac{da}{a} H(1/a) \quad \text{respectively.}$$

The last integral is again convergent and therefore $X_2 \geq cs(t^{1\pm\epsilon})$. This estimate also holds for the sum of the three contributions for t small enough and thus for all $\epsilon > 0$ we have

$$r(t) \geq cs(t^{1\pm\epsilon}) + O(t^n), \quad \text{respectively,}$$

for all $n > 0$.

In the same way we may estimate $s(a) \leq s(t^{1\mp\epsilon})$ in X_2 and hence

$$r(t) \leq cs(t^{1\mp\epsilon}) + O(t^n)$$

for all n.

Therefore, since ϵ is arbitrary the theorem follows from Lemma 9.1.2. □

Since we may exchange the roles of g and h the theorem is proved.
The proof remains the same if $\lim_{a\to 0} \Gamma_g(a, q) < \infty$. □

Remark. The proof needed the high localization of the $\Pi_{g,h}$, which in turn reflects that the wavelets must be very regular with all moments vanishing. However, for a fixed q, and a given $\epsilon > 0$ only some regularity is needed and some moments have to vanish in order to compute the dimensions $\kappa^-(q)$ without error and the dimensions $\kappa^+(q)$ with error at most ϵ.

Theorem 9.3.3
The generalized fractal wavelet dimensions $\kappa(q) = \kappa^\pm(q)$ of the invariant measure of an IFS with length-scale l_k and a priori probabilities p_k, $k = 0, \ldots, n-1$ satisfy

$$\sum_{k=0}^{n-1} p_k^q l_k^{1-q-\kappa(q)} = 1, \qquad q \geq 1.$$

Proof. Thanks to the remark above we may suppose that g is compactly supported. It follows that the partition function (9.3.1) satisfies the following equation

$$G(a, q) = \sum_{k=0}^{n-1} p_n^q l_k^{1-q} G(a/l_k, q).$$

We now may conclude as in the proof of Theorem 9.2.1. We leave the details to the reader. □

10 Time evolution and the dimension $\kappa(2)$

As we saw in Chapter 2 Section 14 for A, a bounded self-adjoint operator with resolvent R_z we have

$$e^{-iTA} = \frac{2\pi}{ic} \int_0^\infty da\, a^{n-1} \int_{-\infty}^{+\infty} db\, \bar{\hat{g}}(aT)\, e^{-iTb}\, R^n_{b+ia}(A),$$

where $c = \int_0^\infty d\omega\, \omega^{n-2} \bar{\hat{g}}(\omega)\, e^{-\omega}$ and the integral over the half-plane is understood as $\lim_{\rho\to\infty} \int_{1/\rho}^{\rho} (da/a) \int_{-\rho}^{\rho} db$. This shows explicitly how the behaviour for large T is determined by the behaviour of R_z at $\Im z \sim 1/T$. Indeed, suppose that the support of $\hat{g}$ is contained in an interval around $\omega = 1$, then the integral over the complex z half-plane actually runs only over a strip around $\Im z \sim 1/T$. Thus the long term behaviour $t \to \infty$ of the time evolution is linked to the small scale behaviour of some wavelet transform of the spectral measure, which in turn, as we have seen, is closely connected with local fractal properties (i.e. local regularity properties) of the measure itself, as we will now see. In particular, we want to establish a relation between the behaviour of the Fourier transform of η and the fractal wavelet dimension $\kappa(2)$. This question is of relevance whenever one considers the long term behaviour $T \to \infty$ of e^{-iTA}, where A is a self-adjoint operator. Indeed, in this case we have

$$\langle \psi \mid e^{-iTA} \phi \rangle = \hat{\mu}_{\psi,\phi}(T),$$

where $\mu_{\psi,\phi}$ is the spectral measure associated with ψ and ϕ. We are mainly interested in time averages of the form

$$\int_0^T d\omega\, |\langle \psi \mid e^{-i\omega A}\, \phi \rangle|^2 = \int_0^T d\omega\, |\hat{\mu}_{\psi,\phi}(\omega)|^2.$$

The following theorem shows how this time average is related to the dimension $\kappa^{\pm}(2)$. We will prove the following theorem not only for measures but also essentially for any tempered distribution that is regular enough at infinity (Holschneider 1994).

Theorem 10.0.1

Let $\eta \in S'(\mathbb{R})$, $\eta \neq 0$, satisfy at $s * \eta \in S(\mathbb{R})$ for all $s \in S(\mathbb{R})$. Suppose that $\eta \notin L^2(\mathbb{R})$. Then $\kappa^-(2) \le 0$ and it follows that

$$-\kappa^+(2) = \liminf_{T\to\infty} \frac{\log \int_0^T d\omega\, |\hat{\eta}(\omega)|^2}{\log T}$$

$$\le \limsup_{T\to\infty} \frac{\log \int_0^T d\omega\, |\hat{\eta}(\omega)|^2}{\log T} = -\kappa^-(2). \qquad (10.0.1)$$

If $\eta \in L^2(\mathbb{R})$, *then* $\kappa^-(2) \geq 0$ *and the large* T *behaviour is trivial in the sense that the integral in* (10.0.1) *tends to a finite constant. The speed of convergence is given by*

$$-\kappa^+(2) = \liminf_{T\to\infty} \frac{\log \int_T^\infty d\omega\, |\hat{\eta}(\omega)|^2}{\log T} \leq \limsup_{T\to\infty} \frac{\log \int_T^\infty d\omega\, |\hat{\eta}(\omega)|^2}{\log T} = -\kappa^2(2).$$

This shows that if the spectral measure of a bounded self-adjoint operator A has a certain two-wavelet dimension then the long term behaviour of e^{-iTA} is governed by this number. In particular, for all $\gamma > -\kappa^+(2)$ there is a constant $c > 0$ such that for T large enough

$$\int_0^T d\omega\, |\langle \psi \mid e^{-i\omega A}\, \phi\rangle|^2 \geq cT^\gamma$$

Proof. Let $g \in S_+(\mathbb{R})$ and consider $\mathscr{W}_g\eta$. As we may, we suppose that $\hat{g}$ is compactly supported. Again we may suppose in addition that $\hat{\eta}(\omega) = O(\omega^m)$ for all m and that therefore the wavelet transform decays fast at large scale.

Suppose $\eta \notin L^2(\mathbb{R})$. This implies that $\int_0^\infty (da/a) \int_{-\infty}^{+\infty} db\, |\mathscr{W}_g\eta(b, a)|^2 = \infty$ and thus

$$\Gamma_g(a, 2) = \int_a^\infty \frac{d\alpha}{\alpha} \int_{-\infty}^{+\infty} db\, |\mathscr{W}_g\eta(b, \alpha)|^2.$$

We therefore now consider $\kappa^-(2) \leq 0$ and thus

$$\int_\alpha^1 \frac{da}{a} \int_{-\infty}^{+\infty} db\, |\mathscr{W}_g\eta(b, a)|^2 \to \infty$$

as $\alpha \to 0$. A direct application of Parseval's equation shows that we have

$$\int_{-\infty}^{+\infty} db\, |\mathscr{W}_g\eta(b, a)|^2 = \int_0^\infty d\omega\, |\hat{g}(a\omega)|^2\, |\hat{\eta}(\omega)|^2.$$

Then let

$$H(\omega) = \int_\omega^\infty \frac{da}{a} |\hat{g}(a)|^2 = \int_1^\infty \frac{da}{a} |\hat{g}(a\omega)|^2.$$

By a simple exchange of integration we have

$$\begin{aligned}\int_{1/T}^\infty \frac{da}{a} \int_{-\infty}^{+\infty} db\, |\mathscr{W}_g\eta(b, a)|^2 &= \int_{1/T}^\infty \frac{da}{a} \int_0^\infty d\omega\, |\hat{g}(a\omega)|^2\, |\hat{\eta}(\omega)|^2 \\ &= \int_0^\infty d\omega\, H(\omega/T)|\hat{\eta}(\omega)|^2.\end{aligned}$$

And finally,

$$\int_0^\infty d\omega\, H(\omega/T)|\hat{\eta}(\omega)|^2 = \int_{1/T}^\infty \int_{-\infty}^{+\infty} db\, |\mathscr{W}_g\eta(b,a)|^2 = \Gamma_g(T^{-1}, 2).$$

Since H is non-negative and of compact support (since $\hat{g}$ is) and $H(0) > 0$ we can find numbers $\lambda > 0$ and $\Lambda > 0$ such that

$$\lambda\chi_{[0,\lambda]}(\omega) \leq H(\omega) \leq \Lambda\chi_{[0,\lambda]}(\omega),$$

where χ_I is the characteristic function of I. Therefore,

$$\lambda \int_0^{\lambda T} d\omega\, |\hat{\eta}(\omega)|^2 \leq \Gamma_g(T^{-1}, 2) \leq \Lambda \int_0^{\Lambda T} d\omega\, |\hat{\eta}(\omega)|^2,$$

and the theorem follows.

Finally, the proof for the case where $\eta \in L^2(\mathbb{R})$ is the same; we only have to use

$$\Gamma_g(a, 2) = \int_0^a \frac{da}{a} \int_{-\infty}^{+\infty} db\, |\mathscr{W}_g\eta(b,a)|^2$$

and to adapt the limits of integration accordingly. □

As an easy corollary of this theorem we have the following quantitative version of the RAGE theorem. Let A be a bounded self-adjoint operator acting in some Hilbert space $\mathscr{H}$ and let dE_λ be its spectral family. Further, let B be a Hilbert–Schmidt operator with singular decomposition

$$B: s \mapsto \sum_{n\in\mathbb{Z}} \gamma_n \langle \phi_n \mid s \rangle \psi_n$$

with respect to the orthonormal sets $\{\phi_n\}$ and $\{\psi_n\}$. Since

$$\|B\, e^{-i\omega A}\, \varphi\|^2 = \sum_{n\in\mathbb{Z}} |\gamma_n|^2\, |\langle \phi_n \mid e^{-i\omega A}\, \varphi \rangle|^2,$$

we have

$$|\gamma_n|^2\, |\langle \phi_n \mid e^{-i\omega A}\, \varphi \rangle|^2 \leq \|B\, e^{-i\omega A}\, \varphi\|^2$$

and thus one can use Theorem 10.0.1 to obtain quantitatively lower estimates for the speed of decay of

$$\frac{1}{T}\int_0^T d\omega\, \|B\, e^{-i\omega A}\, \varphi\|^2$$

in terms of the two-wavelet dimensions of the spectral measures $d\mu_{\phi_n,\varphi}(\lambda) = \langle \phi_n \mid dE_\lambda \varphi \rangle$.

Appendix

11 Local self-similarity and singularities

We will end this section with a short discussion about the relation of local self-similarity, and singularities and fractal dimensions. It turns out that, for the invariant measures that we have constructed, there is an intimate relation between the local singularities, local self-similarity, and the generalized fractal dimensions.

Consider again an IFS ϕ, and suppose that t^* is contained in a limit cycle, that is, t^* is a fixed point of the iterated dynamic $\phi^N(t^*) = t^*$. To fix the ideas, we assume that we have only two scales and two probabilities. The general case follows immediately. So let t^* be coded by a periodic sequence, say

$$t^* = RRLRRL, \ldots,$$

so that t is a fixed point of the iterated mapping $\phi^3 t^* = t^*$. Now consider a sufficiently small interval $I = I(t^*, \epsilon)$ of length ϵ centred around t^*. According to the kneading sequence of the fixed point, this interval is dilated under successive application of the dynamic ϕ, first by a dilation factor l_R^{-1}, then again by l_R^{-1}, and finally by l_L^{-1} before it again becomes a neighbourhood of t. At the same time the probabilities of each of these stretched intervals are re-scaled by p_R, then by p_R and finally by p_L. Therefore, we have shown that

$$\mu_\infty(I(t^*, \epsilon)) = c\mu_\infty(I(t^*, \lambda\epsilon)),$$

where $c = p_R^{-2} p_L^{-1}$ and $\lambda = l_R^2 l_L$. This implies that, around every fixed point, the measure is self-similar in the sense that the measure at a smaller scale looks the same, up to a re-scaling factor. Therefore, we have the local scaling behaviour

$$\mu_\infty(I) \sim |I|^\alpha, \qquad |I| \to 0, I \ni t^*.$$

The number $\alpha = \alpha(t^*)$ is called the *local scaling exponent* or *local fractal dimension.* It is actually a dimension because it measures the relation between the I-scale and the μ_∞-scale.

In analogy we define for any $t \in \Lambda_\infty$ the local scaling exponent as

$$\alpha(t) = \liminf_{|I| \to 0, t \in I} \frac{\log \mu(I)}{\log|I|}.$$

As we have shown, for any point t^* in a limit cycle we have

$$\alpha = \frac{\sum n_k \log p_k}{\sum n_k \log l_k},$$

where n_k is the number of times that the symbol k appears in the kneading sequence of t^*. To put it a little differently, for every $t \in \Lambda_\infty$ we define the local stretching rate of the dynamic via

$$l^{-1}(t) = \lim_{|I| \to 0, t \in I} \frac{|\phi(I)|}{|I|}.$$

In the same way we have a local re-scaling in probability space:

$$p(t) = \lim_{|I| \to 0, t \in I} \frac{\mu(\phi I)}{\mu(I)}.$$

Then we can write for any t in the attracting domain of some limit cycle

$$\alpha(t) = \frac{\langle \log p \rangle}{\langle \log l \rangle},$$

where $\langle \cdot \rangle$ is the mean over the orbit of t under the dynamic ϕ.[13]

11.1 The $f(\alpha)$ spectrum

We have seen so far that local self-similarity and local re-renormalization imply the existence of scaling behaviour, and thus the presence of singularities. We now want to show how the local singularities and the fractal dimensions are related.

We have already defined, for every measure, the local singularity strength via

$$\alpha(t) = \liminf_{|I| \to 0, t \in I} \frac{\log(\mu(I))}{\log|I|}.$$

Let us now consider the set of points that produce the same type of singularity. That is, for every $\gamma \in \mathbb{R}$ we now consider

$$\Psi_\gamma = \{t \in \mathbb{R}: \alpha(t) = \gamma\}.$$

To quantify the importance of this set of points, having all the same type of singularity, we look at the Hausdorf dimension of the set Ψ_α:

$$f(\alpha) = d_{\mathrm{H}}(\Psi_\alpha).$$

This function is called the *singularity spectrum* of the measure μ. The important fact is that—at least for the fractal measures that we have constructed—there is an intimate relation between the singularity spectrum and the generalized fractal dimensions, as stated in the theorem, we give without proof and we refer to Collet *et al.* (1987).

---

[13] If the length of the cycle becomes arbitrarily large, the quantity $\langle \log l \rangle$ is also known as a *Lyapunov* exponent.

Theorem 11.1.1

The singularity spectrum and the generalized fractal dimensions of a balanced measure associated to an IFS are related via a Legendre transform

$$\alpha = \frac{d\tau(q)}{dq}, \qquad f(\alpha) = \alpha q - \tau(q).$$

Instead of giving a proof we will only give a heuristic argument due to Frisch and Parisi (1985). Suppose we cover the fractal measure with cubes Q_n of size ϵ. Then every probability $p_n = \mu(Q_n)$ behaves like ϵ_n^α, where α is the local scaling exponent at the box n. Then we can write

$$G(\epsilon, q) = \sum_n p_n^q \sim \sum_n \epsilon^{\alpha_n p}.$$

We now sum first over all boxes with the same α and then over different alphas to obtain

$$G(\epsilon, q) = \sum_\alpha \sum_{\alpha_k = \alpha} \epsilon^{\alpha_n p} = \sum_\alpha \epsilon^{\alpha_n p} \sum_{\alpha_k = \alpha} 1.$$

The last sum is the number of boxes in which we have the exponent α. This number is heuristically given by $\epsilon^{-f(\alpha)}$ and we have

$$G(\epsilon, q) \sim \sum_\alpha \epsilon^{\alpha q - f(\alpha)}.$$

The main contribution to this sum comes from the point α where the exponent is minimal. On the other hand, by definition, $G \sim e^{\tau(q)}$ and therefore

$$\tau(q) = \min_\alpha(\alpha q - f(\alpha)).$$

This is precisely the definition of the Legendre transform. If we suppose that all functions are convex and smooth we have the following equivalent system of equations:

$$d\tau(q)/dq = \alpha, \qquad \tau(q) = \alpha q - f(\alpha).$$

In Figure 11.1 we show the relations between the singularity spectrum $f(\alpha)$ and the generalized fractal dimensions in the case of the invariant measure of an IFS. Note that the maximum of $f(\alpha)$ corresponds to the Hausdorf dimension of the support of μ_∞. The dimensions $d_{+\infty}$ and $d_{-\infty}$ correspond to the weakest and strongest singularities respectively.

12 On the fractality of orthonormal wavelets

Consider the orthonormal wavelets that we constructed in Chapter 3. As we saw, they all satisfied the so-called two-scale equation, which

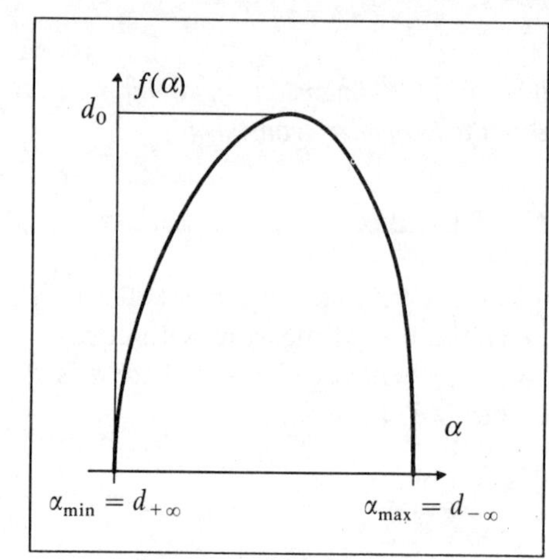

Fig. 11.1 The $f(\alpha)$ spectrum.

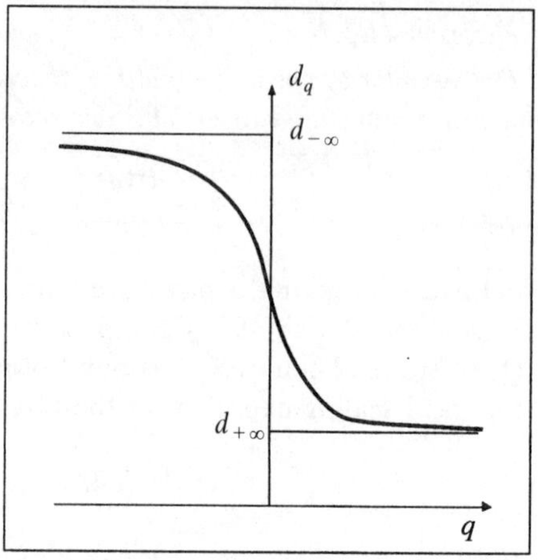

Fig. 11.2 The generalized fractal dimensions.

read

$$\phi(t) = \sum_{k \in \mathbb{Z}} \alpha(k)\phi(2t - k).$$

This is essentially of the same kind as (8.1.1). One important difference, however, is that there are no gaps since the different copies of the scaled ϕ overlap. This overlap disappears if we interpret this equation as the vector-valued analogue for the construction on the multi-fractal measures associated to an IFS. The rescaling with the *a priori* numbers is replaced by the action of some operators. Let us follow the idea of Daubechies and Lagarias (1991) and introduce the vector-valued function $\eta\colon [0, 1) \mapsto \mathbb{R}^{2N-1}$, $t \mapsto \eta(t)_k$, defined through

$$\eta(t)_k = \phi(t + k), \qquad 0 \le t < 1,\ k \in \mathbb{Z}.$$

For $0 \le t < 1/2$ we have $\phi(2t + n) = \eta(2t)_{2n}$, whereas for $1/2 \le t < 1$ we can write $\phi(2t + n) = \phi(2t + 1 + 2n - 1) = \eta(2t - 1)_{2n-1}$. Therefore, we have

$$\phi(t + n) = \sum_{k \in \mathbb{Z}} \alpha(k)\phi(2t + 2n - k),$$

$$\eta(t)_n = \phi(t + n) = \sum_{k \in \mathbb{Z}} \alpha(k)\phi(2t + 2n - k)$$

$$= \sum_{k \in \mathbb{Z}} \alpha(k)\eta(2t)_{2n-k} \quad (0 \le t < 1/2),$$

$$\eta(t)_n = \phi(t + n) = \sum_{k \in \mathbb{Z}} \alpha(k)\phi(2t - 1 + 2n - k + 1)$$

$$= \sum_{k \in \mathbb{Z}} \alpha(k)\eta(2t - 1)_{2n-k+1} \quad (1/2 \le t < 1).$$

Therefore, if we introduce the two linear operators acting on sequences

$$\mathcal{T}_0 : s(k) \mapsto \alpha * (\Sigma_2^{\mathbb{Z}})^* s = D_{\alpha,2}^{\mathbb{Z}} s, \qquad \mathcal{T}_1 : s(k) \mapsto \alpha * (\Sigma_2^{\mathbb{Z}})^* \Sigma s = D_{\alpha,2}^{\mathbb{Z}} \Sigma s,$$

then we have the following vector-valued identity:

$$\eta(t) = \begin{cases} \mathcal{T}_0 \eta(2t) & 0 \le t < 1/2, \\ \mathcal{T}_1 \eta(2t-1) & 1/2 \le t < 1. \end{cases}$$

The right-hand side defines an operator acting in the space of vector-valued measures. It is formally of the same type as the operator (8.1.1) where the numbers p_k/l_k have simply been replaced by operators.

5

GROUP THEORY AS UNIFYING LANGUAGE

In this chapter we will give a very short introduction to the theory of representations of groups and their application to the understanding of the wavelet transform, at least in a Hilbert-space context. The discrete wavelet transform can be understood in an algebraic setting.

1 Some notions of group theory

A set G, together with an operation '$\circ$'

$$\circ : G \times G \to G, \qquad [x, y] \mapsto x \circ y$$

which associates to every pair $x, y \in G$ a new element $x \circ y \in G$ is called a *group* iff

(i) (neutral element): there is $e \in G: e \circ x = x$ for all $x \in G$,
(ii) (inverse element): for all $x \in G$ there is $x^{-1} \in G: x^{-1} \circ x = e$.
(iii) (associative law): $(x \circ y) \circ z = x \circ (y \circ z)$ for all $x, y, z \in G$.

In the following we will recall some facts about groups without proof. The neutral element $e \in G$ is uniquely determined by condition (i). It satisfies, in addition, $x \circ e = x$ for all $x \in G$. The inverse element x^{-1} as defined in condition (ii) is also unique. It satisfies, in addition, $x \circ x^{-1} = e$. Condition (iii) states that, the associative law holds. In general, $x \circ y \neq y \circ x$. A group in which for all $x, y \in G$ we have

(iv) (commutative law): $x \circ y = y \circ x$

is called *Abelian.* In the case of an Abelian group one usually writes $x + y$ instead of $x \circ y$, and, consequently, $-x$ instead of x^{-1}. A subset $H \subset G$ is called a *subgroup* of G if, together with the operation $\circ$ inherited from G, it is a group on its own. For $H \subset G$ to be a subgroup it is necessary and sufficient that, together with x and y, $x \circ y^{-1}$ is an element of H, for all $x, y \in H$.

Example 1.0.1
Consider the set of real numbers $\mathbb{R}$. It is an Abelian group with respect to the usual addition

$$+ : \mathbb{R} \times \mathbb{R} \to \mathbb{R}, \qquad [t, u] \mapsto t + u.$$

The neutral element is clearly the 0. We will write for this group $(\mathbb{R}, +)$ or simply $\mathbb{R}$ if no confusion is possible. The set of integers $\mathbb{Z} \subset \mathbb{R}$ is a subgroup. It is denoted by $(\mathbb{Z}, +)$ or simply by $\mathbb{Z}$ if no confusion is possible.

The positive real numbers $\mathbb{R}^+$ equipped with the multiplication are also a commutative Abelian group, which will be denoted by $(\mathbb{R}^+, \cdot)$ or simply $\mathbb{R}^+$.

Example 1.0.2
Consider the upper half-plane $\mathbb{H}$. It can be made into a group by setting

$$(b, a) \circ (b', a') = (b + ab', aa').$$

The neutral element is $(0, 1)$, and the inverse is given by

$$(b, a)^{-1} = (-b/a, 1/a).$$

We leave it to the reader to verify the associative law. This group is called the affine group. The reason for this is the following. Consider the set of affine transformations of the real line $\mathbb{R}$:

$$A^l_{(b,a)}: \mathbb{R} \to \mathbb{R}, \qquad t \mapsto at + b.$$

These transformations satisfy the composition law

$$A^l_{(b,a)} A^l_{(b',a')} = A^l_{(b+ab',aa')} = A^l_{(b,a)\circ(b',a')}.$$

The affine group will be denoted by $(\mathbb{H}, \circ)$ or simply by $\mathbb{H}$ if no confusion is possible. This group is not Abelian.

Example 1.0.3
This last example of a group is very important for our subsequent considerations. Let $\mathscr{H}$ be a Hilbert space, and let $\mathscr{U}(\mathscr{H})$ be the set of unitary operators on $\mathscr{H}$. Then $\mathscr{U}(\mathscr{H})$ is a group if we take as group operation the composition of maps

$$\circ: \mathscr{U}(\mathscr{H}) \times \mathscr{U}(\mathscr{H}), \qquad [U_1, U_2] \mapsto U_1 \circ U_2 = U_1 U_2.$$

1.1 Direct sum of groups

Let G_1 and G_2 be two groups. The set $G_1 \times G_2$ consisting of all couples $[x, y]$ with $x \in G_1$ and $y \in G_2$, can be made into a group on its own by componentwise group operation:

$$[x, y] \circ [x', y'] = [x \circ x', y \circ y'].$$

This group will be denoted by $G_1 \oplus G_2$. It is called the direct sum of G_1 and G_2. We can identify the set of points $[x, e_2]$ where $x \in G_1$ and e_2 is the neutral element of G_2, with G_1. Therefore, $G_1 \oplus G_2$ contains G_1 as a subgroup. The same is true for G_2, which can be identified with all couples $[e_1, y]$, where $y \in G_2$ and e_1 is the neutral element of G_1.

1.2 Quotient groups

A subgroup $H \subset G$ is called *normal* if for all $x \in G$ we have[1] $x^{-1} \circ H \circ x \subset H$. Note that this does not mean that all elements of H commute with all elements of G. Rather it means that for all $h \in H$ and for all $x \in G$ we can find an $h' \in H$ such that $h \circ x = x \circ h'$. In the case of an Abelian group, every subgroup is normal. We say

$$x = y \bmod H \tag{1.2.1}$$

if there is an $h \in H$ such that $x = y \circ h$, or, what amounts to the same, if there is an h' with $x = h' \circ y$. Relation (1.2.1) is actually an equivalence relation. The set of different classes is denoted by G/H and it is called the *quotient group*. Every element $x \bmod H \in G/H$ may be identified with the set $xH \subset G$ consisting of all elements of the form $x \circ h$ with $h \in H$. The element $x \in G$ is called the representative of the class $x \bmod H$. It is not unique since every other element $x \circ h$ with $h \in H$ could do the job as well.

The set of classes G/H can be made into a group by setting

$$x \bmod H \circ y \bmod H = (x \circ y) \bmod H.$$

If G is an Abelian group, then G/H is Abelian, too.

Example 1.2.1

Consider the circle $\mathbb{T}$. It may be identified with the set of complex numbers $z \in \mathbb{C}$ with unit modulus $|z| = 1$. Together with the multiplication of complex numbers,

$$[z_1, z_2] = [e^{2i\pi\theta_1}, e^{2i\pi\theta_2}] \to z_1 z_2 = e^{2i\pi(\theta_1 + \theta_2)},$$

$\mathbb{T}$ is an Abelian group. We may identify the circle with the periodized real line by setting $z = e^{2i\pi\theta}$ with $\theta \in \mathbb{R}$. However, two points θ_1 and θ_2 that differ by an integral number, $\theta_1 = \theta_2 + n$ with $n \in \mathbb{Z}$, correspond to the same point on $\mathbb{T}$ and we write $\theta_1 = \theta_2 \bmod \mathbb{Z}$. Thus we obtain $\mathbb{T} \leftrightarrow \mathbb{R}/\mathbb{Z}$.

In the same way, we can use the group $\mathbb{Z}$ together with its subgroup $N\mathbb{Z}$ consisting of all points kN with $k \in \mathbb{Z}$ to make the discretized torus $\mathbb{Z}/N\mathbb{Z}$ into a group. We may identify $\mathbb{Z}/N\mathbb{Z}$ with the subgroup of $\mathbb{T}$ that consists of the complex numbers $z_k = e^{2i\pi k/N}$, with $k = 0, 1, \ldots, N-1$, that is, the Nth roots of unity

$$\mathbb{Z}/N\mathbb{Z} \simeq \{z_k \in \mathbb{C}: z_k^N = 1\}.$$

[1] Operations with sets should be understood in the obvious way.

1.3 Homomorphisms

Consider two groups G_1 and G_2. A map between them:

$$\phi: G_1 \to G_2$$

is called a *homomorphism* if it translates the group operations on G_1 into group operations on G_2, that is, if for all $x, y \in G_1$ we have

$$\phi(x \circ y) = \phi(x) \circ \phi(y) \qquad \begin{array}{ccc} G_1 \times G_1 & \xrightarrow{\phi \times \phi} & G_2 \times G_2 \\ \downarrow{\circ} & & \downarrow{\circ} \\ G_1 & \xrightarrow{\phi} & G_2 \end{array}$$

where we have written $\phi \times \phi$ for the map from $G_1 \times G_1$ to $G_2 \times G_2$ that is naturally induced by ϕ by componentwise action

$$\phi \times \phi: G_1 \times G_1 \to G_2 \times G_2, \qquad [x, y] \mapsto [\phi(x), \phi(y)].$$

If ϕ is bijective, then the map ϕ^{-1} exists and is a homomorphism, too. We then cannot distinguish from a purely group-point-of-view between G_1 and G_2. We therefore say that both are isomorphic and write

$$G_1 \simeq G_2.$$

Now consider again an arbitrary homomorphism. The subset of points in G_1 that is mapped onto the neutral element $e_2 \in G_2$ is called the *kernel* of ϕ:

$$\text{kern}\, \phi = \{x \in G_1 : \phi(x) = e_2\} \subset G_1.$$

On the other hand, the image of ϕ is defined as usual by

$$\text{imag}\, \phi = \{\phi(x) : x \in G_1\} \subset G_2.$$

The image of a group under a homomorphism ϕ is always a group and the kernel of ϕ is always a group, too. The last one is even a normal subgroup of G_1 and we may consider the quotient group $G_1/\text{kern}\, \phi$.

Theorem 1.3.1 (*First homomorphy theorem*)
We have

$$\text{imag}\, \phi \simeq G_1/\text{kern}\, \phi.$$

Example 1.3.2
Consider the map

$$\phi: \mathbb{R} \to \mathbb{T}, \qquad t \mapsto e^{2i\pi t}.$$

This mapping is a surjective homomorphism, that is, imag $\phi = \mathbb{T}$. Its kernel is the set of integral numbers kern $\phi = \mathbb{Z}$, since $e^{2i\pi n} = 1$ iff $n \in \mathbb{Z}$, and by what we have said we re-obtain

$$\mathbb{T} \simeq \mathbb{R}/\mathbb{Z}.$$

1.4 Representations

Let $\mathscr{H}$ be a Hilbert space, and let $\mathscr{U}(\mathscr{H})$ be the group of unitary operators acting in $\mathscr{H}$. A homomorphism

$$U: G \to \mathscr{U}(\mathscr{H}), \qquad x \mapsto U(x)$$

from a group G to $\mathscr{U}(\mathscr{H})$ is called a unitary *representation*. This means that for every $x \in G$ we have an associated unitary operator $U(x)\colon \mathscr{H} \to \mathscr{H}$, such that

$$U(x \circ y) = U(x)U(y), \qquad \text{for all } x, y \in G.$$

By the unitarity of $U(x)$ it follows that

$$U(x^{-1}) = U^{-1}(x) = U^*(x).$$

Example 1.4.1
For every $b \in \mathbb{R}$ we have defined the translation operator via

$$T_b\colon L^2(\mathbb{R}) \to L^2(\mathbb{R}), \qquad s(t) \mapsto s(t-b).$$

The translations are unitary operators acting on the Hilbert space $L^2(\mathbb{R})$. It can be immediately verified that

$$T_b T_{b'} = T_{b+b'}$$

and thus

$$U: \mathbb{R} \to \mathscr{U}(L^2(\mathbb{R})), \qquad b \mapsto T_b,$$

is a unitary representation of the group $(\mathbb{R}, +)$.

Example 1.4.2
Consider the unitary[2] dilations:

$$D_a\colon L^2(\mathbb{R}) \to L^2(\mathbb{R}), \qquad s(t) \mapsto \frac{1}{\sqrt{a}}\, s\left(\frac{t}{a}\right), \quad a \in \mathbb{R}_+.$$

Clearly, the map

$$U: \mathbb{R}_+ \to \mathscr{U}(L^2(\mathbb{R})), \qquad a \mapsto D_a,$$

is a unitary representation of the group $\mathbb{R}_+$ equipped with the multiplication as group operation.

[2] Henceforth we use the unitary version of the dilations.

Example 1.4.3

Consider the affine group $\mathbb{H}$. We can use the previous representations to obtain a representation of $\mathbb{H}$ on Hilbert space $L^2(\mathbb{R})$ via

$$U: \mathbb{H} \to \mathscr{U}(L^2(\mathbb{R})), \qquad (b, a) \mapsto U(b, a) = T_b D_a.$$

It can be immediately verified that

$$U(b, a)U(b', a') = U(b + ab', aa'),$$

as it should be.

Let $g \in \mathscr{H}$. The set of vectors that can be reached by the action of the representation is called the *orbit* of g:

$$\text{orbit of } g = \{U(x)g \colon x \in G\}.$$

A subspace $\mathscr{B} \subset \mathscr{H}$ is called *invariant* iff it is mapped into itself under the action of all the operators $U(x)$ with $x \in G$, or, equivalently, iff it contains all its orbits:

$$\mathscr{B} \text{ invariant} \Rightarrow \{U(x)g \colon x \in G, g \in \mathscr{B}\} \subset \mathscr{B}.$$

The null space $\{0\}$ and the whole space $\mathscr{H}$ are trivially invariant spaces. A vector $g \in \mathscr{H}$ is called *cyclic* iff its orbit is dense in $\mathscr{H}$:

$$g \text{ cyclic} \Rightarrow \text{closed span } \{U(x)g \colon x \in G\} = \mathscr{H}.$$

The central definition may now be stated as follows.

Definition 1.4.4

A representation is called irreducible *iff the only closed invariant subspaces are the trivial ones* $\{0\}$ *and* $\mathscr{H}$, *or, equivalently, if every vector* $g \neq 0$ *is cyclic.*

Example 1.4.5

The representation of $\mathbb{R}$ on $L^2(\mathbb{R})$ as translation operators is clearly *not* irreducible. For consider the set of band-limited functions $s \in L^2(\mathbb{R})$ with the Fourier transform supported by some fixed, non-atomic, copact set $K \subset \mathbb{R}$:

$$\operatorname{supp} \hat{s} \subset K.$$

Since the translation operator reads in Fourier space as multiplication by $e^{-ib\omega}$ it does not change the support of the frequency representations. Therefore, for every K we obtain a non-trivial translation invariant subspace of $L^2(\mathbb{R})$. In addition, since K is compact, the space is closed.

Example 1.4.6

The representation of the group $\mathbb{R}_+$ with the multiplication as operation by means of the dilation operator is also not irreducible. For consider a closed cone K with non-empty interior in the upper half-plane with its top in $(0, 0)$. Consider the space Ω of functions $T \in L^2(\mathbb{H})$ that are supported by the cone K:

$$\operatorname{supp} T \subset K.$$

This is a true closed subspace of $L^2(\mathbb{H})$. Now consider an admissible wavelet $g \in S_0(\mathbb{R})$. The projection $\Pi_g \Omega$ of Ω into the image of the wavelet transform is a closed subspace. It is done with the help of the reproducing kernel equation. The kernel is arbitrarily well localized, and therefore it is easy to see that $\Pi_g \Omega$ is a true closed subspace of imag $\mathscr{W}_g$. Therefore, $\mathscr{M}_g \Pi_g \Omega$ is a true closed subspace of $L^2(\mathbb{R})$. By the co-variance properties of the wavelet transform under dilations this subspace is invariant under dilations, and we are done.

Also, neither the representations of the translations alone nor those of dilations are irreducible, their combination is actually irreducible. Consider the set $H^2_+(\mathbb{R})$ and $H^2_-(\mathbb{R})$ consisting of those functions in $L^2(\mathbb{R})$ that are progressive or regressive respectively. By Example 1.4.1 we obtain a unitary representation of the affine group $\mathbb{H}$ on $H^2_+(\mathbb{R})$ and $H^2_-(\mathbb{R})$.

Theorem 1.4.7

The representation

$$U: \mathbb{H} \to \mathscr{U}(H^2_+(\mathbb{R})), \qquad (b, a) \mapsto U(b, a) = T_b D_a,$$

is irreducible.

By symmetry the theorem also holds for $H^2_-(\mathbb{R})$.

Proof. Let $g \in H^2_+(\mathbb{R})$, $g \neq 0$. If the span of the orbit of g were not dense, we could find an $s \in H_+(\mathbb{R})$, $s \neq 0$ with

$$\langle U(b, a) g \mid s \rangle_{\mathbb{R}} = 0, \quad \text{for all } (b, a) \in \mathbb{H}.$$

But this means that the wavelet transform of s with respect to g vanishes identically. We may then find a wavelet $h \in S_0(\mathbb{R})$ that is a reconstruction wavelet for g in the sense that $r = h * \tilde{g}$ is admissible. By Young's inequality we may write $\mathscr{W}_r s(\cdot, a) = h_a * \tilde{g}_a * s$, where the subscript $\cdot_a$ indicates the dilation. By the inversion theorem of Chapter 2, Section 36 we find that $s = 0$ and we are done. □

Consider two unitary representations U_1 and U_2 of the same group G in two different Hilbert spaces $\mathscr{H}_1$ and $\mathscr{H}_2$ respectively. We say that both representations are equivalent and we write $U_1 \simeq U_2$ if there is a continuous, bijective operator $K: \mathscr{H}_1 \to \mathscr{H}_2$ that translates the group translations into each other:

$$KU_1(x)K^{-1} = U_2(x), \quad \text{for all } x \in G.$$

Consider the space $\mathscr{H}_1 \oplus \mathscr{H}_2$ consisting of all vectors $[s, r]$ with $s \in \mathscr{H}_1$ and $r \in \mathscr{H}_2$. This space is a Hilbert-space with the scalar product being componentwise:

$$\langle [s, r] \mid [v, w] \rangle_{\mathscr{H}_1 \oplus \mathscr{H}_2} = \langle s \mid v \rangle_{\mathscr{H}_1} + \langle r \mid w \rangle_{\mathscr{H}_2}.$$

Any operator acting on $\mathscr{H}_1 \oplus \mathscr{H}_2$ can be identified with a matrix of operators acting componentwise:

$$\begin{bmatrix} A & B \\ C & D \end{bmatrix} : \mathscr{H}_1 \oplus \mathscr{H}_2 \to \mathscr{H}_1 \oplus \mathscr{H}_2,$$

$$\begin{bmatrix} s \\ r \end{bmatrix} \mapsto \begin{bmatrix} A & B \\ C & D \end{bmatrix} \begin{bmatrix} s \\ r \end{bmatrix} = \begin{bmatrix} As + Dr \\ Cs + Dr \end{bmatrix}.$$

This can be used to define the direct sum $U_1 \oplus U_2$ of two representations U_1 and U_2 as the representation of G on $\mathscr{H}_1 \oplus \mathscr{H}_2$ defined through componentwise action:

$$U_1 \oplus U_2 : G \to \mathscr{U}(\mathscr{H}_1 \oplus \mathscr{H}_2), \qquad x \mapsto \begin{bmatrix} U_1(x) & 0 \\ 0 & U_2(x) \end{bmatrix}.$$

Now let us consider the other way round. Let $\mathscr{B}$ be a closed invariant subspace $\mathscr{B} \subset \mathscr{H}$. The restriction of the representation to $\mathscr{B}$ is a representation of G in its own. It is called a subrepresentation of U, and we denote it by $U_{\mathscr{B}}$. Together with $\mathscr{B}$, the orthogonal complement $\mathscr{B}^{\perp}$ is also a closed invariant subspace and, once more, the restriction of U to $\mathscr{B}^{\perp}$ is a subrepresentation, called $U_{\mathscr{B}^{\perp}}$. Now $\mathscr{H} = \mathscr{B} \oplus \mathscr{B}^{\perp}$ and therefore we have the following decomposition of U:

$$U \simeq U_{\mathscr{B}} \oplus U_{\mathscr{B}^{\perp}}.$$

A representation is irreducible iff no non-trivial decomposition is possible.

Example 1.4.8

Consider the representation of the affine group on $L^2(\mathbb{R})$. As we saw in Theorem 1.4.7, it can be split into irreducible components

$$L^2(\mathbb{R}) = H^2_+(\mathbb{R}) \oplus H^2_-(\mathbb{R}).$$

1.5 Schur's lemma

Consider once more two representations U_1 and U_2 of the same group G, but on different representation spaces $\mathscr{H}_1$ and $\mathscr{H}_2$. A bounded operator,

$$K : \mathscr{H}_1 \to \mathscr{H}_2,$$

is called an *intertwining* operator iff we have

$$KU_1(x) = U_2(x)K, \qquad \begin{array}{ccc} \mathscr{H}_1 & \xrightarrow{U_1(x)} & \mathscr{H}_1 \\ \downarrow{\scriptstyle K} & & \downarrow{\scriptstyle K} \\ \mathscr{H}_2 & \xrightarrow{U_2(x)} & \mathscr{H}_2 \end{array} \qquad \text{for all } x \in G.$$

Note that we did not suppose K to be bijective. The set of all intertwining operators is an algebra, denoted by $\Gamma(U_1, U_2)$. In the case where $U_1 = U_2$ we have the following theorem.

Theorem 1.5.1

Let U be an irreducible representation. Then $\Gamma(U, U)$ is one-dimensional. It consists of the scalar multiples of the identity.

For a proof of this standard fact see Warner (1972, page 246). We will need the following corollary that characterizes the possible intertwining operators if either U_1 or U_2 is irreducible. In its simplest form it is commonly known as Schur's lemma. In its present form it is essentially due to Mackey.

Theorem 1.5.2

If either of the representations U_1 or U_2 is irreducible and if K is a bounded, injective, intertwining operator whose range is dense, then K is a constant multiple of a surjective isometry.

Proof. The proof is adapted from Warner (1972, page 245). Since $KU_1(x) = U_2(x)K$ we have, upon taking adjoints, $U_1^*(x)K^* = K^*U_2^*(x)$ or, equivalently, $K^*U_2(x) = U_1(x)K^*$. Therefore, $K^*KU_1(x) = K^*U_2(x)K = U_1(x)K^*K$. Let $K = \Phi|K|$ be the polar decomposition of K. Then $|K| = \sqrt{K^*K}$ is a bounded, injective operator from $\mathscr{H}_1$ to $\mathscr{H}_1$, with a dense range, whereas $\Phi\colon \mathscr{H}_1 \to \mathscr{H}_2$ is a surjective isometry. We have shown so far that $|K|^2U_1(x) = U_1(x)|K|^2$. Now $|K|$ can be approximated in operator norm by polynomials in $|K|^2$. Hence we have

$$|K|U_1(x) = U_1(x)|K|. \tag{1.5.1}$$

We now claim that this implies that

$$\Phi U_1(x) = U_2(x)\Phi. \tag{1.5.2}$$

Supposing this to be true it implies that both representations are unitarily equivalent. Hence, in particular, if one of them is irreducible, the other is irreducible, too. This implies, thanks to (1.5.1) and the previous theorem, that $|K|$ is a constant multiple of the identity and the theorem follows.

We still have to prove (1.5.2). We have $KU_1(x) = \Phi U_1(x)|K|$ and $KU_1(x) = U_2(x)\Phi|K|$. Hence $(\Phi U_1(x) - U_2(x)\Phi)|K| = 0$. Since the range of $|K|$ is dense and all operators are bounded, equation (1.5.2) follows. □

1.6 Group action

Let G be a group and S an arbitrary set. We say that G is *acting on S on the left* iff for every $x \in G$ there is a transformation of S into itself, $A^l_x: S \to S$, such that $A^l_{x \circ y} = A^l_x A^l_y$ for all $x, y \in G$. We say that G is *acting on S on the right* iff there is a map $A^r_x: S \to S$ with $A^r_{x \circ y} = A^r_y A^r_x$ for all $x, y \in G$. To every left-action A^l we may associate a right-action via $A^r_x = A^l_{x^{-1}}$ and, vice versa, this defines a left-action if a right-action is given.

Example 1.6.1
Consider the affine group $\mathbb{H}$ acting on the real line $\mathbb{R}$. A left-action is given by

$$A^l_{(b,a)}: \mathbb{R} \to \mathbb{R}, \qquad t \mapsto at + b,$$

whereas a right-action is given by

$$A^r_{(b,a)}: \mathbb{R} \to \mathbb{R}, \qquad t \mapsto \frac{t-b}{a}.$$

We may take $S = G$, that is, we consider the action of the group G on itself. The action of G on the left is given by the left-multiplication

$$A^l_x: G \to G, \qquad x \mapsto x \circ z,$$

whereas the right-multiplication gives rise to a right-action

$$A^r_x: G \to G, \qquad z \mapsto z \circ x.$$

Example 1.6.2
The left-action of the affine group in itself is given by

$$A^l_{(b,a)}: \mathbb{H} \to \mathbb{H}, \qquad (b', a') \mapsto (b + ab', aa'),$$

whereas the right-action in itself is given by

$$A^r_{(b,a)}: \mathbb{H} \to \mathbb{H}, \qquad (b', a') \mapsto (b + a'b, a'a).$$

1.7 Invariant measures

From now on we make the technical assumption that G is a locally compact group. This assumption will be made throughout the rest of this book. Consider a group G that acts on a set S. Suppose, in addition, that S is equipped with some (Borell) measure. We write

$$\int_S d\mu(t) f(t)$$

for the integral of a measurable function f over the S with respect to the measure μ, and $\mu(I)$ will stand for

$$\int_S d\mu(t) \chi_I(t),$$

where χ_I is the characteristic function of the measurable set $I \subset S$. Consider a measurable subset $I \subset S$ and its 'volume' $\mu(I)$. Under the left-action of G on S, the set I is transformed into the set GI, consisting of all points $A_x^l(t)$ with $x \in G$ and $t \in I$, where A_x^l is the left-action of G onto S. We say the measure μ_l is left-invariant, or simply invariant, if the volume remains constant under the left-action of G:

$$\mu_l(GI) = \mu_l(I), \qquad \text{for all measurable } I \subset S.$$

In the same way we may consider the behaviour of I under the right-action. Let IG be the set consisting of all points $A_x^r t$ with $x \in G$ and $t \in S$. A measure is called right-invariant iff

$$\mu_r(IG) = \mu_r(I), \qquad \text{for all measurable } I \subset S.$$

Such non-trivial invariant measures need not always exist. Consider, for example, the affine group acting on $\mathbb{R}$. The only translation invariant measure is the usual Lebesgue measure dt, whereas the dilation invariant measure is $dt/|t|$.

However, if $S = G$ such invariant measures always exist. In addition, the left- and right-invariant measures are uniquely determined by the group up to a scalar multiple. These measures are called the left- and right-invariant Haar measures of G and we denote them by $d\mu_l$ and $d\mu_r$ respectively. In the case where G is compact we can normalize the measures to be probability measures:

$$\mu_l(G) = \mu_r(G) = 1.$$

If the right-invariant measure is equal to the left-invariant measure—up to a scalar multiple—then the group is called *unimodular*. All Abelian groups are trivially unimodular. But, as we will soon see, there are non-unimodular groups. For non-unimodular groups the left- and right-invariant measures are related via

$$\mu_l(I^{-1}) = \mu_r(I), \qquad \text{for all measurable } I \subset G. \tag{1.7.1}$$

Indeed, we have, upon replacing I by $I \circ x$ for every $x \in G$,

$$\mu_l((Ix)^{-1}) = \mu_l(x^{-1} \circ I^{-1}) = \mu_l(I^{-1}),$$

thereby showing the right-invariance of the measure μ_r as defined by the left-hand side of equation (1.7.1). We may choose the equality because of the uniqueness of the measures. For further use we note that relation (1.7.1) takes on the following form if we integrate over functions:

$$\int_G d\mu_l(x)s(x) = \int_G d\mu_r(x)s(x^{-1}). \tag{1.7.2}$$

Example 1.7.1

Let G be a finite group, that is, it contains only a finite number of elements. The unique invariant measure is given by the counting measure, that is,

$$\mu_l(I) = \mu_r(I) = \#\text{elements in } I.$$

Indeed, neither the left- nor the right-action on G changes the number of elements in I. Therefore, all finite groups[3] are unimodular.

Example 1.7.2

Consider the affine group $\mathbb{H}$. By direct computation we can verify that a left-invariant Haar measure on $\mathbb{H}$ is given by

$$d\mu_l(b, a) = \frac{db\, da}{a^2}.$$

A right-invariant Haar measure is given by

$$d\mu_r(b, a) = \frac{db\, da}{a}.$$

Therefore, the affine group $\mathbb{H}$ is not unimodular.

1.8 Regular representations

Consider the Hilbert space $L^2(G, d\mu_l)$ consisting of the square integrable functions over the group with respect to the left-invariant Haar measure

$$\|s\|^2_{L^2(G, d\mu_l)} = \int_G d\mu_l(x)|s(x)|^2 < \infty.$$

The associated scalar product is denoted by $\langle \cdot \mid \cdot \rangle_{L^2(G, d\mu_l)}$. The left-action $A^l_x\colon y \mapsto x \circ y$ of G onto itself induces a group translation operator

$$\mathrm{L}_x\colon L^2(G, d\mu_l) \to L^2(G, d\mu_l), \qquad s(y) \mapsto s(A^{l-1}_x y) = s(x^{-1} \circ y).$$

Lemma 1.8.1

The operator L_y *is unitary for every* $y \in G$.

[3] This result may be extended to compact groups.

Proof. Indeed, if $s \in L^2(G, d\mu_1)$, then $L_x s \in L^2(G, d\mu_1)$, too, for

$$\int_G d\mu_l(y)|L_x s(y)|^2 = \int_G d\mu_1(y)|s(x^{-1} \circ y)|^2$$
$$= \int_G d\mu_1(x \circ y)|s(y)|^2$$
$$= \int_G d\mu_1(y)|s(y)|^2.$$

In the last equation we have used the left-invariance of the measure. □

In the same way, let us introduce $L^2(G, d\mu_r)$ as the Hilbert space of functions such that

$$\|s\|^2_{L^2(G, d\mu_r)} = \int_G d\mu_r(x)|s(x)|^2 < \infty.$$

The right-action of G onto itself defines a right-translation operator

$$R_x: L^2(G, d\mu_r) \to L^2(G, d\mu_r), \qquad s(y) \mapsto s(A^r_x y) = s(y \circ x).$$

Again this operator is unitary.

By construction we have

$$L_x L_y = L_{x \circ y}, \qquad R_x R_y = R_{x \circ y},$$

and therefore we obtain two representations of the group G on $L^2(G, d\mu_1)$ and on $L^2(G, d\mu_r)$, respectively, by setting

$$\left.\begin{aligned} U_{\text{lreg}}&: G \to \mathscr{U}(L^2(G, d\mu_1)), &\quad x &\mapsto L_x, \\ U_{\text{rreg}}&: G \to \mathscr{U}(L^2(G, d\mu_r)), &\quad x &\mapsto R_x. \end{aligned}\right\} \tag{1.8.1}$$

These representations are called the left-regular and the right-regular representations respectively. Note, however, that both representations are acting in different spaces, namely, $L^2(G, d\mu_r)$ and $L^2(G, d\mu_1)$. However, there is a natural isometric map from one space to the other. It is given by[4]

$$\tilde{}: L^2(G, d\mu_1) \to L^2(G, d\mu_r), \qquad s \mapsto \tilde{s}(x) = \bar{s}(x^{-1}). \tag{1.8.2}$$

We claim that this defines a skew linear isometry. Indeed, we have, by (1.7.2),

$$\int_G d\mu_r(x)|\tilde{s}(x)|^2 = \int_G d\mu_r(x)|s(x^{-1})|^2 = \int_G d\mu_1(x)|s(x)|^2.$$

[4] Note that this coincides with the usual '~' operator $\tilde{s}(t) = \bar{s}(-t)$ in the case where $G = (\mathbb{R}, +)$ and μ is the usual Lebesgue measure.

Example 1.8.2
Consider the affine group $\mathbb{H}$. The left-regular representation is acting in the Hilbert space $L^2(\mathbb{H}, da\, db/a^2)$ and is given by

$$U_{\text{lreg}}(b, a)\colon L^2(\mathbb{H}, da\, db/a^2) \to L^2(\mathbb{H}, da\, db/a^2),$$
$$T(b', a') \mapsto T\left(\frac{b'-b}{a}, \frac{a'}{a}\right).$$

The right-regular representation is, instead, acting in $L^2(\mathbb{H}, db\, da/a)$ via

$$U_{\text{rreg}}(b, a)\colon L^2(\mathbb{H}, db\, da/a) \to L^2(\mathbb{H}, db\, da/a),$$
$$T(b', a') \mapsto T(b' + a'b, a'a).$$

1.9 Group convolutions

Let $s \in L^2(G, d\mu_{\text{r}})$ and $r \in L^2(G, d\mu_{\text{l}})$. We define their left-convolution by

$$s *_l r(y) = \int_G d\mu_{\text{l}}(x) s(x^{-1} \circ y) r(x).$$

With the help of (1.8.2) this can also be written

$$s *_l r(y) = \langle \mathrm{L}_y \tilde{s} \mid r \rangle_{L^2(G, d\mu_{\text{l}})},$$

thereby showing that everything is well defined because $s \mapsto \tilde{s}$ is an isometry. In the same way we define the right-convolution of a function $s \in L^2(G, d\mu_{\text{l}})$ with a function $r \in L^2(G, d\mu_{\text{r}})$ by

$$s *_r r(y) = \int_G d\mu_{\text{r}}(x) s(y \circ x^{-1}) r(x).$$

Again this may be written as a scalar product:

$$s *_r r(y) = \langle \tilde{s} \mid \mathrm{R}_y r \rangle_{L^2(G, d\mu_{\text{r}})}.$$

Note that the convolutions are in general not commutative for non-Abelian groups. Instead, we have

$$s *_l r = r *_r s.$$

1.10 Square integrable representations

Consider a representation U of a group G in some Hilbert space $\mathcal{H}$. For fixed $g \in \mathcal{H}$ let us look at the set of scalar products

$$\langle U(x) g \mid g \rangle_{\mathcal{H}}.$$

This defines a function over the group that associates to every group element, x, the 'correlation' between the group-translated vector $U(x)g$ and g itself. Square integrability means that this correlation function is localized. More precisely, we give the following definition.

Definition 1.10.1

We say the representation U of a group G in a Hilbert space $\mathscr{H}$ is square integrable if there exists one vector, $g \in \mathscr{H}$, $g \neq 0$, such that

$$\langle U(x)g \mid g\rangle_{\mathscr{H}} \in L^2(G, d\mu_1). \tag{1.10.1}$$

Equivalently, we may require that $g \neq 0$ satisfies

$$\int_G d\mu_1(x) \mid \langle U(x)g \mid g\rangle_{\mathscr{H}}|^2 < \infty.$$

We need not distinguish between right- and left-square integrability. Indeed, suppose that g satisfies (1.10.1) then we also have

$$\langle U(x)g \mid g\rangle_{\mathscr{H}} \in L^2(G, d\mu_r).$$

This can be proved by direct computation:

$$\begin{aligned}\int_G d\mu_1(x)|\langle U(x)g \mid g\rangle_{\mathscr{H}}|^2 &= \int_G d\mu_r(x)|\langle U(x^{-1})g \mid g\rangle_{\mathscr{H}}|^2\\ &= \int_G d\mu_r(x)|\langle g \mid U(x)g\rangle_{\mathscr{H}}|^2\\ &= \int_G d\mu_r(x)|\langle U(x)g \mid g\rangle_{\mathscr{H}}|^2.\end{aligned}$$

Example 1.10.2

By the results of the preceding sections, the affine group has a square integrable representation acting in $L^2(\mathbb{R})$. Indeed, consider

$$U(b, a)\colon g \mapsto \frac{1}{\sqrt{a}}\, g\left(\frac{t-b}{a}\right).$$

Square integrability now means that there is a function $g \in L^2(\mathbb{R})$ such that

$$\int_{-\infty}^{+\infty} dt \frac{1}{\sqrt{a}}\, \bar{g}\left(\frac{t-b}{a}\right) g(t) \in L^2(\mathbb{H}, da\, db/a^2).$$

But, as we know, every admissible wavelet will do the job.

Example 1.10.3

Consider the discretized real line $\mathbb{Z}$ together with addition as group operation. Let $\mathscr{H}$ be an equidistant sampling space. A representation of the additive group $\mathbb{Z}$ on $\mathscr{H}$ is given by

$$U(n): \mathscr{H} \to \mathscr{H}, \qquad s(t) \mapsto s(t-n).$$

This representation is square integrable. Indeed, there is a translation invariant basis $\{\phi(t-b)\}$. For this ϕ we obtain

$$\sum_{n \in \mathbb{Z}} |\langle U(n)\phi \mid \phi \rangle|^2 = \sum_{n \in \mathbb{Z}} \delta_{n,0} = 1.$$

2 The 'wavelet' analysis associated to square integrable representations

After all these preparations we now come to the analogue of the wavelet transform for arbitrarily locally compact groups. We will essentially follow Grossmann *et al.* (1985). Consider a square integrable representation U of a (locally compact) group G acting in a Hilbert space $\mathscr{H}$.

Definition 2.0.1

The left-transform over G of a function $s \in \mathscr{H}$ with respect to a wavelet $g \in \mathscr{H}$ is given by the set of scalar products

$$\mathscr{L}_g s(x) = \langle U(x)g \mid s \rangle_{\mathscr{H}}, \qquad x \in G.$$

We sometimes write $\mathscr{L}[g; s]$ instead of $\mathscr{L}_g s$. It maps a vector in the Hilbert space $\mathscr{H}$ to a function over the group G.

In the right-transform, instead, the analysing function g is fixed and is 'compared' with all group translated versions of the analysed function s.

Definition 2.0.2

The right-transform is given by the set of scalar products

$$\mathscr{R}_g s(x) = \langle g \mid U(x)s \rangle_{\mathscr{H}}, \qquad x \in G.$$

Again we use the notation $\mathscr{R}[g; s]$ for $\mathscr{R}_g s$.

Example 2.0.3

The wavelet transform $\mathscr{W}_g$ that we considered in previous chapters can be seen as a left-transform associated to the affine group. Indeed, we have

$$\mathscr{W}_g s(b, a) = \sqrt{a}\, \mathscr{L}_g s(b, a).$$

2.1 *A priori* estimates

Because of the Schwarz inequality we have the following estimates of the modulus of the left- and right-transforms:

$$|\mathscr{L}_g s(x)| \le \|g\|_{\mathscr{H}} \|s\|_{\mathscr{H}}, \qquad |\mathscr{R}_g s(x)| \le \|g\|_{\mathscr{H}} \|s\|_{\mathscr{H}}.$$

Therefore, the left- and right-transforms are in general of higher regularity than the analysed function itself. If the representation is strongly continuous, then the left- and right-transforms are continuous functions.

2.2 Transformation properties

By unitarity of $U(x)$ we have the following relation between the left- and right-transformations:

$$\mathscr{L}_g s(x) = \mathscr{R}_g s(x^{-1}).$$

The role of s and g may be exchanged, as we have already seen in the wavelet case,

$$\mathscr{L}[g; s](x) = \mathscr{L}[\bar{s}; \bar{g}](x^{-1}) = \widetilde{\mathscr{L}[s; g]}(x),$$
$$\mathscr{R}[g; s](x) = \mathscr{R}[\bar{s}; \bar{g}](x^{-1}) = \widetilde{\mathscr{R}[s; g]}(x).$$

The behaviour of the left-transform under group translations of the analysed function can be written as

$$\mathscr{L}_g U(y) = \mathrm{L}_y \mathscr{L}_g, \qquad \begin{array}{ccc} \mathscr{H} & \xrightarrow{U(y)} & \mathscr{H} \\ \downarrow \mathscr{L}_g & & \downarrow \mathscr{L}_g \\ L^2(G, d\mu_l) & \xrightarrow{\mathrm{L}_y} & L^2(G, d\mu_l) \end{array} \tag{2.2.1}$$

with L_y the left regular representation. For the right-transform we have, instead,

$$\mathscr{R}_g U(y) = \mathrm{R}_y \mathscr{R}_g, \qquad \begin{array}{ccc} \mathscr{H} & \xrightarrow{U(y)} & \mathscr{H} \\ \downarrow \mathscr{R}_g & & \downarrow \mathscr{R}_g \\ L^2(G, d\mu_r) & \xrightarrow{\mathrm{R}_y} & L^2(G, d\mu_r) \end{array}$$

where R_y is the right regular representation. With respect to the wavelet we have the following co-variance for the left- and right-transforms:

$$\mathscr{L}[U(y)g; s](x) = \mathscr{L}[g; s](x \circ y) = \mathrm{R}_y^* \mathscr{L}[g; s](x),$$
$$\mathscr{R}[U(y)g; s](x) = \mathscr{R}[g; s](y \circ x) = \mathrm{L}_y^* \mathscr{R}[g; s](x).$$

All formulas are easy to verify and we leave the details of the reader.

3 Energy conservation

In this section we will give sufficient conditions on the analysing function that the left- and right-transforms conserve energy. In analogy with the wavelet transform that we have treated in the previous chapters we give the following definition.

Definition 3.0.1
A vector $g \in \mathscr{H}$ is called admissible *if*

$$c_g = \int_G d\mu_l(x) |\mathscr{L}_g g(x)|^2 < \infty. \tag{3.0.1}$$

Note that we do not distinguish between left- and right-admissible vectors since every admissible vector satisfies:

$$\int_G d\mu_r(x) |\mathscr{R}_g g(x)|^2 = \int_G d\mu_r(x^{-1}) |\mathscr{R}_g g(x^{-1})|^2 = \int_G d\mu_l(x) |\mathscr{L}_g g(x)|^2,$$

where we have used equation (1.7.1).

The left- and right-transforms with respect to an admissible analysing vector g conserve energy, as is shown by the following theorem.

Theorem 3.0.2
Let $g \in \mathscr{H}$ be an admissible vector. Then we have

$$\langle \mathscr{L}_g s \mid \mathscr{L}_g r \rangle_{L^2(G, d\mu_l)} = \int_G d\mu_l(x) \overline{\mathscr{L}_g s(x)} \mathscr{L}_g r(x) = c_g \langle s \mid r \rangle_{\mathscr{H}},$$

where $c_g > 0$ is given by (3.0.1). *For the right-transform an analogous result holds:*

$$\langle \mathscr{R}_g s \mid \mathscr{R}_g r \rangle_{L^2(G, d\mu_r)} = \int_G d\mu_r(x) \overline{\mathscr{R}_g s(x)} \mathscr{R}_g r(x) = c_g \langle s \mid r \rangle_{\mathscr{H}}.$$

Proof. We will make use of Schur's lemma for the proof. Consider the graph of the left-transform, that is, consider the set $\mathscr{H}' \subset \mathscr{H} \oplus L^2(G, d\mu_l)$ of couples $[s, T]$ with

(i) $s \in \mathscr{H}$,
(ii) $T = \mathscr{L}_g s \in L^2(G, d\mu_l)$.

We can make $\mathscr{H}'$ a pre-Hilbert space by introducing the following scalar product:

$$\langle [s, T] \mid [s', T'] \rangle_{\mathscr{H}'} = \langle s \mid s' \rangle_{\mathscr{H}} + \langle T \mid T' \rangle_{L^2(G, d\mu_l)}.$$

The associated norm is

$$\|[s, T]\|^2_{\mathscr{H}'} = \|s\|_{\mathscr{H}} + \|T\|_{L^2(G, d\mu_1)}.$$

Lemma 3.0.3
The space $\mathscr{H}'$ is a Hilbert space.

Or, equivalently, we claim that $\mathscr{H}'$ is a closed subspace of $\mathscr{H} \oplus L^2(G, d\mu_1)$, or, equivalently, we claim that $\mathscr{L}_g$ is a closed operator. Recall that an operator is closed iff for all $s_n \to s$ such that Ks_n converges we have $Ks_n \to Ks$.

Proof. It clearly is a pre-Hilbert space but we still have to check for completeness. Therefore, consider a Cauchy sequence $[s_n, T_n]$ in $\mathscr{H}'$. It has a limit in $\mathscr{H} \oplus L^2(G, d\mu_1)$, say (s, T). We now have to show that the limit couple (s, T) actually is in $\mathscr{H}'$, that is, that $T = \mathscr{L}_g s$.

By construction of $\mathscr{H}'$ we have $T_n = \mathscr{L}_g s_n$. By the continuity of the scalar product we therefore have pointwise, for every arbitrarily fixed $x \in G$,

$$T_n(x) = \langle U(x)g \mid s_n \rangle_{\mathscr{H}} \to \langle U(x)g \mid s \rangle_{\mathscr{H}} = \mathscr{L}_g s(x).$$

On the other hand, because $T_n \to T$ in $L^2(G, d\mu_1)$, there is a subsequence $T_{m(n)}$ such that $T_{m(n)}(x)$ converges almost everywhere to $T(x)$ (e.g. Torchinsky, page 214). Therefore, $\mathscr{L}_g s = T$ and $\mathscr{H}'$ is a Hilbert space. □

Now consider the unitary representation $U' = U \oplus \mathrm{L}$ of G acting on $\mathscr{H} \oplus L^2(G, d\mu_1)$ via

$$U'(x)\colon [s, T] \mapsto [U(x)s, \mathrm{L}_x T].$$

By the co-variance of (2.2.1) it is clear that $\mathscr{H}'$ is invariant under U'. Hence the restrictions of U' to $\mathscr{H}'$ is a subrepresentation. Now consider the following linear map:

$$K\colon \mathscr{H}' \to \mathscr{H}, \qquad [s, T] \mapsto s.$$

Clearly, this mapping is continuous because

$$\|K(s, T)\|^2_{\mathscr{H}} = \|s\|^2_{\mathscr{H}} \le \|s\|^2_{\mathscr{H}} + \|T\|^2_{L^2(G, d\mu_1)} = \|(s, T)\|^2_{\mathscr{H}'}.$$

In addition, it is an intertwining operator for the representations U and U':

$$UK = KU'.$$

It is clearly injective and its range is dense since U is irreducible. By Theorem 1.5.2 the map K is a constant multiple of an isometry and therefore with some $\lambda \ge 0$ we have

$$\|s\|^2_{\mathscr{H}} = \|K[s, \mathscr{L}_g s]\|^2_{\mathscr{H}} = \lambda\|s\|^2_{\mathscr{H}} + \lambda\|\mathscr{L}_g s\|^2_{L^2(G, d\mu_1)}.$$

From this it follows that $\lambda > 1$, otherwise $\|\mathscr{L}_g g\|^2_{L^2(G,d\mu_l)} = 0$, which is not true because g is admissible. Now, by construction, $T = \mathscr{L}_g s$ and we obtain

$$\|\mathscr{L}_g s\|^2_{L^2(G,d\mu_l)} = (\lambda - 1)\|s\|^2_{\mathscr{H}},$$

thereby showing that the left-transform is an isometry. Choosing $s = g$ we see that the constant $c_g = (\lambda - 1)$ is given by

$$c_g = \int_G d\mu_l(x)|\mathscr{L}_g g(x)|^2,$$

which shows the theorem for the left-transform. The proof for the right-transform is analogous. □

4 The left- and right-synthesis

Consider a function $T \in L^2(G, d\mu_l)$. The left-synthesis is (formally) given by

$$\mathscr{L}_h^* T = \int_G d\mu_l(x) T(x) U(x) h.$$

The right-synthesis reads

$$\mathscr{R}_h^* T = \int_G d\mu_r(x) T(x) U^*(x) h.$$

These operators allow us to write functions in $\mathscr{H}$ as superpositions of group-translated functions $U(x)h$ or $U^*(x)h$ respectively. The weight of each is given by $T(x)$. In a more precise way, the synthesis is defined by duality, namely, the left- and right-synthesis with respect to an admissible wavelet h are the adjoint operators of the left- and right-transforms with respect to h.

$$\langle s \mid \mathscr{L}_h^* T\rangle_{\mathscr{H}} = \langle \mathscr{L}_h s \mid T\rangle_{L^2(G,d\mu_l)}, \qquad \langle s \mid \mathscr{R}_h^* T\rangle_{\mathscr{H}} = \langle \mathscr{R}_h s \mid T\rangle_{L^2(G,d\mu_r)}. \quad (4.0.1)$$

Therefore,[5] the left- and right-synthesis are bounded linear maps:

$$\mathscr{L}_h^*: L^2(G, d\mu_l) \to \mathscr{H}, \qquad \|\mathscr{L}_h^* T\|^2_{\mathscr{H}} \le c_h \|T\|^2_{L^2(G,d\mu_l)},$$

$$\mathscr{R}_h^*: L^2(G, d\mu_r) \to \mathscr{H}, \qquad \|\mathscr{R}_h^* T\|^2_{\mathscr{H}} \le c_h \|T\|^2_{L^2(G,d\mu_r)}.$$

4.1 Co-variance

With respect to the expansion coefficients T we have the following co-variance:

[5] The proof of Theorem 23.0.1 in Chapter 1 may easily be adapted to this case.

$$\mathscr{L}_h^* \mathrm{L}_x = U(x)\mathscr{L}_h^*, \qquad \begin{array}{ccc} L^2(G, d\mu_l) & \xrightarrow{\mathrm{L}_x} & L^2(G, d\mu_l) \\ \downarrow \mathscr{L}_h^* & & \downarrow \mathscr{L}_h^* \\ \mathscr{H} & \xrightarrow{U(x)} & \mathscr{H} \end{array}$$

and

$$\mathscr{L}_h^* \mathrm{L}_x = U(x)\mathscr{R}_h^*, \qquad \begin{array}{ccc} L^2(G, d\mu_r) & \xrightarrow{\mathrm{R}_x} & L^2(G, d\mu_r) \\ \downarrow \mathscr{R}_h^* & & \downarrow \mathscr{R}_h^* \\ \mathscr{H} & \xrightarrow{U(x)} & \mathscr{H} \end{array}$$

Left- and right-syntheses are finally related by

$$\mathscr{L}^*[h; T(x)] = \mathscr{R}^*[h; T(x^{-1})].$$

5 The inversion formula

As isometry the left- and right-transforms may be inverted by the adjoint. But, as we have found already for the wavelet case, more general inversion formulae exist.

Definition 5.0.1

A vector $h \in \mathscr{H}$ is called a reconstructing vector for the analysing vector g iff the constant

$$c_{g,h} = \frac{\langle \mathscr{L}_h g \mid \mathscr{L}_g h \rangle_{L^2(G, d\mu_l)}}{\langle g \mid h \rangle_{\mathscr{H}}} \tag{5.0.1}$$

satisfies $0 < |c_{g,h}| < \infty$.

We also write $\langle g \mid C \mid h \rangle$ for $c_{g,h}$. This will be justified below. Again, we need not distinguish between left- and right-constructing vectors because

$$\begin{aligned} &\int_G d\mu_l(x) \overline{\langle U(x)h \mid g \rangle}_{\mathscr{H}} \langle U(x)g \mid h \rangle_{\mathscr{H}} \\ &\qquad = \int_G d\mu_l(x) \overline{\langle h \mid U(x^{-1})g \rangle}_{\mathscr{H}} \langle g \mid U(x^{-1})h \rangle_{\mathscr{H}} \\ &\qquad = \int_G d\mu_r(x) \overline{\langle h \mid U(x)g \rangle}_{\mathscr{H}} \langle g \mid U(x)h \rangle_{\mathscr{H}} \\ &\qquad = \langle \mathscr{R}_h g \mid \mathscr{R}_g h \rangle_{L^2(G, d\mu_r)}. \end{aligned}$$

The wavelet synthesis with respect to a reconstruction vector is actually the inversion of the wavelet transform.

Theorem 5.0.2
Let $g \in \mathscr{H}$ be admissible, and let $h \in \mathscr{H}$ be admissible and, in addition, a reconstruction vector for g. Then

$$\mathscr{L}_h^* \mathscr{L}_g = c_{g,h} \mathbb{1}_{\mathscr{H}}, \qquad \mathscr{R}_h^* \mathscr{R}_g = c_{g,h} \mathbb{1}_{\mathscr{H}}.$$

The constant $c_{g,h}$ is given by (5.0.1).

Proof. First observe that everything is well defined because the left- and right-transforms and the left- and right-syntheses are bounded maps. Because of the co-variance of the left- and right-transforms and the respective syntheses, we obtain

$$\mathscr{L}_h^* \mathscr{L}_g U(x) = \mathscr{L}_h^* \mathrm{L}_x \mathscr{L}_g = U(x) \mathscr{L}_h^* \mathscr{L}_g.$$

The analogue is true for $\mathscr{R}_h^* \mathscr{R}_g$. Hence, by Schur's lemma, these operators are a constant multiple of the identity. The value of the constant is obtained by considering $\langle g \mid \mathscr{L}_h^* \mathscr{L}_g \mid h \rangle$ and $\langle g \mid \mathscr{R}_h^* \mathscr{R}_g \mid h \rangle$. □

Note that this theorem implies the following orthogonality relations for admissible g and h and arbitrary $s, r \in \mathscr{H}$:

$$\left.\begin{aligned} \langle \mathscr{L}_g s \mid \mathscr{L}_h r \rangle_{L^2(G, d\mu_l)} &= \langle g \mid C \mid h \rangle_{\mathscr{H}} \langle s \mid r \rangle_{\mathscr{H}}, \\ \langle \mathscr{R}_g s \mid \mathscr{R}_h r \rangle_{L^2(G, d\mu_r)} &= \langle g \mid C \mid h \rangle_{\mathscr{H}} \langle s \mid r \rangle_{\mathscr{H}}. \end{aligned}\right\} \tag{5.0.2}$$

In particular, we have

$$c_{g,h} = \langle \mathscr{L}_g e \mid \mathscr{L}_h e \rangle_{L^2(G, d\mu_l)},$$

where $e \in \mathscr{H}$ is any vector of unit length. In particular, this shows that $(g, h) \mapsto c_{g,h}$ is a quadratic form.

5.1 On the constant $c_{g,h}$

The dependency of the constant $c_{g,h}$ on the two functions g and h can be expressed by the following theorem.

Theorem 5.1.1
There is a closed, densely defined, positive operator C whose domain is identical with the set of admissible vectors such that

$$c_{g,h} = \langle h \mid C \mid g \rangle_{\mathscr{H}}.$$

The domain of C is invariant under the representation U. If the group G is unimodular then $C = \mathbb{1}_{\mathscr{H}}$.

As an immediate consequence we find that for unimodular groups all vectors are admissible if there is at least one vector which is admissible. Recall that C being positive means that

$$\langle s \mid Cs \rangle \geq 0, \qquad \text{for all } s \in \mathscr{H}.$$

Proof. Let $\mathscr{D}$ be the set of admissible vectors in $\mathscr{H}$. Consider the following quadratic form:

$$B(s) = \|s\|^2_{\mathscr{H}} + c_{g,g}. \tag{5.1.1}$$

Its square-root $\|s\| = \sqrt{B(s)}$ is obviously a norm on $\mathscr{D}$ and $s \in \mathscr{H}$ is admissible iff $\|s\| < \infty$. We now want to show that the space of admissible vectors $\mathscr{D}$ is a complete space with respect to this norm. This would show that the correspondence $\mathscr{D} \times \mathscr{D} \to \mathbb{C}, (s, r) \mapsto c_{s,r}$ defines a closed, symmetric, positive bi-linear form. We could now evoke the second representation theorem (e.g. Kato 1966) to see that in this case there exists the desired operator C having the properties as stated in the theorem.

Consider now a Cauchy sequence g_n in $\mathscr{D}$ with respect to the norm of (5.1.1). By definition, it is also a Cauchy sequence in $\mathscr{H}$ and we have $g_n \to g$. We claim that g is again admissible. By the continuity of the scalar product we have pointwise $\lim_{n \to \infty} \mathscr{L}_{g_n} e(x) = \mathscr{L}_g e(x)$, where e is an arbitrary unit vector in $\mathscr{H}$. At the same time we have by hypothesis that $c_{g_n, g_n} = \|\mathscr{L}_{g_n} e\|^2_{L^2(G, d\mu_l)}$ is convergent. By Fatou's lemma,

$$\int_G d\mu_l(x) |\mathscr{L}_g e(x)|^2 = \int_G d\mu_l(x) \liminf_{n \to \infty} |\mathscr{L}_{g_n} e(x)|^2 \leq \liminf_{n \to \infty} c_{g_n, g_n} < \infty.$$

Therefore, g is admissible and we are done. □

5.2 More general reconstruction

As we saw in the wavelet case, we are able to reconstruct the analysed function from its transform even if the analysing function is not admissible. This is in general true, as we will now show. In the following we call a vector $g \in \mathscr{H}$ admissible in the general sense for the admissible reconstructing vector h iff there is a sequence of admissible vectors g_n that tends to g in $\mathscr{H}$ and for which the following limit holds:

$$\lim_{n \to \infty} c_{g_n, h} = c_{g, h}, \tag{5.2.1}$$

where $0 < |c_{g,h}| < \infty$. Note that g need not be admissible, as the case of the wavelet transform over the real line shows.

Theorem 5.2.1
Let $g \in \mathscr{H}$ be admissible in the general sense for some admissible $h \in \mathscr{H}$. Suppose, further, that h satisfies $\mathscr{L}_h h \in L^1(G, d\mu_1)$ in addition. Then for every $r = U(y)h$ in the orbit of h we have

$$\int_G d\mu_1(x)\overline{\mathscr{L}_h r(x)}\mathscr{L}_g s(x) = c_{g,h}\langle r \mid s\rangle_{\mathscr{H}}$$

for all $s \in \mathscr{H}$ and the constant $c_{g,h}$ given by the limit (5.2.1) *above. The integral is absolutely convergent.*

In particular, we therefore know the wavelet transform of s with respect to the admissible wavelet h:

$$\mathscr{L}_h s(y) = \langle U(y)h \mid s\rangle_{\mathscr{H}} = c_{g,h}^{-1}\int_G d\mu_1(x)\overline{\mathrm{L}_y \mathscr{L}_h h(x)}\mathscr{L}_g s(x).$$

Now the first inversion formula may be used to recover s from the left-coefficients $\mathscr{L}_g$ with respect to the non-admissible vector $g \in \mathscr{H}$. Clearly, an analogous theorem holds for the right-transform.

Proof. Let $g_n \to g$ be a sequence of admissible vectors that tend to g in $\mathscr{H}$. By (5.0.2) we have

$$\lim_{n\to\infty}\int_G d\mu_1(y)\overline{\mathscr{L}_{g_n} s(y)}\mathscr{L}_h r(y) = \lim_{n\to\infty} c_{g_n,h}\langle s \mid r\rangle_{\mathscr{H}} = c_{g,h}\langle s \mid r\rangle_{\mathscr{H}}. \quad (5.2.2)$$

By continuity of the scalar product we have pointwise, for every $x \in G$,

$$\lim_{n\to\infty} \overline{\mathscr{L}_{g_n} s(x)}\mathscr{L}_h r(x) = \overline{\mathscr{L}_g s(x)}\mathscr{L}_h r(x). \quad (5.2.3)$$

On the other hand, for n large enough, thanks to the Cauchy–Schwarz inequality, we have, since g_n converges in $\mathscr{H}$,

$$|\overline{\mathscr{L}_{g_n} s(x)}\mathscr{L}_h r(x)| \leq 2\|g\|_{\mathscr{H}}\|s\|_{\mathscr{H}}|\mathscr{L}_h r(x)|.$$

We claim that the right-hand side is absolutely integrable. Indeed, if $T \in L^1(G, d\mu_1)$, then $\mathrm{L}_x T \in L^1(G, d\mu_1)$, too, and we have

$$\int_G d\mu_1(y)\,\|\mathrm{L}_x T(y)| = \int_G d\mu_1(y)|T(x^{-1}\circ y)| = \int_G d\mu_1(y)|T(y)|.$$

By the co-variance we have $\mathscr{L}_h U(x)h = \mathrm{L}_x\mathscr{L}_h h$ and thus $\mathscr{L}_h r \in L^1(G, d\mu_1)$ for every $r = U(x)h$ in the orbit of h. Therefore, the sequence of (5.2.3) is uniformly majorized by an integrable function and we may use the dominated convergence theorem to exchange the limit with the integration in (5.2.2). □

6 The reproducing kernel equation

We now come to the characterization of the image of the left- and right-transforms. Again the image is characterized by a reproducing kernel equation. Let us consider the following operators:

$$\Pi^{\mathrm{l}}_{g,h} = c^{-1}_{g,h} \mathscr{L}_g \mathscr{L}^*_h, \qquad \Pi^{\mathrm{r}}_{g,h} = c^{-1}_{g,h} \mathscr{R}_g \mathscr{R}^*_h.$$

Again, they turn out to be projectors on the image of the left- and right-transforms respectively.

Theorem 6.0.1

The image of the left-transform $\mathscr{L}_g$ consists of exactly those functions $T \in L^2(G, d\mu_{\mathrm{l}})$ for which we have

$$T = P_{g,h} *_l T \tag{6.0.1}$$

or, more explicitly,

$$T(x) = \int_G d\mu_{\mathrm{l}}(y) P_{g,h}(y^{-1} \circ x) T(y).$$

The image of the right-transform $\mathscr{R}_g$ consists of exactly those functions $\mathscr{R} \in L^2(G, d\mu_{\mathrm{r}})$ for which we have

$$\mathscr{R} = P_{g,h} *_r \mathscr{R} \tag{6.0.2}$$

or, more explicitly,

$$\mathscr{R}(x) = \int_G d\mu_{\mathrm{r}}(y) P_{g,h}(y \circ x^{-1}) \mathscr{R}(y).$$

In both cases the kernel $P_{g,h}$ is given by

$$P_{g,h}(x) = c^{-1}_{g,h} \mathscr{L}_g h(x) = c^{-1}_{g,h} \widetilde{\mathscr{L}_h g}(x).$$

For arbitrary functions $T \in L^2(G, d\mu_{\mathrm{l}})$ and $\mathscr{R} \in L^2(G, d\mu_{\mathrm{r}})$ the right-hand side of equations (6.0.1) *and* (6.0.2) *define a projector on the respective image space. For $g = h$ this projector is orthogonal.*

Remark. Note that for $g \neq h$ we have $\bar{P}_{g,h} \notin \operatorname{imag} \mathscr{L}_g$ and $\bar{P}_{g,h} \notin \operatorname{imag} \mathscr{R}_g$. The unique reproducing kernel,[6] whose complex conjugate is in the image of the left- and the right-transforms, is given by $P_{g,g}$.

[6] This is a general feature of Hilbert spaces with reproducing kernel: there is only one reproducing kernel *in* the Hilbert space. See, for example, Meschkowsky (1969). But in some larger spaces there may be some more kernels, of course.

Proof. That the operators $\Pi^{l}_{g,h}$ and $\Pi^{r}_{g,h}$ are projectors, and that $\Pi_{g,g}$ is an orthogonal projector, follows immediately from the inversion formula as in the proof of Theorem 20.0.1 of Chapter 1. We only have to show, that the operators can be written as integral operators, as stated in the theorem. By definition of the left- and right-syntheses of (4.0.1) we have

$$\mathscr{L}_g \mathscr{L}_h^* T(x) = \langle U(x) g \mid \mathscr{L}_h^* T \rangle_{\mathscr{H}} = \langle \mathscr{L}_h U(x) g \mid T \rangle_{L^2(G, d\mu)}.$$

By the co-variance of the right-transform the theorem follows. □

7 Fourier transform over Abelian groups

In the following sections we will use the language of group theory to unify the concept of multi-resolution analysis over the Abelian groups as $\mathbb{R}$, $\mathbb{T}$, $\mathbb{Z}$ and $\mathbb{Z}/N\mathbb{Z}$ that we have considered so far.

In this section we will generalize the Fourier transform over $\mathbb{R}$ to any locally compact Abelian group. However, to avoid unnecessary technicalities we consider mostly only groups that can be written as a finite sum:

$$G = G_0 \oplus G_1 \oplus \cdots \oplus G_{N-1},$$

where each summand is isomorphic to one of the groups out of the set $\{\mathbb{R}^n, \mathbb{Z}^n, \mathbb{T}^n, \mathbb{Z}^n/(N\mathbb{Z})^n\}$, for some $n \in \mathbb{N}$.

We start by recalling the Fourier transform over locally compact groups. Essentially no proofs will be given and we refer to Kirilov (1969).

The elementary functions of Fourier analysis over Abelian groups are the so-called *characters*. A character χ is a continuous homomorphism between the group G and the multiplicative group of complex numbers of modulus one that we identify with the circle $\mathbb{T}$. To put it differently, a character χ is a continuous complex-valued function $\chi\colon G \to \mathbb{C}$ with

$$\chi(x + y) = \chi(x) \cdot \chi(y), \qquad |\chi(x)| = 1.$$

Example 7.0.1

Consider the real line $\mathbb{R}$. All characters are given by the pure oscillations

$$\chi_\omega(t) = e^{i\omega t}.$$

Therefore, there are infinitely many characters. Each can be identified in a unique way with a point $\omega \in \mathbb{R}$, which in this case may be interpreted as a frequency.

Example 7.0.2

Consider the group $\mathbb{T}$ that we identify with the angles $\theta \in \mathbb{R} \bmod 2\pi\mathbb{Z}$. It can easily be seen that the only characters are given by

$$\chi_n(\theta) = e^{in\theta},$$

with some $n \in \mathbb{Z}$. There are still infinitely many characters. However, this time the set of characters is discrete. It turns out that this is true in general if the group is compact.

Example 7.0.3

Consider the discretized circle $\mathbb{Z}/N\mathbb{Z}$. A minute's reflection shows that all characters can be written as

$$\chi_m(n) = e^{2i\pi nm/N},$$

where $m = 0, 1, \ldots, N-1$. Therefore, there are as many characters as elements in the group. This is true for finite Abelian groups in general, as we will see.

The characters form an Abelian group themselves under pointwise multiplication, that is, we have

$$(\chi_1 \circ \chi_2)(x) = \chi_1(x) \cdot \chi_2(x)$$

for every $x \in G$. This group of characters is called the *dual group* of G and we denote it by $\hat{G}$. By convention it is written as a multiplicative group. It may sometimes be useful to write it as an additive group. This is achieved by setting

$$\chi(x) = e^{i\langle \chi' | x \rangle}.$$

We will not distinguish in notation between χ and χ' since it is clear from the context.

Example 7.0.4

By the previous examples we see that we have the following identifications:

$$\hat{\mathbb{R}} \leftrightarrow \mathbb{R}, \qquad \hat{\mathbb{T}} \leftrightarrow \mathbb{Z}, \qquad \widehat{\mathbb{Z}/N\mathbb{Z}} \leftrightarrow \mathbb{Z}/N\mathbb{Z}.$$

The dual group can again be made into a topological group. Let χ_n be a sequence of characters. We say $\chi_n \to \chi$ in (the topology of) $\hat{G}$ iff for every compact set $K \subset G$ we have $\chi_n(x) \to \chi(x)$ uniformly in $x \in K$. Together with this topology, the group $\hat{G}$ is again a locally compact Abelian group. What is its dual group? The answer is given by the following theorem.

Theorem 7.0.5 (Pontryagin)

The dual group of $\hat{G}$ is G.

Example 7.0.6

Consider the integer group $\mathbb{Z}$. Its characters are given by

$$\chi_\theta(n) = e^{in\theta}, \qquad \theta \in \mathbb{T}.$$

Therefore, we find $\hat{\mathbb{Z}} = \mathbb{T}$, as it should be.

7.1 The Fourier transform

As in the cases we have encountered so far it turns out that the characters are a complete set of functions over the group, and arbitrary functions may be expanded into superpositions of characters. The expansion coefficients will be obtained through Fourier analysis.

As usual, we will denote by $L^p(G)$ the equivalence classes of functions such that $\int_G d\mu(x) \mid s(x)|^p < \infty$. If we want to stress the dependence on the measure μ we also write $L^p(G, d\mu)$. Let $s \in L^1(G)$ be a function over the group. Its Fourier transform is defined by taking the scalar products with all characters:

$$F^G s(\chi) = \hat{s}(\chi) = \langle \chi \mid s \rangle_G = \int_G d\mu(x) \overline{\chi(x)} s(x) = \int_G d\mu(x)\, e^{-\langle \chi | x \rangle}\, s(x).$$

Therefore, the Fourier transform is a function over the dual group. Obviously, in all examples of Abelian groups that we have considered so far this concept of Fourier analysis coincides with the one given previously. So it should be no surprise that the Fourier transform over locally compact Abelian groups is an isometry, and hence conserves energy.

Theorem 7.1.1
For $s \in L^1(G, d\mu) \cap L^2(G, d\mu)$ we have $\hat{s} \in L^2(\hat{G}, d\hat{\mu})$. In addition, $d\hat{\mu}$ can be normalized in such a way that

$$\int_G d\mu(x) |s(x)|^2 = \int_{\hat{G}} d\hat{\mu}(\chi) |\hat{s}(\chi)|^2.$$

Therefore, the Fourier transform extends in a unique way to a map from $L^2(G, d\mu)$ to $L^2(\hat{G}, d\hat{\mu})$. The inversion is given by the adjoint. Explicitly, supposing some regularity (e.g. $s \in L^1(G, d\mu)$ and $\hat{s} \in L^1(\hat{G}, d\hat{\mu})$), we have (pointwise)

$$s(x) = \int_{\hat{G}} d\hat{\mu}(\chi) \hat{s}(\chi) \chi(x) = \int_{\hat{G}} d\hat{\mu}(\chi) \hat{s}(\chi)\, e^{i\langle \chi | x \rangle}.$$

For compact groups—since the characters are discrete—it follows that the characters are a complete orthogonal set:

$$\langle \chi \mid \eta \rangle_G = \begin{cases} |G| & \text{if } \chi = \eta, \\ 0 & \text{otherwise,} \end{cases}$$

where $|G|$ stands for the total measure of G.

7.2 Group translations

Consider the following operator:

$$S_y^G\colon L^2(G, d\mu) \to L^2(G, d\mu), \qquad s(x) \mapsto s(x-y), \quad y \in G.$$

The mapping $y \mapsto S_y^G$ is the regular representation of G as defined in (1.8.1). Now consider the operators

$$E_y\colon L^2(\hat{G}, d\mu) \to L^2(\hat{G}, d\mu), \qquad r(\chi) \mapsto \bar{\chi}(y)r(\chi), \quad y \in G.$$

Again, $y \mapsto E_y$ is a representation of G. We have

$$\begin{aligned}\langle \chi \mid S_y^G s\rangle_G &= \int_G d\mu(x)\bar{\chi}(x)s(x-y)\\ &= \int_G d\mu(x)\bar{\chi}(x+y)s(x)\\ &= \bar{\chi}(y)\int_G d\mu(x)\bar{\chi}(x)s(x) = \bar{\chi}(y)\langle \chi \mid s\rangle_G\end{aligned}$$

and thus the Fourier transform is an intertwining operator

$$F^G S_y^G = E_y F^G, \qquad \begin{array}{ccc} L^2(G) & \xrightarrow{S_y} & L^2(G) \\ \downarrow{\scriptstyle F^G} & & \downarrow{\scriptstyle F_G} \\ L^2(\hat{G}) & \xrightarrow{E_y} & L^2(\hat{G}) \end{array} \tag{7.2.1}$$

However, these representations are in general reducible.

7.3 The convolution theorem

We recall that for two functions over the group s, r the group convolution is defined as

$$s * r(x) = \int_G d\mu(y)s(x-y)r(y).$$

The convolution product over commutative groups is commutative itself and associative:

$$s * r = r * s, \qquad (s * r) * v = s * (r * v).$$

In Fourier space the convolution is given by pointwise multiplication.

Theorem 7.3.1
The convolution theorem holds for $s, r \in L^1(G) \cap L^2(G)$:

$$\widehat{s * r}(\chi) = \hat{s}(\chi) \cdot \hat{r}(\chi),$$

for all $\chi \in \hat{G}$.

Proof. We can write the convolution as a scalar product of shifted functions:

$$s * r(x) = \langle S_x^G \tilde{s} \mid r \rangle_G,$$

where $\tilde{s}(x) = \bar{s}(-x)$, as usual. By the isometry property of the Fourier transform (7.1.1) and the intertwining property (7.2.1) we obtain

$$s * r(x) = \langle F^G S_x^G \tilde{s} \mid F^G r \rangle_{\hat{G}} = \int_{\hat{G}} d\hat{\mu}(\chi) \chi(x) \hat{s}(\chi) \hat{r}(\chi).$$

By the inversion formula of the Fourier transform the theorem is proved. □

8 Periodizing, sampling, and M. Poisson

In this and the following section we present the sampling theorems of Chapter 3, Section 1 in a group-theoretic setting. From now on G will be the direct sum of a finite number of groups, each being isomorphic with one of the groups out of the set $\{\mathbb{R}^n, \mathbb{Z}^n, \mathbb{T}^n, \mathbb{Z}^n/(N\mathbb{Z})^n; n = 0, 1, 2, \ldots\}$. Accordingly, the term 'discrete group' will stand for sums of $\mathbb{Z}^n$ and $\mathbb{Z}^n/(N\mathbb{Z})^n$, whereas, 'compact group' means that G is isomorphic to sums of $\mathbb{T}^n$ and $\mathbb{Z}^n/(N\mathbb{Z})^n$.

8.1 Sampling

Let $H \subset G$ be a discrete subgroup of G. There is a natural map from functions over the group G to functions over the subgroup. This map is the perfect sampling operator associated to H:

$$(G) \to (H), \qquad \Sigma^H s(x) = s(x) \quad \text{for } x \in H \subset G.$$

Here (G) stands for some suitable space of functions on G. In the case where, for example, $G = \mathbb{R}$ and $H = \mathbb{Z}$ this definition coincides with the perfect sampling operator that we have considered before.

It is natural to replace the perfect sampling operator by an imperfect sampling operator associated to some filter $\varphi \in L^2(G)$, say

$$\Sigma_\varphi^H : L^2(G) \to L^\infty(H), \qquad \Sigma_\varphi^H s(x) = \langle S_x^G \varphi \mid s \rangle_{L^2(G)}.$$

Note that the imperfect sampling operator is obtained by sampling a smoothed version of s perfectly:

$$\Sigma_{\varphi}^{H} s = \Sigma^{H}(\tilde{\varphi} * s),$$

with $\varphi(x) = \varphi(-x)$, as usual.

8.2 Periodization

Now consider an H-periodic function, that is, a function for which

$$s(x + y) = s(x), \qquad \text{for all } y \in H.$$

Again, this definition coincides with the usual definition of periodicity in the case where $G = \mathbb{R}$ and $H = \mathbb{Z}$. Because an H-periodic function depends only on $x \bmod H$ we may identify the set of H-periodic functions with functions over the quotient group G/H. We will not write the identification map explicitly; it should be clear from the context which space we are working in. Again, the same identification leads us to identify the periodic functions over $\mathbb{R}$ with functions over the circle $\mathbb{T} = \mathbb{R}/\mathbb{Z}$.

There is a natural map for functions over the whole group G to functions over the quotient G/H. It is given by the periodizing operator

$$\Pi^{H}: L^{1}(G) \to L^{1}(G/H), \qquad s(x) \mapsto \sum_{y \in H} s(x - y).$$

It is immediately verified that Π^{H} is H-periodic and hence is actually a function over G/H. Again, this operator is the usual periodization operator in the cases that we considered before.

The quotient group G/H is again a locally compact Abelian group and it has a natural invariant measure inherited from G. In all expressions involving integrals over G/H this measure is to be understood. Therefore, in the case of G/H being compact, $|G/H|$ is the measure in G of a fundamental domain for the action of H, that is, for one measurable set of representatives. The periodization operator allows us to write the scalar product of an H-periodic function s with an arbitrary function r over G as a scalar product over G/H:

$$\langle s \mid r \rangle_{G} = \langle s \mid \Pi^{H} \rangle_{G/H}, \qquad \text{for } s \quad H\text{-periodic.} \tag{8.2.1}$$

The proof is straightforward (compare 9.2.1 of Section 2) and is left to the reader.

8.3 The Poisson summation formula

Consider the set $H^{\perp}$ consisting of those characters $\chi \in \hat{G}$ that satisfy

$$\chi(x) = 1, \qquad \text{for } x \in H. \tag{8.3.1}$$

This set is exactly the set of H-periodic characters. Indeed, let $y \in H$ and $x \in G$. Then

$$\chi(x + y) = \chi(x) \cdot \chi(y) = \chi(x),$$

and vice versa, since $\chi(0) = 1$ we have for an H-periodic character χ that $\chi(x) = 1$ for all $x \in H$. The set $H^\perp$ is actually a subgroup of $\hat{G}$. Indeed, property (8.3.1) is preserved under multiplication and division of characters.

On the other hand, consider the group G/H. We now want to construct its characters $\widehat{G/H}$ from the characters of G. It turns out that $\widehat{G/H}$ can be identified with the set $H^\perp$. Indeed, let $\chi \in \widehat{G/H}$. Then by definition of the characters of $\chi \in \widehat{G/H}$ we should have

$$\chi(x \bmod H + y \bmod H) = \chi(x + y \bmod H) = \chi(x \bmod H) \cdot \chi(y \bmod H).$$

Therefore, χ should only depend on $x \bmod H$, and, vice versa, every such character defines, by the above equation, a character of $\widehat{G/H}$.

Example 8.3.1

Let $G = \mathbb{R}$ and $H = \mathbb{Z}$. The characters χ_ω in $\mathbb{Z}^\perp$ satisfy

$$\chi_\omega(n) = e^{i\omega n} = 1, \qquad \text{for all } n \in \mathbb{Z}.$$

Therefore, we have $\omega = 2\pi m$ with some $m \in \mathbb{Z}$. These are the characters of the circle group $\mathbb{T}$.

We are now ready to state the Poisson summation formula. Again it links sampling to periodizing via the Fourier transform.

Theorem 8.3.2

Suppose H is a discrete subgroup of G such that G/H is compact. On $L^1(G, d\mu) \cap F^{G-1}L^1(\hat{G}, d\hat{\mu})$ we then have

$$\Sigma^{H^\perp} F^G = F^{G/H} \Pi^H, \qquad |G/H| F^H \Sigma^H = \Pi^{H^\perp} F^G.$$

In the definition of the various Fourier transforms G is equipped with some fixed Haar measure, H with the counting measure, and G/H with the measure inherited from G.

More explicitly, this can be written as

$$\sum_{y \in H} s(x + y) = \frac{1}{|G/H|} \sum_{\chi \in H^\perp} \hat{s}(\chi)\chi(x),$$

$$\frac{1}{|G/H|} \sum_{\chi \in H^\perp} \hat{s}(\eta \cdot \chi) = \sum_{x \in H} s(x)\bar{\eta}(x).$$

Graphically, the Poisson summation formula can be written as

$$\begin{array}{ccc} (G) & \xrightarrow{\Pi^H} & (G/H) \\ \downarrow F^G & & \downarrow F^{G/H} \\ (\hat{G}) & \xrightarrow{\Sigma^{H^\perp}} & (H^\perp) \end{array} \qquad \begin{array}{ccc} (G) & \xrightarrow{\Sigma^H} & (H) \\ \downarrow F^G & & \downarrow |G/H| F^H \\ (\hat{G}) & \xrightarrow{\Pi^{H^\perp}} & (\hat{G}/H^\perp) \end{array}$$

For $x = 0$ we obtain

$$\sum_{y \in H} s(y) = \frac{1}{|G/H|} \sum_{\chi \in H^\perp} \hat{s}(\chi).$$

In this form it states that the Fourier transform of a delta-comb is again a delta-comb.

Proof. Let $\chi \in H^\perp$. Then χ is H-periodic. Because of (8.2.1) we may write

$$\Sigma^{H^\perp} \hat{s}(\chi) = \langle \chi \mid s \rangle_G = \langle \chi \mid \Pi^H s \rangle_{G/H} = F^{G/H} \Pi^H s(\chi).$$

The second formula follows immediately from the first. □

9 Sampling spaces over Abelian groups

Let H again be a discrete subgroup of G. We now give the following definition.

Definition 9.0.1

A subspace V of $L^2(G)$ is called a perfect sampling space if the perfect sampling operator

$$\Sigma^H : V \to L^2(H)$$

is an isomorphism and if, in addition, the following diagram is commutative for all $x \in H \subset G$:

$$S_x^H \Sigma^H = \Sigma^H S_x^G, \qquad \begin{array}{ccc} V & \xrightarrow{S_x^G} & V \\ \downarrow \Sigma^H & & \downarrow \Sigma^H \\ L^2(H) & \xrightarrow{S_x^H} & L^2(H) \end{array}$$

A subspace V of $L^2(G)$ is called an (imperfect) sampling space if the perfect sampling operator may be replaced by the imperfect sampling operator

$$\Sigma_\varphi^H : s \mapsto \Sigma^H(\tilde{\varphi} * s)$$

with respect to some $\varphi \in L^2(G)$.

Again, perfect and imperfect sampling spaces can be characterised completely by means of the following theorem.

Theorem 9.0.2

Suppose that H is a discrete subgroup of G such that G/H is compact. Then for V an H-sampling space there is a $\phi \in V$ such that $\{S_x^G \phi\}$, with $x \in H$, is a Riesz basis for V. For a perfect sampling space this generating function ϕ may be chosen to be the unique function in V that satisfies

$$\phi(x) = \begin{cases} 1 & \text{for } x = 0, \\ 0 & \text{for all } x \in H. \end{cases}$$

We then have

$$s(y) = \sum_{x \in H} s(x)\phi(y - x) \qquad \text{for all } s \in V.$$

The proof is as in the case over $\mathbb{R}$ and is left to the reader.

Let V be the closed linear span of the H-translates of some function $\varrho \in L^2(G)$. In Fourier space the functions $s \in V$ can all be written as

$$\hat{s}(\chi) = m(\chi)\hat{\varrho}(\chi),$$

with m an $H^\perp$-periodic function

$$m(\chi \cdot \eta) = m(\eta), \qquad \text{for all } \chi \in H^\perp,$$

where $m \in L^2(\hat{G}/H^\perp, d\nu)$, where $d\nu$ is the measure whose Radon–Nikodim derivative with respect to the invariant measure $d\hat{\mu}$ of $\hat{G}$ is given by the L^1-density

$$d\nu/d\hat{\mu}(\eta) = \sum_{\chi \in H^\perp} |\hat{\varphi}(\chi \cdot \eta)|^2.$$

Again, orthonormality means that this density is constant.

Lemma 9.0.3

A family of functions $\{S_x^G \varphi\}$ with $x \in H$ is an orthonormal set in $L^2(G, d\mu)$ iff

$$\sum_{\chi \in H^\perp} |\hat{\varphi}(\chi \cdot \eta)|^2 \equiv |G/H|.$$

Proof. Orthonormality of the set $\{S_x^G \varphi\}$ is equivalent to

$$\Sigma^H(\tilde{\varphi} * \varphi) = \delta,$$

with $\tilde{\varphi}(x) = \bar{\varphi}(-x)$, as usual. Now we apply the Poisson summation formula to $\widehat{\tilde{\varphi} * \varphi} = |\hat{\varphi}|^2$, and the lemma follows. □

As before, one can show that $\{S_x^G \varrho : x \in H\}$ is a Riesz basis for their linear span iff the above density is equivalent to 1, namely,

$$0 < A < \sum_{\chi \in H^\perp} |\hat{\varrho}(\chi \cdot \eta)|^2 < B < \infty,$$

for all $\eta \in \hat{G}/H^\perp$. An orthonormal base for V can then be obtained from ϱ by means of

$$\hat{\varphi}(\chi) = m(\chi)\hat{\varrho}(\chi), \qquad |m(\chi)|^2 = \frac{|G/H|}{\sum_{\eta \in H^\perp} |\hat{\varrho}(\eta \cdot \chi)|^2}.$$

10 The discrete wavelet transform over Abelian groups

In abstract algebraic language we can generalize the concept of wavelet decomposition over Abelian groups as follows. Let H and J be two discrete subgroups of G that satisfy at

$$J \subset H \subset G.$$

Suppose, in addition, that G/H is compact and that H/J is finite.

Let $V_H \subset L^2(G)$ be an H-sampling space and let $V_J \subset L^2(G)$ be a J-sampling space. Suppose, further, that

$$V_J \subset V_H.$$

In some sense V_H has more small scale features that V_J since it must be sampled over some denser grid. We now want to find a decomposition

$$V_H = \bigoplus_{x \in H/J} W_x = V_J \oplus \bigoplus_{x \in H/J - \{0\}} W_x$$

with $W_0 = V_J$ where all spaces W_x are J-sampling spaces. For $G = \mathbb{R}$, $H = \mathbb{Z}$ and $J = 2\mathbb{Z}$ we have $H/J = \{0, 1\}$, and we thus find the usual wavelet decomposition approach.

Suppose that such a decomposition exists. Let ϕ be a function whose H-translates are an orthonormal basis for V_H. In the same way let ρ be a function whose J-translates are a basis for V_J. In the same way, by hypothesis, there are functions ψ_x whose J-translates are an orthonormal basis for W_x, $x \in G/H$. We will from now on identify W_0 with V_J and, accordingly, ρ with ψ_0. Altogether, we have found a family of functions such that $\{S_y^G \psi_x : x \in G/H, y \in J\}$ is an orthonormal basis for V_H.

Now consider the following family of sampling sequences: $\alpha_x = \Sigma_\phi^H \psi_x$. Since Σ_ϕ^H is a surjective isometry from V_H to $L^2(H)$ that intertwines with the

J-translations we have that

$$\{S_y^G \alpha_x : x \in G/H, y \in J\} \text{ is an orthonormal basis of } L^2(H) \quad (10.0.1)$$

and we give the following definition.

Definition 10.0.1
A family of $\alpha_x \in L^2(H)$, $x \in H/J$, is called a QMF-system iff (10.0.1) *holds.*

Therefore, finding the generators of the translation invariant bases of W_x is equivalent to completing a QMF-system of which one sequence is given. Indeed, let $\alpha_0 = \Sigma_\phi^H \rho$. Then if we can find the missing α_x such that they are a QMF-system we have that

$$\psi_x = (\Sigma_\phi^H)^* \alpha_x$$

are the generators of a translation invariant orthonormal basis of the W_x. In the following we will concentrate on the algebraic content of QMF-systems. In particular, we will show that, again, there is a group structure associated to them. In particular, we will show that our problem of completing a QMF-system always has a solution, at least in a pure L^2-context, not involving any smoothness.

10.1 A group of operators

Consider now the group of unitary operators K acting in the Hilbert space $L^2(H)$ which commute with the translates with respect to the subgrid $J \subset H$:

$$[K, S_x^H] = KS_x^H - S_x^H K = 0 \quad \text{for all } x \in J.$$

Let $\{\alpha_i\}$ be a QMF-system. We claim that the image of a QMF-system under a unitary operator K that commutes with the J-translates is again a QMF-system. Indeed, as an image of an orthonormal basis it is again an orthonormal basis and in addition we have ($i, i' \in H/J$, $x, y \in J$)

$$\begin{aligned}\langle S_x^H K\alpha_i \mid S_y^H K\alpha_{i'}\rangle &= \langle KS_x^H \alpha_i \mid KS_y^H \alpha_{i'}\rangle \\ &= \langle S_x^H \alpha_i \mid S_y^H \alpha_{i'}\rangle = \delta_{i,i'}\delta_{x,y}.\end{aligned}$$

Now consider the quotient H/J that we may identify with a fundamental domain for the action of J on H. It contains $N = |H/J|$ points. Clearly, the family of delta sequences $\{S_k^H \delta : k \in H/J\}$ is a QMF-system. But its image is, by what we have said, again a QMF-system. Therefore, upon setting

$$\alpha_k = KS_k^H \delta, \qquad k \in H/J$$

we obtain a whole family of QMF-systems. Actually, we obtain all of them in this way. Indeed, every unitary operator that commutes with the J-translations is uniquely determined by its image of the delta sequences

$S_k^H\delta$ with $k \in H/J$. Let us call these functions α_k again. We can then recover K explicitly by means of the following: let Π^J be the orthogonal projector on $L^2(J)$ taken in the obvious way as a subspace of $L^2(H)$. Further, let $\Pi_K^J = S_k^H \Pi^J (S_k^H)^*$ be the projection on the coset $S_k^H J$. We now set

$$K: s \mapsto \sum_{k \in J} (\Pi_k^J s) * ((S_k^H)^* \alpha_k). \tag{10.1.1}$$

This operator clearly commutes with the J-translations and its image of the δ_k with $k \in H/J$ are just the α_k. But now the $\{\alpha_k\}$ are a QMF-system and the operator as defined by (10.1.1) is therefore unitary.

10.2 The Fourier space picture

We now come to the Fourier space picture. The Fourier transform of a sequence s in $L^2(H)$ reads

$$\hat{s}(\eta) = \sum_{x \in H} s(x)\bar{\eta}(x).$$

Recall that the Poisson summation formula states that the projection operator Π^J reads in Fourier space

$$\Pi^J: \hat{s}(\omega) \mapsto \frac{1}{|H/J|} \sum_{\xi \in J^\perp} \hat{s}(\omega\xi). \tag{10.2.1}$$

Thus the orthogonality relations of the QMF-system α_k can also be written as

$$\langle \alpha_k \mid S_x^H \alpha_{k'} \rangle_{L^2(H)} = \delta_{k,k'} \delta(x), \qquad x \in J,\ k, k' \in J^\perp$$

$$\Leftrightarrow \Pi_J(\tilde{\alpha}_k * \alpha_{k'}) = \delta_{k,k'},$$

where $\tilde{\alpha}_k(p) = \bar{\alpha}_k(-p)$. We will see in a moment that these conditions are actually equivalent to the QMF-condition, that is, we will see that the orthogonality relations imply completeness of the $S_x^H \alpha_k$. From (10.2.1) it follows that the orthonormality conditions are equivalent to

$$\sum_{\xi \in H^\perp} \bar{\hat{\alpha}}_k(\omega\xi)\hat{\alpha}_k(\omega\xi) = |H/J|\delta_{k,k'}.$$

Or, to put it still differently, the orthonormality condition means that

$$M_{k,\xi}(\omega) = \frac{1}{\sqrt{|H/J|}} \hat{\alpha}_k(\omega\xi), \qquad k \in H/J,\ \xi \in J^\perp, \tag{10.2.2}$$

is unitary for every $\omega \in \hat{H}$.

The unitary operator (10.1.1) in Fourier space reads as follows using the Poisson summation formula:

$$K: \hat{s}(\eta) \mapsto \frac{1}{|H/J|} \sum_{x \in H/J} \eta(x)\hat{\alpha}(\eta) \cdot \sum_{\chi \in J^{\perp}} \chi(x)\hat{s}(\chi\eta).$$

Thus if we introduce the functions $\hat{s}_{\varrho}(\eta) = \hat{s}(\varrho\eta)$, $\varrho \in J^{\perp}$, we obtain

$$K: \hat{s}_{\varrho}(\eta) \mapsto \sum_{\chi \in J^{\perp}} L_{\varrho,\chi}(\eta)\hat{s}_{\chi}(\eta), \tag{10.2.3}$$

with

$$L_{\varrho,\chi}(\eta) = \frac{1}{|H/J|} \sum_{x \in H/J} \hat{\alpha}_x(\varrho\eta)\eta(x)\chi(x). \tag{10.2.4}$$

In this basis the operator multiplication is easily implemented as matrix multiplication. If we introduce the matrix associated to the QMF-system δ_x, $x \in H/J$:

$$F_{x,\chi}(\omega) = \frac{1}{\sqrt{|H/J|}} \bar{\hat{\delta}}_x(\chi \cdot \omega) = \frac{1}{\sqrt{|H/J|}} \chi(x)\omega(x) \tag{10.2.5}$$

then we have the following relation:

$$L = M^t \cdot F, \qquad \text{or, explicitly, } L_{\varrho,\chi}(\omega) = \sum_{x \in H/J} M_{x,\varrho}(\omega)F_{x,\chi}(\omega). \tag{10.2.6}$$

Now since F is unitary it follows that L is unitary, too.

The structure of this function L taking values in the group of unitary matrices $U(|H/J|)$ will be analysed in the sequel.

10.3 QMF and loop groups

A function $A: \hat{H} \to U(N)$, $N = |H/J|$, will be called a loop. In the case where H is isomorphic to the lattice $\mathbb{Z}^n$, the loops are the sections of the unitary bundle over the n-dimensional torus $\mathbb{T}^n$. The loops form a group under pointwise multiplication:

$$A \cdot B(\omega) = A(\omega) \cdot B(\omega).$$

There is an obvious subgroup consisting of those loops that are $J^{\perp}$-periodic:

$$A(\omega\eta) = A(\omega), \qquad \eta \in J^{\perp}.$$

Let $R: J^{\perp} \to U(N)$ be the permutation representation of $J^{\perp}$ defined via $R(\eta) = \delta_{x,\eta\varrho}$. The loops that satisfy at the following condition:

$$B(\omega\eta) = R(\eta)B(\omega)R^*(\eta), \qquad \eta \in J^{\perp},$$

are again a subgroup of all loops. Its elements will be called the twisted loops. The loop L (10.2.4) is a twisted loop as direct computation shows.

With the help of the permutation representation of $J^\perp$ we can also characterize the loops of the type of (10.2.2). They are precisely the loops that satisfy at $M(\omega\rho) = M(\omega)R^*(\rho)$ for $\rho \in J^\perp$. We will call loops satisfying this relation QMF-loops.

The relations between all the objects we have introduced so far become clear through the next theorem.

Theorem 10.3.1

There is an explicit bijection between

(i) *the QMF-filters,*
(ii) *the QMF-loops:* $M(\omega\rho) = M(\omega)R^*(\rho)$, $\rho \in J^\perp$,
(iii) *the group of unitary operators acting in* $L^2(H)$ *with* $[K, S_x^H] = 0$, $x \in J$,
(iv) *the twisted loops:* $B(\omega\rho) = R(\rho)B(\omega)R^*(\rho)$, $\rho \in J^\perp$,
(v) *the untwisted loops:* $A(\omega\rho) = A(\omega)$, $\rho \in J^\perp$.

More precisely, the groups (iii), (iv) *and* (v) *are isomorphic. The set of QMF-loops is a left-coset with respect to* (iv) *and a right coset with respect to* (v). *If in addition, J is isomorphic with H then* (iii), (iv) *and* (v) *are isomorphic to the whole loop group.*

Proof. We have already seen the equivalence between (i), (ii), (iii) and (iv), and the isomorphy between groups (iii) and (iv). To show the equivalence between (iv) and (v) it is enough to observe that $A \mapsto FAF^*$ with F given by (10.2.5) defines an isomorphism between the twisted and the $J^\perp$-periodic loops. On the other hand, it is readily verified that $A \mapsto FA$ maps the twisted loops onto the QMF-loops. They are thus a right-coset with respect to the twisted loops. In addition, $A \mapsto AF^*$ maps the $J^\perp$-periodic loops onto the QMF-loops. They are hence a left-coset with respect to this subgroup.

In the case where J is isomorphic to H it follows that $\hat{H}$ is isomorphic to $\hat{H}/J^\perp$. Indeed, $J^\perp$ is the dual of H/J, which, as the dual of the trivial group, is the trivial group itself. Therefore, if Φ: $\hat{H}/J^\perp \to \hat{H}$ is a fixed isomorphism, then $A(\omega) \mapsto A(\Phi(\omega))$ is an isomorphism between the $J^\perp$-periodic loops and the whole loop group. □

This result shows in particular that our problem of completing the QMF-matrix has an L^2-solution. Indeed, the problem is to complete a unitary matrix-valued function whose first row is given and which satisfies the above compatibility condition. Now in L^2 this can obviously be done by choosing the complement for $\omega \in \hat{H}/J^\perp$ freely and using the condition above to extend it to all of $\hat{H}$. We have thus shown the following theorem.

Theorem 10.3.2

Let $J \subset H \subset G$ *be a chain of Abelian groups. Suppose H is discrete, G/H is* *compact, and H/J is finite. Let $V_J \subset V_H \subset L^2(G)$ be J- and H-sampling spaces. Then there is a decomposition*

$$V_H = \bigoplus_{x \in H/J} W_x = V_J \oplus \bigoplus_{x \in H/J - \{0\}} W_x,$$

where the spaces W_x are J-sampling spaces.

Note, however, that the construction we have given here introduces in general discontinuities. A way to achieve this globally and in a smooth way for specific lattices in $\mathbb{Z}^n$ has been found by Gröchenig (1989) using arguments from algebraic topology.

Example 10.3.3

Let $G = \mathbb{R}$, $H = \mathbb{Z}$ and $J = 3\mathbb{Z}$. This corresponds to the construction of wavelets where the dilation by 2 is replaced by a dilation by 3. Let us denote the sampling space over H by V_0 and that over J by V_{-1}. In addition, we suppose, in analogy with the wavelet case, that V_{-1} is nothing other than a re-scaled version of V_0, that is, $s \in C_0$ iff $s(t/3) \in V_{-1}$. The quotient group $H/J = \mathbb{Z}/3\mathbb{Z}$ contain three elements and thus there are two complement spaces

$$V_0 = V_{-1} \oplus W_{1,1} \oplus W_{-1,2}.$$

By sampling we obtain the associated QMF-systems. They are three sequences over $\mathbb{Z}$. Their Fourier[7] transforms will be denoted by m_i, $i = 0, 1, 2$. If the translates of $\varphi(\cdot - 3k)$ are an orthonormal basis of V_0 then we have the following scaling equation:

$$\hat{\varphi}(3\omega) = m_0(\omega)\hat{\varphi}(\omega).$$

The wavelets ψ_1 and ψ_2 whose translates with respect to $3\mathbb{Z}$ are an orthonormal basis for W_0 and W_1, respectively, can be written as

$$\hat{\psi}_1(3\omega) = m_1(3\omega)\hat{\varphi}(\omega), \qquad \hat{\psi}_2(3\omega) = m_2(3\omega)\hat{\varphi}(\omega),$$

with m_1 and m_2 some 2π-periodic functions. In addition, they satisfy the following generalized QMF-condition:

$$\frac{1}{\sqrt{3}} \begin{bmatrix} m_0(\omega) & m_0(\omega + 2\pi/3) & m_0(\omega + 4\pi/3) \\ m_1(\omega) & m_1(\omega + 2\pi/3) & m_1(\omega + 4\pi/3) \\ m_2(\omega) & m_2(\omega + 2\pi/3) & m_2(\omega + 4\pi/3) \end{bmatrix} \quad \text{is unitary for all } \omega \in \mathbb{T}.$$

[7] The respective Fourier transforms over $\mathbb{R}$ and $\mathbb{Z}$ are defined as usual. In particular, it is written additively again.

10.4 Polynomial loops: the factorization problem

In the case where $H = \mathbb{Z}^n$ the loops are maps from $\mathbb{T}^n$ to $U(N)$. Hence it makes sense to consider the groups of polynomial loops—unitary matrices whose entries are all Laurant polynomials in the elementary frequencies of the torus. This class is of particular importance given that it corresponds to QMF of finite length.

In one dimension we have been able to parametrize all polynomial loops or, equivalently, to construct all QMF-systems of finite length. In more than one dimension the factorization really is a problem. It is equivalent to the question of whether an arbitrary polynomial loop may be factorized into simpler polynomial loops.

Instead of answering this question, which we cannot, we show that there is a general construction for obtaining a large quantity of polynomial loops and hence QMF-systems. It again uses the equivalence between the unitary operators acting in $L^2(H)$ and the loop groups. Let U_u^F be the unitary operator defined by the following: its restriction to some fundamental domain $F = H/J$—consisting of N points non-congruent modulo J—coincides with the unitary map $u \in U(N)$. This definition is extended to all of $\mathbb{Z}^n$ by periodicity. This defines a whole manifold of finite impulse response QMF-systems and hence of polynomial loops. By changing the fundamental domain we obtain another family of such unitary operators. In general, the operators corresponding to different fundamental domains do not commute and thus one can compose them to obtain non-trivial new ones, namely,

$$U = \prod_{p=0}^{m} U_{u_p}^{F_p}, \quad \text{with } u_p \in U(N),\ F_p \text{ some fundamental domain.}$$

Whether or not this family is exhaustive is not yet known.

Appendix

We now want to give some applications of the general framework that has been developed before analysing functions over the two-dimensional plane, with the help of the wavelet transform. In Section 12 we will use the wavelet technique to find an inverse formula for the Radon transform.

11 The wavelet transform in two dimensions

The construction in this section is largely taken from Murenzi (1990). The analysed function will now be a function over the two-dimensional plane

$\mathbb{R}^2$, and the underlying group will be the two-dimensional group of Euclidean motions with dilations. This group will be denoted by ED^2. Its elements may be identified with the vector $[b, \phi, a]$ that parametrizes the Euclidean motions—translation $b = [b_0, b_1]$, rotations $\phi \in \mathbb{T}$ of the plane together with the dilations—parametrized by $a \in \mathbb{R}^+$. The group law is given by

$$[b_0, b_1, \phi, a] \circ [b'_0, b'_1, \chi, a']$$
$$= [b_0 + ab'_0 \cos\phi - ab'_1 \sin\phi, b_1 + ab'_0 \sin\phi + ab'_1 \cos\phi, (\phi + \chi) \bmod 2\pi, aa'].$$

The neutral element is given by

$$e = [0, 0, 0, 1]$$

and the inverse is given by

$$[b_0, b_1, \phi, a]^{-1}$$
$$= [-a^{-1}b_0 \cos\phi - a^{-1}b_1 \sin\phi, a^{-1}b^0 \sin\phi + a^{-1}b_1 \cos\phi, -\phi \bmod 2\pi, a^{-1}].$$

The group ED^2 contains the rotations $\mathbb{T}$:

$$ED^2 \supset \{[0, 0, \phi, 1] : \phi \in \mathbb{T}\} \simeq \mathbb{T},$$

the dilations $\mathbb{R}^+$:

$$ED^2 \supset \{[0, 0, 0, a] : a \in \mathbb{R}^+\} \simeq \mathbb{R}^+,$$

and the translations $\mathbb{R}^2$:

$$ED^2 \supset \{[b, 0, 1] : b \in \mathbb{R}^2\} \simeq \mathbb{R}^2,$$

as subgroups. By direct computation we can verify that a left- and a right-invariant Haar measure is given by

$$d\mu_l(b, \phi, a) = \frac{da\, d\phi\, d^2b}{a^3}, \qquad d\mu_r(b, \phi, a) = \frac{da\, d\phi\, d^2b}{a^2}.$$

Let us introduce the rotation operator

$$R_\phi : L^2(\mathbb{R}^2) \to L^2(\mathbb{R}^2), \qquad s(t) \mapsto s(\mathscr{R}_{-\phi} t),$$

where we have written $\mathscr{R}_\phi$ for the left-action of the rotation group $\mathbb{T}$ on the plane

$$\mathscr{R}_\phi : \mathbb{R}^2 \to \mathbb{R}^2, \qquad [t_0, t_1] \mapsto [t_0 \cos\phi - t_1 \sin\phi, t_0 \sin\phi + t_1 \cos\phi].$$

The translation operator is obtained by the action of the group $\mathbb{R}^2$ on itself via

$$T_b : L^2(\mathbb{R}) \to L^2(\mathbb{R}^2), \qquad s(t) \mapsto s(t - b).$$

Finally, the dilation operator is defined as

$$D_a\colon L^1(\mathbb{R}^2) \to L^1(\mathbb{R}^2), \qquad s(t) \mapsto \frac{1}{a^2}\, s\left(\frac{t}{a}\right).$$

Remark. We do not choose unitary dilation, but, rather, the dilation that leaves the $L^1(\mathbb{R}^2)$ norm invariant. This is purely conventional. The unitary dilation is given by aD_a.

A unitary representation of the group ED^2 is now obtained from these operators via

$$U\colon ED^2 \to \mathscr{U}(L^2(\mathbb{R}^2)), \qquad [b, \phi, a] \mapsto U(b, \phi, a) = aR_\phi D_a T_b.$$

We can now use our machinery to obtain the following analysis of two-dimensional functions.

The associated left- and right-transforms read

$$\mathscr{L}_g s(b, \phi, a) = \langle aT_b R_\phi D_a g \mid s\rangle_{L^2(\mathbb{R}^2)} = \int_{\mathbb{R}^2} d^2t\, \frac{1}{a^2}\, \bar{g}(a^{-1}\mathscr{R}_\phi^{-1}[t - b])s(t)$$

$$\mathscr{R}_g s(b, \phi, a) = \langle g \mid aT_b R_\phi D_a s\rangle_{L^2(\mathbb{R}^2)} = \int_{\mathbb{R}^2} d^2t\, \frac{1}{a}\, \bar{g}(t)s(a^{-1}\mathscr{R}_\phi^{-1}[t - b]).$$

We will not use the left- and the right transforms, but we will use, in analogy with the one-dimensional case, the $L^1(\mathbb{R}^2)$-normalization to define the wavelet transform of an analysed function s with respect to the analysing wavelet g as

$$\mathscr{W}_g s(b, \phi, a) = \langle T_b R_\phi D_a g \mid s\rangle_{L^2(\mathbb{R}^2)} = \int_{\mathbb{R}^2} d^2t\, \frac{1}{a^2}\, \bar{g}(a^{-1}\mathscr{R}_\phi^{-1}[t - b])s(t).$$

Let us introduce the two-dimensional Fourier transform and its inversion formula as

$$\hat{s}(k) = \int_{\mathbb{R}^2} d^2t\; s(t)\, e^{-i\langle k \mid t\rangle}, \qquad s(t) = \frac{1}{(2\pi)^2} \int_{\mathbb{R}^2} d^2k\; s(k)\, e^{i\langle k \mid t\rangle},$$

Then we can re-write the wavelet transform in Fourier space as follows:

$$\mathscr{W}_g s(b, \phi, a) = \int_{\mathbb{R}^2} d^2k\; \hat{g}(a\mathscr{R}_\phi^{-1}k)\, e^{i\langle b \mid k\rangle}\, \hat{s}(k).$$

Theorem 11.0.1

The representation above is irreducible and square integrable. The positive operator C as given by Theorem 5.1.1 *reads, in this case,*

$$\langle g \mid C \mid h\rangle = \int_{\mathbb{R}^2} \frac{d^2k}{|k|^2}\, \bar{\hat{g}}(k)\hat{h}(k).$$

Proof. As we will see in Section 11.3 the wavelet transform has an inverse for arbitrary $g \in L^2(\mathbb{R}^2)$. Therefore, $\mathcal{W}_g s = 0$ implies $s = 0$, and thus there is no non-trivial space orthogonal to the orbit of g under the representation of ED^2.

The rest of the proof follows by pure computation analogous to the one-dimensional case and we leave the details to the reader. □

11.1 Energy conservation

We now want to give easy-to-handle sufficient conditions on the analysing wavelet which ensure the energy conservation of the wavelet transform. As we have seen, everything concerning admissibility relies on the positive operator C. Accordingly, by its explicit form (Theorem (11.0.1)), a wavelet g is called admissible iff the constant

$$c_g = \int_{\mathbb{R}^2} \frac{d^2k}{|k|^2} |\hat{g}(k)|^2$$

satisfies $c_g \neq 0$ and $c_g < \infty$. In consequence, the wavelet transform with respect to an admissible wavelet is a multiple of an isometry

$$\langle s \mid r \rangle_{\mathbb{R}^2} = \frac{1}{c_g} \langle \mathcal{W}_g s \mid \mathcal{W}_g r \rangle_{L^2(ED^2,\, da\, d\phi\, d^2b/a)}.$$

In particular, it conserves energy:

$$\int_{\mathbb{R}^2} dt\, |s(t)|^2 = \frac{1}{c_g} \int_{\mathbb{R}^+} \frac{da}{a} \int_{\mathbb{T}} d\phi \int_{\mathbb{R}^2} d^2b |\mathcal{W}_g s(b, \phi, a)|^2.$$

11.2 Reconstruction formulae

According to our general construction we call a wavelet h a reconstruction wavelet if it is admissible and if the constant

$$c_{g,h} = \int_{\mathbb{R}^2} \frac{d^2k}{|k|^2} \bar{\hat{g}}(k) \hat{h}(k) \tag{11.2.1}$$

satisfies $c_{g,h} \neq 0$ and $|c_{g,h}| < \infty$. The wavelet transform may then be inverted by means of the following integral over the parameter space:

$$s(t) = \int_{\mathbb{R}^+} \frac{da}{a} \int_{\mathbb{T}} d\phi \int_{\mathbb{R}^2} d^2b \mathcal{W}_g r(b, \phi, a) h_{b,\phi,a}(t),$$

where $h_{b,\phi,a} = T_b R_\phi D_a$. Note that this formula holds *a priori* only in a weak sense and only for admissible analysing and re-construction wavelets. However, as we will see in Section 11.3 the range of validity of this formula may, as in the one-dimensional case, be considerably enlarged. In particular, we may choose the δ-function as the reconstruction wavelet, in which case we reconstruct the analysed function by a simple summation over the scales and rotations:

$$s(t) = c_{g,\delta} \int_{\mathbb{R}_+} \frac{da}{a} \int_{\mathbb{T}} d\phi \, \mathcal{W}_g r(t, \phi, a)$$

and the constant $c_{g,\delta}$ is given by

$$c_{g,\delta} = \int_{\mathbb{R}^2} \frac{d^2k}{|k|^2} \bar{\hat{g}}(k).$$

11.3 The inversion formula

In this section we will give the proof of the inversion formula for two-dimensional wavelet transforms. Essentially, we follow the one-dimensional case. In particular, we will give sufficient conditions for pointwise and uniform convergence of the inversion formula.

Let h be an arbitrary function over the plane and let

$$F_{g,h}(t) = \frac{1}{t^2} \int_{|x| \leq t} d^2x (h * \tilde{g})(x), \qquad t > 0, \tag{11.3.1}$$

where $\tilde{g}(x) = \bar{g}(-x)$. This function is well defined whenever $h * \tilde{g}$ is locally integrable. Note that either g or h may be a distribution, as long as their convolution product is regular. A function h is called a uniform reconstruction wavelet for the analysing wavelet g if we have

$$\left.\begin{aligned} &\text{(i)} \quad tF_{g,h}(t) \in L^1(\mathbb{R}^+), \\ &\text{(ii)} \quad \int_0^\infty t \, dt \, F_{g,h}(t) = 1. \end{aligned}\right\} \tag{11.3.2}$$

In the case where condition (i) can be replaced by the stronger one:

i′) $|F_{g,h}(t)| \leq c(1 + t)^{-2-\gamma}, \qquad \gamma > 0,$

we say that h is a pointwise reconstruction wavelet for g.

Condition (ii) can be rephrased in Fourier space as (see equation (11.3.7) for $k = 0$ below)

$$\int_{\mathbb{R}^2} \frac{d^2k}{|k|^2} \bar{\hat{g}}(k) \hat{h}(k) = 1 \tag{11.3.3}$$

and we re-discover the admissibility condition of (11.2.1).

We now want to give a strong meaning to the inversion formula. Consider the following approximation of the reconstruction integral:

$$s_{\epsilon,\rho}(t) = \int_{\epsilon}^{\rho} \frac{da}{a} \int_{0}^{2\pi} d\phi \int_{\mathbb{R}^2} d^2b \mathscr{W}_g s(b, \phi, a) a^{-2} h(a^{-1}\, d\phi\, \mathscr{R}_{\phi}^{-1}[t-b])$$

$$= \int_{\epsilon}^{\rho} \frac{da}{a} \int_{0}^{2\pi} \int_{\mathbb{R}^2} d^2b \mathscr{W}_g s(b, \phi, a)(T_b R_{\phi} D_a h)(x).$$

First note that for fixed direction ϕ and fixed scale a, the integral over the translations is nothing other than a convolution with the dilated and rotated re-reconstruction wavelet h. In the same way, the wavelet transform is, for fixed scale and fixed direction, a convolution. Therefore, we may write

$$s_{\epsilon,\rho} = \left\{ \int_{\epsilon}^{\rho} \frac{da}{a} \int_{\mathbb{T}} d\phi\, h_{\phi,a} * \tilde{g}_{\phi,a} \right\} * s. \tag{11.3.4}$$

In this form the reconstruction integral makes sense for all finite $\epsilon > 0$ and $\rho < \infty$, even if some of the s, g, and h are tempered distributions, provided the repeated convolution is well defined (e.g. all functions have compact support). The reconstruction theorem can now be stated as follows.

Theorem 11.3.1

Let h be a uniform reconstruction wavelet for g. Then for $s \in L^p$ we have that the approximations of (11.3.4) *satisfy*

$$s = \lim_{\varepsilon \to 0, \rho \to \infty} s_{\epsilon,\rho},$$

and the convergence holds uniformly in $L^p(\mathbb{R}^2)$, $1 < p < \infty$. Suppose, further, that h is a pointwise reconstruction wavelet. Then the limit holds pointwise in any point of continuity of $s \in L^p(\mathbb{R}^2)$, $1 < p < \infty$.

Proof. We can write

$$s_{\varepsilon,\rho} = \left\{ \int_{\epsilon}^{\rho} \frac{da}{a} \int_{0}^{2\pi} d\phi\, D_a R_{\phi}(h * \tilde{g}) \right\} * s$$

$$= (B_{\epsilon,\rho} * s)(t),$$

where

$$B_{\epsilon,\rho}(t) = \int_{\epsilon}^{\rho} \frac{da}{a} \int_{0}^{2\pi} d\phi [D_a R_{\phi}(h * \tilde{g})](t).$$

The filter $B_{\epsilon,\rho}$ is actually the difference of two filters:

$$B_{\epsilon,\rho}(t) = B_{\epsilon,\infty}(t) - B_{\rho,\infty}(t),$$

where

$$
\begin{aligned}
B_{\epsilon,\infty}(t) &= \int_{\epsilon}^{\infty} \frac{da}{a} \int_0^{2\pi} d\phi \, \frac{1}{a^2} R_\phi(h * \tilde{g})\left(\frac{t}{a}\right) \\
&= \frac{1}{|t|^2} \int_{\epsilon/|t|}^{\infty} \frac{da}{a} \int_0^{2\pi} d\phi \, \frac{1}{a^2} R_\phi(h * \tilde{g})\left(\frac{t}{|t|a}\right) \\
&= \frac{1}{|t|^2} \int_0^{|t|/\epsilon} a \, da \int_0^{2\pi} d\phi \, R_\phi(h * \tilde{g})(at/|t|) \\
&= \frac{1}{|t|^2} \int_{|y|<|t|/\epsilon} d^2 y h * \tilde{g}(y).
\end{aligned}
$$

Therefore we have, by the definition of $F_{g,h}$, in (11.3.1),

$$
\begin{aligned}
B_{\epsilon,\rho}(t) &= \frac{1}{\epsilon^2} B_{1,\infty}\left(\frac{|t|}{\epsilon}\right) - \frac{1}{\rho^2} B_{1,\infty}\left(\frac{|t|}{\rho}\right) \\
&= \frac{1}{\epsilon^2} F_{g,h}\left(\frac{|t|}{\epsilon}\right) - \frac{1}{\rho^2} F_{g,h}\left(\frac{|t|}{\rho}\right).
\end{aligned}
$$

In particular,

$$
B_{1,\infty}(t) = F_{g,h}(|t|).
$$

By hypothesis (i) on $F_{g,h}$ the function $B_{1,\infty}$ is absolutely integrable and hence the theorem now follows from the last expression in (11.3.5) in connection with the theorems on the approximation of the identity (11.3.5) as in the one-dimensional case. □

The important point is that neither g nor h alone needs to be regular. It is enough that the combination $\tilde{g} * h$ is regular in the sense that we have explained. Therefore, a convenient choice of the reconstruction wavelet h may be used to invert the 'wavelet' transform even if g is only a distribution.

In Fourier space we have

$$
\widehat{s_{\varepsilon,\rho}}(k) = \hat{s}(k)(\widehat{B_{1,\infty}}(\epsilon k) - \widehat{B_{1,\infty}}(\rho k)), \tag{11.3.6}
$$

where

$$
\begin{aligned}
\widehat{B_{1,\infty}}(k) &= \int_1^{\infty} \frac{da}{a} \int_0^{2\pi} d\phi [D_a^* R_\phi^* \hat{h} \cdot \bar{\hat{g}}](k) \\
&= \int_1^{\infty} \frac{da}{a} \int_0^{2\pi} d\phi \, \hat{h} \cdot \bar{\hat{g}}(a \mathcal{R}_\phi k) \\
&= \int_{|k'| \geq |k|} \frac{d^2 k'}{|k'|^2} \hat{h} \cdot \bar{\hat{g}}(k').
\end{aligned} \tag{11.3.7}
$$

From expression (11.3.6) we see that for finite ϵ and ρ only the spatial frequencies of s, which are essentially localized between two concentric circles, are contained in the approximation $s_{\epsilon,\rho}$. As $\epsilon \to 0$ more and more small scales (i.e. high frequencies) are reconstructed, whereas for $\rho \to \infty$ more and more large scales (i.e. low frequencies) are reconstructed.

As a last point observe that we have in addition shown that the relation between s and the reconstruction at one single scale ($a = 1$ say):

$$\tilde{s}(t) = \int_0^{2\pi} \int_{\mathbb{R}^2} d^2b\, \mathscr{W}(b, \phi, 1)(T_b T_\phi h)(t)$$

is given by

$$\tilde{s} = K * s$$

with

$$\hat{K}(k) = \hat{K}(|k|) = \int_0^{2\pi} d\phi\, \hat{h}(\mathscr{R}_\phi k)\bar{\hat{g}}(\mathscr{R}_\phi k).$$

12 A class of inverse problems

The following situation is quite common in the study of inverse problems. Consider an arbitrary function s over $\mathbb{R}^n$. Unfortunately, we are unable to measure s itself, but only a set of (formal) projections:

$$\langle g_\lambda \mid s \rangle, \qquad \lambda \in \Lambda, \tag{12.0.1}$$

onto some states g_λ with $\lambda \in \Lambda$ is known. As an example consider the Radon transform. Here the family g_λ are delta functions concentrated on m-dimensional planes and Λ is some set that parametrizes these planes.

A natural inverse problem might be the following. Given all the 'projections' of (12.0.1) how is it possible to recover s? In this generality the problem is certainly not tractable. Therefore, we consider the following special case. Suppose that Λ is a (locally compact) not necessarily Abelian group. Suppose, further, that the mapping $\lambda \to g_\lambda$ is obtained by a representation U of Λ in the space $\mathscr{U}(L^2(\mathbb{R}^n))$ of unitary operators acting on $L^2(\mathbb{R}^n)$:

$$U: \Lambda \to \mathscr{U}(L^2(\mathbb{R}^n)), \qquad U(\lambda_1\lambda_2) = U(\lambda_1)U(\lambda_2),$$

by setting

$$g_\lambda = U(\lambda)g.$$

Then to know all projections of (12.0.1) amounts to knowing the formal left-transform of s:

$$\mathscr{L}_g s(\lambda) = \langle U(\lambda)g \mid s \rangle. \tag{12.0.2}$$

Therefore, as long as we stay in a Hilbert-space context we already know how to invert this transform. We now want to leave the Hilbert space context that was more or less present in all previous applications. By doing so we will show how the wavelet technique may be used to solve the inverse problem of (12.0.1) in cases where the functions g_λ are no longer Hilbert-space vectors, but may be arbitrary distributions. In this case we speak of singular analysing functions, or, in the case of the affine group, of singular analysing wavelets. By considering such singular functions, inverse problems, as, for instance, the Radon inversion, may be treated with the help of the wavelet technique. But other problems where g_λ becomes singular—or not—may be treated using the wavelet approach.

Once more, however, the existence of an underlying group structure for the set of parameters λ, is crucial; this situation may still be quite common, since it merely reflects a certain symmetry in the way of taking our projections in (12.0.1).

We will not treat the problem for arbitrary groups. Instead, we will show how to work with the wavelet technique by considering the example of the affine group with rotations acting in two dimensions. In particular, we will show how to obtain an inversion formula for the two-dimensional Radon transform. The results in this section are borrowed from Holschneider (1991).

12.1 The Radon transform as wavelet transform

The Radon transform of a function s over the two-dimensional plane is the integral of s over all one-dimensional affine subspaces of $\mathbb{R}^2$, that is, over the set of all straight lines. These lines are the solution of

$$\langle p \mid x \rangle = \alpha,$$

with $p \in \mathbb{R}^2$ a vector of unit length and α a real number. The Radon transform then reads

$$\mathcal{R}\mathcal{A}(p, \alpha) = \int_{\mathbb{R}^2} d^2x \, \delta(\langle p \mid x \rangle - \alpha) s(x).$$

Upon identifying p with an angle ϕ by setting $p = \mathcal{R}_\phi e$ with some reference direction e we have

$$\mathcal{R}\mathcal{A}(\phi, \alpha) = \int_{\mathbb{R}^2} d^2d \, \delta(\langle \mathcal{R}_\phi e \mid x \rangle - \alpha) s(x).$$

Since every straight line is obtained by means of translations and dilations of the line $\langle e \mid x \rangle = 0$ we may view the radon transform as a wavelet transform with respect to the singular wavelet which is a delta distribution concentrated on a line:

$$g(x) = \delta(\langle e \mid x \rangle). \tag{12.1.1}$$

Indeed, we have for this g

$$(T_b R_\phi D_a g)(x) = a\,\delta(\langle \mathcal{R}_\phi e \mid x\rangle - \langle b \mid \mathcal{R}_\phi e\rangle),$$

and thus the wavelet transform with respect to the singular wavelet (12.1.1) is related to the Radon transform by means of

$$\mathcal{W}_g s(b, \phi, a) = a^{-1}\mathcal{R}\mathcal{A}(\langle b \mid \mathcal{R}_\phi e\rangle, \phi).$$

Clearly, this is a highly redundant way to present the Radon transform. Indeed, the wavelet is invariant under translations in the direction of the delta line. Therefore, for every angle ϕ the wavelet transform $\mathcal{W}$ is constant along the parallel lines that are orthogonal to $\mathcal{R}_\phi e$. To put it the other way around, the wavelet transform depends only on the projection of b on the direction $\mathcal{R}_\phi e$:

$$\mathcal{W}_g s(b, \phi, a) = \mathcal{W}(\langle b \mid \mathcal{R}_\phi e\rangle, \phi). \tag{12.1.2}$$

The formal computations have a precise meaning if $s \in L^1(\mathbb{R})$, as follows from Fubini's theorem.

12.2 The Radon-inversion formula

In the case of the singular wavelet of (12.1.1) the condition of (11.5.3) for a possible reconstruction wavelet is reduced to

$$2\pi \int_{-\infty}^{+\infty} \frac{d\lambda}{|\lambda|^2}\,\hat{h}(\lambda e) = 1$$

because the Fourier transform of the delta line is again—up to a factor 2π—a delta line but this time in the direction of e perpendicular to the original delta line. The filter $B_{1,\infty}$ is given by

$$\widehat{B_{1,\infty}}(k) = 2\pi \int_{|\lambda| > |k|} \frac{d\lambda}{\lambda^2}\,\hat{h}(\lambda e) = 1 - 2\pi \int_{-|k|}^{+|k|} \frac{d\lambda}{\lambda^2}\,\hat{h}(\lambda e).$$

Therefore, it is easy to find a function h such that $B_{1,\infty}(x) = F_{g,h}(|x|)$ is sufficiently localized to satisfy conditions (i) and (ii) in (11.3.2). We will give sufficient conditions below.

Therefore, reconstruction wavelets for our singular wavelet exist and we may use the relation between $\mathcal{W}$ and $\mathcal{R}\mathcal{A}$ to write

$$s(x) = \lim_{\epsilon \to 0, \rho \to \infty} \int_\epsilon^\rho \frac{da}{a} \int_0^{2\pi} \int_{\mathbb{R}^2} d^2b\, a^{-1}\mathcal{R}\mathcal{A}(\langle b \mid \mathcal{R}_\phi e\rangle, \phi)\, a^{-2} h(a^{-1}\mathcal{R}_\phi[x - b]).$$

This inversion formula is uniformly convergent for $s \in L^p(\mathbb{R}^2) \cap L^1(\mathbb{R}^2)$ in the topology of $L^p(\mathbb{R}^2)$ and if h is chosen a pointwise reconstruction wavelet it converges pointwise in every point of continuity of s.

We now want to simplify this expression using the fact that the wavelet is translation invariant in one direction. For every ϕ we write $d^2b = d\alpha\, d\alpha^*$, where $\alpha = \langle b \mid \mathscr{R}_\phi e\rangle$ is the coordinate in the direction of $\mathscr{R}_\phi e$ and α^* is the one orthogonal to $\mathscr{R}_\phi e$. By the invariance of (12.1.2) the wavelet transform is independent of α^*, and thus integrating over this variable yields:

$$s(x) = \lim_{\epsilon \to 0, \rho \to \infty} \int_\epsilon^\rho \frac{da}{a} \int_0^{2\pi} \int_{\mathbb{R}} d\alpha\, a^{-1} \mathscr{R}\mathscr{A}(\phi, a) a^{-1} X(a^{-1}\mathscr{R}\mathscr{R}_\phi[x - \alpha e])$$

$$= \lim_{\epsilon \to 0, \rho \to \infty} \int_\epsilon^\rho \frac{da}{a} \int_0^{2\pi} \int_{\mathbb{R}} d\alpha\, \mathscr{R}\mathscr{A}(\phi, \alpha)(\mathscr{R}_\phi T_{\alpha e} D_a X)(x)$$

where X is a function over $\mathbb{R}^2$ that depends only on $\beta = \langle x \mid e\rangle$:

$$X(b) = Y(\beta) = \int_{\mathbb{R}^2} d^2x\, \delta(\langle e \mid x\rangle - \beta) h(x).$$

Therefore, the (one-dimensional) Fourier transform of $Y(\beta)$ is just the value of $\hat{h}$ along e:

$$\hat{Y}(\omega) = \hat{h}(\omega e).$$

Note that the integral over the translation da is nothing other than the convolution of the Radon coefficients $\mathscr{R}\mathscr{A}(\phi, \cdot)$ with the dilated profiles $Y_a(t) = a^{-2}Y(t/a)$. For a given profile Y the partial reconstruction filter $B_{1,\infty}$ is given in Fourier space by

$$\widehat{B_{1,\infty}}(k) = 2\pi \int_{|\lambda| > |\lambda|} \frac{d\lambda}{\lambda^2} \hat{Y}(\lambda) = 1 - 2\pi \int_{-|k|}^{+|k|} \frac{d\lambda}{\lambda^2} \hat{Y}(\lambda).$$

Now suppose that $B_{1,\infty} \in L^1(\mathbb{R}^2)$. We then obtain the following theorem.

Theorem 12.2.1
Let $s \in L^p(\mathbb{R}^2) \cap L^1(\mathbb{R}^1)$ with $1 < p < \infty$. Then s may be recovered from its Radon coefficients by means of

$$s(x) = \lim_{\epsilon \to 0, \rho \to \infty} \int_\epsilon^\rho \frac{da}{a} \int_0^{2\pi} d\phi (\mathscr{R}\mathscr{A}(\phi, \cdot) * Y_a)(\langle \mathscr{R}_\phi e \mid x\rangle). \quad (12.2.1)$$

The convergence is uniformly in $L^p(\mathbb{R}^2)$ and pointwise in any point of continuity of s.

Therefore, for fixed scale a the reconstruction is equivalent to the back-projection technique.

By a theorem of Bernstein (1914), a sufficient condition for $B_{1,\infty} \in L^1(\mathbb{R}^2)$

is given by $\widehat{B_{1,\infty}} \in L^1(\mathbb{R}^2)$ and $\widehat{B_{1,\infty}}$ of Hölder regularity 1/2, namely,

$$|\widehat{B_{1,\infty}}(k) - \widehat{B_{1,\infty}}(k')| \leq c|k - k'|^{1/2}.$$

Therefore, a sufficient condition on the profile Y to ensure the uniform reconstruction of (12.2.1) in $\mathbf{L}^p(\mathbb{R}^2)$, $1 < p < \infty$, is given by

(i) $\hat{Y}(\lambda) = O(\lambda^{3/2})$, $(\lambda \to 0)$,
(ii) $\hat{Y}(\lambda) = O(\lambda^{-\epsilon})$, $(\lambda \to \pm\infty)$,

(iii) $$\int_{-\infty}^{+\infty} \frac{d\lambda}{\lambda^2} \hat{Y}(\lambda) = \frac{1}{2\pi}.$$

In cases where we use (12.2.1) at one single scale (say $a = 1$) we obtain

$$\tilde{s}(x) = \int_0^{2\pi} d\phi \, (\mathscr{R}\mathscr{A}(\phi, \cdot) * Y_a)(\langle \mathscr{R}_\phi e \mid x \rangle) = K * s(x)$$

with

$$\hat{K}(k) = \hat{Y}(|k|)/|k|.$$

6

FUNCTIONAL ANALYSIS AND WAVELETS

In the first section we saw that the space of C^{∞} regular, localized functions with all moments vanishing is characterized by a rapid decay of the modulus of the wavelet coefficients at small and large scales and at large positions so that the wavelet transform is localized over the position-scale half-plane. By duality the space of tempered distributions was characterized by an at most polynomial growth of the wavelet coefficients at small and large scales and large positions. A third space, the Hilbert space of square integrable functions, was characterized by a square integrable distribution of the energy over the position-scale half-plane. In all three cases this characterization was independent of the analysing wavelet. It therefore seems natural to introduce spaces that are characterized by their distribution of the modulus of the wavelet coefficients over the position-scale half-plane. This will be done in this chapter. For supplementary information we refer the reader to the literature. In particular in the case of the wavelet transform the books by Meyer (1985) and Coifman and Meyer are certainly standard references. On the other hand, many ideas that are now unified under the wavelet approach can be found in great detail in the book by Triebel (1992). Again, all results can be generalized in the language of group theory, as was shown by Feichtinger and Gröchenig (1989).

1 Some function spaces

In this section we will introduce a family of Banach spaces that are all characterized by the localization of the absolute value of the wavelet coefficients over the half-plane. We show that this family contains most of the function spaces that one encounters in day-to-day functional analysis. We will work in $\mathbb{R}^n$ right from the beginning since this causes no additional problems. We will use the wavelet transform over $\mathbb{R}^n$ in the same way as in Section 30 of Chapter 1 to which we refer for the notation. The wavelet transform is then a map from functions over $\mathbb{R}^n$ to functions over the half-space $\mathbb{H}^n = \{(b, a) : b \in \mathbb{R}^n, a \in \mathbb{R}_+\}$.

In order to quantify different types of localization of a function over the position-scale half-space we consider two weight functions $\kappa(b)$—responsible for the localization over the positions—and $\phi(a)$—characterizing the localization over the scales.

Following Hörmander (1982) we say that a non-negative function κ is a (tempered) weight function over $\mathbb{R}^n$ if it satisfies at

$$\kappa(x + y) \leq O(1)(1 + |x|^2)^{\alpha/2}\kappa(y)$$

for all $x, y \in \mathbb{R}^n$ with some $\alpha \in \mathbb{R}$. An obvious example is $\kappa(x) = (1 + |x|^2)^{\alpha/2}$. Analogously, we say that a positive function is a (tempered) weight function over $\mathbb{R}_+$ if it satisfies at

$$\phi(xy) \leq O(1)(x + 1/x)^\alpha \phi(y)$$

for all $x, y \in \mathbb{R}_+$ with some $\alpha \in \mathbb{R}$. Again, the most frequently used example is $\phi(a) = a^\alpha/(1 + a)^{\alpha+\beta}$. One can easily show that the sum and the product of two weight functions is again a weight function. In addition, κ^δ and ϕ^δ are weight functions, too, for any real δ.

It is, then, natural to consider the space of functions T over the half-space which are localized in such a way that we have

$$\|T\|_{LP^{p,q}_{\kappa,\phi}(\mathbb{H}^n)} = \left\{\int_{\mathbb{R}^n} d^n b \kappa^q(b) \left(\int_0^\infty \frac{da}{a} \phi^p(a)|T(b, a)|^p\right)^{q/p}\right\}^{1/q} < \infty$$

for some $p, q > 0$ and some weight functions κ: $\mathbb{R}^n \to \mathbb{R}_+$ and ϕ: $\mathbb{R}_+ \to \mathbb{R}_+$. Here the symbol 'LP' stands for 'Littlewood–Paley' since it will turn out that this kind of localization over the half-space is typical for the characterization of L^p-spaces with the help of the Littlewood–Paley theory. Note that this complicated expression may be rephrased as follows: let $L^q(\mathbb{R}^n, d^n b\kappa^q(b))$ be the space of functions over $\mathbb{R}^n$ that satisfy at $\|\rho\|^q_{L^q(\mathbb{R}^n, d^n b\kappa^q(b))} = \int d^n b\kappa^q(b)|\rho(b)|^q < \infty$ and let $L^p(\mathbb{R}_+, da\, \phi^p(a)/a)$ be the analogue weighted L^p space over $\mathbb{R}_+$. Then we have

$$\|T\|_{LP^{p,q}_{\kappa,\phi}(\mathbb{H}^n)} = \|\rho\|_{L^q(\mathbb{R}^n, d^n b\kappa^q(b))},$$

where

$$\rho(b) = \|T(b, \cdot)\|_{L^p(\mathbb{R}_+, da\, \phi^p(a)/a)}.$$

This shows that the expression above actually defines a seminorm. By identifying all functions that agree almost everywhere we obtain norms and we no longer distinguish between the norms and the seminorms.

It is also possible to look at these norms in terms of vector-valued functions: let B be a Banach space with norm $\|\cdot\|_B$, and let (M, μ) be a positive measure space. Consider the set of functions over M with values in B. For s such a vector-valued function we denote by $|s(x)|$ the function that associates to $x \in M$ the non-negative number $\|s(x)\|_B$. Consider now the set $L^p(M, \mu, B)$ of vector-valued functions for which

$$\|s\|_{L^p(M,\mu,B)} = \left\{\int_M d\mu(x)\, |s(x)|^p\right\}^{1/p} < \infty.$$

The right-hand side defines a norm which makes $L^p(M, \mu, B)$ a Banach space if we identify functions that agree almost everywhere (see Treves 1967 for a proof). For $p = \infty$ the usual modifications apply. With this notation we have

$$\|T\|_{LP^{p,q}_{\kappa,\phi}(\mathbb{H}^n)} = \|\mathscr{R}\|_{L^q(\mathbb{R}, db\kappa^q(b), L^p(\mathbb{R}_+, daa^{-1}\phi^p(a)))},$$

where $\mathscr{R}(b) = T(b, \cdot)$ is the function that associates to a position $b \in \mathbb{R}^n$ the whole zoom perpendicular to that point. This interpretation as the norm of a space of vector-valued functions shows that $LP^{p,q}_{\kappa,\phi}(\mathbb{H}^n)$ is a Banach space provided we identify functions that agree almost everywhere.

A second way of parametrizing the localization of a function over the half-space is given by requiring that the following expression be finite $(p, q > 0)$:

$$\|T\|_{B^{p,q}_{\kappa,\phi}(\mathbb{H}^n)} = \left\{\int_0^\infty \frac{da}{a}\,\phi^p(a)\left(\int_{\mathbb{R}^n} d^n b\kappa^q(b)\,|T(b, a)|^q\right)^{p/q}\right\}^{1/p} < \infty$$

for some weight functions $\kappa\colon \mathbb{R}^n \to \mathbb{R}_+$ and $\phi\colon \mathbb{R}_+ \to \mathbb{R}_+$. The label '$B$' stands for 'Besov' since this kind of localization over the position-scale half-space characterizes Besov spaces (see Meyer 1990 or Triebel 1992). Note that, again, p quantifies the localization over the scales, that is, it is the 'smoothness' index.

Again this expression may be interpreted as

$$\|T\|_{B^{p,q}_{\kappa,\phi}(\mathbb{H}^n)} = \|\sigma\|_{L^p(\mathbb{R}_+, da\,\phi^p(a)/a)}, \quad \text{where} \quad \sigma(a) = \|T(\cdot, a)\|_{L^q(\mathbb{R}^n, d^n b\kappa^q(b))}.$$

Again this shows that this actually defines a norm provided we identify functions that agree almost everywhere. As before, there is a vector-valued interpretation, namely,

$$\|T\|_{B^{p,q}_{\kappa,\phi}(\mathbb{H}^n)} = \|\mathscr{R}\|_{L^p(\mathbb{R}_+, daa^{-1}\phi^p(a), L^q(\mathbb{R}^n, db\kappa^q(b)))},$$

where $\mathscr{R}(a) = T(\cdot, a)$ is the function that associates to a scale $a \in \mathbb{R}_+$ the whole voice at scale a. Thus $B^{p,q}_{\kappa,\phi}(\mathbb{H}^n)$ is a Banach space.

The interpretation as norms of vector-valued L^p-spaces allows us to obtain their dual spaces. We have that the dual space of $L^p(M, d\mu, B)$ is $L^{p'}(M, d\mu, B')$ with $p^{-1} + p'^{-1} = 1$, $1 \le p < \infty$, and B' the dual of B, provided the double dual B'' is separable (e.g. Gajewski *et al.* 1974, or Reed and Simon 1972). This means that every continuous linear functional of $L^p(M, d\mu, B)$ may be written as

$$s \mapsto \int_M d\mu(x)(\eta(x) \mid s(x)),$$

with some $\eta \in L^{p'}(M, d\mu, B')$ and, clearly, every such η gives rise to a linear functional in $L^p(M, d\mu, B)$. We agree to write $(\eta \mid s) = \eta(s)$ for the action of η on s. We therefore have the following theorem.

Theorem 1.0.1

Let $1 < p, q < \infty$. Then the dual space of $B^{p,q}_{\kappa,\phi}(\mathbb{H}^n)$ is $B^{p',q'}_{1/\kappa,1/\phi}(\mathbb{H}^n)$ and the dual space of $LP^{p,q}_{\kappa,\phi}(\mathbb{H}^n)$ is $LP^{p',q'}_{1/\kappa,1/\phi}(\mathbb{H}^n)$ with $(1/p) + (1/p') = 1$ and $(1/q) + (1/q') = 1$. The pairing is understood as

$$r: s \mapsto (r \mid s) = \int_0^\infty \frac{da}{a} \int_{\mathbb{R}^n} d^n b\, r(b, a) s(b, a).$$

Therefore, since for α large enough we have

$$(a + 1/a)^{-\alpha}(1 + |b|^2)^{-\alpha/2} \in B^{p,q}_{\kappa,\phi}(\mathbb{H}^n) \cap LP^{p,q}_{\kappa,\phi}(\mathbb{H}^n),$$

the next theorem follows.

Theorem 1.0.2

The following inclusions holds for $1 < p, q < \infty$ and arbitrary tempered weight functions κ and ϕ:

$$S(\mathbb{H}^n) \subset B^{p,q}_{\kappa,\phi}(\mathbb{H}^n), \quad LP^{p,q}_{\kappa,\phi}(\mathbb{H}^n) \subset S'(\mathbb{H}^n).$$

The inclusions also hold in the topological sense, that is, the inclusion maps are continuous.

The spaces $B^{p,q}_{\kappa,\phi}(\mathbb{R}^n)$ and $LP^{p,q}_{\kappa,\phi}(\mathbb{R}^n)$ are characterized by requiring that the wavelet coefficients are localized such that the norms we have considered above are finite. We give the following definition.

Definition 1.0.3

Let $g \in S_0(\mathbb{R}^n)$. A distribution $s \in S_0'(\mathbb{R}^n)$ is in $B^{p,q}_{\kappa,\phi}(\mathbb{R}^n)$ (with respect to g) if its wavelet transform satisfies

$$\|s\|_{B^{p,q}_{\kappa,\phi}(\mathbb{R}^n)} = \|\mathcal{W}_g s\|_{B^{p,q}_{\kappa,\phi}(\mathbb{H}^n)} < \infty.$$

We say a distribution $s \in S_0'(\mathbb{R})$ is in $LP^{p,q}_{\kappa,\phi}(\mathbb{R})$ (with respect to g) if

$$\|s\|_{LP^{p,q}_{\kappa,\phi}(\mathbb{R}^n)} = \|\mathcal{W}_g s\|_{LP^{p,q}_{\kappa,\phi}(\mathbb{H}^n)} < \infty.$$

It seems that the definition of spaces we have given depends on the wavelet g. Actually, essentially it does not, since for admissible wavelets all norms are equivalent.

Theorem 1.0.4

Let $g, h \in S_0(\mathbb{R}^n)$ be two admissible wavelets, that is

$$c_g = \int_0^\infty \frac{da}{a} |\hat{g}(ak)|^2 \neq 0, \qquad c_h = \int_0^\infty \frac{da}{a} |\hat{h}(ak)|^2 \neq 0.$$

Then there is a constant $c > 1$ *such that*

$$c^{-1}\|\mathcal{W}_g s\|_{B^{p,q}_{\kappa,\phi}(\mathbb{H}^n)} \leq \|\mathcal{W}_h s\|_{B^{p,q}_{\kappa,\phi}(\mathbb{H}^n)} \leq c\|\mathcal{W}_g s\|_{B^{p,q}_{\kappa,\phi}(\mathbb{H}^n)},$$
$$c^{-1}\|\mathcal{W}_g s\|_{LP^{p,q}_{\kappa,\phi}(\mathbb{H}^n)} \leq \|\mathcal{W}_h s\|_{LP^{p,q}_{\kappa,\phi}(\mathbb{H}^n)} \leq c\|\mathcal{W}_g s\|_{LP^{p,q}_{\kappa,\phi}(\mathbb{H}^n)},$$

for all $s \in S_0'(\mathbb{R})$ *for which the left-hand side (and hence the right-hand side, too) is finite.*

Proof. Recall that the convolution over the half-space reads

$$Q * T(b, a) = \int_0^\infty \frac{d\alpha}{\alpha} \int_{\mathbb{R}^n} d^n\beta \frac{1}{\alpha^n} Q\left(\frac{b-\beta}{\alpha}, \frac{a}{\alpha}\right) T(\beta, \alpha).$$

Since the passage from the wavelet transforms with respect to g to the one with respect to h is given by the convolution over $\mathbb{H}^n$ with $\Pi = c_g^{-1}\mathcal{W}_h g$, which is in $S(\mathbb{H}^n)$, the rightmost inequalities follow from the next theorem. The missing inequality follows upon exchanging the roles of g and h.

Theorem 1.0.5
For $Q \in S(\mathbb{H}^n)$ *the convolution operator*

$$T \mapsto Q * T$$

is continuous in $LP^{p,q}_{\kappa,\phi}(\mathbb{H}^n)$ *and* $B^{p,q}_{\kappa,\phi}(\mathbb{H}^n)$ *with* $1 < p, q < \infty$.

Proof. Since all weight functions are tempered it follows by simple changing of integration variables that for m large enough we have

$$\|T(\cdot\alpha + \beta, \cdot\alpha)\|_{LP^{p,q}_{\kappa,\phi}} \leq O(1)(\alpha + 1/\alpha)^m(1 + |\beta|^2)^{m/2}\|T\|_{LP^{p,q}_{\kappa,\phi}},$$
$$\|T(\cdot\alpha + \beta, \cdot\alpha)\|_{B^{p,q}_{\kappa,\phi}} \leq O(1)(\alpha + 1/\alpha)^m(1 + |\beta|^2)^{m/2}\|T\|_{B^{p,q}_{\kappa,\phi}}$$

Now by simple superposition,

$$\|Q * T\|_{LP^{p,q}_{\kappa,\phi}} = \left\| \int_0^\infty \frac{d\alpha}{\alpha} \int_{\mathbb{R}^n} d^n\beta Q(\beta, \alpha) T(\cdot\alpha + \beta, \cdot\alpha) \right\|_{LP^{p,q}_{\kappa,\phi}}$$
$$\leq O(1)\|T\|_{LP^{p,q}_{\kappa,\phi}} \int_{\mathbb{R}^n} d^n\beta \int_0^\infty \frac{d\alpha}{\alpha} |Q(\beta, \alpha)| (\alpha + 1/\alpha)^m(1 + |\beta|^2)^{m/2}.$$

This last integral is finite, thanks to the high localization of Q, and we are done. The proof for $B^{p,q}_{\kappa,\phi}$ is the same. □

This proves that the norms are well defined. □

Now that we have proved that the norms we consider are well defined we have the following theorem, as expected.

Theorem 1.0.6
The spaces $B^{p,q}_{\kappa,\phi}(\mathbb{R}^n)$ *and* $LP^{p,q}_{\kappa,\phi}(\mathbb{R}^n)$ *with* $1 < p, q < \infty$ *are Banach spaces. Their dual spaces are* $B^{p',q'}_{1/\kappa,1/\phi}(\mathbb{R}^n)$ *and* $LP^{p',q'}_{1/\kappa,1/\phi}(\mathbb{R}^n)$ *respectively with* $(1/p) + (1/p') = 1$ *and* $(1/q) + (1/q') = 1$. *The pairing is understood as*

$$r: s \mapsto (r \mid s) = \int_0^\infty \frac{da}{a} \int_{\mathbb{R}^n} d^n b \, \mathscr{W}_g s(b, a) \mathscr{W}_{\bar{g}} r(b, a)$$

with some admissible wavelet $g \in S_0(\mathbb{R}^n)$. *The following inclusion holds in the topological sense:*

$$S_0(\mathbb{R}^n) \subset B^{p,q}_{\kappa,\phi}(\mathbb{R}^n),\ LP^{p,q}_{\kappa,\phi}(\mathbb{R}^n) \subset S'_0(\mathbb{R}^n).$$

Proof. We argue for $B^{p,q}_{\kappa,\phi}(\mathbb{R}^n)$. The case $LP^{p,q}_{\kappa,\phi}(\mathbb{R}^n)$ is analogous. By definition, we may identify $B^{p,q}_{\kappa,\phi}(\mathbb{R}^n)$ with the functions T over the half-space that are in $B^{p,q}_{\kappa,\phi}(\mathbb{H}^n)$ and that satisfy the reproducing kernel equation

$$T = \Pi * T, \qquad \Pi = \frac{1}{c_g} \mathscr{W}_g g.$$

Since the right-hand side varies continuously with T it follows that the space is complete. Indeed, by definition, a Cauchy sequence in $B^{p,q}_{\kappa,\phi}(\mathbb{R}^n)$ has a wavelet transform that is a Cauchy sequence in $B^{p,q}_{\kappa,\phi}(\mathbb{H}^n)$. It therefore has a limit which, by continuity of the convolution above, again satisfies the reproducing kernel equation. Hence it is the wavelet transform of some function in $B^{p,q}_{\kappa,\phi}(\mathbb{R}^n)$, as was to be shown. □

Remark. In all theorems we have supposed that h and g are arbitrarily regular and have all moments vanishing. This implied an arbitrarily high localization of the reproducing kernel over the half-space. Actually, for fixed p, q, κ, ϕ, only some localization of the reproducing kernel is needed. This can be achieved with some regularity and some vanishing moments. In particular, we may suppose that h and g are compactly supported (which is not possible if all moments vanish).

2 Wavelet multipliers

We now wish to study a large class of operators and their continuity properties with respect to the Banach spaces we have just defined. Recall that for any polynomially bounded function σ over the half-space we have defined the operators

$$K_\sigma: s \mapsto \mathscr{M}_h(\sigma \cdot \mathscr{W}_g s)$$

that are essentially multiplication in wavelet space with σ. Henceforth, we call σ a wavelet multiplier. Polynomially bounded multipliers give rise to continuous operators between spaces of the Besov or Littlewood–Paley scales. More precisely, suppose that σ is some function over $\mathbb{H}^n$ that satisfies at

$$|\sigma(b, a)| \leq \rho(a)\eta(b)$$

for some weight functions ρ, η. Then K_σ is continuous from $LP^{p,q}_{\kappa,\phi}(\mathbb{R}^n)$ to $LP^{p,q}_{\kappa\eta,\phi\rho}(\mathbb{R}^n)$ and from $B^{p,q}_{\kappa,\phi}(\mathbb{R}^n)$ to $B^{p,q}_{\kappa\eta,\phi\rho}(\mathbb{R}^n)$ for all weight functions κ, ϕ, as follows trivially from the definitions.

In the following sections we study a class of continuous operators that leave the spaces $LP^{p,q}_{\kappa,\phi}(\mathbb{R}^n)$ and $B^{p,q}_{\kappa,\phi}(\mathbb{R}^n)$ invariant. These are the so-called Calderón–Zygmund operators that we have already encountered several times throughout this book.

3 The class of highly regular Calderón–Zygmund operators (CZOs)

Here we wish to consider a class of Calderón–Zygmund operators (i.e. CZO) that are bounded operators simultaneously in all the Banach spaces we have defined so far. This is due to the fact that the operators we consider are almost diagonal in wavelet space: the image of a regular wavelet is again something wavelet-like having the same regularity, and is located at roughly the same position and scale. We now go into the details.

Consider an integral kernel B: $\mathbb{H}^n \times \mathbb{H}^n \to \mathbb{C}$. We can associate with this an operator acting on functions on the half-space $\mathbb{H}^n$ by setting

$$M_B\colon T \mapsto \int d\mu(z')B(z; z')T(z'),$$

where we have used the notation $z = (b, a)$ and $d\mu(z) = da\, d^nb/a$. This defines an operator for functions on $\mathbb{R}^n$ if we set

$$K_B\colon s \mapsto \mathcal{M}_h M_B \mathcal{W}_g s \tag{3.0.1}$$

for some $g, h \in S_0(\mathbb{R}^n)$. Recall that a function over the half-plane is highly localized if it satisfies at

$$(a + 1/a)^m(1 + |b|^2)^{m/2}|\Pi(b, a)| \leq O(1)$$

for all $m > 0$.

Definition 3.0.1

We call an operator defined by (3.0.1) *a highly regular Calderón–Zygmund operator in* $\mathbb{R}^n$ *iff the kernel* B *with respect to some wavelets* g, $h \in S_0(\mathbb{R}^n)$

satisfies at

$$|B(b, a; b', a')| \leq \frac{1}{a'^n} \Pi\left(\frac{b - b'}{a'}, \frac{a}{a'}\right)$$

for some highly localized function Π.

This class of operators includes the previously considered wavelet multipliers $s \mapsto \mathcal{M}_h(F \cdot \mathcal{W}_g)$, where F is a bounded function over the half-space. Indeed, its action is given by the following kernel:

$$B(b, a; b', a') = F(b, a) \frac{1}{a'^n} P\left(\frac{b - b'}{a'}, \frac{a}{a'}\right)$$

where P is one possible reproducing kernel for $\mathcal{W}_g$.

Since there are many different reproducing kernels, this example shows that B is not uniquely determined by K_B. For fixed g and h, admissible, one possible B is obtained from a general operator K through ($c = c_g c_h$)

$$B(b, a; b', a') = \frac{1}{c} (\mathcal{W}_h(K g_{b', a'}))(b, a) = \frac{1}{c} \langle h_{b,a} \mid K \mid g_{b', a'} \rangle .$$

Therefore, it is not difficult to see that the highly regular Calderón–Zygmund operators are precisely those for which the matrix elements with respect to two admissible wavelets g, h, satisfy

$$|\langle h_{b,a} \mid K \mid g_{b',a'} \rangle| \leq \frac{1}{a'^n} \Pi\left(\frac{b - b'}{a}, \frac{a}{a'}\right) \tag{3.0.2}$$

for some highly localized function Π. We recall that in this context a wavelet is admissible iff

$$0 < \int_0^\infty \frac{da}{a} |\hat{g}(ak)|^2 = \text{const} = c_g < \infty,$$

which implies that the wavelet transform with respect to g has a continuous inverse. Thus, roughly speaking, a highly regular Calderón–Zygmund operator maps a wavelet that is localized at some position b and at some scale a into something that is again wavelet-like and localized at some position β' and scale α' such that $(b - \beta)/\alpha$ and a/α stay close to 1. From this it follows immediately that they leave invariant all spaces that are characterized by the localization of the wavelet coefficients over $\mathbb{H}^n$. We leave the details to the reader.

Theorem 3.0.2

The highly regular Calderón–Zygmund operators map X continuously into X, where X is any of the spaces out of the set $\{S_0(\mathbb{R}^n), LP^{p;q}_{\kappa,\rho}(\mathbb{R}^n), B^{p;q}_{\kappa,\rho}(\mathbb{R}^n), S'_0(\mathbb{R}^n)\}$

for $1 < q, p < \infty$ *and all tempered weight functions* κ, ρ.

3.1 The dilation co-variance

Consider the Riesz transforms defined in Fourier space via

$$R_\xi \colon \hat{s}(k) \mapsto \hat{s}(k) \frac{\langle \xi \mid k \rangle}{|k|}.$$

As we will see below, this is a highly regular Calderón–Zygmund operator since the multiplier is smooth and dilation invariant. We thus have the following translation and dilation invariance of the Riesz transforms:

$$T_b R_\xi T_b^{-1} = R_\xi, \qquad D_a R_\xi D_a^{-1} = R_\xi.$$

In the one-dimensional case the Riesz transform is the Hilbert transform which is given as the principal value of the singular integral kernel $1/\pi(t - u)$.

For general Calderón–Zygmund operators the above commutation relations no longer hold in the strict sense, but approximately in the sense that if the left-hand side is applied to a test-function in $S_0(\mathbb{R}^n)$ the set of image vectors is bounded in $S_0(\mathbb{R})$.

To be more precise, let K be a continuous operator in $S_0(\mathbb{R}^n)$ and consider the family of translated and dilated operators

$$T_b D_a K D_a^{-1} T_a^{-1}, \qquad b \in \mathbb{R}^n,\ a > 0.$$

Then we have the following theorem.

Theorem 3.1.1

The operator K is a highly regular Calderón–Zygmund operator iff the above family of operators is equi-continuous in $S_0(\mathbb{R}^n)$.

Corollary 3.1.2

The set of highly regular Calderón–Zygmund operators is an algebra.

Proof. Let K be a highly regular Calderón–Zygmund operator. Clearly, it is then continuous from $S_0(\mathbb{R}^n)$ to $S_0(\mathbb{R}^n)$. The matrix elements of the above family read

$$\langle h_{b,a} \mid T_\beta D_\alpha K D_\alpha^{-1} T_\beta^{-1} \mid g_{b',a'} \rangle = \alpha^{-n} \langle h_{(b-\beta)/\alpha, a/\alpha} \mid K \mid g_{(b'-\beta)/\alpha, a'/\alpha} \rangle.$$

By (3.0.2) this is bounded in absolute value by

$$\frac{1}{a'^n} \Pi\left(\frac{b - b'}{a'}, \frac{a}{a'}\right),$$

which shows that the family above is equi-continuous.

On the other hand, suppose that the family above is equi-continuous. Then for a fixed, admissible $g \in S_0(\mathbb{R}^n)$ by Theorem 28.0.1 of Chapter 1 there is a highly localized function Π such that for all $(\beta, \alpha) \in \mathbb{H}^n$ we have

$$|(\mathscr{W}_g T_\beta D_\alpha K D_\alpha^{-1} T_\beta^{-1} g)(b, a)| \leq \Pi(b, a).$$

By the co-variance of the wavelet transform we obtain once more the estimation of (3.0.2). □

3.2 Fourier multipliers as highly regular CZO

We now focus on the case where K commutes with all the translations. The next theorem shows that this corresponds to highly regular Fourier multipliers with same scaling invariance.

Theorem 3.2.1

Let K be a highly regular Calderón–Zygmund operator. Then K commutes with the translations

$$T_b K = K T_b$$

iff there is a smooth function $m\colon \mathbb{R}^n \to \mathbb{C}$ satisfying

$$|\partial^\alpha m(k)| \leq \frac{c_\alpha}{|k|^{|\alpha|}} \tag{3.2.1}$$

for all multi-indices[1] *α such that K is given by the Fourier multiplier*

$$K\colon \hat{s} \mapsto m \cdot \hat{s}.$$

Proof. We first show the necessity part of the theorem. Since K commutes with the translations, it is given by some Fourier multiplier m. We still have to show that it satisfies (3.2.1). Suppose, as we may, that $g \in S_0(\mathbb{R}^n)$ is such that $\hat{g}$ is real valued, compactly supported and that $\hat{g}(k) = 1$ for $1 \leq |k| \leq 2$. By hypothesis on K we have

$$a^n |\langle g_{ba,a} | K | g_{0,a} \rangle| = \frac{1}{(2\pi)^n} \left| \int d^n k |\hat{g}(k)|^2 m(a^{-1}k)\, e^{i\langle b|k\rangle} \right| \leq f(b),$$

for some rapidly decaying function f. By the choice of g it follows that for fixed α we have $|\partial^\alpha(m(a^{-1}k))| = a^{-|\alpha|}|(\partial^\alpha m)(a^{-1}k)| \leq O(1)$ for $1 \leq |k| \leq 2$ uniformly in a. This implies (3.2.1).

[1] That is, the standard notation $\partial^\alpha = \partial_0^{\alpha_0} \partial_1^{\alpha_1} \cdots \partial_{m-1}^{\alpha_{m-1}}$ and $b^\alpha = b_0^{\alpha_0} b_1^{\alpha_1} \cdots b_{m-1}^{\alpha_{m-1}}$ with $\alpha_0 + \alpha^1 \cdots + \alpha_{m-1} = |\alpha|$.

To show the sufficiency note that in Fourier space we have

$$\langle g_{b,a} \mid K \mid g_{0,a'} \rangle = \frac{1}{(2\pi)^n} \int d^n k\, \bar{\hat{g}}(ak)\, e^{i\langle k \mid b \rangle}\, m(k) \hat{g}(a'k)$$

$$= a^{-n} \frac{1}{(2\pi)^n} \int d^n k\, \bar{\hat{g}}(k)\, e^{i\langle k \mid b/a \rangle} m(k/a) \hat{g}(ka'/a).$$

We now suppose, as we may, that $\hat{g}$ is supported by some compact set, say $\{1/2 \leq |k| \leq 2\}$. Then this integral is 0 whenever a/a' is not between 1/4 and 4. For fixed a and a' the expression is clearly rapidly decreasing as b tends to ∞ and, as we will see, thanks to the condition on m, this decay is uniform in a and a'. Note, first, that it is uniformly bounded by $O(a^{-n})$, as follows by estimating the L^1-norm of the integrand. We therefore only have to show that the integral can be estimated by

$$O((|b|/a)^{-\varrho}), \qquad (|b| \to \infty)$$

for all $\varrho > 0$ uniformly in a and a'. By writing

$$e^{i\langle k \mid b/a \rangle} = \frac{a^{|\alpha|}}{b^{\alpha}} \partial^{\alpha}\, e^{i\langle k \mid b/a \rangle}$$

and integrating by parts we obtain derivatives of m and $\hat{g}$. By hypothesis on m, we have, for all multi-indices γ,

$$\tfrac{1}{4} \leq |k| \leq 4 \Rightarrow |\partial^{\gamma} m(k/a)| \leq c_{\gamma}.$$

The partial derivatives $\partial^{\beta} \hat{g}(ka'/a) = (a'/a)^{|\beta|} \hat{g}^{(\beta)}(ka'/a)$ are bounded, too, for all multi-indices β since $1/4 \leq a'/a \leq 4$. Therefore, we have that the integral is of order $O((a/|b|)^{|\alpha|})$ for all multi-indices α. □

In the case where K commutes with dilations and translations it is now an easy matter to verify that K is given by a smooth Fourier multiplier m that is homogeneous of degree 0:

$$m(\lambda k) = m(k).$$

Hence, this subalgebra of highly regular CZO can be identified with the algebra of smooth functions over the $n-1$-dimensional sphere. Typical examples are precisely the Riesz transforms.

3.3 Singular integrals as highly regular CZO

The next theorem shows what the generalization of the Hilbert transforms typically look like. The integral kernels are those one likes to call singular integrals. To illustrate this let us consider the Hilbert transform. In order to regularize the integral

$$\int_{-\infty}^{+\infty} dt \, \frac{s(u)}{t-u},$$

one can replace $1/t$ by a sequence $f_\epsilon \in S_0$ that converges in $S_0'(\mathbb{R})$ to $p.v.t^{-1}$. Each approximation $f_\epsilon * s$ is in $S_0(\mathbb{R})$. The following theorem may then be used to show that the limit actually is in $S_0(\mathbb{R})$ since the family $\{f_\epsilon\}$ may be chosen such that the family of operators $s \mapsto f_\epsilon * s$ is equi-continuous.

Theorem 3.3.1
Suppose $\{B \in S_0(\mathbb{R}^n)\}$ is a family of functions; each of them satisfies at

$$|\hat{B}(k)| \leq c,$$

and, in addition,

$$|\partial^\alpha B(x)| \leq \frac{c_\alpha}{|x|^{|\alpha|+n}}, \qquad x \neq 0$$

for all multi-indices α. Then the family of operators

$$K = K_B \colon s \mapsto B * s$$

is an equi-continuous family of highly regular Calderón–Zygmund operators in the topology of $S_0(\mathbb{R}^n)$.

Note that the first and the second conditions are independent as shown by the example of the function $1/|x|$.

Proof. First note that we may limit ourselves to matrix elements taken between wavelets that are not in $S_0(\mathbb{R}^n)$ but that are $C^m(\mathbb{R}^n)$ with compact support and m-vanishing moments, $|\hat{g}(k)| = o(|k|^m)$ $(k \to 0)$. Indeed, let us call this class of functions Λ_m. Then consider an operator $M \colon \Lambda_m \to S_0'(\mathbb{R}^n)$, $m \in \mathbb{N}$. Suppose that for all m we can find a pair of admissible wavelets g°, h° in some $\Lambda_{m'}$ with $m' \geq m$ and a non-negative function Π with

$$\Pi(b, a) \leq C_m(a + 1/a)^{-m}(1 + |b|^2)^{-m/2}$$

such that

$$|\langle h^\circ_{b,a} \mid M \mid g^\circ_{b',a'} \rangle| \leq \frac{1}{a'^n} \Pi\left(\frac{b-b'}{a'}, \frac{a}{a'}\right).$$

We claim that M is a highly localized Calderón–Zygmund operator. Indeed, the matrix elements taken between two fixed admissible wavelets g and h in $S_0(\mathbb{R}^n)$ can be obtained from the previous ones by the action of some transition kernels $\Pi_{h \to h^\circ} = c_{h,h}^{-1} \mathscr{W}_{h^\circ} h$ and $\Pi_{g^\circ \to g} = c_{g^\circ,g^\circ}^{-1} \mathscr{W}_g g^\circ$. Let us call $|\Pi_{h \to h^\circ}|$ and $|\Pi_{g^\circ \to g}|$ their absolute values. Then the matrix elements between g and h can be estimated by

$$|\langle g_{b,a} \mid M \mid g_{b',a'} \rangle| \leq \frac{1}{a'^n} |\Pi_{h \to h^\circ}| * \Pi * |\Pi_{g^\circ \to g}| \left(\frac{b - b'}{a'}, \frac{a}{a'} \right).$$

Since the kernels $\Pi_{h \to h^\circ}$ and $\Pi_{g^\circ \to g}$ have some localization that grows arbitrarily large with m, it follows that the matrix elements above are arbitrarily well localized, as claimed.

We now return to the proof of the theorem. Observe that the conditions on B are scaling invariant. More precisely, the kernel $D_a B$ corresponding to the operator $D_a K D_a^{-1}$ satisfies the same estimates as B with the same constants c and c_α for all multi-indices α. Therefore, it is enough to estimate the localization of

$$X(b, a) = |\langle h_{b,a} \mid K \mid g \rangle|.$$

More precisely, we show that for every $m \geq 0$ and some admissible g, h in Λ_m we have

$$X(b, a) \leq C_\beta (a + 1/a)^{-\beta}, \quad \text{and} \quad X(b, a) \leq C_\beta (1 + |b|^2)^{-\beta/2}$$

with some localization exponent $\beta > 0$ and some constant C_β that depends only on g, h, and the constants c and c_α above. In addition, we will show that this exponent β may become arbitrarily large with m, thereby proving the theorem.

Now pick $g, h \in \Lambda_m$. The first condition on B ensures that X is localized over the scales

$$X(b, a) \leq C_\beta (a + 1/a)^{-\beta}$$

with some exponent β that may become arbitrary large as $m \to \infty$, and a constant C_β depending only on g and c above. Indeed, the localization over the scales depends, as we know, only on the behaviour of the Fourier transform of the analysing wavelet and the analysed function around 0 and at ∞. But by the first hypothesis on B this behaviour is not altered by K.

The second condition on B ensures the localization in b. We have that $g \in \Lambda_m$ is compactly supported, say by $I = [-\delta, +\delta]^n$, and we wish to estimate the localization of

$$Kg(x) = \int_{\mathbb{R}^n} d^n y g(x - y) B(y).$$

Clearly, by the first condition on the Fourier transform of B, we have globally

$$|Kg(x)| \leq c \|\hat{g}\|_1$$

and we may limit ourselves to $|x| \geq 4\delta$. We can now choose $g \in \Lambda_m$, admissible such that $g = (-\Delta)^m \phi$ for some $\phi \in C_c^\infty(I)$. Using partial integration we write for $|x| \geq 4\delta$

$$|Kg(x)| = \left| \int_{\mathbb{R}^n} d^n y \, \phi(x-y) \Delta^m B(y) \right| \leq \frac{C}{|x|^{2m}} \|\phi\|_1,$$

where in the last inequality we have used the fact that the integral actually only runs over $|x - y| \leq 2\delta$. The constant C depends only on the c_γ with $|\gamma| = 2m$ and g. In addition, together with m, the localization in x becomes arbitrarily large, thanks to Theorem 11.0.2 of Chapter 1 which proves the theorem. □

4 Pointwise properties of highly regular CZO

Let us consider the one-space dimension. The delta functional and the kernel $1/x$ are both homogeneous of degree -1. Taken as a convolution operator the first acts like the identity and therefore clearly preserves the local singularity structure. The second kernel is *a priori* de-localized and thus there is no obvious reason why it should preserve the local nature. However, there is a family of spaces on which it acts like δ, namely, the Hardy spaces H^p_+ given by the traces on $\mathbb{R}$ of analytic functions in the upper half-plane. Therefore, there is perhaps something that is preserved under the singular integral operators. On the other hand, as we have seen, local singularities are characterized through the scaling behaviour of their wavelet coefficients. Now Calderón–Zygmund operators are, roughly speaking, dilation invariant operators: they see all scales the same. Therefore, it is no surprise that many of the local regularities are preserved under the action of the highly regular Calderón–Zygmund operators. In particular, we have the following theorem (For notations see Chapter 4.).

Theorem 4.0.1
Let $\eta \in S'(\mathbb{R}^n)$ at x be locally of regularity W^α. Then $K\eta$ with K any highly regular Calderón–Zygmund operator has the same regularity at x.

Proof. This follows trivially from the characterization of this regularity through the decay of the wavelet coefficients. This decay is not altered by K and we are done. □

For exactly the same reason we have that uniform global regularity is preserved under the action of a highly regular Calderón–Zygmund operator.

Theorem 4.0.2
The space of tempered distributions in $S'(\mathbb{R})$ of regularity Λ^α is invariant under the action of the highly localized Calderón–Zygmund operators.

5 Littlewood–Paley theory

Consider the space $LP^{2;2}_{1,1}(\mathbb{R}^n)$. As we have already shown, this is exactly the space of square integrable functions over the n-dimensional space $\mathbb{R}^n$. In this section we wish to show that the $L^q(\mathbb{R}^n)$ can be identified with $LP^{2;q}_{1,1}(\mathbb{R}^n)$. If we write this statement more explicitly we wish to prove the following theorem.

Theorem 5.0.1

Let $1 < q < \infty$. Let $g \in S_0(\mathbb{R})$ be admissible. For any $s \in S_0'(\mathbb{R})$ define the auxiliary G-function pointwise via

$$(Gs)(b) = \left\{ \int_0^\infty \frac{da}{a} |\mathscr{W}_g s(b, a)|^2 \right\}^{1/2}$$

for all points b where this integral is finite. Then $s \in L^q(\mathbb{R}^n)$ if and only if Gs is defined almost everywhere and $Gs \in L^q(\mathbb{R}^n)$. In addition, there is a constant $c > 0$ only depending on q and g such that

$$\frac{1}{c} \|s\|_q \leq \|Gs\|_q \leq c\|s\|_q.$$

The main tool that we need to prove this is the theorem concerning the continuity of the singular integral operators in $L^q(\mathbb{R}^n)$. We will not prove it here and refer to Stein (1979), from which the central idea for this section is taken.

Theorem 5.0.2

Let $\{B\}$ be a family that satisfies the conditions of Theorem (3.3.1). *Then $K: s \mapsto B * s$ is equi-continuous for the topology of $L^q(\mathbb{R}^n)$.*

The important point for us now is that this theorem also holds if K is not a scalar-valued function but takes values in some Banach space. In particular, let $\mathscr{H}_1$ and $\mathscr{H}_2$ be two separable Hilbert spaces. By $B(\mathscr{H}_1, \mathscr{H}_2)$ we denote the Banach space of bounded linear operators from $\mathscr{H}_1$ to $\mathscr{H}_2$. Let K be a function over $\mathbb{R}^n$ with values in $B(\mathscr{H}_1, \mathscr{H}_2)$ and let s be a function over $\mathbb{R}^n$ with values in $\mathscr{H}_1$. Then the convolution is defined in the usual way:

$$K * s(x) = \int_{\mathbb{R}^n} d^n y K(x - y) s(y)$$

and defines a function over $\mathbb{R}^n$ with values in $\mathscr{H}_2$. This formal expression has a meaning whenever $K \in L^q(\mathbb{R}^n, d^n t, B(\mathscr{H}_1, \mathscr{H}_2))$ and $s \in L^p(\mathbb{R}^n, d^n t, \mathscr{H}_1)$, in which case $K * s$ is an element of $L^r(\mathbb{R}^n, d^n t, \mathscr{H}_2)$, $1/p + 1/q = 1 + 1/r$.

The Fourier transform of a vector-valued function is defined in the obvious way.

With these preliminaries we have the following theorem.

Theorem 5.0.3
Theorem 5.0.2 *holds in the analogous sense if K takes its values in $B(\mathscr{H}_1, \mathscr{H}_2)$ and s takes its values in $\mathscr{H}_1$.*

We are now ready to prove the Littlewood–Paley theorem concerning the characterization of $L^q(\mathbb{R}^n)$ through the modulus of the wavelet coefficients.

Proof. Consider the two Hilbert spaces $\mathscr{H}_1 = \mathbb{C}$ and $\mathscr{H}_2 = L^2(\mathbb{R}_+, da/a)$. For given $\epsilon > 0$ let K_ϵ be the function over $\mathbb{R}^n$ that takes values in $B(\mathscr{H}_1, \mathscr{H}_2)$ defined by setting

$$K_\varepsilon(x)\colon \mathscr{H}_1 \ni t \mapsto t\tilde{g}_\alpha(x)\chi_\epsilon(a) \in \mathscr{H}_2, \qquad \tilde{g}_a(x) = \bar{g}(-x/a)/a^n,$$

where χ_ϵ is some smooth, non-negative cut-off function tending monotonically to 1 as $\epsilon \to 0$. Note that with this ugly notation we have that the zoom of the wavelet transform of s is obtained by convolution with K_ϵ, namely,

$$\mathscr{W}_g s(b, \cdot)\chi_\epsilon(a) = K_\epsilon * s(b).$$

The G function now reads

$$(Gs)(b) = \lim_{\epsilon \to 0} |K_\varepsilon * s(b)|,$$

where $|\cdot|$ stands for the norm of a vector-valued function. The theorem we wish to prove can now be rephrased as estimations of the norm $L^q(\mathbb{R}^n, \mathscr{H}_2)$ of $K_\epsilon * s$, that is, we want to show first that

$$\|K_\epsilon * s\|_{L^q(\mathbb{R}^n, \mathscr{H}_2)} \leq c\|s\|_{L^q(\mathbb{R}^n)},$$

with some c that does not depend on ϵ. We therefore only have to check the requirements for the singular integral to apply. In particular, note that

$$|\partial^\alpha K_\epsilon(x)| = \left\{\int_0^\infty \frac{da}{a} |\chi_\epsilon(a)\partial^\alpha \tilde{g}_a(x)|^2\right\}^{1/2} \leq c_\alpha |x|^{-n-|\alpha|}.$$

In addition, the Fourier transform of K_ϵ is obtained via

$$\hat{K}_\epsilon(k)\colon \mathscr{H}_1 \ni t \mapsto t\cdot\bar{\hat{g}}(ak)\chi_\epsilon(a) \in \mathscr{H}_2$$

and therefore satisfies for all ϵ at

$$|\hat{K}_\epsilon(k)| \leq c.$$

We may therefore invoke Theorem 5.0.2, together with its modification,

Theorem 5.0.3, to see that for $1 < q < \infty$ we have

$$\|G_g(s)\|_q \leq c\|s\|_q,$$

where c only depends on the wavelet g and on q but not on s.

To prove the converse inequality we make, as usual, use of the fact that, for $s_1, s_2 \in L^2(\mathbb{R}^n)$ we have

$$\int d^n x \overline{s_1(x)} s_2(x) = c_g^{-1} \int d^n x \overline{G_g(s_1)(x)} G_g(s_2)(x).$$

Therefore, if we suppose, in addition, that $s_1 \in L^q(\mathbb{R}^n)$ and $s_2 \in L^p(\mathbb{R}^n)$, $(1/p) + (1/q) = 1$ then, by Hölder's inequality and thanks to the previous estimation, we have

$$\int d^n x \overline{s_1(x)} s_2(x) \leq c_g^{-1} \|G_g(s_1)\|_q \|G_g(s_2)\|_p \leq c' \|G_g(s_1)\|_q \|s_2\|_p.$$

Now using the well-known fact that

$$\|s_1\|_q = \sup \frac{|\langle s_1 \mid s_2 \rangle|}{\|s_2\|_p},$$

where the sup runs over a dense subset of $L^p(\mathbb{R}^n)$, we may conclude that the converse inequality holds, at least for $s_1 \in L^2(\mathbb{R}^n) \cap L^q(\mathbb{R}^n)$. For general, $s_1 \in L^q(\mathbb{R}^n)$, the same argument applied to some Cauchy sequence $L^2(\mathbb{R}^n) \cap L^q(\mathbb{R}^n) \ni s_n \to s_1$ allows us to conclude. □

6 The Sobolev spaces

As an easy application of the preceding section we show how to characterize the Sobolev spaces in terms of the modulus of the wavelet coefficients. A function belongs to the Sobolev space $L^{m,p}(\mathbb{R}^n)$ if it and all its partial derivatives up to order m belong to $L^p(\mathbb{R}^n)$. Here the derivatives have to be taken in the weak sense, that is, $\partial^\alpha s = r$ iff for all $\rho \in S(\mathbb{R}^n)$ we have $\langle s \mid \partial^\alpha \rho \rangle = (-1)^{|\alpha|} \langle r \mid \rho \rangle$ for all multi-indices α with $|\alpha| \leq m$. Equivalently, we may require that s and $(-\Delta)^{m/2} s \in L^p(\mathbb{R}^n)$, where Δ stands for the Laplace operator in n dimensions. In wavelet space the partial derivative $(-\Delta)^{\gamma/2}$ is given, up to a highly regular Calderón–Zygmund operator, by the Töplitz operator with symbol $1/a^\gamma$. Indeed, we have

$$(-\Delta)^{\gamma/2} = K_\gamma \mathcal{M}_h a^{-\gamma} \mathcal{W}_g,$$

where the operator K_γ is necessarily translation and dilation invariant. By direct computation it can be seen that it is given by the Fourier multiplier

$$K_\gamma\colon \hat{s}(k) \mapsto m_\gamma(k)\hat{s}(k), \qquad m_\gamma(k)^{-1} = \int_0^\infty \frac{da}{a}\,\bar{\hat{g}}(ak/|k|)\hat{h}(ak/|k|).$$

Clearly, we may choose g and h such that the right-hand side is never 0 and hence, applying Theorem 3.2.1, we see that K_γ is a highly regular CZO.

We therefore define $L^{\gamma,p}(\mathbb{R}^n)$, $\gamma > 0$, to be $LP^{2;p}_{1,1+a^{-\gamma}}(\mathbb{R}^n)$. Explicitly, the associated norm reads

$$\|s\|_{L^{\gamma,p}(\mathbb{R}^n)} = \left\{\int d^n b\left(\int_0^\infty \frac{da}{a}(1 + a^{-\gamma})^2|\mathcal{W}_g s(b, a)|^2\right)^{p/2}\right\}^{1/p}.$$

Note that the weight given to the large scales ensures that $s \in L^p(\mathbb{R}^n)$. By the above remarks these spaces actually coincide with the classical Sobolev spaces.

BIBLIOGRAPHY

Arnéodo, A., Grasseau, G., and Holschneider, M. (1988), On the wavelet transform of multifractals, *Phys. Rev. Letters*, **61**, 2281–4.

Arnéodo, A., Grasseau, G., and Holschneider, M. (1988), Wavelet transform analysis of invariant measures of some dynamical systems, in *Wavelets*, Combe *et al.*, eds., Springer-Verlag.

Aslasken, E. W. and Klauder, J. R. (1968), Unitary representations of the affine group, *J. Math. Phys.*, **9**, 206–11.

Avron, J. and Simon, B. (1981), Transient and recurrent spectrum, *J. Funct. Anal.*, **43**, 1.

Barnwell, Th. P. (1982), Subband coder design incorporating recursive quadrature filters and optimum ADPCM coders, *IEEE Trans ASSP*, **30/5**, 751–65.

Barnwell, Th. P. and Smith, M. J. T. (1986), Exact reconstruction techniques for tree structured subband coders, *IEEE Trans ASSSP*, **34/3**, 434–41.

Battle, G. (1987), A block spin construction of ondelettes, Part I, Lemarié functions, *Comm. Math. Phys.*, **110**, 601–15.

Battle, G. (1988), A block spin construction of ondelettes, Part II, The QFT Connection, *Comm. Math. Phys.*, **114**, 93–102.

Beltrami, E. J. and Wohlers, M. R. (1966), *Distributions and the Boundary Value of Analytic Functions*, Academic Press, New York.

Cavareta, A. S., Dahmen, W., and Miccelli, C. (1991), Stationary subdivision, *Mem. Amer. Math. Soc.*, **93**, 1–186.

Chui, C. K. (1992), *An Introduction to Wavelets*, Academic Press, New York.

Chui, C. K. (1992b) (ed.) *Wavelets: A Tutorial in Theory and Applications*, Academic Press, New York.

Cohen, A. (1990), Ondelettes, analyses multirésolutions et filtres miroirs en quadrature, *Ann. Inst. H. Poincaré, Anal. non linéaire*, **7**, 439–59.

Cohen, A. and Daubechies, I. (1992), A stability criterion for biorthogonal wavelet bases and their related subband coding schemes, *Duke Math. J.*, **68**(2), 313–35.

Cohen, A., Daubechies, I., and Vial, P. (1992), Biorthogonal bases of compactly supported wavelets, *Comm. Pure Appl. Math.*, **45**, 485–500.

Coifman, R. and Meyer, Y. (1990), *Ondelettes et opérateurs III*, Hermann, Paris.

Collet, P., Lebowitz, J., and Porzio, A. (1987), The dimension spectrum of some dynamical systems, *J. Stat. Phys.*, **47**, 609–44.

Combes, J. M., Grossmann, A., and Tchamitchian, Ph. (1989) (eds.) *Wavelets*, Springer-Verlag, Berlin.

Daubechies, I. (1987), The wavelet transform, time–frequency localization and signal analysis, *IEEE, Trans. Inform. Theory*, **36**, 961–1005.

Daubechies, I. (1988), Orthonormal bases of compactly supported wavelets, *Comm. in Pure and Applied. Math.*, **41/7**, 909–96.

Daubechies, I. (1992), *Ten Lectures on Wavelets*, CBMS–NSF Regional Conference Series in Applied Mathematics, Philadelphia (1992).

Daubechies, I., Grossmann, A., and Meyer, Y. (1986), Painless nonorthogonal expansions, *J. Math. Phys.*, **27**, 1271–83.

Deslauriers, G. and Dubuc, S. (1987), *Interpolation dyadique et Fractals, Dimensions non entières et Applications*, Masson, Paris, pp. 44–5.

Dubuc, S. (1986), Interpolation through an iterative scheme, *J. of Math. Anal. and Appl.*, **114/1**, 185–204.

Esteban, D. and Galand, C. (1977), *Application of Quadrature Mirror Filters to split-band voice coding schemes*, Proc. of IEEE International Conf. ASSP, Hartford, Connecticut.

Falconer, K. J. (1985), *The Geometry of Fractal Sets*, Cambridge University Press, Cambridge.

Federbush, P. (1987), Quantum field theory in ninety minutes, *Bull. Amer. Math. Soc.*, **17**(1), 93–103.

Feichtinger, H. G. and Gröchenig, K. (1989), Banach spaces related to integrable group representations and their atomic decompositions I., *J. Funct. Anal.*, **86**, 107–15, II. *Monatsh. Math.*, **108**, 129–48.

Flornes, K., Grossmann, A., Holschneider, M., and Torresani, B. (1994), Wavelet analysis over finite fields, *J. Appl. Comp. Harm. Anal.*, **1**, 137–46.

Franklin, P. (1928), A set of continuous orthogonal functions, *Math. Annalen*, **100**, 522–9.

Frisch, U. and Parisi, G. (1985), Turbulence and predictability. In *Geophysical Fluid Dynamics and Climate Dynamics*, Ghil, M., Benzi, R., and Parisi, G. (eds.), North-Holland, Amsterdam.

Gajewski, H., Gröchner, K., and Zacharias, K. (1974), *Nichtlineare Operatorgleichungen under Operatordifferentialgleichungen*, Akademie-Verlag, Berlin.

Goupillaud, P., Grossmann, A., and Morlet, J. (1984), Cycle-octave and related transforms in seismic signal analysis, *Geoexploration*, **23**, 85–102, Elsevier.

Grochenig, K. (1987), Analyse multi-échelle et bases d'ondelletes, *C.R. Acad. Sci. Paris*, Série, I, 13–17.

Grossmann, A. (1984), Wavelet transforms and edge detection, in *Stochastic Processes in Physics and Engineering*, Blanchard, Ph., Streit, L., and Hazewinkel, M. (eds.), Reidel Publishing Co.

Grossmann, A., Holschneider, M., Kronland-Martinet, R., and Morlet, J. (1987), Detection of abrupt changes in sound signals with the help of wavelet transforms, in *Advances in Electronic and Electron Physics*, Suppl. 19, *Inverse Problems*, Academic Press.

Grossmann, A., Kronland-Martinet, R., and Morlet, J. (1989), Reading and understanding continuous wavelet transforms, in *Wavelets*, Combes, J. M., Grossmann, A., and Tchamitchian, P., (eds.), Springer-Verlag, Berlin.

Grossmann, A., Morlet, J., and Paul, T. (1985), Transforms associated to square integrable group representations I: general results, *J. Math. Phys.*, **26**, 2473–9.

Grossmann, A., Morlet, J., and Paul, T. (1986), Transforms associated to square integrable group representations II: examples, *Ann. Inst. Henri Poincaré Physique théorique*, **45**, 293–309.

Grossmann, A. and Morlet, J. (1984), Decomposition of Hardy functions into square integrable wavelets of constant shape, *S.I.A.M., J. Math. Ann.*, **15**, 723–36.

Grossmann, A. and Morlet, J. (1985), Decomposition of functions into wavelets of constant shape, and related transforms, in *Mathematics and Physics, Lectures on Recent Results*, L. Streit (ed.), World Scientific Publishing (Singapore).

Grossmann, A. and Paul, T. (1984), Wave functions on subgroups of the group of affine canonical transformations, in *Resonances, Models and Phenomena, Lectures Notes in Physics*, Vol. 211, Springer-Verlag, Berlin.

Haar, A. (1910), Zur Theorie der orthogonalen Funktionensysteme, *Math. Ann.*, **69**, 331–71.

Hardy, G. H. (1916), Weierstrass's nondifferentiable function, *Trans. Am. Math. Soc.*, **17**, 301–25.

Hardy, G., Littlewood, J. E., and Pólya, G. (1952), *Inequalities*, Cambridge University Press, Cambridge.

Hentschel, H. G. E. and Procaccia, I. (1983), The infinite number of generalized dimensions of fractals and strange attractors, *Physica*, **8D**, 435–44.

Hernandez, E. and Weiss, G., *A First Course of Wavelets*, CRC Press. (In press.)

Hille, E. and Philips, R. S. (1957), Functional Analysis and Semi-groups, *AMS*, vol. XXXI, Providence.

Hoffmann, K. (1962), *Banach Spaces of Analytic Functions*, Prentice-Hall, Englewood Cliffs, N.J.

Holschneider, M. (1988), L'analyse d'objets fractals et leur transformé en ondelettes, Thèse de doctorat, Université de Provence (Aix-Marseille I).

Holschneider, M. (1988a), On the wavelet transformations of fractal objects, *J. Stat. Phys.*, **50**, 953–93.

Holschneider, M. (1990), Wavelet analysis on the circle, *J. Math. Phys.*, **31/3**, 39–44.

Holschneider, M. (1991), Inverse Radon transforms through inverse wavelet transforms, *Inverse Problems*, **7**, 853–61.

Holschneider, M. (1993a), General inversion formulas for wavelet transforms, *J. Math. Phys.*, **34**(9), 4190–8.

Holschneider, M. (1993b), Localization properties of wavelet transforms, *J. Math. Phys.*, **34**(7), 3227–44.

Holschneider, M. (1994), Fractal wavelet dimensions and localization, *Comm. Math. Phys.*, **160**, 457–73.

Holschneider, M. (1994b), Functional calculus using wavelet transforms, *J. Math. Phys.*, **35**(7), 3745–52.

Holschneider, M. (1994c), More on the analysis of local regularity through wavelet transforms, *J. Stat. Phys.*, **77**, 807–400.

Holschneider, M. (1995), Wavelet analysis over Abelian groups, *J. of Applied and Comput. Harm. Anal.*, **2**, 52–60.

Holschneider, M. and Pinkall, U. (1993), *Quadratic Mirror Filters and Loop-groups*, Preprint.

Holschneider, M., Kronland-Martinet, R., Morlet, J., and Tchamitchian, P. (1988), *The algorithme 'à trous'*, CTP-88/P2.115.

Holschneider, M., Kronland-Martinet, R., Morlet, J., and Tchamitchian, P. (1989), A real-time algorithm for signal analysis with the help of the wavelet transform, in *Wavelets*, Combes, J. M., Grossmann, A., and Tchamitchian, P. (eds.), Springer-Verlag, Berlin.

Holschneider, M. and Tchamitchian, Ph. (1991), Pointwise regularity of Riemann's 'nowhere differentiable function', *Inventiones Mathematicae*, **105**, 157–75.

Hörmander, L. (1982), *The Analysis of Linear Partial Differential Operators I*, Springer-Verlag, Berlin.

Jaffard, S. (1989), Estimations Hölderiennes ponctuelles des fonctions au moyen de leurs coefficients d'ondelettes, *C.R. Acad. Sci., Paris*, Sér. I, **308**, 7.

Jaffard, S. (1991), Pointwise smoothness, two-microlocalisation and wavelet coefficients, *Publications Mathematics*, vol. 35, 155–68.

Kato, T. (1966), *Perturbation Theory for Linear Operators*, Springer-Verlag, New York.

Kirilov, A. A. (1969), *Elements of the Theory of Representations*, Springer-Verlag, Berlin.

Kronland-Martinet, R., Morlet, J., and Grossmann, A. (1987), Analysis of sound patterns through wavelet transforms, *Int. J. Pattern Recognition and Artificial Intelligence*, **1**, 273–302.

Lawton, W. (1991), Necessary and sufficient conditions for constructing orthonormal wavelet bases, *J. Math. Phys.*, **32**(1), 57–61.

Lemarié, P. G. (1988), Ondelettes à localisation exponentielle, *Journal de Math. Pures et Appl.*, **67**, 227–36.

Lemarié, P. G. and Meyer, Y. (1986), Ondelettes et bases hilbertiennes, *Revista Matematica IberoAmericana*, **2**, 1–18.

Mallat, S. G. (1989), Multiresolution approximations and wavelet orthonormal bases of $L^2(\mathbb{R})$, *Trans. Amer. Math. Soc.*, **315**, 69–88.

Mallat, S. G. (1989b), A theory for multiresolution signal decomposition: the wavelet representation, *IEEE Trans. PAMI*, **11**, 674–93.

Marr, D. (1982), *Vision*, Freeman and Co., San Francisco.

Meschkowsky, H. (1962), *Hilbertsche Räume mit Kernfunctionen*, Springer-Verlag, Berlin.

Meyer, Y. (1985), *Principe d'incertitude, bases hilbertiennes et algèbres d'opérateurs*, Séminaire Bourbaki, 662, Asterisque (Société Mathématique de France).

Meyer, Y. (1990), *Ondelettes et opérateurs I, II*, Hermann, Paris.

Murenzi, R. (1990), Doctoral Thesis, Louvain la Neuve.

Paley, R. and Wiener, N. (1934), *Fourier Transforms in the Complex Domain*, Repr. 1960, Amer. Math. Soc. Colloqu. Publ. XIX.

Paul, T. (1984), Functions analytic on the half-plane as quantum mechanical states, *J. Math. Phys.*, **25/11**.

Paul, T. (1985), Ondelettes et Mécanique Quantique, Thèse de doctorat d'état, Université d'Aix-Marseille II.

Polya, G. and Szegö, G. (1971), *Aufgaben und Lehrsätze aus der Analysis II*, Springer-Verlag, Berlin.

Pressley, A. and Segal, G. (1986), *Loop Groups*, Oxford Mathematical Monograph.

Queffelek, M. (1971), Dérivabilité de cerains sommes de Fourier lacunaires, *CRAS*, **273**, A, 291–3.

Reed, M. and Simon, B. (1972), *Methods of Modern Mathematical Physics I*, Academic Press.

Rudin, W. (1991), *Functional Analysis* (2nd edn), McGraw-Hill, New York.

Saracco, G. (1994), Propagation of transient waves through a stratified medium: wavelet analysis of a homoasymptotic decomposition of the propagator, *J. Acoust. Soc. Am.*, **95**(3), 1191–205.

Schaefer, H. H. (1970), *Topological Vectorspaces*, Springer-Verlag, Berlin.

Stein, E. (1979), *Singular Integrals and the Differentiability Property of Functions*, Princeton University Press, Princeton, NJ.

Strichartz, R. (1990), Fourier asymptotics of fractal measures, *J. Funct. Anal.*, **89**, 154.

Stromberg, J. O. (1982), A modified Haar system and higher order spline systems, *Conference in Harmonic Analysis in Honor of Antoni Zygmund, II*, pp. 475–93, W. Beckner *et al.* (eds.), Wadsworth Math. Series.

Torchinsky, A. (1988), *Real Variables*, Addison-Wesley, New York.

Treves, J. F. (1967), *Topological Vector Spaces Distributions and Kernels*, Academic Press, New York.

Triebel, H. (1992), *Theory of Function Spaces II*, Monograph in Mathematics 84, Birkhäuser Verlag, Basel.

Volkmer, H. (1992), On the regularity of the wavelets, *IEEE Trans. Inform. Theory*, **38**, 872–6.

Warner, G. (1972), *Harmonic Analysis on Semi-Simple Lie Groups*, Springer-Verlag, Heidelberg.

Young, R. M. (1980), *An Introduction to Nonharmonic Fourier Series*, Academic Press, New York.

Zygmund, A. (1968), *Trigonometric Series*, 2nd edn, Cambridge University Press, Cambridge.

INDEX